INTEGRATED SUSTAINABLE URBAN WATER, ENERGY, AND SOLIDS MANAGEMENT

INTEGRATED SUSTAINABLE URBAN WATER, ENERGY, AND SOLIDS MANAGEMENT

Achieving Triple Net-Zero Adverse Impact Goals and Resiliency of Future Communities

Vladimir Novotny

WILEY

Registered Office
John Wiley & Sons, Inc., 111 River Street, Hoboken, NJ 07030, USA

Editorial Office
111 River Street, Hoboken, NJ 07030, USA

For details of our global editorial offices, customer services, and more information about Wiley products visit us at www.wiley.com.

Wiley also publishes its books in a variety of electronic formats and by print-on-demand. Some content that appears in standard print versions of this book may not be available in other formats.

Library of Congress Cataloging-in-Publication Data
Names: Novotny, Vladimir, 1938- author.
Title: Integrated sustainable urban water, energy, and solids management : achieving triple net-zero adverse impact goals and resiliency of future communities / Vladimir Novotny, Northeastern University.
Description: Hoboken, NJ, USA : John Wiley & Sons, Inc., 2020. | Includes bibliographical references and index.
Identifiers: LCCN 2019045370 (print) | LCCN 2019045371 (ebook) | ISBN 9781119593652 (hardback) | ISBN 9781119593690 (adobe pdf) | ISBN 9781119593669 (epub)
Subjects: LCSH: Municipal engineering. | Environmental policy. | Urban ecology (Sociology) | Water-supply. | Planned communities. | Sustainable development.
Classification: LCC TD159 .N68 2020 (print) | LCC TD159 (ebook) | DDC 363.6—dc23
LC record available at https://lccn.loc.gov/2019045370
LC ebook record available at https://lccn.loc.gov/2019045371

Cover image: Wiley
Cover design: Courtesy of Bullitt Center

Set in 10/12pt and TimesTenLTStd by SPi Global, Chennai, India

V10016630_010220

CONTENTS

6 BIOPHILIC SUSTAINABLE LANDSCAPE AND LOW IMPACT DEVELOPMENT **141**

7 BUILDING BLOCKS OF THE REGIONAL INTEGRATED RESOURCES RECOVERY FACILITY (IRRF) **175**

8 INTEGRATING GASIFICATION AND DEVELOPING AN INTEGRATED "WASTE TO ENERGY" POWER PLANT **211**

PREFACE

This book is not a reminiscence of the past; it is an outlook into a near future. It ventures there with suggested solutions to avoid the dark consequences of dangerous trends that the Earth is now undergoing and that are expected to increase if nothing is done to reduce emissions of greenhouse gases, impacts of excessive plastics use, and waste generation and pollution by chemicals, and to control the devastating effects of these trends.

I was born at the onset of World War II in a small industrialized country in central Europe – Czechoslovakia (today the Czech Republic). I lived through some of the environmental and economic apocalypses of the war, terrible environmental pollution, and economic and political struggles of my native country that for many years, along with the rest of the industrialized world, embarked on unrestricted economic development and disregarded the environment and sound environmentally friendly economics. After the suffering and devastation of World War II, in the second half of the twentieth century many countries experienced terrible pollution catastrophes, which resulted in mercury poisoning in Japan, rivers on fire (the Cuyahoga River in Cleveland, Ohio), the dying of Lake Erie and some other Great Lakes due to excessive use of industrial fertilizers, unrestricted industrial expansion resulting in contaminated brownfields, and the *Silent Spring* (as described in Rachel Carson's book on the disappearance of birds due to toxic pollution) caused by contamination of the environment by chemicals.

After that, pushed by the strong environmental movement in the early 1970s, the US Congress passed the Clean Water and Clean Air Acts, followed at the end of the century by similar legislative acts in most developed countries. These developments opened the path to the cleanup of the environment. However, despite this progress during the last three decades of the century, it was realized that the fragmented water/stormwater/solid waste management system still threatened the use of resources for future generations. In addition, methane emissions from landfills were still very high, uncontrolled, and now exceed those related to water supply and sewage disposal. Enormous quantities of waste plastics have accumulated in water resources and oceans and the mass of discarded plastics is expected to increase if nothing is done to control it. Furthermore, at the end of the previous century and the beginning of the new millennium, societies realized the danger of global warming that, if not addressed and abated, would by the end of this century change the ecology of the Earth and threaten life as we know it in many parts of the world.

After publishing several books over several decades (the first appeared in 1980), this book is special to the author. Most of the previous books described the state-of-the-art knowledge on water quality, nonpoint (diffuse) pollution, urban drainage, and water conservation. However, in this millennium, scientists, politicians, industrialists, and the informed public have realized that the rise of global warming gas concentrations in the atmosphere must be stopped and eventually reversed. Among other social and economic changes, this will require a change in the paradigm of water, stormwater, solid waste, air pollution control, and energy management. Furthermore, the change will not only have beneficial effects on the environment, but it will also bring tremendous economic benefits. Large-scale use of renewable energy that ten years ago was expensive and unrealistic is exploding in Europe (Germany, Austria, Poland, France, Sweden, and others) and Asia (Singapore, China, and Republic of Korea), which already have communities that claim to have sustainable net-zero greenhouse gas (GHG) impact by implementing green (energy from waste, vegetation biomass, and sludge digestion methane) and blue (hydro, solar, wind) energy. Israel, Singapore, and Australia are leaders in water conservation and reuse. However, the efforts are still fragmented, and environmental engineering is still divided between water supply, drainage, and liquid and solid waste resorts and these resorts rarely cooperate. Urban landscape architects and city and industrial planners urgently need guidance of what is wrong with the past and current systems, what to improve, and what is economically possible in the near and more distant future to achieve urban sustainability.

Implementing water and materials reuse, recycling resources, currently considered waste, and deriving a large portion of community energy and some resource needs from local green and blue sources are now becoming near-future reality. There is a synergy between resources in water and solid waste systems that can be harnessed. The new technologies leading to zero GHG emissions rapidly evolving in the automobile industries are highly applicable and already being applied to other sectors but not yet to urban water/stormwater/solid waste management. These technologies have a potential not only to dramatically reduce urban GHG emissions to net-zero or better (negative GHG emissions) but they could also make cities self-reliant on green and blue energy sources that to a large degree can be derived from waste resources.

Switching to hydrogen as an energy source and carrier opens a real possibility of generating electricity by ultra-clean and very efficient hydrogen fuel cells, which today are opening new possibilities and revolutionizing many segments of the economy. *Fuel cells are an energy user's dream: an efficient, combustionless, virtually pollution-free power source, capable of being sited in downtown urban areas or in remote regions that runs almost silently and has few moving parts* (US Department of Energy). The author of this book argues and documents that these new concepts are highly compatible and adaptable to used water and waste solids disposal.

Ideas and examples of the sustainable and resilient urban landscape providing water storage wherein storm and combined sewers become obsolete and sanitary sewers much smaller, are also presented. Stormwater is not waste; it is the third best source of water for communities, and reusing it locally also saves energy. The "triple net-zero" goal (no waste of water, net-zero GHG emissions, and no waste to landfills) featured in this book is a challenge leading to integration of water, used (waste) water, solids waste, and energy management that is a foundation of the future sustainable urban and suburban areas. This conceptual idea was also a challenge to the author, who attempted to use his 50 years of experience to prove or disprove whether these goals are realistic.

In the late 1980s, the author led a small international team of experts to prepare an English-language adaptation of Karl Imhoff's *Taschenbuch der Stadtentwässerung (Pocketbook of Urban Sewage)*. Founder of European sewage management, Karl Imhoff wrote the first edition of this book in 1906 for a growing audience of planners, students, and engineers embarking on implementing an emerging and challenging paradigm of sewage of cities plagued at that time by extreme pollution of urban waters and polluting industrial activities (mining and steel mills in Germany's Ruhr industrial area). Even city streets were severely impacted by pollution from solid wastes and horse manure. The original

Taschenbuch has been translated and used by engineers of urban water/wastewater management of many countries for decades as a tried and tested work aid. In the US it was first introduced by the pioneer of US environmental engineering (at that time known as sanitary engineering), Gordon M. Fair (K. Imhoff and G.M Fair, *Sewage Treatment*, John Wiley & Sons, 1956) and revised and adapted again in 1989 (V. Novotny, K.R. Imhoff et al., *Handbook of Urban Drainage and Wastewater Disposal*, Wiley). This compact but comprehensive guide has been continuously updated and republished, most recently the 32nd German edition in 2007 (prepared by Klaus Imhoff with coauthors). It has provided fundamental engineering/planning guidance in Europe and in many translations elsewhere to generations of students and professionals.

This book attempts a similar mission: to provide fundamental information and guidelines to students, planners, politicians, and other stakeholders to implement the new paradigm of sustainable integrated water/stormwater/solid waste and energy infrastructures concepts. This shift is as much or more revolutionary as the shift from the uncontrolled waste disposal at the beginning of the twentieth century to the wastewater "fast conveyance – end of pipe treatment paradigm" and engineered solid waste landfilling that dominated the end of the last century. The revolution toward the new paradigm of how cities and homes are built is already in progress. In the next 20–30 years, most homes and commercial houses in countries adhering to the Paris 21 Agreement will have photovoltaic solar panels. Very soon France, Sweden, Denmark, Norway, Singapore, Austria, Israel, Iceland, Germany, and possibly other countries will soon be net-zero GHG-impact countries. Large wind energy farms and solar energy power plants are already ubiquitous in some EU countries, China, India, Israel, Australia, the US, and other countries.

Since 2005 I have been a member of the international team and a founding member of the International Water Association (IWA) committee of scientists, which under the auspices of the IWA organized and promoted the "Cities of the Future" program. Consequently, this book provides information in the international context. The uniqueness of this book is in integration of used (waste) water treatment, municipal solid waste, and suburban agriculture organic waste collection, and disposal into one system of water, solid waste, energy, and resources management and recuperation. In this system, gasification of a wide range of waste organic solids, including plastics, producing syngas, replaces landfilling and environmentally damaging and inefficient incineration. Reforming syngas and methane to hydrogen and subsequent power production by hydrogen fuel cell power plants (2030–2050 horizon) produces with high efficiency blue electricity, heat, resources from waste, and even some ultraclean water.

This book presents the concepts and designs of the integrated sustainable water, energy, solids management, and resource recovery in the "Cities of the Future." Beginning with defining urban sustainability, the text presents historic scientific geological and ecological reasons why humans must reduce the present levels of greenhouse gases in the atmosphere to avoid catastrophic consequences of global warming. The book introduces the concepts of sustainable triple net-zero adverse impact communities and describes methodologies toward meeting sustainability goals. It guides users through the latest emerging technologies of urban water, energy, and solid waste management disposal to the realization that water must be conserved, and that stormwater and solid waste are resources. This book does not directly cover the water/energy nexus of the industrial, agricultural, or even transportation sectors. However, the circular economy of cities is expected to have significant direct and indirect effects on the production sector and switching to hydrogen energy will affect transportation and other sectors, as is already happening in Iceland, Germany, and elsewhere.

The 2018 reports of the International Panel on Climatic Change (IPCC, 2018) and the 13 US government scientific agencies (US Global Change Research Program, 2018a, b) confirmed what scientists and scientific research have been finding and warning about for decades: that if water, solids, and energy management and other economic practices continue practicing business as usual, atmospheric temperature will continue to rise and the damages to the planet's ecology – the frequency of catastrophic storms, floods, droughts, and wildfires, and the melting of artic ice and glaciers – will continue to magnify and within a generation will reach a point of no return. The innocent people greatly impacted by these adverse changes and catastrophes includes the author's children, grandchildren, and future generations. To avoid the serious and catastrophic consequence of global climate change in the near future (from now to 2040), societies must implement radical, fast, yet very logical changes of the urban paradigm from fragmented waste and excessive water and energy use and solid waste disposal to integrated systems that save water, produce excess green and blue electricity, and recover resources from used water and solid waste without landfilling. Recovered water, energy, and resources, including hydrogen and high concentration of CO_2 gas, will have a commercial value, and income and savings achieved with the new paradigm may pay for a great part of the cost of implementation.

The main technical objective of the book is to systematically prove that, within a generation, the move to the urban sustainability paradigm with triple net-zero adverse impacts is feasible both in new and historic communities and to prove that this new paradigm is not a utopia, but it is a realistic goal and even a necessity because of global warming, population increases, and other stresses. The book does not claim that the paths toward these sustainability goals outlined in it are the only way. New technologies unknown in the last century are rapidly emerging. The author trusts that this book will help in the movement toward the sustainable Cities of the Future.

This book is in no way a product of only one person. It contains the knowledge and ideas of many experts and visionaries, personally starting with the late professor Peter Krenkel, a renowned expert on water quality management, who invited me – at that time a young scientist just starting out – to Vanderbilt University in Nashville, Tennessee, during the time of crisis in the author's native country 50 years ago. Also significant is the late Professor W. Wes Eckenfelder, Jr., who taught me the craft of wastewater treatment and disposal design as a teacher and first US employer. Work on adaptation of the *Taschenbuch* with Ing-Dr Klaus R. Imhoff, Director of Ruhrverband Water and Sewage Management Agency in Germany, introduced me to regional urban integrated water and sewage management. Past presidents of the International Water Association, Professors Petr Grau, László Somlyody, and Glen Daigger, and IWA past Executive Directors Anthony Milburn, Paul Reiter, and current Executive Director Professor Kala Vairavamoorthy, with whom I collaborated, created the IWA "Cities of the Future" program and should also be acknowledged, along with many international scientists now participating in the Cities of the Future movement. These scientific leaders are a part of many thousands of scientists of all ages who discovered the danger of overuse of fossil fuels, the ensuing global warming, and the unsustainability of the way cities and industries use energy and resources. Thousands of visionaries of all ages and professional organizations are committed to the goals of achieving the triple net-zero adverse-impact communities, transportation, and industrial production and are working on solutions. And finally, I must mention my former graduate students, who are now university professors and department leaders, presidents, vice presidents, and leaders in large engineering and research companies and city governments.

However, this book is mainly dedicated to the current and next generation of scientists, engineers, political leaders, and all those who will be saving the Earth from the damages of global warming and environmental degradation caused by their ancestors. Thanks and appreciation go also to my wife and life-long partner for her support and collaboration.

Working under a different paradigm of economic development, these ancestors eliminated famine and poverty in most countries, have flown to the Moon, cleaned up the water and air in some cities, and kept the world mostly at peace now for more than 75 years. The realization that unlimited development and urbanization are not sustainable and may severely damage the Earth's ecosystems and humanity occurred later. Now is the time to make the world sustainable and liveable for this and future generations. The goals of sustainable Cities of the Future are realistic and achievable in a generation.

Vladimir Novotny, February 2019

Integrated Sustainable Urban Water, Energy, and Solids Management

Vladimir Novotny

BOOK VISION

Provide guidance on achieving sustainable integrated water, energy, and resource recovery in urban areas. There is discussion of drainage infrastructures connected to receiving waters and protecting or mimicking nature, and being resilient to natural and anthropogenic stresses, including extreme events. It outlines how to reduce emissions of greenhouse gases to net zero level by water conservation, recycling, and generating blue and green energy from waste, complemented by installing solar power in houses and wind power in communities, with the goal of providing good quality of natural and reclaimed water for diverse uses and blue and green energy to present and future generations. Urban sustainability will consider municipal solid waste as a source of energy and resources and eliminate the need for landfills.

Achieving Triple Net-Zero Adverse Impact Goals and Resiliency of Future Sustainable Communities

1

SUSTAINABILITY GOALS FOR URBAN WATER AND SOLID WASTE SYSTEMS

1.1 INTRODUCTION TO URBAN SUSTAINABILITY

This treatise proposes and presents an integrated and sustainable system of urban water, used (waste) water, waste solids management that would save and protect quality of water, recover energy and other resources from waste solids and minimize or eliminate the need for landfills. The system, because it promotes providing more storage of water and resources during the times of excess and safe conveyance during catastrophic events, also enhances the resiliency of urban systems against extreme events such as flooding and severe droughts, which are expected to become more severe and frequent. Technologies to achieve this goal are available and their use will be profitable.

Urban development and, specifically, economic concepts in most of the twentieth century and before followed the idea that development and use of resources should be maximized at minimum cost without considering the impact on future generations and impact of the adverse effects on the environment. In some countries, smoking industrial stacks were a sign of progress. This unrestricted development and disregard for the impact on the environment and city population led to environmental and human catastrophes such as epidemics due to dirty air, poisoned water, and devastating soil losses (dust bowl conditions).

The gravity of the plight of the water and other resources in the world cities and their future under the business as usual (BAU) scenario were felt and known for more than a century but the response was either to do nothing or to convert urban streams to underground sewers out of sight. Resistance, inertia, and the tradition of past urbanism paradigms based on hard infrastructures and pavements still persist and are difficult to overcome. But no matter how many billions will be spent under the last-century paradigm of building new hard water/wastewater infrastructure and/or fixing the old, the ecological goals of the Clean Water Act in the US, the Water Framework Directive in European Community countries, and similar laws in many other countries may not be met. There is also a need to build many new cities; especially in Asia and Latin America, to accommodate anticipated population increase and flux of people from rural areas to the cities. The problems have been magnified by the ongoing global climatic changes, and gradual exhaustion of water and other resources.

There is an ongoing discussion on what is "sustainability" and how it differs from "economic progress." Is it increasing living standards by maximizing economic gains, even though it results in unsustainable use of natural resources? Or is it the naturalistic view that preserving nature in its original state is superior to economic development? The author remembers that in the 1950s, signs of "progress" in totalitarian "socialist" countries were smoking stacks of expanding heavy industries, cultivation of forests, draining wetlands, and irrigating virgin lands and deserts for agriculture, all with heavy adverse impacts on natural resources. In the US the influx of European immigrants in the nineteenth century changed Midwest forests and prairies into agricultural lands and, more recently, industrialization and population increase rapidly changed pristine and agricultural lands to urban lands and sprawl, often without considering environmental consequences.

The advocates of economy only consideration argue that in a free capitalist society a company should have no "social responsibility." "The only corporate responsibility a company should have is to maximize profits for itself and for its shareholders" (Friedman, 2002). Unrestricted development was ubiquitous in the US and other developed countries and in many larger developing countries before World War II and lasted until three decades after the war. But it was also a characteristic of totalitarian socialistic societies of the twentieth century in Europe and Asia, although the economy therein was planned and dictated by the government and not by the market. Unrestricted economic development focusing only on maximization of profits led – and still leads – to health and ecological calamities and widespread damages to air and water quality and public health. Hence, there is a "social" cost of pollution caused by unrestricted development and production focusing only on economic gains.

At the beginning of the new millennium, countries around the world almost universally agreed that environment sustainability and economic development are interconnected. However, arguments that "environmental prosperity" overrides "environmental sustainability" are still being discussed and advocated in few countries rich with fossil fuel resources. Unrestricted economic development would create "environmental diseconomy" through excessive air and water pollution, and would accelerate global warming. In reality, there cannot be economic prosperity with environmental sustainability unless the country does not have an environment to begin with or would accept a scenario that part of a large country can be environmentally devastated so that another part could be prosperous. In the doom scenario of the future under the business as usual (BAU) scenario, Rees (2014) argues that under the current global threat of climate change and limited nonrenewable resources these "resources from honeybees through petroleum to songbirds slip down the scale from abundance to scarcity." Worldwide, people have yet to acknowledge that on a planet already overburdened, there is no possibility of raising even the present world population to developed country material standards sustainably with known technologies and available resources. This situation leads to conflicts and unsustainable migration of people. Rees also warned, "No country, however virtuous, can be sustainable on its own or remain insulated from global turmoil."

Environmental economists argue that maximizing economic profits without considering environmental consequences results in "external diseconomy," which is an economic activity that imposes a negative economic effect such as air and water pollution, including carbon emissions, on an unrelated third party. Market forces alone cannot reduce pollution by the unrestricted production and urban development. Without regulations, upstream polluters would disregard downstream impact on ecology and users of water. The problem with external diseconomy effects of production could be resolved by public pressure, litigation, taxation, and enforcement of government regulations, which is most effective.

The issues, problems, and solutions of the environmental diseconomy related to pollution were extensively presented and summarized in Novotny (2003a).

The situation of unrestricted economic thinking persisted almost to the middle of the second half of the last century while the discourse was continuing. The starting point of the new sustainability concepts was the work of the World Commission on Environment, chaired by then Prime Minister of Norway G. Brundtland, which defined sustainability as follows (Brundtland et al., 1987):

> *Humanity has the ability to make development sustainable – to ensure that it meets the needs of the present without compromising the ability of future generations to meet their own needs.*

The Brundtland commission's definition, which has been generally accepted since the end of the last century, does not differentiate between sustainable development and sustainability. The report expresses the desire of society to use and manage resources on the bases of economic sustainability, social equity, intergenerational justice, and intrinsic value of nature (Dilworth, 2008) and makes the intergenerational preservation of resources as paramount as economic development. It is now generally agreed that these values are not mutually exclusive; they overlap to some degree.

Howarth (2007) discussed the views that led to formulation of sustainability. One definition of "sustainability" that tried to modify unrestricted development of land and use of natural resources for economic "progress" was the so-called economic *Net Investment Rule*, stating:

> *A dynamic economy will maintain a constant or increasing level of per capita utility only if value of natural resources investments in manufactured capital exceed the monetary depletion on an economy-wide basis.*

Incorporating the loss of natural resources into economic thinking was a step forward from strictly economic monetary benefit/cost analyses. "Environmental impact statements," along with economic analyses, were attached to strictly monetary assessment of benefits and costs of development projects. This concept was included in the US in the National Environmental Policy Act (NEPA), passed by the Congress in 1969. NEPA was one of the first laws ever written that established a broad national framework for protecting our environment. NEPA's basic policy is to assure that all branches of government consider the environment prior to undertaking any major federal action that significantly affects the environment. Unfortunately, NEPA was mandatory only for federally financed projects, but some states passed their own NEPA-like rules. Large private developments were excluded. The weakness of the NEPA assessments was the inability to include intangible environmental gains and losses into the economic analyses.

As indicated previously, sustainability is evolving from gradual merging and discourse of population groups ascribing to two social views (Novotny et al., 2010a):

- The anthropogenic views regard nature as a resource that should be used and developed for economic gains.
- The biocentric views regard preserving and restoring nature as the goal for humans.

Most people subscribe to both views, meaning that they want to increase their living standard, yet they do not want to live in a polluted or severely damaged environment and want to preserve the nature and resources for the future generations. Protection of the environment and public health is a cognitive value. In the second half of the last century this dual adherence to two seemingly contradicting principles could have been one of the reasons for urban sprawl benefiting those who could afford to leave behind the deteriorating cities with their polluted air and water and be, for a limited time, closer to the nature. Yet, the same people participated in the economic activities in the cities that created the problem. This process of affluent people leaving behind degraded and abandoned urban zones and poor inhabitants is a social problem known as *environmental injustice,* which is still rampant in shantytowns of some developing countries and can also be seen in the US and other developed countries There may be other reasons for such intragenerational injustice, such as crime or quality of education.

Howarth (2007) presented the arguments and various views in the discourse leading to the present thinking and rules. The new millennium views have been expressed in the Millennium Economic Assessment (2005):

> *Fair Sharing Principle – Cateris paribus (latin for "all other things remaining equal") – each member of present and future society is entitled to share fairly the benefits derived from environmental resources. Specific stocks of environmental resources should not be depleted without rendering just compensation to members of the future generations.*

Hence, the term "sustainability" and adherence to its principles is the historical shift from "a maximum economic use model" that understood resources to be merely raw materials for production and sinks for the disposal of waste (a purely anthropogenic view), to a more biocentric optimal model that recognizes the environment as a finite resource that needs to be conserved through public stakeholder involvement and governmental regulation in order to create a long-term relationship between economy and nature. The "sustainability" is also salient to the "land ethic" (including water and watershed) expressed by Leopold (2001), who emphasized a balance between preserving nature and development.

Sustainability as an interdisciplinary science is still emerging; nevertheless, stronger calls for sustainable development and a "paradigm shift" from the public, public officials, and in the media have been heard since the 1990s and intensified in the new millennium. People who are well informed can align themselves with the movement toward sustainability, which, however, may not mean the same thing to different people. The cognitive values of sustainability are related to:

Preservation of human societies today and in the future

Preservation of nature and restoration where nature is damaged

Achieving and maintaining good economic status of present and future generations

Minimizing or eliminating risks to public health and providing healthy and green urban environment

Sustainability has three interacting dimensions: environmental, economic, and societal, and sustainability can be achieved only when these components are balanced (Figure 1.1). A change in one compartment of the trinity of sustainability affects the other two compartments. If they are not balanced, the outcomes can be numerous, such as pollution and environmental degradation but also social injustice or unsustainable development (Novotny, 2003a).

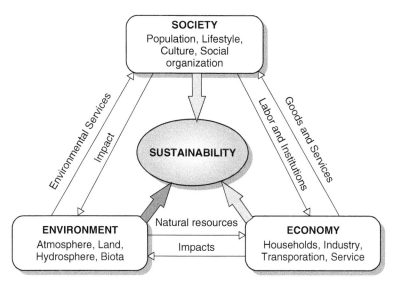

Figure 1.1. Trinity of factors and impacts determining sustainability. *Source:* Adapted from Brundtland (1987), Novotny (2003a). and Allan (2005).

It turned out that today there are hundreds of definitions of sustainability (Maršálek at al., 2007). Mihelcic et al. (2003) pointed out that sustainability is not merely a preference for economic development with some environmental protection (an anthropogenic development view), nor preserving nature with "green" development (a biocentric view). Sustainability, however, is not limited only to water or resources that would benefit humans. The serious threats from climate change to humans and other species add other components of integrity – atmosphere and marine life – that are now impacted by greenhouse gases (GHG), air pollution emissions that cause global warming of the atmosphere. In addition to global warming, GHGs in the atmosphere cause increased acidity of oceans that has been associated with dying coral reefs, loss of glaciers, and disappearing polar ice. These ecological changes will adversely affect both nonhuman organisms and human beings. As a hypothesis, this author (Novotny et al., 2010a), in defining sustainability reflecting water, resources, and society, suggested that the concept of water body integrity defined in the Clean Water Act (PL 92-500, US Congress 1972) is universal and applies to the environment in general (i.e., air, water and soil). In addition to nonhuman organisms, this also considers damages to humans, substitutes "environment" for "water," and redefines environmental integrity as:

> *the ability of the environment (air, water and soil) to support on regional or global scale balanced communities of organisms comparable to that of natural biota of the region and also provide conditions for unimpaired wellbeing of present and future human generations.*

Sustainability would then mean "restoration, preservation and maintaining integrity of the ecology, environment and resources for present and future generations." This simple definition recognizes the fact that humans are a part of the ecological system and, in addition to biota, they can also impose a damage upon themselves. According to the definitions in the US Clean Water Act (Section 5), "pollution" is caused by humans or their actions and is differentiated from changes of the quality (integrity) of the environment due to natural causes such as natural CO_2 content of the atmosphere, natural erosion, weathering of rocks, volcanic eruptions, and CO_2 emissions from natural forest fires or from natural biomass growth and respiration processes converting organic carbon to CO_2. However, converting organic carbon from fossil sources to energy and industrial products by humans resulting in GHG emissions is clearly pollution that impairs integrity.

1.2 HISTORIC AND CURRENT URBAN PARADIGMS

Loosely defined, a *paradigm* is a set of theories and associated rules that explain the way a particular subject or behavior is understood at a particular time (Cambridge dictionary). It is a set of assumptions, concepts, values, and practices based on science and/or tradition that constitutes a way of viewing reality shared by a community.

Human paradigms have changed over millions of years and could be linked to "revolutions" defined in this context as a major change in the way people gathered and produced food, built settlements, produced energy, and interacted as societies. In the seminal book *Engineering Response to Climate Change,* Watts (2013) quoted from a 1953 book by Galdon Darwin (*The Next Million Years),* which identified five revolutions, each having different paradigms:

Use of fire – separated humans from other primates about 400,000 years ago

Invention of agriculture – allowed humans to stop being nomadic hunters

Urban revolution – organized larger human settlements, center of commerce 7,500 years ago

Scientific revolution – understanding the nature and onset of Industrial Revolution

The first four revolutions of human development were fueled by energy, which was originally derived by burning wood or even dried dung of domestic animals, but at the onset of the industrial revolution humans started to use fossil fuel – coal, oil, and natural gas. The fossil fuel energies in the industrial period fueled unprecedented population increases, which became exponential, with rapid expansion of agriculture, building cities and industries, and fast transportation by trains and aviation.

More than sixty years ago, Galdon Darwin and others quoted in Watts's book (Watts, 2013, Chapter 1) envisioned the "fifth revolution" during which the fossil sources of fuel would be exhausted. They anticipated that this revolution would occur centuries from current time and would slow down the development and reduce population. They did not anticipate the global climatic changes that the burning of fossil fuels would cause and the consequences. They predicted that energy during and after the fifth revolution would be provided by sun, wind, tides, hydropower, and geothermal sources. Before 1950, nuclear power was known but its use for energy instead of for making bombs was fuzzy. Today, the consequences of fossil fuel burning and other industrial uses (e.g., making fertilizers, steel, plastics, tires, chemicals) and transportation has become a serious problem and a cause of global climate change. It has become evident that humans as well as the entire ecosystem on the Earth are at the onset of the fifth revolution, which will lead to dramatic reductions of the use of fossil fuel not because of their immediate exhaustion, but because of damaging impacts they have on life on the Earth.

Paradigms of Urbanization

This book focuses on paradigms of human urban development and settlements, specifically cities in the industrial period. There are at least four recognizable historical models or paradigms (Table 1.1) that reflect the evolution and development of urban water resources.

While humans were detected by archeologists to exist about between 1.3 to 1.8 million years ago, the first urban settlements were found in Mesopotamia around 7500 BCE. Early cities also arose in the Indus Valley and ancient China (Wikipedia). Ancient and medieval cities, from ancient Egypt and Rome 2000–4000 years ago and even until the end of the nineteenth century, were relatively filthy and miserable walled places to live for ordinary people. Even today, living in some large settlements and shantytowns in Africa or Latin America is difficult and lacking the amenities of modern cities in developed countries. After the walls surrounding the cities were removed in the nineteenth century, urban dwellers desired to live near nature, which led to developing parks. Later, in the first part of the twentieth century, "garden cities" emerged in the outskirts and wealthy suburbs. With automobiles and freeways, "green" but unsustainable suburban and even distant subdivision type communities identified as *urban sprawl* dotted with thirsty lawns were emerging, mainly in the US. This development led in the US to the reduction of forests and agricultural lands, the demise of older historic city centers, and the dramatic increase of energy use for daily long-distance commuting by fuel-inefficient cars and very high use of water for irrigating lawns. European and Asian cities were growing too but mostly kept and improved their mass transport systems and avoided urban sprawl. Meanwhile, urbanization, industrialization, and change from family farms to industrialized agriculture resulted in heavy pollution in and around cities, extended in the US as far as the Gulf of Mexico.

The physical connections (both structural and natural) between cities and their water resources and nature have changed through the centuries. At the same time, our conceptual models of these systems and our understanding of how they should function and relate to one another have changed as well. The first paradigm of water management of ancient cities was characterized by the utilization of local shallow wells for water supply; exploitation of easily accessed surface water bodies for transportation, washing, and irrigation; and the shared use of streets and roadside channels for the conveyance of people, waste products, and precipitation. Urban runoff of ancient and medieval cities was not clean; it carried feces from horses and other animals and sometimes from people, in spite of the street sweeping (by brooms) that most likely was practiced in some cities by people in front of their houses and shops.

As water demand increased and easily accessible local groundwater and surface supplies became insufficient to support life and commerce (Rome two thousand years ago had a population approaching one million), a second paradigm emerged in growing ancient and medieval cities – the engineered capture, conveyance, and storage of water. The aqueducts of ancient Roman and Byzantine cities brought water to fountains, public baths, and villas from mountains as far away as fifty kilometers to Rome and up to several hundred kilometers to the Byzantine capital Constantinople. In many ancient cities, rural castles, and villas, rainwater was collected and stored in underground cisterns, a practice still prevalent in some communities in the Mediterranean. Similarly to the ancient cities, street surfaces were polluted by fecal matter and trash and the smell must have been nauseating. The Roman sewer Cloaka Maxima has been functioning for more than two thousand years but sewers were installed centuries later in other European cities. Sewers were built mostly for conveyance of urban runoff polluted by garbage and fecal matter, but in Roman and some medieval

Table 1.1. Historical paradigms of water and solids management.

Paradigm	Time Period	Characterization	Quality of Receiving Waters
I. Basic water supply	BCE to the Middle Ages, still can be found in some developing countries	Wells and surface waters for water supply and washing, streets and street drainage for stormwater and wastewater, fecal matter from animals and sometimes humans disposed onto streets and into surface drainage, privies and outhouses for black waste, most street surface pervious or semi-permeable, thatched roofs.	Excellent in large rivers, in small and medium urban streams, poor during large rains, good in between the rains. Pollutants of concern: most likely pathogens.
II. Engineered water supply and polluted urban runoff conveyance	Ancient Crete, Greece, and Rome, Middle Ages, cities in Europe till the beginning of Industrial Revolution in the 19th century	Wells and long distance aqueducts for public fountains, baths (Rome), and some castles and villas, some treatment of potable water, wide use of capturing rain in underground cisterns, medium imperviousness (cobblestones and pavers), many roofs covered with tiles, sewers and surface drainage for stormwater, some toilets flushing in homes of aristocracy discharging into sewers, otherwise privies and outhouses for black waste, fecal matter from animals and sometimes humans disposed onto streets and nearest streams and into surface drainage, no wastewater treatment.	Excellent to good in large rivers, poor to very poor in small and medium urban streams receiving polluted urban runoff contaminated with sewage, widespread epidemics from waterborne and other diseases. Pollutants of concern: pathogens, lead (in Roman cities because of widespread use of lead, including pipes), BOD of runoff.

III Fast conveyance with no minimum treatment	From the second half of 19th century in Europe and US, later in Asian cities, till the second half of the 20th century in advanced countries, still persisting in many countries	Wells and aqueducts for water supply, potable water mostly from surface sources treated by sedimentation and filtration, wide implementation of combined sewers in Europe and North America, beginning of widespread use of flushing toilets, conversion of many urban stream into underground conduits, initially no or only primary treatment for wastewater. Secondary treatment installed in some US and German cities after 1920s. After 1960 some smaller communities built lower-efficiency secondary treatment, paving the urban surface with impermeable (asphalt and concrete) surfaces. Swimming in rivers became unsafe.	Poor to very poor in rivers receiving large quantities of untreated or partially treated wastewater discharges, runoff discharged into sewers and combined sewer overflows. Some rivers were devoid of oxygen with devastating effects on biota. Cuyahoga River on fire in Cleveland, but waterborne disease epidemics diminishing due to treatment of potable water. Pollutants of concern: BOD, DO, sludge deposits, pathogens
IV Fast conveyance with the end of pipe treatment	From the passage of the Clean Water Act in the US in 1972 to the beginning of the 21st century	Gradual implementation of environmental constraints resulting in mandatory secondary treatment of biodegradable organics, regionalization of sewerage systems, additional mandatory nitrogen removals required in European Community, recognition of nonpoint (diffuse) pollution as the major remaining problem. There were increasing concerns with pollution by urban and highway runoff as a source of sediment, toxins, and pathogens, increasing focus on implementation of best management practices for control of pollution by runoff, and emphasis on nutrient removal from point and nonpoint sources, beginning of stream daylighting and restoration efforts in some communities.	Improved water quality in places where point source pollution controls were installed. Due to regionalization, many urban streams lost their natural flow and became effluent dominated, major water quality problems shifted to the effects of sediment, nutrients, toxins, salt from deicing compounds, and pathogens. Biota of many streams recovered; however, new problems with eutrophication and cyanobacteria harmful blooms have emerged.

Source: Updated from Novotny et al. (2010a).

Arab cities they also received polluted flows from baths and continuously flushing public and aristocratic toilets.

The third paradigm for urban water and wastewater starting with the era of industrialization could be linked to the invention and widespread use of flushing toilets that were connected to existing streams and storm sewers, which then became combined sewers. The cities added a massive investment in sewage collection and later some controls and treatment of point sources of pollution resulting from sewer systems, and provided increased treatment of potable water supplies. These later improvements that emerged at the end of the nineteenth century were driven by epidemics of waterborne diseases caused by the cross-contamination of water supplies with raw sewage discharges and by leaking sewers contaminating wells. In Chicago, the course of the heavily polluted Chicago River was diverted from Lake Michigan to the watershed of the Mississippi River to prevent devastating cholera epidemics by contaminated lake water used for drinking. At the same time, impervious surfaces in cities were also increasing, resulting in higher volumes of stormwater runoff and more frequent flooding but also in diminishing base flow in streams. Consequently, many streams were lined with masonry or concrete and often buried underground. The aim of these fast conveyance urban drainage systems was to remove large volumes of polluted water as quickly as possible from the premises, protecting both public safety and property, and discharging these flows into the nearest receiving water body. Solid waste, including feces of horses from the streets were collected and dumped in unsanitary dumps. The third paradigm introduced first primary and then secondary wastewater treatment but did not address the overall, uncontrolled water-sewage-water cycle (Imhoff, 1931; Lanyon, 2007; Novotny, 2007a) in which water in an upstream community is converted to sewage, discharged into receiving water body, and reused downstream as potable water by another community.

The worst period of environmental pollution and degradation was from the end of the World War II till several years after the passage of the CWA in 1972. This was the period of fast and unrestricted economic development during which the industries that had been manufacturing war weapons and chemicals during the war were rapidly developing pesticides, industrial chemicals, and fertilizers; expanding the manufacture of machines and transformers that used polychlorinated biphenyls (PCBs) for cooling, rapidly increasing energy production mainly through sulfur-containing coal-fired power plants in the eastern US and Europe, which emitted sulphur oxides that resulted in acid rain. Water pollution was reaching catastrophic levels, as seen in the US with rivers on fire, such as the Cuyahoga River in Cleveland (Figure 1.2) and anoxia from excessive discharges of biodegradable pollutants from sewers. The situation was as bad or worse in Japan, where, among other catastrophes, fish were poisoned by mercury and cadmium. In Europe, pollution catastrophes occurred both in democratic capitalist and in totalitarian socialist countries. Rachel Carson (2002) described this situation in her seminal book *Silent Spring* (first published in 1962); she stated, "The most alarming of all man's assaults upon the environment is the contamination of air, earth, rivers, and sea with dangerous and even lethal materials." Carson described the situation resulting from excessive application of chemicals that led to the disappearance of birds because they stopped reproducing.

In the United States, the third paradigm period ended with the passage of the Water Pollution Control Act Amendments of 1972 (Clean Water Act, CWA), which made treatment of pollution mandatory. It established the Environmental Protection Agency, which was required to develop, implement, and enforce environmental protection standards and criteria. The period from the enactments of the CWA in the US and the Water Framework Directive in European Community countries until the present time comprises the fourth

Figure 1.2. Cuyahoga River in Cleveland on fire, 1952. *Source:* Cleveland Press Collection, Cleveland State University Library.

paradigm of urban water management and protection, in which increasingly diffuse sources of pollution were considered and addressed in many separate and discreet initiatives. These are also characterized as "end-of-pipe control" because the predominant point of control occurs where the polluted discharge enters the fast conveyance system (sewer or lined channel) or the receiving water body. Pollution by urban runoff and other diffuse sources was recognized as a problem only about 30–40 years ago and was included in the CWA. At the end of the twentieth century the European Parliament enacted the Water Framework Directive (WFD).

In the US after the passage of the Clean Water Act, the new massive building program of treatment plants considered mainly cost-based on "economy of scale" that preferred large regional treatment facilities with long-distance transfers of wastewater over smaller local plants. Local treatment plants built before 1970 were mostly rudimentary primary-only plants or low-efficiency trickling filter facilities or aerobic/anaerobic lagoons. In most cases these plants were unable to meet the goals of the Clean Water Act. The new large-scale activated sludge treatment facilities offered better efficiency capable of meeting the more stringent effluent standards and were managed by highly skilled professionals. As some cities have reached their limits of water supply, reuse has been considered but not widely implemented.

The long-distance transfers of water and wastewater dramatically changed the hydrology of the impacted surface water bodies, which became flow deficient after withdrawal; after receiving the effluent downstream, these water bodies then became effluent dominated. Under the fourth paradigm the problems with combined and sanitary sewer overflows (CSOs and SSOs) have not been and most likely will not be fully mitigated in the near future due to the high cost of capture, storage, and treatment in expensive, mostly underground storage facilities. For example, many kilometers of twelve-meter-diameter interceptors known as "deep tunnel" were built in Milwaukee and Chicago, storing millions m^3 of a mixture of stormwater and wastewater from CSOs and SSOs. The underground pumping stations – with several pumps that are among the largest ever built – pump the sewage/stormwater mixture into the regional treatment plants, and pumping energy use is high. Long-distance water/wastewater transfers from source areas over large distances also require electric energy for pumping, treatment (e.g., aeration), and transporting treatment residuals to their point of disposal. Large volumes of "clean" groundwater infiltration and illicit inflows (I-I) into sanitary sewers must be pumped and treated with the sewage, which uses more energy. During wet weather the I-I inputs can more than triple the volume of dry weather wastewater flows in sewer systems and overwhelm treatment plants (Metcalf & Eddy, 2003; Novotny et al., 1989).

1.3   GLOBAL CLIMATE CHANGES

The Earth is now undergoing accelerated global atmospheric warming, which could become the most serious threat to humans and ecology on the Earth unseen in geologic times for thousands, maybe millions of years. The Committee on Geoengineering Climate of the National Academies (National Research Council, 2015) summarized the forecasts by scientists who have predicted that doubling the carbon dioxide in Earth's atmosphere from preindustrial levels would warm Earth's surface by an average of between 1.5°C and 4.5°C (about 3°F to 8°F). The CO_2 concentration at the onset of the Industrial Age was stable at 280 ppm. It started to increase with expanded use of fossil fuels, first coal (heating, steam engines, making steel and syngas) in the nineteenth century, then oil (automobiles and other transportation) at the beginning of the twentieth century and natural gas after 1950. Figure 1.3 shows that in 2013 the CO_2 concentration measured by the National Oceanic and Atmospheric Administration (NOAA) exceeded 400 parts per million (ppm). If the rate of atmospheric CO_2 concentration increased linearly as indicated on the figure, the doubling of the preindustrial carbon dioxide concentrations would be reached by the end of this century. Similar increases have been measured in atmospheric methane, which is a more potent GHG than CO_2. The report from the Intergovernmental Panel on Climate Change (IPCC, 2013, 2018) concluded that, if current emissions trends continue, by the end of this century if the planet experiences a warming of up to 4.5°C, sea level will rise by as much as 1 meter and, during some years, the Arctic will be ice free in the summer by midcentury. Thermal warming increases air temperatures; 80% of the heat added by GHG gases is absorbed by oceans. Water warming causes thermal expansion and about half of the past century's rise in sea level is attributable to warmer oceans simply occupying more space. The other half of the sea water level rise is due to the melting of the permafrost over Greenland, and the Antarctic will lose a great portion of its ice cover. The melting of Arctic and Antarctic ice has been ongoing over the last 100 years, during which sea levels have risen by 10 to 20 cm. The rise is accelerating and the rate has doubled (https://www.nationalgeographic.com/environment/global-warming/sea-level-rise/). Because of physics, melting of North Pole sea ice does not increase the water level of oceans.

As part of this climatic change, society will experience an increase in the frequency and severity of heat waves (including forest fires), droughts, and heavy precipitation events. Figure 1.3 also shows the atmospheric concentrations of methane, which is also very potent greenhouse gas. Its global warming potential is 25, while that of carbon dioxide is 1.

CO_2 remains in the atmosphere for decades to centuries. Excess carbon dioxide from the atmosphere is absorbed chemically by the terrestrial biosphere by forming carbonates in water; reacting with calcium, magnesium, and silicates to form rocks; and biologically by photosynthesis. Most CO_2 emitted today will still be in the atmosphere, land biosphere, or

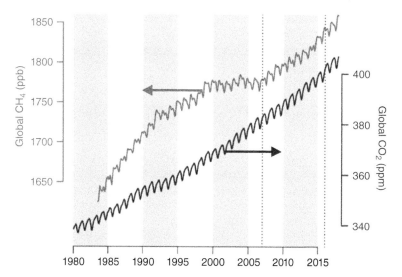

Figure 1.3. Increase of the atmospheric carbon dioxide (CO_2) and methane (CH_4) concentrations in the atmosphere between 1980 and 2018 in gas volume per million volumes (ppm) or per billion volumes (pbb) of dry air of the of the US National Oceanic and Atmospheric Administration's global monitoring network. The annual increases in atmospheric carbon dioxide over the past 60 years are about 100 times faster than previous natural increases in the last 800000 years *Source:* National Oceanic and Atmospheric Administration (NOAA), US Government Data, reprinted from USGCRP, 2018b.

ocean tens of thousands of years later, until geologic processes can form rocks and deposits that would incorporate this carbon (Archer et al., 2009; NRC, 2015). See Chapter 4 for more discussion.

The NRC (2015) report divides the atmospheric carbon-reducing actions into the following climate intervention categories intended to produce a targeted change in some aspect of the climate (e.g., global mean or regional temperature):

Carbon Dioxide Removal (CRD), which also includes other greenhouse gases (GHG), covers intentional efforts to remove GHGs from the atmosphere, including land management strategies, accelerated weathering, ocean iron fertilization, bioenergy with carbon capture and sequestration, and direct air capture and CO_2 sequestration. CDR techniques complement carbon capture and sequestration methods that primarily focus on reducing CO_2 emissions from point sources such as fossil fuel power plants. Reducing or eliminating CO_2 and methane emissions from urban water and waste solids management systems fall in this category.

Albedo (radiation reflectivity) modification concerns intentional efforts to increase the amount of sunlight that is scattered or reflected to space, thereby reducing the amount of sunlight absorbed by Earth, including injecting aerosols into the stratosphere, marine cloud brightening, and efforts to enhance surface reflectivity. Snow and polar ice have a high albedo.

The large-scale geo-engineering actions are described in the NRC (2015) report and in Watts (2013). The NRC scientists concluded, "there is absolutely no substitute for slashing fossil fuel emissions in order to prevent catastrophic disruption of the Earth's climate," but also recommended that the other large-scale geoengineering actions should be researched because in the final outcome they could help in reducing the atmospheric CO_2 to acceptable levels.

1.4   NEED FOR A PARADIGM SHIFT TO SUSTAINABILITY

Since the beginning of this millennium, experts have agreed that the current paradigms of urban water, stormwater, and wastewater management – along with disposal of municipal solid waste and used water solid residuals onto landfills and related infrastructures – are unsustainable. The fast-conveyance drainage infrastructure, most of it conceived in Roman times to eliminate unwanted, highly polluted urban runoff and sewage, has produced great gains in protecting public health and safety but, after the US has spent billions of dollars since the passage of the CWA in 1972, the water resources are still damaged or vulnerable to damage by various water quality and habitat impairment and hydrological stresses; this results in water wasting on one side and shortages on the other side. Stresses such as infestation with cyanobacteria are reappearing and spreading with increasing frequency and intensity worldwide. The present water/stormwater/solid waste management system still threatens the use of resources by future generations. The centrally managed infrastructure of underground pipes for drainage and water delivery is not resilient to extreme events such as flooding that are expected to increase because of global warming. It excessively uses energy in spite of the fact that used water and, by the same reasoning, other organic wastes, contain recoverable energy in a form of heat and electricity. The current urban stormwater paradigm could also be called the "fast conveyance/end-of-pipe control."

In developed countries, the infrastructure is crumbling and requires massive investments for repairs. Traditional subsurface stormwater drainage can only handle smaller storms with the recurrence intervals ranging from less than two years (e.g., Tokyo) to five to ten years (standards for storm sewer designs in many US and European communities). Combined sewer overflows (CSO) are still allowed, which again is unacceptable to the public because, after each storm, beaches and bathing areas are closed or permanently unsafe for swimming. These systems are not resilient to extreme events. The current unsustainable situation is being further exaggerated by global climate warming (increasing sea levels, changes in drought and water availability patterns) (IPCC, 2007, 2013) and has now become the most serious worldwide threat to sustainability (Figure 1.3). Other adverse effects and challenges are:

- *Population increase:* urban population is expected to increase by 50% in the next 20–30 years and continue to grow at a reduced rate till 2100; many new cities and megacities will be built in Asia and other parts of the world (United Nations, 2014, 2016).
- *Migration of people* from rural areas to the cities and from war-torn and poverty-ridden countries of the Middle East, Latin America, and Africa to cities in Europe and North America; according to the United Nations (UN Water, 2014) by 2050, 40% of the global population is projected to be living in areas subjected to severe water stress, defined as less than 500 $m^3 cap^{-1}$ $year^{-1}$ for all uses (residential, commercial, industrial, and agricultural) and for personal residential use the minimum use is 36 $m^3 cap^{-1}$ $year^{-1}$, which is 100 L cap^{-1} day^{-1} (Damkjaer and Taylor, 2017).
- *Increasing living standards,* resulting in more demand for food and, consequently, water and other resources

- *Aging and crumbling water infrastructure* (often built more than 100 years ago) magnified by the unwillingness of public officials to increase funding for rehabilitation and adaptation to the changes
- *More frequent and severe harmful algal (cyanobacteria, red tide) bloom (HAB) outbreaks*
- *Emerging pollutants* (endocrine disruptors, pharmaceutical residuals)
- *Conversion of urban waters* into effluent-dominated water will require management of the urban water hydrological cycle and decentralization of sewerage (Novotny, 2007a, b).
- *Severe and life-threatening air* pollution in many megalopolises such Beijing in China, Cairo in Egypt, and large cities in India. New Delhi (India) is the most air-polluted city; forest fires in 2018 made some Californian cities (San Francisco-Oakland; Los Angeles) the worst air polluted urban areas in the world.
- Increased flooding due to global warming effects, imperviousness and other land use changes through urbanization in the watershed.
- Plastics accumulation in the oceans, lakes, and environment

Table 1.2 lists other worldwide threats. A large portion of water shortages, flooding, and pollution in urban areas are caused by the typical characteristics of the urban landscape: a preference for impervious over porous surfaces; fast "hard" conveyance infrastructure rather than "softer" approaches like ponds and vegetation; and rigid stream channelization instead of natural stream courses, buffers, and floodplains. Under the paradigm of the last century, the hard conveyance and treatment infrastructure was designed to provide only five- to ten-year protection, but because maximum temperatures have reached record levels in almost all years in this century and hot seasons are longer, these systems are usually unable to cope safely with extreme events. In fact, they sometimes failed with serious consequences, as with Hurricanes Katrina in New Orleans in 2005, Sandy in New York and New Jersey in 2012, and Harvey in Houston and Maria in Puerto Rico in 2017. It has become evident that despite billions (trillions worldwide) of US dollars spent on the current paradigm of water management measures and solid waste disposal, these systems are inherently incapable of becoming sustainable and resilient. Meanwhile, global temperatures have been rising in this century and average high temperatures of summer months exceeding 30°C (86°F) were measured in central and northern Europe. If this trend continues, parts of the world (India, Pakistan, Bangladesh Africa, the Middle East, parts of Australia, and the southwestern US) will become uninhabitable.

Urban developments and population migration from distant suburbs back to cities are not necessarily bad for the environment; human urban habitats can mimic nature and preserve it, as demonstrated by urban ecotones and parks designed to mimic nature, as built by Frederick Law Olmsted in New York, Boston, Chicago, Milwaukee, and other cities almost 150 years ago (Hill, 2007, Heaney, 2007). The precipitation–runoff–groundwater recharge balance in future cities (both new developments and retrofitted historic communities) can approach the natural hydrologic cycle. Even today, a typical urban dweller in the US uses about half of the energy of an average American. In some cities, such as New York and San Francisco, about half the city dwellers use public transportation for commuting and many do not own an automobile. In the first two decades of this century, the initiatives of communities toward sustainability have been local. There is even a contest among city mayors for a title of most sustainable city. Some "green" organizations and magazines provide sustainability city ranking, based, for example, on the portion of renewable energy used by the

Table 1.2. Current global threats.

Earth Process	Boundary Threatening Earth and Restricting Development	Current Status and Threat
Climate change	Atmospheric concentration of CO_2 300 ppm as a boundary preventing catastrophic changes of Earth's climate.	Significantly exceeded, CO_2 concentration is rising and use of fossil fuels is still increasing.[*]
Loss of biodiversity	10 species per million species extinct per year.	Significantly exceeded.
Phosphorus cycle	11 million tons of phosphorus flowing from land to oceans. Most of the lost P is replaced by mined P.	Mineral P resources could be exhausted by the end of this century.
Nitrogen cycle	35 million tons of N are removed from the atmosphere each year for human use. This process requires a lot of energy and organic fossil fuel carbon and emits GHGs.	The limit has been significantly breached.
Global freshwater use	$4000\,km^3$ of freshwater is consumed each year by humans.	Regional water shortages are severe and some regions and states are irreversibly exhausting water resources, especially fossil groundwater sources. Reliance on desalination – a process that uses a great deal of energy – is increasing.
Chemical pollution	The levels of both atmospheric and water pollution in rapidly developing countries are reaching unhealthy and even life-threatening levels. New pollutants (pharmaceuticals, endocrine disruptors) are emerging in developed countries, along with the reappearing threats of excessive nutrients loads causing cyanobacteria blooms.	Atmospheric pollution in many cities in China, India, and other industrial countries has reached levels several times higher than the World Health Organization limits, to the point that breathing the urban air is dangerous. Extremely dangerous air pollution also comes from more frequent and more intensive forest fires.
Plastic	Plastic accumulation in oceans and some inland lakes is reaching close to catastrophic mass proportions.	This has severe impact for several reasons: fish and aquatic mammals consume plastics, coral reefs are damaged, plastics manufacturing causes GHG emissions, and large floating masses of plastic are now found in oceans and coastal lagoons.

Source: Adapted from Schoon et al. (2013)
[a] IPCC, 2013.

population, minimization of emissions from transportation, green living (parks, urban agriculture), recycling of solid waste, and other sustainability parameters (Svoboda et al., 2008).

1.5 POPULATION INCREASE, URBANIZATION, AND THE RISE OF MEGALOPOLISES

In the late eighteenth century, before the Industrial Revolution, world population was relatively small, growing at a rate of about 0.05% per year and reaching the first billion in the year 1800. World population in 2016 was 7.4 billion. Figure 1.4 shows world population recorded and forecasted from 1950 through 2050. World population in 2016 was growing at a rate of around 1.13% per year, more than 20 times faster than before the preindustrial age, and the current average population change is estimated at around 80 million per year. The rate of increase has almost halved since its peak of 2.2% reached in 1963. The annual growth rate is currently declining and is projected to continue to decline in the coming years. Currently, it is estimated that it will become less than 0.5% by 2050 (United Nations. 2017).

Future population growth is highly dependent on the path that future fertility will take, because relatively small changes in fertility behavior, when projected over several decades, can generate large differences in total population. While some earlier forecasts predicted that world population will level off after 2050 due to decreasing birth rates and diminishing resources, the revised 2015 United Nations (2017) report states that continued population growth after 2050 is almost inevitable, even if the decline of fertility accelerates. There is an 80% probability that the world population will be between 8.4 and 8.6 billion in 2030,

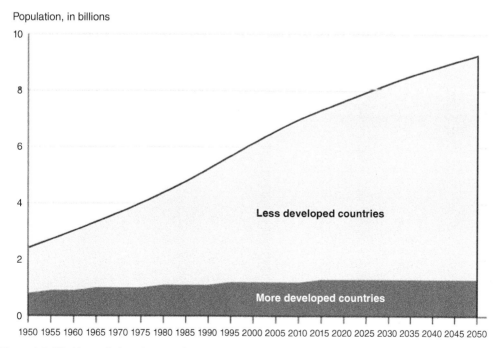

Figure 1.4. World population changes since 1950 and expected future growth, based on the optimal model. *Source:* United Nations, *World Population Prospects* (2017).

between 9.4 and 10 billion in 2050, and between 10 and 12.5 billion in 2100. The world population growth presented in Figure 1.4 represents the medium option.

The UN population statistics state that more than half of global population growth between now and 2050 is expected to occur in Africa. Thus, Africa's share of global population is projected to grow from 16% in 2016 to 25% in 2050 and 39% by 2100, while the share of population residing in Asia will fall from 59% in 2015 to 54% in 2050 and 44% by 2100. The rapid population growth in Africa will occur in the poorest 33 countries and could give rise to social upheavals and accelerated migration to Europe and wealthy Middle Eastern countries (Turkey and Arabic countries). This population increase and migration will affect mostly urban areas of the receiving countries. Less rapid but still significant growth will occur in Latin America, which will put pressure on migration to the US and Canada. As a matter of fact, most of the future population growth therein will be attributed to immigration and not to birth rates.

Population growth remains especially high in the group of 48 countries designated by the United Nations as the least developed countries (LDCs), of which 27 are in Africa. The population of this group is projected to double in size from 954 million inhabitants in 2015 to 1.9 billion in 2050, and further increase to 3.2 billion in 2100. In contrast, Europe is projected to have a smaller population in 2050 than in 2015. Several Eastern European countries and Japan are expected to see their populations decline by more than 15% by 2050 and due to the expected increase in longevity, this will represent a reduction in the proportion of the active working population over older retirees relying on Social Security payments and pensions. Forty-eight countries or areas are projected to experience population decline between 2015 and 2050. Table 1.3 shows population history and degree of urbanization.

It should be noted that the population of India is expected to surpass that of China. In 2015, the population of China was approximately 1.38 billion, compared with 1.31 billion in India. By 2022, both countries are expected to have approximately 1.4 billion people. Because of a high population growth in India and the policy of one child (changed to two children in 2015) per family in China, the population in China will stabilize around 2030 and begin to decrease thereafter. With unrestricted growth, after 2030 India will become the country with the largest population. India's population is projected to continue growing for several decades to 1.5 billion in 2030 and 1.7 billion in 2050 and then perhaps to stabilize in the second half of this century. Based on the median model, US population was expected to increase by 2050 by 67 million; however, demographers attribute this increase mainly to immigration. The effect of the 2018 US government policy changes regarding immigration is not known.

Table 1.3. Total and urban population in selected countries, in millions.[*]

Year	World	United States		United Kingdom		China[**]		Japan		India	
		Total	Urban	Total	Urban	Total	Urban	Total	Urban	Total	Urban
1950	2,525	159		51		544		83		376	
1990	5,268	255	191	60	45	1,184	308	122	94.5	853	222
2015	7,349	322	266	65	52	1,376	786	126.6	118	1,311	418
2030	8,500	356	300	70	57	1,415	902	120	114	1,527	607
2050	9,775	390	345	75	60	1,348	1,050	107	105	1,705	814
2100	11,200	550		82		1,004		83		1,659	

Source: United Nation population statistics.
[*]Values for some years were interpolated from the UN data.
[**]Not including Hong Kong and Macao.

The fraction of the population living in cities is expected to rise in almost all countries. McKinsey Global Institute (Woetzel et al., 2009) analytically described and forecasted the remarkable rapid but planned urbanization in China, which is building and expanding cities at a planned rate of almost 10 million each year until 2050. Urbanization is forecasted to increase therein from 26% in 1990 to 76% in 2050. By 2030 the policies of rapid urbanization will result in China having more than 220 cities with more than one million people. At that time (2030) the US will be about 87% urbanized. As shown on Figure 1.5, after 2015 all population increase will occur in urban areas. The economy of China has been developing rapidly in this century and cities are the places where most job opportunities are found. The rapid growth of urban areas has given a rise to megacities and megalopolises.

Examples of the megacities and megalopolises in 2015 and projected to 2030 are presented in Table 1.3. Arbitrarily defined, a megacity is a city with closely connected suburbs that has a population greater than 10 million. United Nation (2017) statistics reported that in 1990 there were 10 megacities with more than 10 million inhabitants, totaling 153 million people and representing less than 7% of the global urban population. By 2015 the number of megacities had tripled to 28, and the population they contain had grown to 453 million, accounting now for 12% of the world's urban dwellers. In 2030, there will be approximately 40 megacities and megalopolis agglomerations with a population of more than 10 million, of which 15 will be in China.

It was pointed out in the preceding section that urbanization and movement of rural people to the cities from urban sprawl suburbs in developed countries, if done right, may be beneficial for the energy and resource use and can lead to more sustainable living. Figure 1.6 shows the relation of carbon GHG emissions to population density. Similar relation can be developed for water use. Cities in the US, Canada, and Australia have the largest per capita $CO_{2\text{-eq}}$ emissions because of low density and the often still sub-standard public transportation in the US, which forces people to rely on commuting by private automobile and causes traffic jams, resulting in high fuel use. US cities also have the highest water use (footprint) because of lawn irrigation, especially in the suburbs, although there are exceptions.

Because of relatively high density, good public transportation by electric light rail, street cars, subway, and buses, and a great portion of electricity being provided by hydropower, both San Francisco and Seattle have the lowest per capita GHG emissions, at the level of

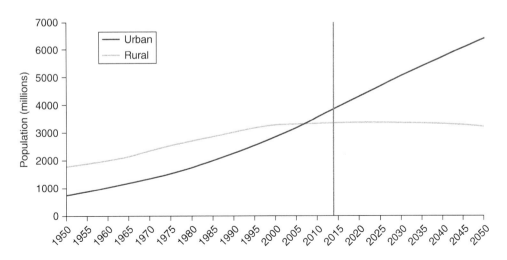

Figure 1.5. World urban and rural population. *Source:* United Nations.

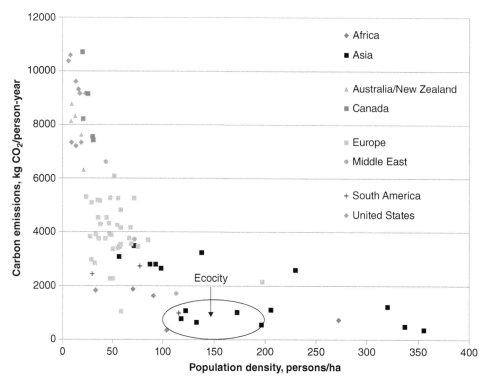

Figure 1.6. Relation of the annual GHG emissions to population density in selected world cities. *Source:* Data from various sources referenced in Novotny and Novotny (2009) and Novotny et al. (2010a).

a typical European city. Water use in some European countries is already approaching the sustainability level of 100 L/cap-day, which should also be a goal for the US.

With respect to carbon emissions, the optimum density for sustainable communities is between 100 and 200 people per hectare (Figures 1.6 and 1.7). High-density megacities with high-rises require more energy for pumping water to high elevations and for operating city

Figure 1.7. Hammarby Sjöstad (Stockholm), one of the first sustainable communities. Population density is 133 people/ha. *Source:* Wikipedia Creative Commons; picture by Peter Kulper.

lights, and more fuel for vehicular traffic in traffic jams. The low- and medium-density urban sections and suburbs will implement the low-impact development (LID) practices to reduce water use and pollution (see Chapter 3).

A McKensey Institute report (Woetzel et al., 2009) points out that by moving rural population to cities by planned urbanization and transfer, in 2025 China could cut its public spending requirement by 2.5% of GDP, or 1.5 trillion renminbi (Chinese currency; in 2018 1 renminbi = $0.16) a year, reduce SO_2, CO_2, and NOx emissions by 35%, halve its water pollution, and deliver private-sector savings equivalent to 1.7% of GDP, mainly through reduced consumption of natural resources.

Waste Accumulation

Only a fraction of waste discharged into water, emitted into air, or deposited onto land is biodegradable and eventually will be converted into harmless by-products. Many conservative and persistent pollutants entered the food web, biomagnified, and caused toxic contamination of flora, fauna, and humans. Persistent organic pollutants (POPs) possess toxic properties, resist degradation, bioaccumulate, and are transported, through air, water, and migratory species, across international boundaries and deposited far from their place of release, where they accumulate in terrestrial and aquatic ecosystems (Norstrom, 2002). Most of the chemicals that have been identified as POPs requiring elimination or restriction of use, or measures to prevent their formation as byproducts, either biomagnify in the food web of flora and fauna as unchanged chemicals (chlordane, DDT, dieldrin, hexachlorobenzene, mirex, toxaphene, PCBs, PCDDs, PCDFs), and/or form persistent, biomagnifying metabolites (aldrin, chlordane, DDT heptachlor). A US Geological Survey research report (Krabbenhoft, 1996) identified coal combustion in coal-fired power plants and waste incineration as main sources of mercury into Florida's Everglades, where fish contamination by methyl mercury is very high. Mercury can be converted in organic aquatic sediments into toxic methyl-mercury that biomagnifies in the aquatic food web from algae to small fish, to large fish, and finally to fish-eating birds and animals. Today most tuna and other large fish meats carry warning about mercury and other contamination. Because people are at the end of the food web, some contaminants are detected in humans.

There are numerous examples of "legacy" pollution, which links to the third paradigm of unrestricted discharges and emissions. The Earth is dotted with water bodies containing contaminated sediments and hazardous soils (brownfields) and abandoned but still polluting landfills. Large masses of plastics (bottles, packages, and many other items) are floating in oceans. The largest, the Great Pacific Garbage Patch, contains 80,000 tons of plastic and other garbage and in 2018 was twice the size of Texas or three times the size of France (https://www.theoceancleanup.com/). It severely interferes with aquatic life and it is expected to increase. Other important water bodies, such as the Venetian Lagoon, have the same issue. The problem with plastics has not yet been solved because they should not be incinerated and they decompose very slowly into particles (e.g., polyvinyl chloride, or PVC) that are even more harmful to the aquatic life and to humans. Nonetheless, solutions are available (see Chapter 8).

Brief Outlook Toward the Future

The revolution toward the new fifth paradigm of how cities and homes are built is already in progress. In the next 20 years, most homes and commercial houses in countries adhering to the Paris 21 agreement will have photovoltaics on the roof. The cost has already reached the breakeven level of affordability. But this is not just in the US or European countries; photovoltaic panels can now be seen on the roofs in the shantytowns of the underdeveloped world to provide electricity for TVs and other appliances when the dwellings are not connected to the electric grids, as well as on rural houses in China. Within a generation, France, Sweden, Denmark, Norway, Singapore, Austria, Israel, Iceland, Germany, New Zealand, Costa Rica, and possibly many others will be net-zero GHG impact countries.

Regarding the progress toward green technology and reduction of GHG emissions, "Global Trends in Renewable Energy Investment 2017" found that in 2016, wind, solar, biomass and waste-to-energy, geothermal, small hydro, and marine energy sources added 138.5 gigawatts to global power capacity, up 8% from the 127.5 gigawatts added the year before. The additional generating capacity roughly equals that of the world's 16 largest existing power-producing facilities combined.

In Massachusetts and some other states, subsidies are provided to homeowners to install solar power without paying any upfront costs, followed by monthly payments that are significantly smaller than the payment to the power providers before solar power installation. The state also is installing large-capacity off-shore wind power, and in 2017 it decommissioned its last regional coal power plant. In 2018, batteries were provided to communities and water utilities to store the excess energy generated during off-peak periods and save money by using it during peak hours.

1.6   WHAT IS A SUSTAINABLE ECOCITY?

Toward the end of the last century, after the major cleanup of water resources, avant-garde urbanization architects and developers realized the economic value of the intangible benefits of living closer to nature and they began to plan and implement new communities, sometimes adapting utopian visions or the ideas of early urban development giants like architects Frank Lloyd Wright or of Frederick Law Olmsted, father of the healthy urban landscape. The concept of the ecocity evolved at the end of last century and was first defined by Richard Register (1987), who coined the term *ecocity*. The following variant of Richard Register's ecocity definition has been accepted and promoted by the Ecocity Builders (2014):

> *An Ecocity is a human settlement modeled on the self-sustaining resilient structure and function of natural ecosystems. The ecocity provides healthy abundance to its inhabitants without consuming more (renewable) resources than it produces, without producing more waste than it can assimilate, and without being toxic to itself or neighboring ecosystems. Its inhabitants' ecological impact reflects planetary supportive lifestyles; its social order reflects fundamental principles of fairness, justice and reasonable equity.*

Following these developments at the end of the last century, the concepts of sustainable ecocities had been encompassed into criteria for some new and retrofitted old communities, which have been evolving as prototypes of sustainable (eco-) cities proposed, conceptualized, or built in Dongtan, Chongming, Tianjin, and Caofeidian (China); Forest City (Malaysia); Västra Hamnen and Hammarby Sjöstad (Sweden; Figure 1.7); Saadiyat Island and Masdar (UAE); Songdo International Business District (South Korea); Eko Atlantic (Nigeria); Ørestad (Denmark); Dockside Greens (Victoria, BC, Canada; Figure 1.8); and many other locations. The *sustainable city* is the most frequently occurring category, linked closely to the *ecocity* and *green city* concepts (de Jong et al., 2015). Advice and inspiration for ensuring that nature is included in the ecocity is covered in a book by Beatley (2017) that defines and describes *biophilic urbanism* (see Chapter 6), promotes well-being, creates an

Figure 1.8. Dockside Green ecocity in Victoria, British Columbia. Note 10 kW size wind turbines. *Source:* Copyright Dockside Green, Victoria, BC, Canada.

emotional connection to the Earth among urban residents, and describes urban biophilic innovations in more than a dozen global cities. The fundamental concepts of connecting urban areas with the nature and ecocities are also described by Ahern (2007, 2010), Farr (2008), Wheeler and Beatley (2014), Roberts, Newton, and Pearson (2014), and others.

Large cities are also undertaking transformation to a more suitable future. The uniform goal is to reduce CO_2 emissions by at least 80% by 2050, which also includes electrification of transportations, black and gray water reuse in coastal cities, using sea water for heating and cooling energy, recycling, and other approaches. The top cities listed in several rankings are Reykjavik, Iceland, where very soon all energy will be provided by geothermal and hydropower; Singapore, known to be greenest city-state in the world that recycles its water; Copenhagen, Denmark; Zurich, Switzerland; Stockholm, Sweden; and the three North American cities of San Francisco, California; Portland, Oregon; and Vancouver, British Columbia.

In addition to these new developments, in a more recent book dealing with ideas of cities that would be in balance with nature, Richard Register hypothesized that if "zero-car cities" like Venice, Italy, and Zermatt, Switzerland, had some solar energy retrofitting and more biodiversity in or adjacent to them, they would be very close to becoming ecological cities and villages (Register, 2006). A number of carless communities already exist, and based on this hypothesis, carless island cities such as Hydra Island, Greece; historic Venice and Capri Island, Italy; Fez, Morocco; and Mackinac Island, Michigan, would allegedly become perfect ecocities by installing blue energy and expanding local agriculture, although this might be difficult or impossible in those places. In historic Venice, cars, buses, and trucks were replaced by gondolas, water taxis, and boats (vaporeti); on Mackinac Island, by horse-drawn carriages and wagons; and on Hydra Island and Fez by donkeys, like in the Middle Ages.

These communities are more island tourist attractions than fully livable and functioning cities. Hence, such exaggeration would lead to a dead end if applied to current cities and urban zones. This idealistic striving toward biophilic future cities – which would not only be dense (thereby claiming avoiding environmentally damaging sprawl) but would also require people to accept a collectivist political philosophy of living in close neighborhoods, to use walking, bicycles (both highly recommended), and transportation incorporating donkeys and horses (highly questionable) instead of fossil-fuel vehicles – would not fit the new ecocity concepts. Furthermore, based on the experience of living and working in historic Venice for a long time, this author knows first-hand that this pearl of history and architectural beauty has serious water and sewage disposal problems, is sinking due overdraft of groundwater (reduced by bringing water from large distance by a long pipeline from the mainland), and in the last century had no sewers and only rudimentary sewage treatment mostly in the canals, as well as problems with disposal of solid waste exhibited by severe pollution by plastics and algae of the Venetian Lagoon surrounding the city, and lack of places to dispose of solid waste. Historic Venice is rapidly losing its permanent population and becoming a park where tourists come in the morning and leave at night, leaving their garbage behind.

However, the recent proliferation of expensive high-rise condominiums and apartment houses in major US cities, abandoned during the second half of the last century, proves that even well-to-do populations like "dense" but convenient locations close to work, especially when social services like daycare, schools, health clinics and hospitals, shopping, culture, areas for walking, good public transportation, and green areas with bike paths are nearby. The traditional large, dense cities like New York, Boston, Milwaukee, Chicago, Portland, Seattle, Minneapolis/Saint Paul, San Francisco, Copenhagen, Stockholm, London, and Paris are very desirable but very expensive places to live.

Based on the analysis of a census of 178 cities said to be ecocities in the first decade of this century, Joss et al. (2013) noted significant global proliferation of ecocity initiatives that have become mainstream and fashionable at various levels of government in many regions of the world. However, Grydehøj and Kelman (2016) and Caprotti (2014) noted that some of the ecocities mentioned above that are already built or are being built, including showcase Masdar in the United Arab Emirates, are indeed "island" developments that are good for testing and conceptualization of the ideas and/or socioeconomic and environmental relations, but they have not been as successful in becoming the flagship developments that would spread widely to other old and new urban developments. Some were abandoned (Dongtan near Shanghai), some slowed down or are on hold (Masdar) due to various reasons such as politics and reduced oil prices. However, in 2017, China was on the way to building many new urban communities designated as ecocities (Williams, 2017; Shepard, 2017). Shepard reported that ecocities are now being built across China, from the eastern seaboard to the fringes of Central Asia, from Inner Mongolia to the jungle hinterlands of the south. It has been estimated that in the near future, over 50% of China's new urban developments will be stamped with labels such as "eco," "green," "low carbon," or "smart."

Meanwhile, in addition to the attractiveness of the ecocity ideas, other more urgent reasons presented in this chapter have emerged in the twenty-first century, including global warming, loss of resources, water shortages, and population growth, which the urbanization paradigm of the last century cannot handle. In addition, catastrophes such as unusually strong hurricanes and heat waves have been occurring and are expected to intensify in the future. Therefore, at the beginning of this century, it was realized that the founding blocks of the ecocity are not only a desirable green architecture but also, and above all, water resources, infrastructure, water and energy management, used and stormwater disposal, and, by the same reasoning, also treating solid and liquid wastes as resources and sources of energy that would lead to saving water, reaching net-zero GHG emissions, sustainable suburban agriculture, avoidance of landfills, and several other key sustainability components. This leads to the following expansion of the Register's original ecocity definition:

A sustainable city or ecocity is a city or a part thereof that balances social, economic, and environmental factors (triple bottom line) to achieve sustainable development. A sustainable city or ecocity is a city designed with consideration of environmental impact, inhabited by people dedicated to minimizing required inputs of energy, water, food, and waste and outputs of heat, air pollution (CO$_2$, methane), and water pollution. Ideally, a sustainable city powers itself with renewable sources of energy, creates the smallest possible ecological footprint, and produces the lowest quantity of pollution. It also uses land efficiently, composts and recycles used materials, and converts waste to energy. If such practices are adapted, overall adverse contributions of the city to climate change will be none or minimal below the resiliency threshold. The cities of the future (COF) will have frugal energy and water (green) infrastructure, resilient and hydrologically functioning landscape, and protected and restored interconnected natural features and water bodies within their zones of influence. Urban (green) infrastructure, resilient and hydrologically and ecologically functioning landscape, and water resources will constitute one system.

Many commonalities would characterize sustainable (eco) cities, such as:

- Many sustainable new (eco)cities are built on lands that might not be suitable for development using traditional criteria. The land is either a brownfield (Hammarby Sjöstad) or an arid land (Tianjin Ecocity in China).
- These sustainable communities are not like low-impact developments (LID) in the US. Typically, they are medium-density multiple-family mixed-use developments (for example, Hammarby Sjöstad in Sweden or Dockside Greens in Victoria, British Columbia).
- They are all frugal with their water use.
- They all use less energy and aspire to reach a net-zero carbon footprint.
- They all recycle.
- Vehicular traffic within the city is restricted; people in the cities walk or use bikes and have convenient public transportation.
- Many ecocities are water-centered (water-centric) developments surrounded by or near a body of water and restoring the natural drainage and old water courses (see Chapter 6).

The Ecocity Builders (2014) define urban ecosystem as a biological environment consisting of all the organisms living in a particular area, as well as all the nonliving physical components of the environment with which the organisms interact, such as air, soil, water, and sunlight. The urban entities (cities, towns, and villages) are urban ecosystems that are also part of larger systems that provide essential services such as regulation (climate, floods, nutrient balance, water filtration), provision (food, medicine), culture (science, spiritual, ceremonial, recreation, aesthetic), and support (nutrient cycling, photosynthesis, soil formation). A bioregional cluster of ecocities with integrated waterways and agricultural lands and connected by public transit and bikeable greenways would constitute an ecological metropolis, or "ecopolis."

Impact of Global Warming and Continuing Overuse of Resources

Professor Rees, founder of the concept of "sustainability footprints" (see Chapter 3) in his seminal essay (Rees, 2014) stated:

> Environmental and earth scientists have shown that human demands on the ecosphere exceed its regenerative capacity and that global waste sinks are overflowing. The accumulation of anthropogenic greenhouse gases, particularly carbon dioxide, is perhaps the best-known example, but is just one symptom of humanity's frontal assault on the ecosphere—climate is changing, the oceans are acidifying, fresh waters are toxifying, the seas are overfished, soils are eroding, deserts are expanding, tropical forests are shrinking, biodiversity is plummeting. The growth of the human enterprise continues at the expense of depleting self-producing natural capital and polluting life support systems.

The worldwide plastic and garbage accumulation in the ecosystem overload that was realized with horror at the end of the second decade of 2000s and the global warming records at the beginning of this century confirm Professor Rees's postulate that economic expansion has its limits and the world has already surpassed its limits on fossil energy and resources and is at the onset of the fifth revolution regarding sources of energy. Rees has also pointed out that ecological damage and resource scarcity is largely the result of production and consumption to satisfy just the wealthiest 20% of the world's population. Roughly half the fossil fuel ever burned has been consumed in just the past 30 years. These are serious challenges to humans. Essentially, cities must reduce consumption and overuse of resources; the general term for overuse is *mining*, which means overuse of resources that are not renewable without replenishing them or fully mitigating their adverse impact (groundwater and fossil fuels mining, deforestation, etc.). This means that society must move to a more "sober" economic regime to limit and reduce the consumption of water, energy, and natural resources (including nonrenewables) and limit waste and air pollution discharges as outlined in a French study on urban integrated management (Lorrain et al., 2018).

The UN 2015 Resolution of Sustainability

In 2015 the United Nations General Assembly (UN, 2015a) adopted a resolution establishing the sustainability goals for 2030, which expressed the intent of countries "to protect the planet from degradation, including through sustainable consumption and production, sustainably managing its natural resources and taking urgent action on climate change, so that it can support the needs of the present and future generations." This charter addressed many important topics such as poverty, global warming, and use of resources. The UN recognizes that sustainable urban development and management are crucial to quality of life and promised to work toward reducing the negative impacts of urban activities and of chemicals that are hazardous to human health and the environment, including through the environmentally sound management and safe use of chemicals, the reduction and recycling of waste, and the more efficient use of water and energy. They also made a commitment to minimize the impact of cities on the global climate system. The goals pertinent to the topics and issues covered in this book are:

- Ensure availability and sustainable management of water and sanitation for all.
- Ensure access to affordable, reliable, sustainable, and modern energy for all.

- Promote sustained, inclusive, and sustainable economic growth, full and productive employment, and decent work for all.
- Build resilient infrastructure, promote inclusive and sustainable industrialization, and foster innovation.
- Reduce inequality within and among countries.
- Make cities and human settlements inclusive, safe, resilient, and sustainable.
- Ensure sustainable consumption and production patterns.
- Take urgent action to combat climate change and its impacts.
- Conserve and sustainably use the oceans, seas, and marine resources for sustainable development.
- Protect, restore, and promote sustainable use of terrestrial ecosystems, sustainably manage forests, combat desertification, halt and reverse land degradation, and halt biodiversity loss.

By 2030:

- Achieve universal and equitable access to safe and affordable drinking water for all.
- Implement integrated water resources management at all levels, including through transboundary cooperation as appropriate.
- Protect and restore water-related ecosystems, including mountains, forests, wetlands, rivers, aquifers, and lakes.
- Ensure universal access to affordable, reliable, and modern energy services.
- Substantially increase the share of renewable energy in the global energy mix.
- Double the global rate of improvement in energy efficiency.
- Enhance international cooperation to facilitate access to clean energy research and technology, including renewable energy, energy efficiency, and advanced and cleaner fossil-fuel technology, and promote investment in energy infrastructure and clean energy technology.
- Expand infrastructure and upgrade technology for supplying modern and sustainable energy services for all in developing countries.
- Reduce the adverse per capita environmental impact of cities, including by paying special attention to air quality and municipal and other waste management.
- Provide universal access to safe, inclusive, and accessible green public spaces, in particular for women and children, older persons, and persons with disabilities.
- Achieve sustainable management and efficient use of natural resources.
- Halve per capita global food waste at the retail and consumer levels and reduce food loss along production and supply chains, including post-harvest losses.
- Achieve the environmentally sound management of chemicals and all wastes throughout their life cycle, in accordance with agreed international frameworks, and significantly reduce their release to air, water, and soil in order to minimize their adverse impacts on human health and the environment.
- Substantially reduce waste generation through prevention, reduction, recycling, and reuse.

These goals represent the most important UN sustainability agenda goals.

THE NEW PARADIGM OF URBAN WATER, ENERGY, AND RESOURCES MANAGEMENT

2.1 THE SEARCH FOR A NEW PARADIGM

The process of defining the new management paradigm for water, stormwater, energy, and solids begins with defining the criteria classifying the future sustainable community that will arise from the necessity to save the world from the severe impacts of global warming and provide a pleasant and sustainable living environment for current and future generations, which is the core of sustainability. In this millennium, this new paradigm, emerging from the past successes and failures of providing water and controlling urban pollution and floods, offers a promise of adequate amounts of clean water for all beneficial uses and avoids adverse effects on present and future resources, the environment, and climatic conditions (Novotny et al., 2010a).

The emerging paradigm is based on the premise that urban waters are the lifeline of cities and the focus of the movement toward more sustainable "green" cities. These new concepts emphasize that all people – including city dwellers – are participants in an ecosystem and ultimately depend on the resilience and renewability of the ecosystem and on maintaining its services for present and future generations. Hence, communities must find ways to live adaptively within the loading (assimilative) capacity of the environment and the availability of resources afforded to them by the ecosystems of which they are a part (Rees, 1992, 1997, 2014). These planning considerations and designs must consider safe assimilation of residual greenhouse gases (GHGs) in the world ecosystems (e.g., oceans, forests, crops, aquifers) without causing global warming and other damages to the atmosphere and land on which we live. The linkages between socioeconomic and ecological systems mean that people must pay attention to the protection and, if necessary, restoration of resilient, self-organizing ecosystems that have the capacity for self-renewal in the wake of extreme disruptions, the frequency of which is expected to increase.

However, urban sustainability does not involve only closeness to nature, water, and energy. By the end of the twentieth century, communities had produced enormous volumes of refuse (garbage) that had been unsafely disposed in dumps in developing countries and in "engineered" landfills, emitting the potent greenhouse gas methane. Even in engineered landfills, capturing, and flaring methane collect only about 60 to 80% of the emitted

CH_4 (Bogner and Mathews, 2003). At the end of the previous century, less than 20% of municipal solid waste (MSW) in the US was recycled and only a small fraction of landfills collected methane to produce energy (US EPA, 2014a). Landfills also produce highly polluted liquid leachate that may overload wastewater treatment plants (called water reclamation plants at that time) and polluted groundwater in failed designs. In developing counties, unsanitary garbage dumps, often of enormous size, have been scavenged by poor people, who in a primitive way recycled anything that had some value (food, metals, clothing, carboard, wood for fuel). Commonly, children are involved in this.

Today, suitable landfill sites are difficult to find. Dumping solid waste, especially plastics, into surface waters has created massive "islands" of plastic mass floating throughout the oceans, harming and killing marine life. An article in the *Washington Post* (Phillips, 2018) reported finding a dead sperm whale in Indonesia that had 30 kg of plastic in the stomach and intestines, including trash bags, polypropylene sacks, ropes, and net segments, in such quantities that the whale was unable to expel the garbage from its digestive system. Since the start of 2018, more than 30 dead sperm whales have been found beached in the UK, the Netherlands, France, Denmark, and Germany with masses of plastic and garbage in their stomach. Today, MSW is considered a resource of a great value and, in the future, its management should not be separated from water, energy, and resource recovery.

The goals of urban sustainability must be developed by considering the overall situation in which our planet finds itself in the new milieu. The UN (2015b) agenda provides the topics but is only a guideline. It is up to the individual countries whether the goals become reality. This century is seeing a rapid rise in development in countries such as China, India, Vietnam, Brazil, and the Middle East, and in some countries in Africa (Kenya, Nigeria, South Africa) that in the last century were classified as Third World low-economy nations. These countries are now striving to achieve the same living standard as that of developed countries after the industrial revolution, but available global resources cannot sustain this development and reception of waste. If all countries were to achieve the same living standard as that of Europe, the amounts of resources would have to be three times greater than all resources currently available on the planet, and five times greater if all countries were to archive North American levels (Daigger, 2009; Schoon et al., 2013; Rees, 2014).

The rapid development of formerly undeveloped countries – specifically China, India, Brazil, and Indonesia – is creating pollution levels similar to those the developed countries of Europe, Japan, and North America experienced in the middle of the last century. As a matter of fact, their pollution may be worse because of much larger population numbers today and ongoing global warming. For example, as of 2016, because of exceedingly rapid development on the way to perhaps becoming the largest world economy, the air pollution in Beijing and other Chinese megalopolises, as well as in New Delhi, India, is as life threatening as it was in 1950s and earlier in London when thousands died, and it is far more widespread. Water pollution there is as bad as that 60–80 years ago in Japan or Ohio (for example, the Cuyahoga River fires). Because of overpopulation and global warming, Delhi is becoming unlivable for poor people during several months before the monsoon season. Air pollution in Beijing and water pollution in the second-largest freshwater in China (Lake Tai) have been widely publicized in the media. The impacts of current rapid development and resources exploitation were shown in Table 1.2.

Great advances in developing compact and highly efficient water and energy reclamation plants for wastewater, landscape designs based on the efficient best management practices to control and buffer diffuse pollution, and water conservation call for a fundamental change in the way water, stormwater, and wastewater are managed. A paradigm of integrated water/used water and energy management has been emerging for the last 15 years, but it has not yet been implemented on a large scale. Integrating municipal solid waste into this new paradigm is the next logical step.

Allan (2005) pointed out that to achieve the fifth paradigm, Integrated Resource Management and Recovery, requires a new holistic approach and a high level of political cooperation. The inclusive political process of the fifth paradigm necessitates including the interests of civil society, hierarchy (government), social movements (NGOs), and the private sector in the policy-making discourse (Thompson et al, 1990). To attain the fifth paradigm goals will require unprecedented levels of political cooperation and discourse. The new paradigm cannot rely on increasing mining of resources, which results in more unmanageable waste (e.g., plastics) and increased energy demands. It must take into account the threats of global warming, which require a number of considerations: implementing significant reductions in fossil fuel–derived energy and greenhouse gas (GHG) emissions, recovering resources such as nutrients from waste, and reusing and recycling waste, in addition to the traditional goals of providing water and safe management of residuals as a resource and not as waste.

Rees (2014) stated, "in theory, the global community is capable of deliberately planning and executing a 'prosperous way down' and still has the resources to do so. The goals would be to restore and maintain the ecosphere while ensuring social order and reasonable economic security for all. This approach requires a complete transformation of national and global development paradigms."

2.2 FROM LINEAR TO HYBRID URBAN METABOLISM

Water, ecological resources, carbon/energy, and economic sustainability footprints are all linked in an expression of urban metabolism, defined as the "sum of the technical and socioeconomic processes that occur within in the cities, resulting in growth, production of energy, and elimination of waste" (Kennedy et al., 2007). Cities and interconnected surroundings are complex systems consisting of nonliving infrastructure, machinery, roads, developed land, and ecosystems, which are accumulated, cycled, attenuated, and transformed within the system. They are intended to produce desirable outputs and should avoid or minimize undesirable outputs (air, water and soil pollution, waste, diseases, irreversible land losses). Currently, a great portion of key resources are needed for sustaining life and the development living organisms. Urban systems require inputs (food, energy, water, chemicals, materials) that are nonrenewable and some may even be exhausted in this century (Schoon et al., 2013) or soon thereafter, for example, mineral phosphorus for fertilizer production (Vaccari, 2009).

Urban metabolism can be linear, cyclic, or hybrid (in between), as shown in Figure 2.1. The balance or imbalance between the inputs, accumulation and growth, and waste – resulting in emissions of undesirable pollutants – determine the sustainability of the city. Current urban systems have usually been linear in terms of urban metabolism. Daigger (2009), Novotny (2008), and others agree that the current "linear" approach, sometimes called the *take, make, use, and waste* approach, has become unsustainable and cannot continue. The linear system discourages water reuse because the source of reclaimed water is far downstream from the city. The past and sometimes current economic benefit/cost or minimum cost evaluations have not considered important social – and in many cases environmental – costs and benefits of environmental protection and reuse and resource recovery, which have traditionally been considered intangible. Linear urban water and energy management exert highly unsustainable demands on resources and inputs (water, energy, food, chemicals, and materials). In the current linear system, economic *production*

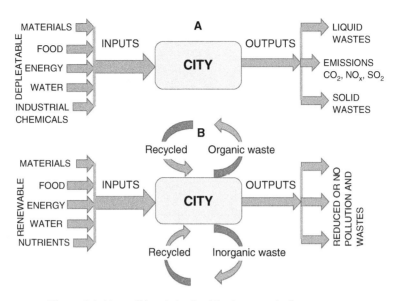

Figure 2.1. Linear (A) and circular (B) urban metabolism systems.

is mostly a *consumptive* process. Manufacturing, for example, irreversibly transforms large quantities of useful energy and material into an equivalent mass of useless waste (even the smaller quantity of product eventually joins the waste stream). Economic activity inexorably dissipates resources and increases the entropy (randomness, disorder) of the ecosphere (Rees, 2014).

Currently, most linear urban systems rely on unrestricted availability of resources and energy, which today does not exist. Linear communities, regions, or states cannot fully handle or may even disregard the adverse impacts of waste and GHG emissions on the environment and society. Often, the linear economy has led to environmental catastrophes during the mid-twentieth century in Europe and Japan and to similar apocalyptic events in twenty-first century China, India, and other rapidly developing countries, especially those located in areas of mining, heavy industry, and deforestation. Waste carbon dioxide being discharged into the atmosphere, mainly from energy production, is linked to the devasting forest and urban fires in California, which in 2018 also occurred in other countries (Sweden, Greece).

In the prevailing linear water systems that use the concept of *fast conveyance end-of-pipe treatment,* water taken from upstream or even from distant other watershed sources (as in ancient Rome and Byzantine times) is delivered to urban areas by underground conduits or surface lined canals (the California Aqueduct, the Arizona Canal, the South-to-North Water Transfer Project in China), is used and polluted, and is then delivered by underground conduits to a regional wastewater treatment facility typically many miles downstream from the points of potential reuse, and finally overwhelms the receiving water body with the partially treated effluent discharge and/or untreated overflows and bypasses. Similarly, electric energy is provided mostly by power plants that use nonrenewable fossil fuels and emit GHGs.

Rainwater and urban stormwater runoff are not waste products that must be discharged into underground pipe systems and conveyed to the nearest water resource or, combined with sewage to cause bypasses and overflows of untreated sewage into receiving waters.

Rainwater and clean (treated) stormwater are the second or third best and sometimes the only source of quality raw water and as such should be managed properly on site and in the neighborhood. Restoring the original hydrology and/or mimicking natural drainage, even in urban settings, is the goal and the solution not only to eliminating pollution but also to providing a good source of water and even some energy. These new resource recovery systems (RRS) are progressing in the following three domains (Guest et al., 2009):

1. *Water conservation and recovery.* Water conservation is achieved by switching from water-demanding household and commercial appliances and by significantly reducing outdoor water use. Hence, implementing water recycling should be universally accepted, first for nonpotable uses to further reduce water demand in water stressed areas. There are numerous examples throughout the world and today the list of remarkable case studies is already long (Novotny et al., 2010a).

2. *Reduction energy use and achieving net-zero GHG emissions.* The most common but still not widely used method of recovering energy from used water is digestion of residual organic solids, which produces methane, and recovering methane to produce energy instead wastefully flaring it. Methane can be also cleaned and sold to local natural gas distributors or cleaned and used as fuel for utility vehicles or heating reactors. It can also be reformed to hydrogen. To increase energy yield of digestion toward a net-zero utility operation, some water reclamation plants look for other sources of digestible organic carbon that can be found in wasted food, food production waste liquids, landfill leachate, or glycol-based airport deicing compounds (for example, Milwaukee Metropolis Sewerage District in Wisconsin or Biorefineries in Denmark; Veolia, 2018). Sensible (latent) heat can also be recovered by heat exchangers and heat pumps. However, combustion of methane derived from used water does not provide enough energy even to cover the energy needs of a conventional aerobic water reclamation plant. Including municipal solid waste (MSW) and other sources (manure, vegetation residues) in the integrated energy recovery system as well as on-site local renewable wind and solar energy production would not only provide enough energy for utilities (both water and solid waste) but would also cover a part of the community energy (per capita) needs. It could also result in net-zero GHG emissions and carbon sequestering for the utility and, in the future, for the community.

   This treatise also affirms that a dramatic reduction of fossil energy use can be achieved by switching, wherever possible, from the "black" fossil fuels to green and blue energy sources and carriers (see Chapter 4). Instead of plants powered by fossil fuel and traffic powered by gasoline produced by energy-demanding refineries, energy can be provided by renewable sources that include hydrogen, solar power, wind, the potential and kinetic energy of water resources, and sustainable electrolytic splitting of water molecules by excess green and blue energy to produce hydrogen. Hydrogen can also be produced biologically by fermentation of waste sludge and other organic waste matter, and abiotically by gasification of organic solids produce syngas and its subsequent steam reforming to H_2, or producing hydrogen from biomass by photocatalytic reactions through light. Several countries (France, Scandinavia, Germany, New Zealand) and California, Oregon, Washington, and some New England states will fully cover their energy demand by renewable (green and blue) sources by 2030.

   The ultimate goal of integrated water/solids/energy management is to reach within a generation (by 2050) a status wherein management of the above sectors and other economic sectors in the sustainable urban areas would be negative CO_2 emitters using

Negative Emissions Technologies (NETs) (NAC, 2018) that remove carbon from the atmosphere and sequester it. Removing CO_2 from the atmosphere and storing it has the same impact on the atmosphere and climate as simultaneously preventing an equal amount of CO_2 from being emitted. (Negative emission technologies are covered in Chapter 4.)

3. *Material recovery.* Producing processed biosolids and using them as fertilizer is still the most common, and is basically the only method of nutrient recovery from used water solids in typical water reclamation plants. Concurrently, recycling of some fraction of MSW has also been practiced and is expanding. Under the fifth sustainability paradigm, the nutrient recovery from used water and MSW will expand and will include production of phosphate (struvite) and, potentially, commercial-grade nitrogen fertilizers. The system will also produce biogas, biofuel, and hydrogen, and enhance traditional solid waste recycling of paper, carboard, aluminum, glass, and metals. Instead of incineration, combustible MSW can produce biogas by gasification that can be reformed to hydrogen and produced concentrated CO_2 can be sequestrated or even reused.

The author (Novotny, 2008) summarized the state of the art in the first decade of the 2000s and outlined the concepts of the emerging new sustainable urban water management system and the criteria by which their performance will be judged as follows:

- Integration of water conservation, stormwater management, and wastewater disposal into one system managed on a principle of a closed-loop hydrologic balance (Heaney, 2007).
- Consideration of designs that reduce risks of failure and catastrophes due to the effects of extreme events and that are adaptable to future anticipated increases of temperature and associated weather and sea level changes (IPCC, 2007).
- Decentralization of water conservation, stormwater management, and reuse of used water treatment into drainage and water/used water management clusters to minimize or eliminate long-distance transfers and enabling water reclamation and energy recovery near the point of use (Heaney, 2007). The municipal stormwater and used water sewage management is expected to be decentralized into urban clusters rather than regionalized. Decentralized management clusters with a simple water reclamation facility (e.g., a primary treatment followed by a wetland and/or a pond followed by disinfection) are especially suitable for megacities in developing countries.
- Incorporating green LEED-certified buildings (US Green Building Council, 2014) that will reduce water use through water conservation and storm runoff with best management practices, including green roofs, rain gardens, and infiltration.
- Possible recovery of heat and cooling energy, biogas (suitable in developing countries) and fertilizing nutrients from sewage in the cluster water reclamation and energy recovery facilities (Engle, 2007; Barnard, 2007).
- Implementing new innovative and integrated infrastructure for reclamation and reuse of highly treated effluents and urban stormwater for various purposes, including landscape irrigation and aquifer replenishment (Hill, 2007; Ahern 2007; Novotny, 2007a; US Green Building Council, 2014) that would also control and remove emerging harmful pollutants such as endocrine disruptors, trihalomethane (THM) precursors, and pharmaceutical (drug) residues.

- Minimization or even elimination of long-distance subsurface transfers of stormwater and used water and their mixtures (Heaney, 2007).
- Practicing environmental flow enhancement of effluent-dominated and flow-deprived streams (Novotny, 2007b), and ultimately providing a source for safe water supply.
- Implementing surface stormwater drainage in hydrologically and ecologically functioning landscape, and making the combined structural and natural drainage infrastructure and the landscape far more resilient to extreme meteorological events than the current underground infrastructure. The landscape design will emphasize interconnected ecotones connecting ecologically with viable interconnected surface water systems. Surface stormwater drainage is also less costly than subsurface systems and enhances aesthetic and recreational amenities of the area (Hill, 2007; Ahern, 2007; Beatley, 2017).
- Considering maximum pollution loading capacity of the receiving waters as the limit for residual pollution loads (Rees, 1992 and 2007; Novotny, 2007b) as defined in the Total Maximum Daily Load (TMDL) guidelines (US EPA, 2007), and striving for zero pollution load systems (Asano et al, 2007).
- Adopting and developing new green urban designs through new or reengineered resilient drainage infrastructure and retrofitted old underground systems interlinked with the daylighted or existing surface streams (Novotny et al., 2010a).
- Reclaiming and restoring floodplains as ecotones buffering the diffuse (nonpoint) pollution loads from the surrounding human habitats and incorporating BMPs that increase attenuation of pollution such as ponds and wetlands (Novotny et al., 2010a).
- Considering and promoting changes in transportation in the future cities, relying on clean fuels (hydrogen, electricity) and public transportation by electric or hydrogen-powered streetcars, buses, subways (driverless in the near future), and trains.
- Developing surface and underground drainage infrastructure and landscape that will store and convey water for reuse and providing ecological flow to urban flow-deprived rivers, and safe downstream uses; treat and reclaim polluted flows; and integrate the urban hydrologic cycle with multiple urban uses and functions to make it more sustainable. The triple bottom line accounting (economy, environment, and society) is the foundation for developing the sustainable systems.

Integrated resources recovery (IRR) is extensively covered in this treatise and is not limited only to water and used water domain. The concepts of co-digestion of biodegradable organic solids and liquids from MSW (solid and liquids) with residual organics from used water treatment and energy and resource recovery facilities makes sense and will substantially expand the resource and energy recovery, leading to net-zero (or better) goals. In addition, instead of low-efficiency incineration (a GHG and toxic air pollution producer) and landfilling, considerations will be given to processing combustible nonbiodegradable organic fractions of MSW by pyrolysis and gasification that would also accept residual waste solids from the water domain, instead of landfilling them. One of the key objectives in the integrated resource recovery is to minimize now and eliminate the need for landfilling in the future.

Circular Economy

A circular economy is an alternative to a traditional linear economy, in which resources are kept in use for as long as possible, extracting the maximum value from them while in use, then recovering and regenerating products and materials at the end of each service life. The system should be waste-free and resilient by design. In this model, the economy is organized in a way that is restorative of ecosystems, and ambitious with its innovation and impact on society. Circular economy closes resource loops, and mimics natural ecosystems in the way we organize our society and businesses (Circle Economy, 2015). Figure 2.2 presents the concepts and principles of circular economy, which are:

1. Design out waste and pollution.
2. Keep products and materials in use.
3. Regenerate natural systems.

Figure 2.2 distinguishes between technical and biological cycles (Ellen McArthur Foundation, 2017). Consumption happens only in biological cycles, where food and biologically based materials (such as cotton or wood) are designed to feed back into the system through processes like composting and anaerobic digestion. These cycles regenerate living systems, such as soil, which provide renewable resources for the economy. Technical cycles recover and restore products, components, and materials through strategies like reuse, repair, remanufacture, or (at last resort) recycling.

Reuse and recycling are the most efficient processes to reduce GHG emissions by preserving forests that assimilate CO_2 instead of burning wood for energy and making paper

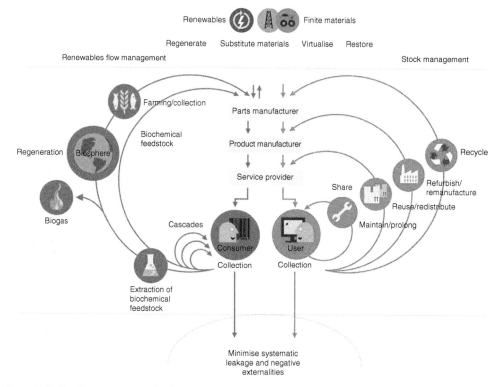

Figure 2.2. Circular economy cycle diagram. *Source:* Copyright ©Ellen MacArthur Foundation (2018); http://www.ellenmacarthurfoundation.org/circular-economy/interactive-diagram.

and cardboard and by saving energy by reusing materials and products. For example, reusing cardboard to make new cardboard from billions of cardboard boxes used in the US for mailing merchandise reduces water use by 99% and energy by 50%. Furthermore, cardboard and corrugated papers can be recycled up to eight times and the final residual solids can be composted or, better, gasified to methane by anaerobic digestion (see Chapter 7) or to synthetic gas (syngas: carbon monoxide and hydrogen) by gasification (see Chapter 8).

The circular economy goes beyond recycling. It is aimed at eliminating or reducing the need for landfills, making solid and liquids into resources, and reducing water and energy demand. Implementing circular economic systems crosses borders between water and sanitation (including municipal waste management), technology, food, agriculture, finance, and governance.

However, in 2018 the US recycling system was not working as hoped. It was not "economical" in the US to recycle and reprocess recyclables; consequently, every day roughly 3,700 shipping containers full of recyclables were trucked to US ports, loaded onto ships, and sent to China. The items in those containers included plastics, metal, paper, cardboard, and textiles, which, as reported by Bloomberg News (Arlington, 2017; Mosbergen, 2018), Chinese manufacturers used as raw materials. For China, imported garbage had replaced raw materials that this large country did not have (e.g., large forests to provide wood pulp for paper making). However, in 2017–2018, China restricted garbage imports and put very strict limits on the percentage of nonrecyclable materials in imported garbage (Mosbergen 2018). This has resulted in California and other locations sending recyclables to landfills.

The application of circular economy to the plastics problem has been addressed in the World Economic Forum (2014) report spearheaded by the Ellen MacArthur Foundation (2017). This report pointed out that 72% of plastic packaging is not recovered at all: 40% is landfilled, and 32% leaks out of the collection system; that is, either it is not collected at all and is dumped, or it is collected but then illegally dumped or mismanaged. Plastics use and management has an inherent design failure: its intended useful life is typically less than one year; however, the disposed material persists for centuries, which is particularly damaging if it leaks outside collection systems. The costs of such after-use externalities for plastic packaging, plus the costs associated with greenhouse gas emissions from its production, are very high, conservatively estimated at US$40 billion.

The WEF-EMF report proposed several circular economy solutions, including chemical recycling, which breaks down plastic polymers into individual monomers or other hydrocarbon products that can then serve as building blocks or feedstock to produce polymers again. However, the report also found that chemical recycling technologies are not yet widespread and/or not yet economically viable for most common packaging plastics. It will be documented that incineration of plastics for energy should not be applied because of air pollution and toxins. The WEF-MEMF report also lists pyrolysis as a possible method for breaking down the plastics into oil but points out that refinement of the pyrolysis oil or wax is costly and has a low market value. Chapter 8 of this book proposes as a feasible solution for breakdown of plastics along with other municipal solid waste and residual solids by indirect gasification to biofuel and syngas for energy in an integrated approach from waste to energy. The proposed integrated resource recovery facility (IRRF) will produce biomass, methane, syngas ($CO + H_2$), and concentrated CO_2, all of which can be used to produce plastic polymers.

Another approach to the resolution of the plastic problem featured in the WEF-EMF report is switching from fossil-based plastics to renewable (biomass) plastics sources. In the production of fossil plastics, petrochemical companies distill crude oil in different fractions, of which the naphtha fraction is the main feedstock for plastics production.

This fraction is cracked into monomer building blocks (e.g. ethylene, propylene). In producing renewable sources, plastics biomass or greenhouse gasses (methane, carbon monoxide and dioxide) in biorefineries are converted into the same or different monomers as the ones derived from fossil feedstock. The carbon footprint of bio-based PE (polyethylene), for example, has been found to be $-2.2\,CO_{2eq}$ (negative emissions – CO_2 is sequestered) per kilogram compared to positive $1.8\,CO_{2eq}$ emitted per kilogram of fossil-based PE produced (World Economic Forum, 2016). Petrochemical industry production of plastics today is the largest industrial black energy consumer and third-largest industrial GHG emitter. In a business as usual (BAU) scenario, GHG emissions from plastic manufacturing are expected to increase 30% by 2050 (International Energy Agency, 2016).

In general, attaining sustainability goals requires a change of the linear water/energy paradigm and metabolism to a hybrid circular (reuse, recycling) system, following the concepts of circular economy. Avoidance of the catastrophic and irreversible effects resulting from the ongoing global climate changes caused by excessive greenhouse gas (GHG) emissions is also a human survival and biota preservation issue that must be addressed. Because these climatic changes are not fully avoidable (IPCC, 2013), communities must become resilient and adaptable to the increased frequency of future extreme events.

2.3   URBAN RESILIENCE AND ADAPTATION TO CLIMATE CHANGE

The most common definition of resilience is the capacity of a system to recover from a damaging disturbance. By this definition, resilience can be measured by how fast the variables of the system's optimal conditions return to predisturbance levels (Walker, 1995). Urban resilience is the capacity of individuals, communities, institutions, businesses, and systems within a city to anticipate, plan, and mitigate the risks to adapt, survive, and grow no matter what kinds of chronic stresses and acute shocks they experience (Roberts, Newton, and Pearson, 2014).

While there are other potential catastrophes that could be detrimental (volcanic eruptions, earthquakes, forest fires, war, large terrorist attacks), the current increasing awareness of climate change has brought to the forefront large-magnitude stresses that were not anticipated in past designs of urban communities and that have already resulted in catastrophes not considered in past designs and planning.

Examples include increasingly stronger hurricanes with larger seawater tidal surges reaching more northern latitudes, such as the Hurricane Sandy's devastating impact on northeastern US coastal areas in 2012 (Figure 2.3) or the danger of increasing sea level rises due to glacier melting that, if not remedied, could result in losses of low-lying urban zones in Venice, Miami and other US coastal cities, large parts of Holland, many urban areas of Bangladesh, Pacific Island countries, and other places. It also includes more frequent catastrophic flooding on one side and drought and water shortages on the other side. Drainage infrastructure of cities, especially those in coastal areas, must adapt and become more resilient to the increased frequency and magnitude of extreme events. The unusually strong record-setting sequences of category 4 and 5 hurricanes Harvey, Irma, and Maria in 2017 devastated the Caribbean islands of Puerto Rico, Virgin Islands, and areas in Texas and Florida. Hurricanes Florence and Mathew brought catastrophic destruction and flooding to North and South Carolina and Florida in 2018.

Figure 2.3. A New York City subway station during Hurricane Sandy, 2012 *Source:* US Department of Transportation.

Resilience of socioecological systems to extreme events is related to (a) the magnitude of shock that the system can absorb and remain within a given state or a magnitude of increasing stressors (e.g., increasing temperature or nutrient loads) with respect to the assimilative capacity of the system to absorb them, (b) the degree to which the system is capable of self-organization during and after the shock, and (c) the degree to which the system can build the capacity for learning and adaptation (Folke et al., 2002). Urban resilience thus refers to the ability of the urban system to withstand a wide array of shocks (extreme events) and return to its more balanced and desirable pre-event state. Important as it is, climate change is then only one of the possible stresses cities will face in the future (Island Press and Kresge Foundation, 2015; Leichenko, 2011).

More resilient socioecological systems can withstand large shocks without changing in fundamental ways. Folke et al. (2002), quoting others, stipulated that the future is arriving so quickly that it cannot be accurately anticipated in planning. This leads to *adaptive planning and management,* wherein planning is a continuing repetitive process and corrective measures are included when more knowledge about the future stresses become available through better data and forecasting methodologies.

The pace of climate change and its severity is greater and more rapid than any other global changes in the existence of the human race. Ice ages arrived between one hundred thousand to ten thousand years ago over a period of thousands of years. Most severe and rapid climatic changes will occur in this century if the goals of the 2015 Paris Agreement are not met. In these cases, two useful tools may help to be prepared: (1) developing alternative structural (infrastructure) scenarios, and (2) adaptive management.

Folke et al. (2002) and Leichenko (2011) identified the following categories of urban resilience:

1. Infrastructure and hazard and disaster (engineering) resilience
2. Ecological resilience
3. Economic resilience
4. Governances or social resilience

To a great degree, these categories are interlinked. For example, hurricane hazards and disasters affect ecology, economy, and governance.

Engineering and Infrastructure Hazards and Disaster Resilience

Engineering resilience is a measure of the rate at which a system approaches a steady state following a perturbation (Folke et al, 2002). Flooding, hurricanes, tornadoes, and droughts would fall under this category. These events are generally expressed in terms of magnitude and frequency or probability of occurrence, which are interlinked. Figure 2.4 shows the concept of resilience to extreme events. The upper part depicts associated rainfall or flood flow. The system control variable on the lower part of the graph could be the amount of damage such as loss of habitable homes, infrastructure damage, aquatic flora and fauna habitat destruction, perhaps even loss of life. Very often what makes the system not resilient (vulnerable) are the policies that allowed development in areas that are highly vulnerable to repeating extreme events. The graph in Figure 2.4 represents the magnitude of the extreme stress, for example, the strength of a hurricane events such as development of sand dune islands and coastlines or areas vulnerable to extreme high tide surges. New Orleans has still not fully recovered from the 2004 destruction of Hurricane Katrina and large parts of the city probably will not recover to pre-Katrina conditions in the foreseeable future. Likely because the sections of the city that were most severely impacted by the flooding were inhabited by disadvantaged population, funds for recovery have been minimal. On the other hand, after similar effects from Hurricane Sandy in 2012, New York City and Hoboken, New Jersey, recovered their basic services in days and fully in few months after the hurricane.

Traditionally, because we deal herein with catastrophic events, the hydrologic and meteorological data from which design parameters for communities have been developed have been, at minimum, fifty-year-long records. To obtain a probabilistic relationship, annual maxima (flood, rainfall, snow depth, hurricane wind speed or category, high tide) or minima (droughts, annual rainfall) from the series of past years are correlated or plotted versus

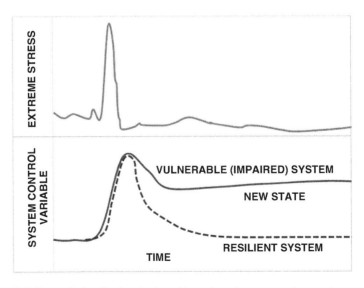

Figure 2.4. Concept of resilient and vulnerable systems in response to an extreme event.

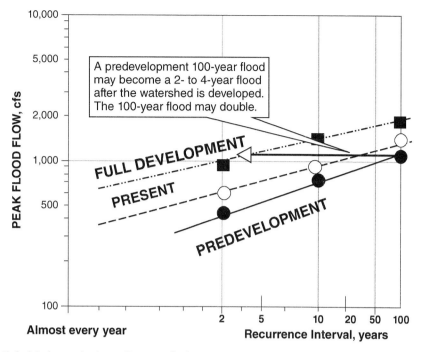

Figure 2.5. A plot of annual extreme flow magnitude versus recurrence interval for a suburban creek. To convert from cfs (cubic feet per second) to m^3 (cubic meters), divide the flows by 35.3. *Source:* From Novotny et al. (2010a).

probability of occurrence of the event being less or equal for floods or greater or equal for droughts. Instead of probability of occurrence (frequency), the magnitudes can be linked to the recurrence interval, which is the average number of years during which the magnitude of the event can be equaled or exceeded only once. Such relations for flood flows in an urban creek are shown in Figure 2.5.

The relationship between the probability of the event, P, in percentage of annual occurrence and recurrence interval, T, in years is:

$$P(\%) = 100/T.$$

For example, if the probability of a high flow being equaled or exceeded in any given year is 1%, then such flow on average would occur once in a hundred years. This leads to definition of extreme events in terms of years (e.g., 100-year flood). Hence, urban planners define the probabilistic annual period for design and hazard protection in terms of the probabilistic interval in years during which the event is equaled or exceeded once. Generally, residential development should not take place in areas that would be in the 100-year flood plain, defined as the area that has 1% or greater probability of being covered by flood in any year.

However, occurrences of catastrophic annual meteorological and hydrological events are not perfectly random; they may occur in lasting cycles of several years, such as those related to periodic warming and cooling of the Pacific Ocean (El Niño and La Niña). The most recent California drought lasted five years (2010–2015) and in 2016 the state was still recovering. Full recovery from water shortages occurred in the winter of 2016–2017 due to unusually large rainfall, snowfall, and mud slides that in some places were catastrophic, magnified by urban forest fires. The hydrologic cyclic phenomenon also implies that there is a greater probability of an extremely wet year being followed by another wet year than an extreme year being followed by a random dry year. This is also true for hot and cold years. This is known in climatic or hydrologic science as the Hurst phenomenon, derived from ancient thousand-year-long flow records of the Nile River in Egypt. It is also known in folk history as sequences of seven good or bad years. However, climate change is not random or a sudden shock; it is now due to continuous global warming, year after year, an increasing trend underlying seemingly random and cyclic fluctuations.

Figure 2.5 also shows that flood hazard probability can change by development, which increases watershed perviousness and/or by the climatic change that increases the magnitude and frequency of flood-causing rainfalls. However, such plots were developed using data gathered mostly in the twentieth century. Before 2018 it was extremely difficult to reconstitute what the precipitation records would be if global temperatures rise by one or two degrees Celsius in this century, yet the infrastructure and systems designed with past data will be dealing with hazards of the future. By the same reasoning, the current drainage built a century ago is incapable of handling current and future extreme hydrological events. Ongoing climate change is making past records and risk-based probabilities unreliable for predicting future hazard risks. Climate changes due to global warming are already occurring, seawater levels are rising, magnitudes and frequency of flood-causing rainfall are increasing, and occurrence of winter blizzards in the northeastern and western US has doubled since 1995. Communities must adapt to these changes and hope that stabilization may occur after 2050 because of the Paris COP 21 agreement. Moddemeyer (2016) and others pointed out:

- Without equilibrium or quasi-steady state, it is difficult to measure probabilities, although statistical methods that can separate trend, cyclic, and random statistical variations do exist.
- Without probabilities there is no measure of risk (risk = the probability of an unwanted event that may or may not occur).
- Without probabilities there is no measure of reliability and resilience.

Urban ecological resilience is related to the amount of change a system can undergo and remain within the same state of domain of attraction, withstanding perturbations and other stressors and maintaining its structure and functions. Examples of ecological functioning include maintaining biological, chemical, and physical integrity of the ecological system (required by the Clean Water Act in the US and the Water Framework Directive in the EU), nutrient cycling, biodiversity, avoidance of dissolved oxygen dead zones, and erosion control. Resilience also describes the degree to which the system is capable of self-organization, learning, and adaptation (Resilience Alliance, 2015; Holling, 1973; Walker et al., 2004).

The flip side of resilience is *vulnerability* (Walker et al., 2004; Novotny, 2003b), which identifies the weakness of the system to withstand and recover from adverse effects when a social or ecological system loses the resiliency to events that would otherwise be previously absorbed, or a trend has reached the ceiling or saturation that results in a shift to a less manageable status. In this context, like resilience, vulnerability can be economical and/or ecological susceptibility to extreme meteorological events, or weakness of governance. An example of governance vulnerability and weakness was the highly publicized lead contamination in Flint, Michigan, in 2015–2016, wherein state government action changed the source of drinking water, which resulted in lead dissolution from old lead pipes into potable water. City and state government could not handle this emergency and thousands were adversely affected by lead poisoning for years.

Adverse effects of ecological disruptions from global warming include loss of biota, shift from less-tolerant temperate species to more warm-weather and heat-tolerant species. In 2015–2017, when the adverse impact and dangers of global warming was finally widely recognized by scientists and then the public (but resisted by some public representatives and the fossil fuel industry), news media began reporting on the already occurring adverse changes to biota, increasing drought and catastrophic flood frequencies, dangers to coastal communities due to increasing sea levels, damages to Artic ecology in Canada, and the accelerated melting of glaciers in Antarctica and Greenland causing higher seawater levels. Coral reefs have been disappearing because of global warming and the increasing acidity of the oceans.

Vulnerability is linked to the *assimilation capacity* or resistance of the system to the change. In hydrological/ecological systems there are thresholds below which the system is resisting the change or can avoid the damage. Systems become vulnerable when they approach the assimilation capacity or threshold, and they become impaired when the threshold is exceeded. For example, a natural river channel can safely contain an annual high flow, which has a probability of less than or equal to 50% of being exceeded (Leopold, Wolman, and Miller, 1992) in an annual time series of maximal high flows. This, based on the above equation linking probability of exceedance to the period between the events, would represent the recurrence interval of two years. Flows exceeding the natural channel flows are considered floods. A severe flood occurring once or a few times in a century or centuries may cause severe damages to vulnerable watersheds that may take years to recover from. Figure 2.5 showed that urbanization increases the probability of flooding, as does global warming, which increases the frequency of extreme rainfall events. Hence, what was a hundred-year flood based on the hydrological data of the last century becomes more frequent flooding in this century because of global warming and more intensive urbanization. The resilience of an urban watershed against flooding can be increased by improving infiltration by implementing low-impact development (LID or SUDS or Sponge city concepts; see Chapter 6) and flood plain management and, above all, by planning the development and adaptation that would consider the uncertainties with the future adverse factors.

Good quality of inland fresh water resources is crucial to urban living. The quality of these resources was severely impaired in the US in the first 70 years of the last century (and before), but it has dramatically improved after the passage of the Clean Water Act (Public Law 92-500) by the US Congress in 1972. However, some problems deemed to be solved 30 years ago – for example, the poor quality of Lake Erie caused by excessive nutrient input stimulating algal growth – are reappearing in this century with a vengeance. Reappearance of cyanobacterial harmful algal blooms (HAB) in Lake Erie is suspected to be caused, at least partially, by warmer waters of the lake. Cyanobacteria HABs have become ubiquitous in some European countries (e.g., Czech Republic), in China (Lake Tai; see Figure 2.6), and in other countries (Novotny, 2011b). And the 2018 trophic collapse of Lake Okeechobee in Florida from cyanobacteria blooms and meteorological extremes caused large releases of nutrients into the Atlantic and the Gulf of Mexico, which resulted in devastating red tide flagellate blooms and harmful cyanobacteria blooms in the rivers connected the lake with the seas.

Cyanobacteria prefer warmer waters. As a matter of fact, the oceans when these microorganisms appeared for the first time three and half billion years ago were much warmer than lakes and coastal waters are now. The increased vulnerability of fresh and marine water bodies caused by climate change will require continuous increased use of economic resources to anticipate, adapt to, and mitigate the expected ecological threats. The water quality models of the last century were generally linear and often steady state, wherein algal growth was

Figure 2.6. The cyanobacteria outbreak in 2007 in Lake Tai (Taihu) in China, which provides potable water to 7 million people. This photo was taken by an anonymous Chinese student, and similar pictures of algal contamination were published in 2010 and thereafter by media for the coastal zones of Lake Erie and in 2016 for southern Florida coastal zones.

linearly related to the concentration of the limiting nutrient. However, Muradian (2001), Walker et al. (2004), and others pointed out ecological discontinuities of an ecological system as a consequence of the smooth and continuous change of an independent variable (a stressor, e.g., temperature or nutrient; Figure 2.7). Essentially, the system can persist in a more or less steady state continuum when subjected to a slowly increasing stress until a critical value or threshold is reached and the assumed continuous, often linear relationships between the stressors and environmental response system collapse in a relatively short period into another less desirable or undesirable state. Recent ubiquitous sudden emergences of harmful algal blooms (HAB) of cyanobacteria over a period of only few years in many water bodies of the world, including coastal areas of Florida and Mississippi, can be attributed to gradual increase of nutrient loads caused by agricultural sources (fertilizers and excessive manure spreading) and excessive suburban use of fertilizers (Novotny, 2011b), magnified by global warming and possible exhaustion of the soil assimilative capacity to absorb and retain the phosphates (Novotny, 2003a).

The same concept of ecological discontinuity expressed in Figure 2.7 can be applied to global warming. Because of gradually increasing releases of greenhouse gases, atmospheric temperature has been gradually rising since the beginning of the industrial age, with accompanying adverse effects on ecology, ice melting, increased sea levels, and several other effects (see Chapter 4). However, a study by the Stockholm Resilience Center, the University of Copenhagen, Australian National University, and the Potsdam Institute for Climate published in the *Proceedings of the National Academy of Sciences* (Steffen et al., 2018) revealed that there is strong evidence that the Earth is progressing to a planetary threshold that, if crossed, could push the Earth system toward an irreversible "Hothouse Earth," which could prevent stabilization of the climate and places would become inhabitable even as human emissions continue to be reduced. Thereafter, reducing or even stopping GHG emissions would not return the planet to a pre-Hothouse Earth status for hundreds of years.

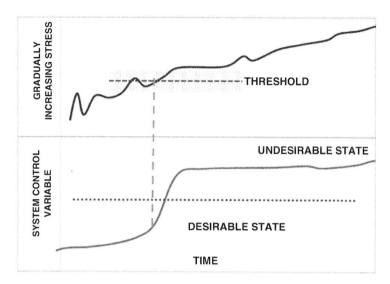

Figure 2.7. Ecological system nonlinear response to a gradually increasing stress and threshold of system adverse change.

In ecological systems, the thresholds are usually expressed as standards or criteria. Standards almost always include a *margin of safety* that accounts for uncertainties. Assimilative (loading) capacity of the system is then back-calculated by models from the standard and other pertinent variables defining the status of the system as it is being done in the US in the Total Maximum Daily Load Process (US EPA, 1991; Novotny. 2003). A criterion is equivalent to the threshold, while a standard includes a safety factor. The difference between the standards and criteria is legality. A criterion is based on scientific knowledge and a standard issued or approved by an authority has a legal binding and should contain a margin of safety to keep the system in the desirable state. The threshold standard for global warming agreed on by the IPCC and many other scientists is to keep the global atmospheric warming increase below 2°C.

Once the system transforms itself into a new state when the threshold is exceeded, reversal to the older "better" state is not easy, and simple reduction of the stressor(s) below the threshold may not suffice. For example, the change of impoundments providing water or recreational opportunities to urban areas from *mesotrophic/low eutrophic status* dominated by diatoms, green algae, and healthy fish population to *hypereutrophic status* exhibited by harmful algal blooms (HAB) dominated by toxin-producing cyanobacteria (Figure 2.6) that are lethal to animals, is relatively rapid. HABs are also characterized by low-quality bottom-feeding fish (carp, buffalo fish), zero oxygen during late night and early morning, and very low transparency. This undesirable state of the water body is caused by increased loadings of phosphorus from urban and agricultural sources, also exasperated by increased temperatures, and the threshold for the change to this undesirable state is reached when vernal concentrations of phosphorus exceed about 100 μg/L. See Table 2.1 for lake quality characterization and criteria.

However, once the hypereutrophic state is reached, remediation of HABs simply by reducing phosphorus input to the water body to a pre-HAB level is insufficient. Remediation is costly and requires multiple actions, including reductions of P loads to significantly lower levels than the threshold (typically down to 25 *μg*/L, which includes some margin of safety), preventing phosphorus inputs from sediments or even sediment dredging,

Table 2.1. Quality characteristics of eutrophication categories of impounded surface waters.

Trophic Index	Chlorophyll a µg/L	Total P µg/L	Transparency Meters[*]	Trophic Class	Characteristics
<30–40	≤8	≤10	≥3	Oligotrophic	Low primary productivity, low algae, high clarity
40–50	8–25	10–35	3–1.5	Mesotrophic	Intermediate level of productivity, submerged aquatic vegetation, good clarity, balanced and diversified aquatic biota
50–70	25–75	35–100	1.5–0.7	Eutrophic	High productivity, dominated by algae and aquatic plants
>70	>75	>100	<0.7	Hypertrophic	Harmful (toxic) blooms dominated by cyanobacteria, loss of submerged aquatic vegetation, zero oxygen, low-quality fish, loss of swimming

Source: Based on Carlson (1977) Trophic Index and OECD (1982) criteria (annual mean values).
[*]Transparency measured by Secchi disc visibility, which can be related to turbidity in NTU units.
Secchi disc (m) $\approx 0.01 * (NTU)^{-0.689}$ http://www.dakotaswcd.org/pdfs/mm_watclar_fs.pdf

restoration of balanced biota, installation of protective buffers (wetlands), biofilters, and vegetative shading to reduce temperature rise. And there is no guarantee that the undesirable eutrophication will not return, as occurred in Lake Erie.

Socioecological or Governance Resilience

Socioecological or governance resilience is the ability of human communities and their ecological systems to withstand external shocks or perturbations to their infrastructure, such as environmental variability or social, economic, or political upheaval and to recover from such perturbations (Folke et al, 2002). Restoring and protecting biodiversity of urban areas were outlined by Parris et al. (2018), Ahern (2010), Beatley (2017), and Novotny et al. (2010a).

Moddemeyer (2016) compiled the key principles of governance resilience, defined in Carpenter et al. (2012), Biggs et al. (2012), and Parris et al. (2018). The principles of biodiversity resilience can be summarized as follows:

1. *Diversity.* Systems can increase adaptive capacity with multiscale diversity. Horizontally distributed systems consisting of units at multiple vertical scales are more resilient than large homogenous systems. Also, higher values for resources that are climate-change proof can be designated, such as reuse of treated used water for non-potable purposes, including toilet flushing and irrigation, both of which typically use potable water.

   Urban landscape architects must re-create high- and low-biodiversity areas in and around the cities that would not only be viable but would be resilient to extreme stresses. It is almost impossible to re-create ecosystems once they are lost. Interconnected parks, nature preserves, wetlands, and golf courses are the places that can provide and maintain high biodiversity. These efforts have a long tradition, from the replacement of medieval walled zones with municipal gardens or to the city parks designed by landscape architect Frederick Law Olmsted.

2. *Modularity* means using cost-effective, modular, repeatable strategies. Modularity increases diversity of connectivity between systems providing additional buffers in times of shock or disruption. Nested semi-autonomous district-scale energy and water systems that use onsite renewable sources have different drivers and overlapping functions that can survive an impact to the larger centralized system. Centralized systems may continue to provide the backbone levels of service, but district-scale systems can relieve the peak demands on the centralized systems and provide additional buffers against extreme events.

   Urban biodiversity landscapes should also include planned and constricted biotic area such as green roofs, raingardens, biofilters, vertical forests, and so on. Water bodies such as creeks and streams that have been lost through construction that turned them into concrete-lined channels or underground sewers can be restored. (These urban landscape features are covered in Chapter 6.)

3. *System connectivity* is about the structure and strength of links between nodes and scales. By connecting most nodes of the system, the flow of information and resources can increase as access to insight, information, and feedback is expanded. For example, if one neighborhood or cluster of the urban system is severely damaged by an extreme event (hurricane), power resources, aid, water, energy and temporary habitation of the affected population can be provided by the connected, less-impacted, and/or more resilient urban units.

   The same aspects of connectivity can be applied to ecological systems. Urban aquatic and terrestrial habitat areas should be interconnected to allow movement of animals, fish, fungi, seeds, and pollen from one area to another. Infrastructure that prevents movement such as dams or simply a freeway cutting though the ecosystem should provide a passage for animals. For example, if biota is severely damaged in one unit by a toxic shock or an extreme climatic event, it can be repopulated and may recover through migration of flora and fauna from surrounding connected ecological units of the system. If a dam cannot be removed, fish ladders should be installed (Novotny, 2003a). Connectivity increases resiliency (see Chapter 6).

4. *Storage* necessitates restoring capacity of reserves at each scale so that isolated elements can survive for a period on their own. The overall ability of a system to absorb shocks or disruption is increased if essential reserves of energy, food, and water are stored at each scale. Peak energy demand can be attenuated if nonpeak uses can be shifted. Domestic hot water can also be stored at the individual building scale where peak demand for hot water can be shifted to off-peak demand by installing redundant hot water systems that are operated centrally by the energy grid provider. This creates additional buffering for wind and other intermittent renewables. In future systems, hydrogen can be produced and stored during times of excess energy availability and used as a source of energy during the times of green energy shortage. Similarly, batteries for electric energy storage at the district or even home levels are now available and becoming affordable (Cuthbertson, 2018).

   Urbanization eliminated the natural hydrology of the area by imperviousness and building fast conveyance underground drainage that interrupted natural infiltration and rainfall recycling and eliminated storage. These hydrological modifications made urban areas more vulnerable to flooding, erosion, or landslides. Without storage and recycling, biotic systems are susceptible to droughts as well as destruction of habitat and biota by flooding. Retaining and recycling water and nutrients are necessary for sustaining ecosystem services and biodiversity and for removal of pollutants that can adversely impact the biotic integrity of the ecosystem.

5. *Feedback* implies incorporating extensive monitoring and feedback loops to enable systems to moderate behavior and adapt as conditions change. Monitoring and understanding signals of impending change can provide feedback that helps to maintain stability in the face of shocks or surprise or long-term underlying variables. Remote sensing and ubiquitous monitoring tools are enhancing the ability to monitor and adapt to change. Early warning of changes underway enhances the ability of a system to accommodate change, whether it is in leak detection, identification of the use of energy from specific devices, water stress levels in urban vegetation, post-disaster mapping, or surveys of affected populations.

6. *Self-organizing* creates systems and subsystems that are semiautonomous and have capacity for self-governance and the ability to adapt to feedback. Self-organizing systems have greater capacity to self-correct given new insight, information, and feedback. They do not require extensive command and control and are the source of innovation that can bolster large systems. Nested self-organizing systems can create novel adaptations to dynamic change. Self-organizing systems and subsystems that have operational autonomy are more capable of making decisions in the field to address emergent issues in a timely manner.

GOALS AND CRITERIA OF URBAN SUSTAINABILITY

3.1 REVIEW OF EXISTING SUSTAINABILITY CRITERIA

To measure suitability progress and impact toward sustainability, the society and planners need measures of urban metabolism and sustainability criteria. A compendium of chapters selected from literature dealing with ecological footprints, urban metabolism, resource availability, comparing resources availability with demand, and resources and services from distant ecosystems (virtual water and resource use) is covered in a book by Munier (2006).

A "footprint" is a quantitative measure of urban metabolisms showing the appropriation of natural resources by human beings (Rees, 2014; Hoekstra and Chapagain, 2008). Footprints can be local or regional or global. Footprints related to urban metabolism have been identified in the literature:

- The water footprint
- The carbon footprint
- The ecological footprint

Large-scale (regional, global) footprints focus on sustainability of fresh water availability, energy and greenhouse emissions, and ecology. The criteria can also be local, such as the widely accepted LEED (Leadership in Energy and Environmental Design) (USGBC, 2014) rating restricted today to buildings, individual developments, or neighborhoods. The One Planet Living (OPL) narrative criteria originally developed by the WWF (formerly known as World Wildlife Fund, 2014) and other nonprofit environmental foundations in Europe address most of the concerns and establish sustainability goals for the communities (Table 3.1). Today One Planet Living is based on a framework of ten guiding principles developed subsequently by Bioregional, an international nonprofit environmental organization that also certifies OPL communities (https://www.bioregional.com/one-planet-living). At the end of the second decade of this century, 600,000 people are living in OPL communities such as Grow Community on Bainbridge Island, Washington (already developed), and Sonoma Mountain Village in California (construction begun 2019) in the US to developments in Great Britain and South Africa (Novotny et al., 2010a).

Table 3.1. One planet living criteria for communities, WWF: Bioregional.

Global Challenge	OPL Principle	OPL Goal and Strategy
Climate change due to human-induced buildup of carbon dioxide (CO_2) in the atmosphere.	**Zero Carbon**	*Achieve net CO_2 emissions.* Implement energy efficiency in buildings and infrastructure; supply energy from on-site renewable sources, and new off-site renewable supply where necessary.
Waste from discarded products and packaging create a huge disposal challenge while squandering valuable resources.	**Zero Waste**	*Eliminate waste flows to landfill and for incineration.* Reduce waste generation through improved design; encourage reuse, recycling and composting; generate energy from waste cleanly; eliminate the concept of waste as part of a resource-efficient society.
Travel by car and airplane can cause climate change, air and noise pollution, and congestion.	**Sustainable Transport**	*Reduce reliance on private vehicles and achieve major reductions of CO_2 emissions from transport.* Invest in transport systems and infrastructure that reduce dependence on fossil fuel use (e.g., by cars and airplanes). Neutralize carbon emissions from unavoidable air travel and car travel.
Destructive resource exploitation and use of nonlocal materials in construction and manufacture increase environmental harm and reduce gains to the local economy.	**Local and Sustainable Materials**	*Transform materials supply to the point where it has a net positive impact on the environment and local economy.* Where possible, use local, reclaimed, renewable, and recycled materials in construction and products, which minimizes transport emissions, spurs investment in local natural resource stocks, and boosts the local economy.
Industrial agriculture produces food of uncertain quality and harms local ecosystems, while consumption of nonlocal food imposes high transport impacts.	**Local and Sustainable Food**	*Transform food supply to have a net positive impact on the environment, local economy, and people's well-being.* Support local and low-impact food production that provides healthy, quality food while boosting the local economy in an environmentally beneficial manner; promote low-impact packaging, processing, and disposal and the benefits of a low-impact diet.
Local supplies of fresh water are often insufficient to meet human needs due to pollution, disruption of hydrological cycles, and depletion of existing stocks.	**Sustainable Water: Net-zero adverse impact**	*Achieve a positive impact on local water resources and supply.* Implement water use efficiency measures, reuse and recycling; minimize water extraction and pollution; foster sustainable water and sewage management in the landscape; restore natural water cycles.

Table 3.1. *(Continued)*

Global Challenge	OPL Principle	OPL Goal and Strategy
Loss of biodiversity and habitats due to development in natural areas and overexploitation of natural resources.	**Natural Habitats and Wildlife**	*Regenerate degraded environments and halt biodiversity loss.* Protect or regenerate existing natural environments and the habitats they provide to fauna and flora; create new habitats.
Local cultural heritage is being lost throughout the world due to globalization, resulting in a loss of local identity and wisdom.	**Culture and Heritage**	*Protect and build on local cultural heritage and diversity.* Celebrate and revive cultural heritage and the sense of local and regional identity; choose structures and systems that build on this heritage; foster a new culture of sustainability.
Some in the industrialized world live in relative poverty, while many in the developing world cannot meet their basic needs from what they produce or sell.	**Equity and Fair Trade**	*Ensure that a community's impact on other communities is positive.* Promote equity and fair-trading relationships to ensure a community has a beneficial impact on other communities both locally and globally, notably on disadvantaged communities.
Rising wealth and greater health and happiness increasingly diverge, raising questions about the true basis of well-being and contentment.	**Health and Happiness**	*Increase health and quality of life of community members and others.* Promote healthy lifestyles and physical, mental, and spiritual well-being through well-designed structures and community engagement measures, as well as by delivering on social and environmental targets.

Source: Adapted from https://www.bioregional.com/one-planet-living.

LEED Criteria for Buildings and Subdivisions

The US Green Building Council has proposed and is continuously developing standards for "green" buildings and neighborhoods (USGBC, 2014) that are becoming a standard for building and development. Each federal, state, and city-owned building in Chicago is expected to comply as closely as possible with LEED standards wherever possible (for example, installing a green roof) and to implement water conservation. Green roofs reduce runoff and provide substantial savings on energy use, which again reduces greenhouse emissions. New green tall buildings are being showcased in New York, the Republic of Korea, Berlin, Milan, Singapore, and elsewhere throughout the world. Most consultants and city planners have tried as best as they can to adhere to LEED's concepts and standards (USGBC, 2014). "Green" subdivisions and satellite cities are now sprouting throughout the world and in the design studios of urban landscape architects. The concept and designs of ecocities of up to several hundred thousand inhabitants are now being implemented in the United Kingdom, Canada, Sweden, Singapore, China, Australia, Abu Dhabi, Saudi Arabia, and elsewhere. The USGBC standards for "green" certification was formulated for homes, neighborhood development, and commercial interiors. The new construction and reconstructions standards include the following categories:

- *Sustainability* of the sites such as site selection and development, brownfield (previously developed and contaminated land) development, transportation, or stormwater design
- *Water efficiency* in landscape irrigation, innovation in wastewater technologies, and reuse and water use reduction
- *Energy and atmosphere*
- *Material and resources* such as construction materials and waste reuse and recycling
- *Indoor environmental quality*
- *Innovation and design*

The pilot LEED Neighborhood Rating System (USGBC, 2014) added the following categories:

- *Smart location and linkage,* including required indices of proximity to water and wastewater infrastructure, floodplain avoidance, endangered species; protection, wetland, and water body conservation, and agricultural land conservation
- *Neighborhood pattern and design* such compact development, diversity and affordability of housing, walkable streets, transit facilities, access to public spaces, or local food production
- *Green construction and technology,* essentially LEED building certification
- *Innovation and design process*

The LEED standards are aimed at buildings and small neighborhoods. They are not *a priori* related to natural resources and the value (total number of points) for the natural resource protection and water resources conservation is relatively small; only about 15% of the points are credited for reducing water use and potential contribution to improving integrity of waters and natural resource. There are no credits for restoration of water bodies or wetlands as a part of the neighborhoods. Maximum two points are available for implementing sound stormwater management strategies and diffuse pollution controls.

Triple Net-Zero (TNZ) Goals

The TNZ goals publicized and promoted by builders and developers (Li et al, 2013; Goldsmith et al., 2011; Torcellini and Plees, 2012) were defined as:

- *Zero waste water from buildings and sites* by limiting the consumption of fresh water resources, treating and returning clean water back to the same watershed so that the groundwater and surface water resources of that region are not depleted in quantity and quality over the course of a year and resources from used water such as nutrients are reclaimed
- *Net-zero (negative) GHG emissions* from building and sites by producing as much blue and green energy on site as they used over the course of a year, recovering energy from used water and waste solids

- *Zero solid waste to landfills* in a community or a site by reducing, reusing, and recovering waste streams and converting them to resource values, including energy, with zero landfilling over the course of a year

Considerable attention in the literature and at professional meetings has been devoted to definitions of the parameters and metrics of the TNZ goals (Marszal et al, 2011; Torcellini and Plees, 2012). The question to be addressed is what "net-zero" represents and how it differs from "absolute zero" (Hernandez and Kenny, 2010). The TNZ goals, as worded above, are striving to achieve no adverse impact of a development or community on nonrenewable diminishing water resources (by overuse and pollution), preventing global warming, and loss of productive and natural land by urban developments such as buildings, campuses, military bases, and subdivisions (Goldsmith et al., 2011; Gerdes, 2012) or by pollution, including liquid and solid wastes disposal on land. The triple net-zero goals are not as comprehensive as the OPL goals, but they are as ambitious and, essentially, include the most important aspects of sustainability. However, achieving TNZ goals does not necessarily imply achieving the sustainability goals and vice versa.

However, there is a general consensus among scientists that the net-zero carbon goal after 2050 may not be enough and will have to be strengthened to achieve overall *negative CO_2 emissions* defined in the National Academies Committee on Developing a Research Agenda for Carbon Dioxide Removal (NAC) (2018) as a situation wherein more CO_2 is taken from the atmosphere by implementing CO_2 removing technologies (e.g., photosynthesis by agriculture and forest management, and CO_2 sequestering) than is emitted from human activities. This concept is more stringent than that advocated by the IPCC, which does not categorize CO_2 emissions from non-fossil fuels as "greenhouse gases." The section "Carbon Dioxide Sequestering and Reuse" in Chapter 4 describes negative emission technologies in more detail.

It is also becoming clear that focusing on these goals separately will give rise to doubts about the reality of achieving these goals. The goal to reach net-zero carbon emissions by 2030 or 2050 has a lot of sceptics, many supported by the fossil fuel industries. There is no doubt that achieving this goal will require a lot of funds, which in a society running large deficits are simply not available. At the same time, landfilling garbage will continue because there are no other alternatives (in 2018) for disposing garbage. So, society is paying large sums of money for disposing garbage with no benefits. Realizing that solid waste is a resource and source of green and blue energy, recovering energy synergistically from municipal solid wastes with other biodegradable and combustible waste solids (manure, vegetation residues, sludge, organic waste deicing liquids) can generate a lot of energy from other resources (biofuel, fertilizers) by using already available and increasing funding that today is going to solid waste landfilling. This idea of addressing the net-zero goals symbiotically as triple net-zero interlinked goals is a mantra of this book.

The triple net-zero water, energy, and wastewater sustainability management of military installations and small cities was in 2015 the subject of a NATO workshop (Goodside and Sirku, 2017). Speakers discussed how to implement and integrate triple net planning and energy, water, and waste sustainability strategies into broad installations of operational management, how to arrive at the best decision, how to create policy, and how to communicate effectively to stakeholders. The workshop explored current and emerging technologies, methods, and frameworks for energy conservation, efficiency, and renewable energy within the context of triple net-zero implementation practice.

The three dimensions of sustainability – economic, environmental, and equity (Novotny et al., 2010a) – are better expressed by the OPL criteria and should also include consideration of resilience and preserving resources for future generations (Brundtland, 1987). Equity considerations are not included in the TNZ criteria and economic criteria are only *a priori* assumed. On the other hand, sustainability may be achieved without reaching absolute TNZ. The environment has an ability to provide water without depleting the sources, to accept residual waste discharges without damaging the integrity of the water resources as assessed by the Total Maximum Daily Load Planning concept included in Section 301 of the Clean Water Act (National Research Council Committee, 2001; Novotny, 2003a), and by anthropogenic and natural sequestering to accept certain emissions of GHGs without increasing atmospheric CO_2 concentrations, to withstand the stresses, and to avoid damaging public health. Estimating this assimilative capacity is the most important aspect of the sustainability assessment.

Water Footprint

Water availability stresses are caused by population increase, overuse, and increased pollution. The US is the largest user of water for industrial, agricultural, and municipal uses wherein the average per capita municipal water uses before 2010 was around 550 Liters/capita-day. In the US, indoor domestic water use is relatively constant among the major urban areas (Heaney, Wright, and Sample, 2000), averaging 350 Liters capita^{-1}day^{-1} for a household without water conservation, and 136 Liters capita^{-1}day^{-1} for a household practicing water conservation. Figure 3.1 shows the water footprint for some representative countries of the world in the first decade of this century. Overuse, mining, exhaustion of

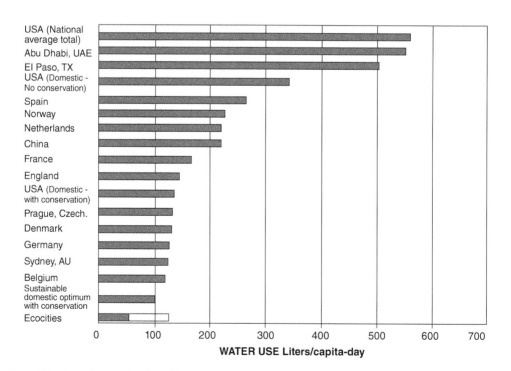

Figure 3.1. Water demand in selected countries and cities, 2010. *Source:* Updated from Novotny et al. (2010a).

fossil underground aquifer resources, pollution, and climate change are the major threats and causes of unsustainable water uses. 100 L cap^{-1}day^{-1} represents adequate minimum water consumption. Less than that calls for recycling.

The water availability situation has become critical in many parts of the world for various reasons and is expected to worsen. It is obvious that the severe water shortages in California and the southwestern US during 2011–2015 were due less to random chance and were more the result of climate change and gross overuse, underlined by the El Niño/La Niña climatic cycles. But in many parts of Africa, the Middle East, and northeast China, adequate water resources are simply not there. Providing water to rapidly increasing urban populations of these regions is done through highly energy-demanding seawater desalination (1.5 to 5 kW-h/m^3 in California, more if water is taken from high-salinity seas such as the Persian/Arabian Gulf) (Asano et al., 2007), and this source is only available to wealthy coastal states. Looking for water in deep salty aquifers (remnants of ancient seas) may not be feasible because of high energy use, high cost of treatment and desalting, increasing radioactivity of deep aquifers, and problems with disposal of high-salinity polluted water rejected from the desalination process.

Some avant-garde green building architects intended to design and build showcase absolute-zero water buildings such as, for example, the Kenton Living Building in Portland, Oregon (City of Portland, 2014), which would not need an outside water supply. The original intent of the design was to derive all water needs from rain falling on the premises and recycled gray water. Conceivably, this would be nothing new insofar as there are some large inhabited islands in the Adriatic and Mediterranean Seas (e.g., the Croatian island of Hvar in the Adriatic), where subsurface karst geology could not provide fresh groundwater and all surface small streams are ephemeral; hence, most of the water supply used to be provided by rainfall captured in communal and house wells and cisterns. Today, water for Hvar Island is provided by a pipeline from the mainland and on some other islands by desalination. Using house and communal cisterns for rainwater capture dates to Roman (100 CE) and Minoan (2500 BCE) civilizations (Novotny et al., 2010a). The British colony of Gibraltar (pop. 29,000) during the blockade by Franco's Spain (1939–1975) derived its fresh water by capturing rainwater, and even today uses this method.

However, it does not make much sense to accept this absolute zero goal in areas that have adequate fresh water supply and/or adequate aquifer recharge. The net-zero goal is then understood as prevention of aquifer drawdown and protection of all beneficial surface water uses, including providing adequate ecological flow to support indigenous biota in receiving water bodies, and implementing water conservation in all cases because of the connection of water use to energy and GHG emissions. Hence, the first step toward the net-zero water use impact is fresh water conservation, followed by or combined with gray water recycling and rainwater (stormwater) capture. Because of high energy use, cost, and problems with reject water disposal, desalinization is the source of last resort but there are now many coastal communities in the Middle East relying solely on desalted water. In the TNZ concept, any energy associated with providing water, recycle treatment, and pumping and disposal of residuals must be derived from or compensated by blue or green renewable energy sources.

Virtual water refers to the water use outside of the city area that is used to produce food, materials, and other goods to satisfy the needs of the people living in the city. Such water-demanding activities outside of the city include agriculture, production of electricity, and making construction materials, paper, and, today, biofuel from corn or sugar cane or mining oil derived from tar sands. Virtual water footprint in cities in developed countries is about three times as large as the household on-site use (Hoekstra and Chapagain, 2008) and should be included in the life-cycle assessment of the net-zero water footprint goal. Virtual-water flows are calculated by multiplying, per trade commodity, the volume of trade by the respective average water footprint per ton of product. Extensive lists of water footprints and virtual water use is presented in the UNESCO report by Mekonnen and Hoekstra (2011).

GHG (Carbon Dioxide) Net-Zero Footprint Goal

It is now generally accepted that the Earth is undergoing a period of adverse global climatic changes, also known as global warming, caused by excessive emissions of greenhouse gases (GHG) from power plants, traffic, industrial operations, home heating, landfills, and so forth, including more frequent forest fires, and heat trapped in the atmosphere. GHGs include carbon dioxide (CO_2), methane (CH_4), nitrous oxides (N_xO), fluorinated gases, and some other gaseous industrial compounds (IPCC, 2013), defined in Chapter 2. In the US, on average, 1 kW-hour of energy produced in 2010 represented about 0.6 kg of CO_2 emissions (Novotny et al., 2010a). This ratio should be decreasing because some US power-generating utilities are switching from coal to natural gas, which has a lower ratio of GHG emissions per kW-h of electric energy produced. Also, the use of blue energy (solar, wind) is increasing. However, the focus in North America on deriving new energy fuel (both gas and oil) from shale by fracking or from (Canadian) tar sands – which requires large amounts of water and energy for extraction and pipeline transportation – may have either no effect on reducing GHG emissions or make them worse. This fact must also be considered in the life-cycle sustainability analyses.

On the other hand, in some European countries and in San Francisco and the Pacific Northwest (Portland and Seattle), the carbon emissions per kW-hour and proportionally the per capita carbon emissions are significantly less than the US national average. In 2015 in France only 10% of energy was produced from fossil fuel, 79% was nuclear, and 11% was hydro energy, which resulted in the ratio of $CO_{2eq.}$ emissions per kW-hour energy produced of less than 0.1. The 2014 French Energy Act anticipates France becoming a truly net-zero GHG country by 2025 by reducing nuclear power from the current 79% to less than 50% and replacing this loss and that of fossil fuel by shifting to renewable wind and solar sources and 30% to energy conservation and recovery (Carre, 2014). Other advanced European countries (Germany, UK, Sweden) are also extensively implementing renewable energy production projects. Today in Norway, Italy, Brazil, Paraguay, and Argentina, most electric energy is derived from hydropower. In small country of Iceland, all energy is geothermal or hydro. Massachusetts and possibly other states are planning large purchases of electricity produced by hydropower plants in Quebec, Canada, and installing their own off-shore wind power turbines. As of 2017, Massachusetts abandoned coal as a source of energy. Hawaii plans to become net-zero carbon emissions territory by 2030.

In the US Pacific Northwest, most energy is provided by hydropower, with an increasing portion also generated by blue and green energies. Consequently, GHG emission per energy used in 2013 in the State of Washington was 0.15 kg CO_2/kW-h (Puget Sound Energy, 2014) and Seattle is planning to be the first large net-zero energy US city. Based on the data from the US Energy Information Administration (2013), the average per capita residential and commercial energy use reported by USEIA (2013) was 10 837 kW-h capita^{-1} year^{-1}.

The energy issue and the goal of reducing the GHG emissions are included in the LEED criteria from the US Green Building Council (2014). Getting gold or platinum LEED certification requires energy savings but achieving this certification does not automatically mean achieving net-zero energy status. Following the efforts already in place in most advanced European countries (UK, France, Scandinavia, Germany), there were calls in the US to achieve the net-zero energy within a generation. President Obama's Executive Order 13514 required all agencies to meet several energy, water, and waste reduction targets in government buildings (more than 500,000) and operations, including implementation of the 2030 net-zero-energy building requirement in all commercial buildings as set by the Energy Independence and Security Act of 2007. However, the interest in large-scale implementation of sustainable energy by the US administration after 2016 is at best uncertain, but private US building industries are slowly accepting this goal and have embarked on developing guidelines for net-zero energy goals. Several net-zero energy buildings have already been built in the US and in the UK, Germany, Japan, Korea, and elsewhere. Green net-zero buildings and cluster (ecoblocks/subdivisions) developments will usually include (Novotny et al., 2010a):

- Passive architectural features for heating and cooling (see Chapter 4)
  - Southern exposure with large windows regulated by shutters
  - Cross ventilation
  - Green roofs
  - A lot of insulation
  - Energy-efficient lighting
  - Possibly vertical forest on large buildings (see Chapter 6)
- Landscape features
  - Shading trees (planting trees also sequesters carbon), urban forests
  - Pavement cooling by reclaimed water to reduce urban heat island effects and heating by waste heat from cooling water in winter to control ice (instead of using salt)
- Renewable energy sources
  - Solar photovoltaic and concentrated solar thermal roof panels
  - Wind turbines
  - Heat in used water (especially in gray water and sludge), stormwater, and municipal solid waste
- Water conservation and reuse, addressing the entire water (hydrologic) cycle within the development, including rainwater harvesting and storage
- Distributed stormwater and used (waste) water management to enable efficient water reuse and renewable energy production
- Xeriscape of the surroundings that reduces or eliminates irrigation and collects and stores runoff from precipitation
- Energy-efficient appliances (e.g., water heaters), treatment (e.g., reverse osmosis), and machinery (e.g., pumps, aerators)
- Connecting to off- and on-site renewable energy sources such as solar power and wind farms
- Organic waste solids and liquids management for energy recovery
- Connection to low or no GHG net emissions heat/cooling sources such as heat recovered from used water or from ground or air or geothermal
- Smart metering of energy and water use and providing flexibility between the sources of water and energy
- Sensors and cyber infrastructure for smart real-time control

Water/Energy Nexus

US EPA (2013) estimated that municipal drinking water and wastewater systems account for approximately 3% to 4% of energy use in the United States (7–15 in California) and are adding over 45 metric tons of greenhouse gases annually to the atmosphere. Furthermore, potable water and wastewater treatment plants are typically the largest energy consumers of municipal governments, accounting for 30–40% of their total energy consumption. Energy as percent of operating costs for drinking water systems can also reach as high as 40% of the total water energy needs and, under the scenario of retaining the current paradigm technologies (business as usual), is expected to increase 20% in the next 15 years due to population growth and tightening drinking water regulations. This will yield the following current water/ energy nexus value for the urban water management:

Energy for extracting, treating, delivering potable water and treating and disposal of used water is:

$$E_W = 0.01 * 3.5 \ (\%) * 10{,}837 \ (\text{kW-h h capita}^{-1}\text{day}^{-1})/365 = 1.03 \ \text{kW-h capita}^{-1} \ \text{day}^{-1}$$

Using the average municipal water demand of the end of the last century (550 Litre capita^{-1}day^{-1}), the energy for delivering and disposing 1 m^3 becomes:

$$E_{WV} = 1.03 \ (\text{kW-h/c-day})/0.55 \ (\text{m}^3/\text{c-day}) = 1.87 \ \text{kW-h/m}^3 \ \text{of water delivered.}$$

US EIA (2013) data also show that in 2009 (based on the 2010 census), 17.7% of the total domestic energy use was for water heating. This percentage remains constant between the states. The rest amounting to 82.3% is for air conditioning (average 6.2%), appliances, electronics, lighting (34.6%), and space heating (41.5%). The values for air conditioning and space heating vary among the states. Hence, the amount of energy for water heating becomes:

$$E_{WH} = 0.01 * 17.7(\%) * 10{,}837 \ (\text{kW-h capita}^{-1}\text{day}^{-1})/365 = 5.24 \ \text{kW-h capita}^{-1}\text{day}^{-1}$$

The AWWA-RF (1999) study also indicates that only about 15% of the water heating energy can be reduced by shower-restricting fixtures; however, a larger portion (~75%) can be recovered by a heat pump installed on gray water or total used water flow pipes (Novotny et al., 2010a). Most of the water heat energy is in gray water flow, which includes bath (shower, bathtub) and laundry drain water. Kitchen sink water and dishwasher waste in the US are considered black water because of food waste disposal through the drain.

Ecological Footprint

Ecological footprint (eco-footprint) is defined as the area of productive land and water ecosystems required, on a continuous basis, to produce the renewable resources that the population consumes and to assimilate its carbon wastes (Rees, 1992, 2014). Rees estimated that the average person on Earth needs 2.7 gha (global hectare) of productive land to satisfy the life needs, but the average citizen in developed countries of Europe requires the productive and carbon assimilative capacities of 4–5 gha per capita to support current levels of consumption. The North American eco-footprint is 7 gha/capita. There are about 12 billion gha of productive land on the Earth and the global population in 2018 was 7.7 billion, which implies that only 1.6 gha/person of productive land was available at the end of 2018. In addition, the area will keep diminishing with increasing population and loss of productive land due to desertification, increasing droughts resulting from global warming, and conversion of productive land to urban paved and built-over zones. This means that the eco-limit has already been exceeded by three to five times. Canada, with its relatively small population and large territory, is one of only a handful of countries whose domestic bio-capacity (15 gha/capita) is more than adequate to satisfy domestic demand (6.5 gha/capita) (Rees, 2014). The same is true for the world largest country by area, the Russian Federation, which has a smaller eco-footprint than that for North America or Europe.

The root causes and solutions are known and to some degree covered in this book but not yet followed by societies, which means that for some time the disparities will be increasing, and ecological and social catastrophes like massive migrations and starvation will be more frequent and more devastating. Rees (2014) highlighted that working cooperatively for the common good of sustainability and resilience will require the ardent exercise of several intellectual and behavioral qualities that are unique (or nearly so) to our species:

- High intelligence, the capacity to reason logically from available facts and data
- The ability to plan ahead, to direct the course of events toward desired ends
- An unequalled array of socio-behavioral means and mechanisms for cooperation
- The capacity for moral judgment, the ability to distinguish right from wrong
- The ability to empathize with other people and even nonhuman species and to exercise compassion toward "the other"

The goals would be to restore and maintain the ecosphere while ensuring social order and reasonable economic security for all. This approach requires a complete transformation of national and global development paradigms.

3.2   ZERO SOLID WASTE TO LANDFILL GOAL AND FOOTPRINT

The sources of municipal solid wastes (MSW) in a community are numerous, and achieving net-zero waste input into landfills must address the problem in an integrated holistic manner. In the US, the quantity of generated MSW grew steadily from 80 Mt (metric tons) in 1960 and increased to 238 Mt (10^6 tons) in 2015, an increase of 195%. During this fifty-five year stretch, the US population increased by 71% (321 million in 2015) on per capita basis, MSW generation increased from 1.2 kg capita^{-1}day^{-1} in 1960 to 2.03 kg capita^{-1}day^{-1} in 2015 (US EPA, 2018a) and will further increase because of rapidly expanding web shopping and mailing.

In 2015, 34.7% of the generated 238 Mt MSW was recycled or composted, nearly 13% combusted with energy recovery, and 52% discarded, mostly into landfills (US EPA, 2014a, 2016, 2018a). Disposal of MSW to landfills from 1960 to 2012 dropped from 89% to 52%, but on the mass basis the drop was from about 145.3 Mt in 1990 to 124 Mt in 2015. This decline was due to a significant increase in the amount of waste recovered for recycling and composting as well as that incinerated. The relatively slow but nevertheless steady drop of MSW going to landfill under the current scenario would meet the zero waste to landfill goal in a very distant future – more than one hundred years. The European Community is asking nations to increase recycling to reduce landfilling to 10% of MSW by 2035.

Despite far better and more sanitary disposal of MSW in the US than that in a typical megalopolis in developing countries, in 2005 US landfills emitted 132 Mt of CO_{2eq} in methane. (Kwon et al., 2010), which decreased to 100 Mt of CO_{2eq} in 2015 (91 Mt from

MSW landfills, 9 Mt from industrial landfills) due to increasing implementation of landfill gas (LFG) capture and flaring and energy production. Municipal solid waste (MSW) landfills are the third-largest source of human-related methane emissions in the US, accounting for approximately 14.1% of these emissions in 2016 (US EPA, 2017). Larger sources of methane are the production and transport of natural gas, coal, and oil, and livestock and other agricultural practices.

Figure 3.2 presents the municipal solid waste production and composition in the US. Out of the 2.03 kg of MSW capita^{-1}day^{-1} of the total mass produced in the US in 2015, 0.86 kg capita^{-1}day^{-1}, or 52%, is biodegradable organic waste (food 15.1%, paper and paperboard 25.6%, and yard trimmings 13.2%) that can be preferably digested to produce methane or composted (including about half of paper, such as toilet paper, paper diapers, or paper cups and plates). Composting emits non-GHG CO_2. Of this, 28.6%, or 0.57 kg person^{-1}day^{-1}, is nonbiodegradable or difficult to biologically degrade combustible solid waste, which can be better gasified (plastics 13.1%, rubber and leather, 3.2%, textiles 6.1%, and wood 6.2%) and some incinerated. Gasification can process plastics to biogas; however, less efficient incineration cannot because combusting plastics emits toxins. In 2015 17%, or 0.37 kg person^{-1}day^{-1}, was inorganic waste, of which a great portion can be recycled (such as construction waste, metals, and glass), as well as waste that is not recyclable (such as fireplace or boiler furnace ash). Landfill disposal of household coal and wood ash waste has mostly been eliminated in the US and the majority of European countries by switching to natural gas for heating. Gasification leaves less than 10% ash residue, which can be used as soil conditioner or in construction. The deposition of the MSW into landfills is shown in Figure 3.3, which documents that the net-zero solid waste to landfill goal is very challenging and at the present rate of the reduction of MSW landfilling, the net-zero goal may not be achieved in this century. However, switching to MSW

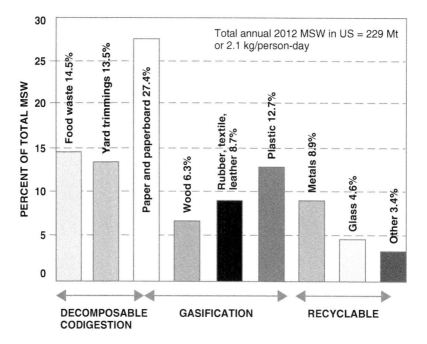

Figure 3.2. Municipal solid waste production in the US, 2012. *Source:* US EPA, (2014a).

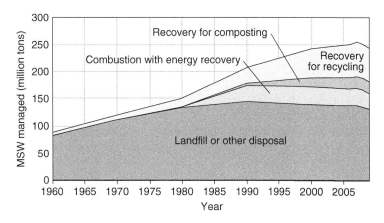

Figure 3.3. Trend in municipal solid waste generation in the US, 1960–2009. *Source:* US EPA, (2014a).

gasification (not incineration) could make a difference toward achieving the goal much earlier, producing clean energy, and recovering other resources.

Solid waste can be separated and collected on a house level. Some "cities of the future" (e.g., Hammarby Sjöstad in Sweden) installed an underground vacuum system delivering sorted solid waste from buildings to district processing and to the energy and resource recovery centers (Figure 3.4). MSW has a great potential as a gasifier feedstock because it has a high heating value (HHV) (dry basis), about 7 kW-hour/kg ds MSW, which is nearly as high as most conventional biomass feedstocks (Valkenburg et al., 2008). It is readily available in the near-term, and it has a preexisting collection/transportation infrastructure and fee provided by the supplier that does not exist for conventional biomass resources such as

Figure 3.4. Solid waste separation in Hammarby Sjöstad (Stockholm). The deposition inlets are connected to an underground vacuum conveyance system that brings refuse to recycling and recovery district facilities. *Source:* Courtesy Malena Karlsson (GlashusEtt) Hammarby Sjöstad.

woodchips and agricultural residues (Valkenburg et al, 2008). The resources value of MSW is tremendous. MSW high-calorific value for energy production and recycling has also a high virtual energy saving value. Per unit of mass, aluminum production has the highest energy requirement; therefore, recycling just one ton of aluminum cans is equivalent to conserving 44 MW-hr electricity or saving 6,295 liters (1,665 gallons or 4.65 Mt) of gasoline (US EPA, 2014a). However, most aluminum in the US is produced in the Pacific Northwest, which has an abundance of blue hydropower with very low or zero CO_2/Kw-h ratio, but recycling aluminum makes this low GHG energy available for use elsewhere.

The problems with deposition of solid waste into landfills include finding the suitable landfill areas, problems with groundwater contamination by highly concentrated and contaminated landfill leachate seepage and its disposal, and loss of land. Solid waste, far more than liquid used water, is a resource and the loss of these resources represents a great portion of the negative ecological and carbon (methane emission) footprints and opportunity cost.

Currently (2015), there are 86 facilities in the United States that recover energy from the combustion of municipal solid waste. These facilities are located in 25 states, mainly in the Northeast. Although no new facilities have opened in the country since 1995, some expanded to handle additional waste and create more energy. These facilities have the capacity to produce 2,720 MWatts of power per year by processing more than 28 million tons of waste, which in 2015 represented about 12.7% of the total MSW. A typical waste to energy plant generates about 550 kilowatt hours (kWh) of energy per ton of waste. EU countries incinerate 20% of their MSW (Quina et al., 2011), but Sweden incinerates most of their combustible MSW.

The adverse effects of MSW incineration on people are both local and regional. Emissions from MSW incineration contain harmful and polluting acidic gases (SO_x, HCl, HF, NO_x, etc.), volatile organic compounds (VOCs), dioxins from burning plastics such as PVC (McKey, 2002), polyaromatic hydrocarbons (PAH), polychlorinated biphenyls (PCBs), polychlorinated dibenzo-p-dioxine/-furans (PCDD/Fs), and leachable toxic heavy metals (He et al., 2009). Dioxins in higher concentrations are lethal persistent organic pollutants with irreparable environmental health consequences on affected populations. Because of this pollution, no MSW incinerators were built in the US after 1995 but they are now becoming ubiquitous in China, often with serious adverse environmental consequences.

Landfill Gas (LFG)

Landfill gas (LFG) is a byproduct of the decomposition of organic materials in MSW in the anaerobic conditions of landfills. By volume, LFG typically contains 45% to 60% methane and 40% to 60% carbon dioxide. It also includes small amounts of nitrogen, oxygen, ammonia, sulfides, hydrogen, and carbon monoxide, and less than 1% non-methane organic compounds and trace amounts of inorganic compounds such as nitric oxides (Lee, Han, and Wang, 2017). Because the GWP of methane is 25, most of the GHG total value will originate from methane. The amount of created LFG primarily depends on the quantity of waste and its composition and moisture content, as well as the landfill design and management practices.

LFG can be collected and combusted in flares or energy recovery facilities to reduce emissions. MSW landfills receive approximately 69% of the total waste generated in the United States and produce 94% of landfill emissions. The remainder of the emissions is released by industrial waste landfills. The number of active MSW landfills in the United States has decreased from approximately 7,900 in 1988 to 1900 in 2009 (US EPA, 2014b) to 1540 in 2015. The situation with inactive landfills and dumps in the US is not good. There are

Figure 3.5. Landfill gas collection. *Source:* Wisconsin Department of Natural Resources, Creative Commons, flickr.

about 10,000 of such sites and many may still be emitting landfill gas (LFG). Some inactive landfills have been capped, but not all emitted methane is captured and flared.

Obtaining energy from MSW at the end of the last century was by capturing LFG from landfill (Figure 3.5) and, after processing to increase its methane content, generating energy by methane combustion. Captured methane used for electricity and related heat generation emits CO_2 but the produced electricity and heat provide CO_2 green energy credit. However, in the second decade of this millennium only 650 landfills (42%) captured landfill gas (LFG) and 75% (480 or 31% of total landfills) of those generated energy, which represented a power capacity of 2,160 MW. The rest has been flared. LFG collecting facilities on average can capture 60% to 90% of formed and released methane. In the most engineered landfills, LFG is extracted from landfills by drilling wells, and then using a blower or vacuum system LFG is routed into a central gas collecting where it is "flared" (burned) to meet strict environmental requirements or sent to energy facility. The flaring practice requires expensive collection and flaring systems with absolutely no return on investment once LFG is withdrawn from the landfill. Environmentally, flaring changes GHG gas methane that has a GWP of 25 of equivalent molar mass of carbon dioxide with a GWP of one. At poorly engineered landfills and inactive landfills in the United States and elsewhere, a portion of greenhouse gases seep back into our environment.

A more advanced and profitable alternative is to treat the collected LFG to high-energy-content biogas that consists of about 95% methane, 4% nitrogen, and 1% carbon dioxide. After removing all contaminants and volatile organic compounds (VOCs) LFG can be sold to a natural gas pipeline, replacing the fossil natural gas, thus eliminating the environmental consequences of solid waste decomposition. Milwaukee Metropolitan Sewerage District (MMSD), East Bay Municipal Utility District in Oakland (CA), Veolia Co in Denmark, and a few other wastewater treatment utilities co-digest food waste with sludge and other biodegradable solid and liquid waste to increase their methane production. MMSD also accepts landfill biogas gas to become a net-zero energy (GHG) treatment facility (Sands, 2014).

Estimating GHG emissions from landfills requires complex modeling. Lee, Han, and Wang (2017), following the IPCC (2008) landfill GHH estimates, pointed out that the GHG emissions from a given amount of landfilled organic solid wastes depend on the fate of the carbon in the waste. Organic carbon in the deposited MSW in landfills is (1) C sequestered in landfills to become, after many years, a precursor to solid organic carbon (about 50% per IPCC); (2) C in waste is decomposed to CO_2; (3) C in waste is decomposed to CH_4 methane; (4) produced methane can be oxidized in the aerobic surface layer to CO_2; and (5) in managed landfills CH_4 is collected and combusted either to convert potent GHG methane (GWP = 25) to CO_2, which has GWP of one, or collected and combusted to produce electricity. About 10% to more than 30% of the CH_4 is oxidized to CO_2 in the aerobic landfill cover. The amount of organic carbon sequestered in the landfill could be larger than 50% used in the IPCC report (Lee, Han, and Wang, 2017).

Using the approximate chemical formula representing wet anaerobic MSW decomposition in a landfill and knowing that the CO_2 and CH_4 gases produced by decomposition of one-half (or less) of the MSW biomass deposited in the landfill, the total emissions can be approximately estimated. The "chemical" composition of MSW is extensively covered in Chapter 8 in the section "Gasification of Municipal Solid Wastes (MSW)." The formula describing the chemical composition of combustible (volatile) New York City MSW is $C_6H_{9.4}O_{3.9}$, or more generally $C_6H_{10}O_4N_{0.06}$ (Themelis, Kim, and Brady, 2002). The approximate formula for generation of the landfill gases is:

$$C_6H_{10}O_4 \rightarrow 1.6\ CH_4 + 1.4\ CO_2 + C_3H_{3.6}O_{1.8}$$

where the last term on the right side represents the mix of the solid/liquid carbonic residuals (Organic C, H_2, H_2O or CO) of MSW decomposition. Simply, one "mole" of MSW has molecular weight of $6 * 12 + 10 + 4 * 16 = 162$ grams. 1 kg of organic (volatile) MSW solids has 1 (kg) $* 1000$ (g/kg)/162 g = 6.17 "moles" of MSW, which, after multiplying by molecular weights of CH_4 and CO_2, emit $1.6 * 6.17 * 16 = 158$ grams of CH_4 and $1.4 * 6.17 * 44 = 380$ grams of CO_2 per 1 kg of volatile MSW solids during the period of decomposition that may last dozens of years. These two emitted gases would have about the same volume or, by dividing with specific density (1 atm and 20°C) in grams/liter:CH_4 volume = 158/0.72 = 219 litres/kg VSS (48% by volumes) and CO_2 volume = 380/1.84 = 207 litres/kg VSS (51%).

Because VSS content of MSW is about 50%, the CH_4 emissions per 1 kg of dry solids would be:

$$CH_4\ \text{emission} = 0.5 * 0.158 = 0.08\ \text{kg}\ CH_4/\text{kg ds MSW} = 0.08 * 25$$

$$= 2\ \text{kg}\ CH_4 GHG CO_{2eq}/\text{kg ds}$$

$$CO_2\ \text{emission} = 0.5 * 0.380 = 0.19\ \text{kg}\ CO_2/\text{kg ds} \approx 0.2\ \text{kg}\ CO_{2eq}/\text{kg ds}$$

Since the GWP of CH_4 is 25 and that of CO_2 is 1, the above estimates could be converted to CHG CO_2 equivalent of landfill emissions, which would yield:

$$\text{Total GHG}\ CO_{2eq} = 0.08 * 25 + 0.19 \approx 2.2\ \text{kg}\ CO_{2eq}/\text{g ds MSW}$$

Obviously, these values are highly variable and are presented herein mainly for illustration and background information. Nevertheless, by researching and quoting older sources (Bingemer and Crutzen, 1987; Halvadakis et al., 1983), Bogner and Mathews (2003) reported that, using a typical municipal solid waste degradable organic carbon (DOC) composition for developed countries, the methane yield from landfills was about 0.10 kg CH_4/kg (dry) solid waste, which is very close to that estimated from the MSW chemical composition calculated above as 0.084 kg CH_4/kg ds MSW. Because the volatile content of dry MSW is about 50% (Tchobanoglous, 1993), the above calculation of CH_4 emissions would coincide with the Bogner and Mathews (2003) compilation. Also, carbon in plastics is generally nonbiodegradable. Bogner and Mathews then proposed CH_4 yields ranging from 0.07 to 0.13 kg CH_4/kg (dry) solid waste. The above analysis implies that landfills are net sinks of organic carbon, similar to the formation of peat in anaerobic deposits of swamps, which was recognized by IPCC; however, formation of the high GWP methane, and to a lesser degree nitrous oxide, makes landfill a large source of $GHGC_{eq}$.

A point has to be made that the decomposition of the MSW in landfills and formation of methane lasts a long time, ranging from 40 to more than 100 years. The annual formation of CH_4 and CO_{2eq} emissions is commensurately a smaller fraction of the total yield. The decomposition is a first-order process, which means that the first year emits the largest amount of CH_4 and CO_2 and the decomposition CH_4 and CO_2 emissions decrease year by year over the decomposition time. Because MSW is added year after year in active landfills, at some point the total emissions may reach a steady state. Annual (2015) GHG emissions can be calculated using information from US EPA (2018b) website. It is interesting to note that the US EPA (2018b) estimates of the total $GHGCO_{2eq}$ emissions loads from US landfills reported CO_{2eq} = 10.8 Mt and CH_4 $GHGCO_{2eq}$ = 100.5 Mt, which exhibit almost the same ratios (1:10) of the CO_{2eq} : CH_4 $GHGCO_{2eq}$ as the estimates derived from the simple MSW formula presented above. If these total loads are divided by the total MSW load deposited into landfills (111 Mt), then the values of the annual gas emissions in kg/kgVSS would be approximately one half of the total emissions, or:

$$\text{GHG } CH_4 \text{ emission (EPA)} \approx 100.5 \text{ Mt}/111 \text{ Mt} \approx 1 \text{ kg } CH_4 GHGCO_{2eq}/\text{kg ds MSW}$$

$$= 0.04 \text{ kg } CH_4/\text{kg ds MSW}$$

and

$$CO_2 \text{ emissions (EPA)} \approx 10.8 \text{ Mt}/111 \text{ Mt} \approx 0.1 \text{ kg } CH_4/\text{kg ds MSW}$$

$$\text{Total landfill GHG (EPA)} \approx 1.1 \text{ kg } CO_{2eq}/\text{kg ds MSW}$$

GHG CO_{2eq} of nitrous oxide was reported by the EPA (2018b) as GHG CO_{2eq} = 0.4 Mt, which is small when compared to CO_{2eq} and CH_4 GHG CO_{2eq}.

Landfills also produce a highly polluted liquid *leachate*. Typically, in engineered managed landfills, leachate collection systems remove leachate from the landfill as it collects on the bottom liner, using perforated collection pipes placed in a drainage layer (e.g., gravel). Waste is placed directly above the leachate collection system in layers. Collected leachate can be treated on-site or transported to off-site treatment facilities. Although traditional landfills tend to minimize the infiltration of rain into a landfill using liners, covers, and caps (sometimes referred to as dry tombs), some landfills recirculate all or a portion of leachate collected to increase the amount of moisture within the waste mass. This practice of leachate recirculation results in a faster anaerobic biodegradation process and increased rate of LFG generation. Similarly, landfills may accept liquids other than leachate, such as sludge and high organic industrial wastewater. Conventional landfills typically have in-situ moisture contents of approximately 20%, whereas landfills recirculating leachate or other liquids may maintain moisture contents ranging from 35% to 65% (US EPA, 2014 ab). The moisture content of collected MSW is 40–50% (Chapter 8).

In an integrated approach, residual solids (sludge) from used water treatment and reclamation should be put into perspective of the overall solid waste problem and solution. The amount of water sludge produced is approximately 0.07 kg ds person^{-1}day^{-1} and represents less than 10% of the total biodegradable MSW. About 55% of the total residual sludge solids receive additional treatment and are applied as biosolids onto land (US EPA, 2014a). The sludge generation in the European Union in the early 2000s was about 10 Mt ds/year, of which about 55% was recycled for agricultural use (applied to land), 24% was incinerated, and 19% went to landfill (Smith, 2002). Land application of residual used water solids has some GHG reduction benefits. As in landfills, not all COD of the solids is converted in soils to CO_2; some is retained as the organic content of the soil, and the nutrients are used for photosynthesis by crops and other vegetation to capture atmospheric CO_2 to grow. Hence, in the context of the integrated approach leading to net-zero solid waste to landfills, the sludge contribution from current and future water reclamation facilities is small.

Biodegradable MSW (food waste, vegetation and yard waste, and nowadays increasing portions of paper and cardboard) can be processed either aerobically by composting, or anaerobically by bio methanation digestion, or in the future, converted to hydrogen in H_2-producing digestion/fermentation processes. Organic solids can also be added directly to pyrolysis/gasification of all combustible solids to produce syngas (hydrogen, carbon monoxide, and several other low-carbon combustible gases), biofuel, and char. After further gasification, char and tar can be gasified to hydrogen-enriched syngas (more than 50% H_2). The CO_2 emitted from composting is not counted as GHG; however, uncontrolled emissions of methane from landfills and other anaerobic decomposition processes are contributing to global warming as a GHG gas.

Exporting Garbage

The disposition of solid waste in the US (Figure 3.2) and other developed countries had been grossly distorted by "exporting garbage," mainly to China. In 2016, China processed half of the world's exports of waste plastic, paper, and metals. The US exported 16 million tons of waste to China that year, worth about $5.2 billion (Mosbergen, 2018), which represented about one-third of all its recycling. In 2016, China imported a total of 45 million tons (or about $18 billion worth) of scrap metal, waste paper, and plastic from many countries.

However, in 2017, China abruptly closed its borders to this trash influx and banned import of 24 categories of solid waste, including scrap plastic and mixed paper. Starting in 2018, more than 100 million tons of garbage, including most plastics, had to be landfilled in the US, Europe, and Japan instead being sent to China, India, Vietnam, or Turkey for recycling and energy production (by polluting incineration). Some recycling companies in the US are stockpiling recyclables, hoping for some solution, but eventually most of it will go to landfill (Mosbergen, 2018). A sustainable solution to this dilemma will be presented in Chapter 8.

Swedish Recycling Revolution

Surprisingly, in 2018 Sweden was still accepting some imported garbage, and in 2014 the country even imported 2.7 million tons of waste from other countries (Fredén 2018). Sweden today has reached the goal of zero solid waste to landfills and, over time, developed a large capacity for and skill in efficient and profitable solid waste treatment. The core of the recycling and energy recovery is incineration that processes 51% of presorted solid waste and turns it into energy; 32% of solid waste is recycled,16% is biologically decomposed, and 1% is the ash residue deposited. Apparently, incineration in Sweden is not a controversy, which it would be in the US. However, this implies that Swedish municipal solid waste incineration emits a lot of CO_2, but IPCC may not count the emitted CO_2 as GHG because solid waste is not a fossil fuel and emitted CO_2 may be compensated by CO_2 not emitted because of recycling.

In Sweden, newspapers are turned into paper mass, bottles are reused or melted into new items, plastic containers become plastic raw material, and food is composted and becomes soil or biogas through a complex chemical process. Sanitation trucks are often run on recycled electricity or biogas. Waste water is purified to the extent of being potable. Special sanitation trucks go around cities and pick up electronics and hazardous waste such as chemicals. Pharmacists accept leftover medicine. Swedes take their larger waste, such as a used TV or broken furniture, to recycling centers on the outskirts of the cities (Fredén, 2018). This is the Swedish Recycling Revolution.

3.3 IMPORTANCE OF RECYCLING VERSUS COMBUSTING OR LANDFILLING

Studies in the US and Canada (ICF International, 2005) have shown that recycling saves several times more virtual energy than combusting and producing energy from combustible solid wastes. This is one of the founding blocks of the circular economy. The ICF report summarized the state of the art and effort of the life cycle of disposing, reusing, and recycling the municipal solid waste and efforts to develop and refine life-cycle GHG emission factors for specific materials commonly occurring in residential, industrial, commercial, and institutional waste stream. Figure 3.6 presents the MSW life cycle/circular economy and hybrid metabolism.

The ICF report documented that recycling is a better and more sustainable alternative for reducing GHG emissions than producing energy from waste as shown in the table.

	Recycle Energy Savings	**Energy Output from Incineration**
Material	**MJoule/ton (kW-hour/kg ds MSW)**	
Paper products	6 to 10 (1.6 to 2.8)	2.2 to 2.6 (0.6 to 0.72)
Plastics	50 to 80 (13.8 to 22.2)	3.2 to 6.3 (0,88 to 1.75)

Incineration is not very efficient "energy from waste" methodology. In a report for the US Environmental Protection Agency, ICF International (2015) developed a methodology for life-cycle assessment of the impact of solid waste by the model of Greenhouse Gas Emission and Energy Factors Used in the Waste Reduction Model (WARM).

The ICF report and US EPA identified the following circumstances affecting the net GHG emissions of materials:

- Through *source reduction* (for example, "lightweighting" a beverage can use less aluminum), GHG emissions throughout the life cycle are avoided. When paper products are source reduced, additional carbon is sequestered in forests through reduced tree harvesting.
- Through *recycling,* the GHG emissions from making an equivalent amount of material from virgin inputs are avoided. In most cases, recycling reduces GHG emissions because manufacturing a product from recycled inputs requires less energy than making the product from virgin inputs.
- *Composting* with application of compost to soils results in carbon storage and small amounts of CH_4 and N_2O emissions from decomposition.

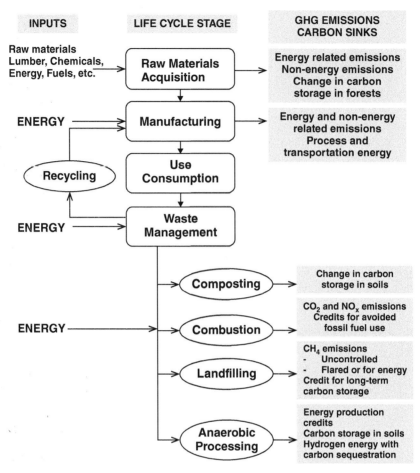

Figure 3.6. Hybrid life-cycle metabolism for integrated material and solid wastes (municipal solid waste) (ICF, 2005). *Source:* Courtesy ICF, Toronto.

- *Landfilling* results in both CH_4 emissions from biodegradation and biogenic carbon storage. If captured, CH_4 may be flared, which simply reduces CH_4 emissions (since the CO_2 produced by flaring is biogenic in origin, it would not be counted by the IPCC as GHG).
- *Combustion* of waste may result in an electricity utility emission offset if the waste is burned in a waste-to-energy facility, which displaces fossil-fuel-derived electricity.
- Changes in *forest carbon storage* are due to paper and wood products (lumber, furniture, etc.) and GHG implications of forest carbon storage. Recycling affects changes in timber harvest on forest carbon stocks, changes in the in-use product carbon pool, the net change in carbon storage, and the limitations of the forest carbon storage factors.

Until recently, the biggest problem of waste recycling has been that it was labor intensive and had problems with plastics. Despite residential requirements for sorting and separation, recycled waste had to be manually sorted in the processing centers. Consequently, overpopulated countries like China and India have been accepting garbage from developed countries, packaged in containers. However, as stated, as of January 2018 import of garbage to China has been severely limited, leaving the US, Britain, and other countries with restricted possibilities to dispose of solid waste, which has led to waste accumulation and filling existing landfills by recyclable MSW.

Combustion of MSW with or without added residual solids from the IRRF requires drying. Evaporating water from moist solids requires a lot of heat, which is captured from the incineration process, and using a part of produced heat for drying solids reduces the energy yield. Sometimes natural gas is used. The heat balances of incineration in the overall "waste to energy" process will be discussed in Chapter 8, where it will be also documented that hydrogen-based energy from waste (used water, residual solids and MSW) recuperation is a more efficient methodology, with energy outputs that are comparable to recycling. However, it can be concluded that recycling is the most acceptable and economic managerial alternative, with beneficial environmental and circular economic life-cycle impacts.

4

ORIGIN OF HYDROGEN ENERGY, GHG EMISSIONS, AND CLIMATIC CHANGES

4.1 INTRODUCTION TO ENERGY

There is no doubt that global climate change is caused to a great degree by excessive anthropogenic emissions of greenhouse gases (GHG) from burning coal, oil, and, to a lesser degree, natural gas fossil fuels, which emit carbon dioxide (IPCC, 2007, 2013, 2018; NRC 2015). Emissions of methane and other GHG gases also contribute to this serious problem facing the planet Earth. GHG emissions have a secondary but also significant effect on the acceleration of global warming by melting polar ice and mountain glaciers, including massive accumulation of ice in Greenland and Antarctica, which decreases the Earth's sun radiation albedo (less heat is returned to space) and by melting permafrost in tundra, which releases methane. Volcanos also emit GHGs (CO_2 and some geologic methane). Methane is 25 times more potent than carbon dioxide. Hence, following the Kyoto 1997 and Paris 2015 international agreements, scientists and nations are continuing to focus on increasing production of blue and green energy, including solar and wind, and to look for alternate sources of energy that would reduce or eliminate dependence on the use of fossil fuels. The requirement for sustainable "Cities of the Future" is not just to reduce GHG emissions but to become carbon neutral or better (i.e., GHG from nonrenewable fossil fuels released into the atmosphere must be reduced and the remaining emissions must be compensated by carbon dioxide sequestration somewhere within the community). GHG emissions are mainly a consequence of generating energy, but other processes also contribute, such as production of industrial fertilizers, deposition and treatment of waste organic materials in landfills, traffic emissions, natural emissions from volcanoes, and methane emission from natural wetlands. However, burning fossil fuels for energy is the main anthropogenic cause of the problem.

Energy Definitions and Units

The units of energy, work, and power are best defined and calculated in SI (System International) units. In the field of energy engineering and science, the old US or British units are no longer used, even in the US.

Force is defined as mass times acceleration and its unit is N (Newton) = kg * m/sec^2. Gravity acceleration is g = 9.81 m/sec^2. Then the unit of weight is a force exerted by gravity of 1 kg mass in vertical direction, or G = mass * g, which is 9.81 N. Specific density of a substance is expressed as ρ = mass/volume and the unit is kg/m^3, which has the same value as the physical unit of grams/liter. Specific gravity is the weight of a unit of volume of a substance, or γ = $\rho \times$ g, and the unit is N/m^3. The maximum specific gravity of water is 9 810 N/m^3 at 4°C. Both specific density and gravity of water change with the temperature.

Pressure is a force applied over a unit of area and the unit of measurement is the pascal: Pa = Newton/m^2. One atmosphere is the weight of a column of air under standard conditions from the zero–sea level elevation to the upper border of the stratosphere exerted on 1 m^2. It is P$_{atm}$ = 1.01325 × 10^5 Pascal (Pa). An approximate unit is 1 bar 100 000 Pa.

Energy is defined as the capacity of a physical system to perform work.

Work is done when force moves 1 kg of mass by a vertical distance of 1 meter. The unit of energy or work is the joule, or J = N * m = 1 kg * m^{2}sec^{-2}.

Power is work done or energy exerted in a unit of time. The unit is the watt, or 1W = J/sec.

$$1 \text{ kW (kilowatt)} = 1000 \text{ W} \qquad 1 \text{ MW (megawatt)} = 10^6 \text{ W}$$

The common unit of energy is also 1 kW-hour which is energy used or work done by 1 kWatt power capacity over one hour, or 1 kW-hour = 3 600 000 joules = 3.6 MJ (megajoules)

The types of energy are:

Kinetic energy is the energy of motion that drives the machines, automobiles, or anything that moves. Kinetic energy is expressed as E$_k$ = m v^2/2, where m is mass and v is velocity.

Potential energy is a physical unit of energy related to overcoming or gaining force by gravity of an object, which is related to the position of the object. The potential energy is E$_{grav}$ = m * g * h, where m is mass in kilograms, g is gravity acceleration, and h is a height in meters above a reference elevation. Potential energy is converted to kinetic energy and vice versa.

Hydropower energy is expressed as P$_h$ = γ Q H, where γ = specific gravity of water, Q is the flow through the power plant, and H is hydraulic depth. Work done by a pump (Efficiency = Eff) delivering volume V(m^3) of water to elevation of H(m) is:

$$W = g * \rho * V * H/Eff = 9810 * V * H/Eff \text{ (joules)} = \frac{9810 \text{ V H}}{3600 * \text{Eff}}(W\text{-}h)$$

Heat or thermal energy is energy from the movement of atoms or molecules, which is related to temperature. The unit of heat energy is the *calorie,* defined as heat energy to increase temperature of one cm^3 (1 milliliter = 1 cm^3) of water by 1°C from temperature of 4°C. 1 Cal = 4.19 joule. In cooling and heating of water as well as liquifying or dry ice making from carbon dioxide, the latent heat of vaporization, freezing, and sublimation must also be considered. Latent heat of water that evaporates 1 cm^3 of water from liquid to gas (steam or vapor) or vice versa is about 570 calories/cm^3. Heating energy is also expressed in kW-hours. 1 Kilocalorie is 0.00116222 kW-hr.

Sensible heat energy is energy obtained by changing a temperature of an object or liquid. It can be calculated as:

$$E_s \text{ (kJoule)} = \Delta T \text{ (°K or °C)} * V(m^3) * \rho \text{ (kg/m}^3) * C_p(kJ/(°K \text{ kg}))$$

where C$_p$ is specific heat. ΔT's in degrees Kelvin and Celsius are same.

Mechanical energy is the sum of the kinetic and potential energy of a body.

Light energy is based on the photon, a particle that transmits light, which provides the energy needed for photosynthesis by flora, including algae and cyanobacteria. Electricity is also generated in photovoltaic panels.

Electric energy is the movement of charged particles, such as protons, electrons, or ions. The most common unit of electric energy is kW-hour, but joules are also used.

Chemical energy is released or absorbed by chemical reactions. It is produced by breaking or forming chemical bonds between atoms and particles. Chemical energy of a reaction is also known as Gibb's ΔG, which is energy released or used by a chemical or biological reaction. An *exothermic* reaction is a chemical or physical reaction that releases heat and has a negative ΔG. It gives net energy to its surroundings. The energy released by this reaction is in the form of electrical or thermal energy. If ΔG is positive, then energy needs to be added for the reaction to proceed and the reaction would be *endothermic*. Chemical energy is also released or used by bacteria growth when reduced organic compounds are oxidized during decomposition or oxidation. Microorganisms converting nitrate to nitrogen gas derive their energy from a parallel exothermic chemical reaction of anaerobic decomposition (oxidation) of organic biodegradable carbon compounds (see Chapter 8). Energies derived from electrolysis of hydrogen with oxygen that produces water (exothermic) or is needed to break water into hydrogen and oxygen (endothermic) are also important. For example, the well-known high school chemistry experiment reaction of electrosynthesis of water from gaseous hydrogen and oxygen is:

$$2H_2 \text{ (g)} + O_2 \text{ (g)} \rightarrow 2H_2O \text{ (g)}$$

$$\Delta G = -483.6 \text{ kJ/mole of O}$$

(g signifies gaseous forms; 1 mole of oxygen is 16 grams and that of hydrogen is 1).

Nuclear energy comes from the interactions of protons and neutrons in the nucleus of an atom. It is the energy that holds neutrons and protons together, which is a very strong force. Examples are energy released by fission of uranium in nuclear power plants and fusion of hydrogen to helium on the sun or in hydrogen atomic bombs.

Geothermal energy is heat energy released by sensible heat from a geological layer, including soil below frost level for home heating or cooling or heat released by magma for large geothermal plants, such as in Iceland or Hawaii.

Energy can be renewable or nonrenewable. Renewable energy is generally defined as energy that is collected from resources that are naturally replenished on a human timescale, such as sunlight, wind, rain, tides, waves, and geothermal heat, or energy from crops and other vegetation, wood, and waste. Renewable energy often provides energy in four important areas: electricity generation, air and water heating/cooling, transportation, and rural (off-grid) energy services (Wikipedia). Another important renewable source of energy is using microorganisms to produce biogas and biofuels from biodegradable urban solids and liquids, from energy crops (corn or colza), from manure and vegetation residues and woodchips, or by gasifying combustible solid wastes that today may also include plastics. An untapped or only marginally tapped source of energy is deposited as solid waste into landfills. Some countries like Sweden are now beginning to incinerate or gasify imported solid wastes and Denmark and Holland are digesting (gasifying) manure on a large scale.

Nonrenewable energy is derived from fossil fuels (coal, oil, natural gas) that were formed millions or billions of years ago and are mined today. Converting these mineral fuels to energy generates GHG emissions that contribute to global atmospheric and sea water warming.

All kinetic energy–producing facilities (e.g., power plants) and energy-using devices (generators, engines) that convert electric and chemical energy to kinetic energy also produce heat, or:

$$\text{Energy coming in} \rightarrow \text{kinetic energy} + \text{heat}$$

In the traditional systems, kinetic energy has been converted to electricity and heat, which is generally wasted in the cooling system and/or emitted into the environment. Typically (except for hydropower), waste heat energy is greater than the electric energy produced, which is exhibited by the power plant efficiently defined as:

$$\text{Eff} = \text{electric energy produced/energy in fuel}$$

and:

$$\text{Waste heat} = (1-\text{Eff}) * \text{energy in fuel}$$

Efficiency of thermal fossil fuel power plants is about 45% and that for nuclear power plants is 35%. The rest is waste heat emitted by cooling systems and flue gases into the environment. The efficiency of a hydroelectric power plant depends mainly on the type of water turbine employed and can be as high as 95% for large installations. The efficiency of converting solid waste to electric energy by incineration (waste to energy systems) is 20–25%, but not all waste heat is recoverable. With introduction of heat pumps and heat exchangers, potential waste heat energy reuse and conversion to high-quality heat energy (temperature greater than 100°C) has been increasing (see Chapters 5 and 7). Heat pumps also provide more opportunity for homeowners and commercial establishments to derive their sensible heat and cooling from air and geothermal sources. Typical efficiency of air-to-air heat pumps is about 80%, slightly more for water-to-air or geothermal-to-water or air cooling and heating systems.

Table 4.1 presents the sources of energy and their color designation. The energy sources that do not emit GHG and have the highest beneficial impact on global climate changes are ranked as "blue." Green energy sources generally release some carbon dioxide, but overall they have a positive impact on atmospheric CO_2 concentrations by taking more CO_2 from the atmosphere through photosynthesis than is released during energy production. If the emitted CO_2 is sequestered and/or heat is recovered, the energy source would be neutral. "Black" energy sources are mostly derived from fossil carbonaceous sources without sequestering CO_2, which increases GHG in the atmosphere. Gray classification applies to fossil sources that emit less GHG CO_2 per unit of energy (e.g., natural gas) that can be sequestered and/or waste heat recovered for a CO_2 emission credit.

The color classification could be extended to cities, regions, and countries. A *green city* designation would go to a community or a region that reached the net-zero carbon goal. A *blue city* is a negative CO_2 emitter, which means that in balance, the city sequesters and removes from the atmosphere more CO_{2eq} gas than it emits.

Greenhouse Gases (GHGs)

As the shortwave energy in the visible and ultraviolet portion of the spectra heats the surface, longer-wave (infrared) energy (heat) is reradiated to the atmosphere. Greenhouse atmospheric gases absorb this energy, thereby allowing less heat to escape back to space, and trapping it in the lower atmosphere. The most important GHGs are carbon dioxide (CO_2), methane (CH_4), nitrous oxide (N_2O), and fluorinated gases such as hydrofluorocarbons, per-fluorocarbons, sulfur hexafluoride, and nitrogen trifluoride. The first three gases

Table 4.1. Energy sources and theirs GHG potential classification.

Energy source	Energy carrier /conversion	Greenhouse gas (GHG) emissions	Energy classification
Solar radiation photovoltaic or heat collecting panels	Electricity or heat	None	Blue
Hydrogen from electrolysis of water	Hydrogen	None if electrolyzed by blue energy	Blue
Hydro- and wind energies	Electricity	None	Blue
Fossil methane burned for energy	Electricity or heat or methane gas	0.55 kg CO_2/kW-h of energy produced[*]	Moderately black or gray
Recent methane energy from wood, vegetation residues, sludge and food waste, naturally grown algae	Electricity and/or heat, hydrogen by reforming methane	CO_2 emission same as fossil methane but less than CO_2 consumption by antecedent photosynthesis	Moderately green
Any significant anthropogenic source of methane (e.g., landfills) without flaring	NA	25 * kg $CO_{2\,eq.}$/kg CH_4	Strongly black
Oil	Electricity, heat	0.74 kg CO_2/kW-h[*]	Black
Coal bituminous	Electricity and/or heat, syngas	0.94 kg CO_2/kW-h[*]	Strongly black
Biofuel from energy crops and algae grown for energy	Biofuel	CO_2 emission from combusting produced biofuel smaller than the CO_2 sequestered by photosynthesis	Neutral or moderately grey if virtual energy use is high
Geothermal energy	Heat and electricity from heat	None	Blue
Nuclear energy	Heat, electricity	None	Blue/green[**]
Sensible heat by heat pumps and heat exchangers; air to air or soil to air	Heating or cooling	About 0.1 kg CO_2/kW-h if heat pumps are used; none with heat exchangers	Moderately blue because of high efficiency

[*]US Energy Information Administration (2017)
[**]Nuclear power plants do not emit GHGs, some virtual fossil fuel energy use occurs in uranium mining, enriching, and processing.

can act like a blanket, insulating the Earth. Different GHGs can have different effects on the Earth's warming. Sources of the GHGs will be subsequently presented. Two key ways in which these gases differ from each other are their ability to absorb energy (their "radiative efficiency"), and how long they stay in the atmosphere (also known as their "lifetime"). IPCC panel assigned a value of the Global Warming Potential (GWP) to each of the GHG as follows IPCC, 2007; US EPA, 2015):

- CO_2, by definition, has a GWP of 1 regardless of the time period. It is used as the reference because it remains in the climate system for a very long time. CO_2 emissions cause increases in atmospheric concentrations of CO_2 that will last thousands of years.
- Methane (CH_4) is estimated to have a GWP of 25–36 over 100 years. CH_4 emitted today lasts about a decade on average, which is much less time than CO_2 but it absorbs significantly more energy than CO_2. The effect of the shorter lifetime and higher energy absorption is reflected in the GWP, which is 25. CH_4 is a precursor to formation of ozone, which itself is a GHG.
- Nitrous oxide (N_2O) has a GWP 265–298 times that of CO_2 for a 100-year timescale. N_2O emitted today remains in the atmosphere for more than 100 years, on average.
- Chlorofluorocarbons (CFCs), hydrofluorocarbons (HFCs), hydrochlorofluorocarbons (HCFCs), perfluorocarbons (PFCs), and sulfur hexafluoride (SF_6) are sometimes called high-GWP gases because, for a given amount of mass, they trap substantially more heat than CO_2. (The GWPs for these gases can be in the thousands or tens of thousands.)

Assigning GWP value to each gas enables reporting and comparing each gas contribution in CO_{2eq} values; hence 1 kg of CH_4 emitted would have a global warming effect of 25 kg of CO_2. Expressed in CO_2 equivalents, US EPA (2015) estimated the 2015 US GHG emissions as 6,580 Mt (megatons = millions of ton) proportioned in CO_{2eq} as follows:

CO_2 equivalent

Carbon dioxide	82%	Nitrous oxides	5%
Methane	10%	Fluorinated gases	3%

In 2014, the top carbon dioxide (CO_2) and other GHG emitters were the United States, China, the European Union, India, the Russian Federation, and Japan (US EPA, 2015), with the US having the highest per capita emission. Note that current composition of the atmosphere (Gano, 2016) is:

nitrogen gas (N_2) – 78%, oxygen – 21%, argon – 0.93%, carbon dioxide – 0.041%

with little vapor, geologic and anthropogenic sulfur oxides (acid rain), and methane in low elevations. This is a result of billions of years of changing the atmospheric composition.

There was no oxygen in the primeval atmosphere at the origin of Earth (Samson, 2016) and the atmosphere was hot, in hundreds of °C, composited mainly of carbon dioxide, vapor, ammonia, gaseous sulfur, and little abiotic methane. It took one billion years to cool down the atmosphere below 100°C so that vapor could precipitate as rain that formed the oceans. Water also could originate from icy comets and asteroids bombarding warm early Earth.

The current GHG content of the atmosphere is relatively low when compared to primeval geological times but it has almost doubled since the preindustrial age. Therefore, releasing only a small fraction of carbon dioxide from fossil fuels combustion, reaching, for example, 0.5% of the atmospheric content, would be detrimental and catastrophic. The atmosphere of the planet Venus is more than 90% CO_2 and its temperature is 490°C (914°F).

4.2 HYDROGEN ENERGY

The future is in blue and green energies and hydrogen.

Blue and Green Sources of Hydrogen on Earth

Considering the damaging effects of reliance on GHGs generating fossil fuels responsible to a large degree for global warming, the US Department of Energy research and scientists in the first decade of this century focused on switching from the reliance on carbonic fossil fuels to green energy and to blue hydrogen as an important energy source and carrier. In the first decade of this century this effort received enthusiastic support from President George W. Bush and excited many scientists and citizens concerned with the dangerous consequences of excessive GHG emissions from the use of carbonic fossil fuels. The focus on hydrogen energy is now worldwide but has met resistance from fossil fuel energy producers in the US and some other countries (China, India) relying on coal (called "black gold") or oil as a main source of energy.

Hydrogen is an efficient carrier of energy. Today, hydrogen is used primarily in ammonia manufacturing to produce industrial fertilizers, petroleum refining, and in synthesis of methanol. Strangely, current hydrogen is manufactured by gasification and reforming of fossil fuels by processes that emit GHGs; consequently, there is an obvious contradiction that creates some resistance against its use as energy source. It will be argued herein that that there are opportunities to generate "blue" or "green" hydrogen without GHG emissions. Hydrogen has been fueling rockets in space programs in the US, Russia, and today also in China, European space projects, and India, and in the fuel cells hydrogen provides heat, electricity, and drinking water for astronauts and energy in automobiles It generates power using a chemical reaction rather than combustion, and in doing so, produces only water and heat as byproducts. Hydrogen fuel cells are already being used in cars, in houses, for portable power production, and in many more applications. In 2018 hydrogen fuel cell power plants up to 100-MWatt power capacity had been installed and greater planned. There is no limit on the size, just the market demand.

Linking Hydrogen to Energy on the Earth: Big Bang Origin. Some opponents of hydrogen energy claim that hydrogen is not a source of energy; they say it is an energy carrier because there are no natural hydrogen sources on the earth and manufacturing hydrogen takes more fossil energy than it produces. (See the response against the claims of "hydrogen hoax" in Lovins, 2005.) This is incorrect. Hydrogen is the original cosmic energy since the first moments after the "big bang" that created universe as we know it thirteen and a half billion years ago. It is the most abundant (75%) element in the universe. During the first seconds after the big bang, mini-particles containing enormous energy began to organize into protons, neutrons, electrons, and photons and the simplest organization was an unstable combination of one proton and one electron, which is the atom of hydrogen. Two hydrogen atoms linked together then made the stable cosmic primal hydrogen gas.

When stars were formed from the stellar gas nebulas, certain favorable conditions like extreme gravity of large cosmic bodies triggered nuclear fusion of two hydrogen molecules that then combined to make the stable gas helium. This nuclear fusion has been releasing enormous energy in the form of radiation for billions of years in our Sun and billions of other stars in our Milky Way and billions of other galaxies. The radiation occurs in a wide spectrum of wavelengths that include ultraviolet (short wave), light, infrared (heat), and charged mini-particles. This radiation energy emitted from the hydrogen fusion is the primeval source of energy on planet Earth and can be directly harnessed as blue energy and converted into electric or heat energies using solar photovoltaic panels that capture ultraviolet and light energy and heat energy recovery panels that absorb infrared radiation.

Water as a Source of Hydrogen.   The second hydrogen-related source of energy is water that can be split into hydrogen and oxygen. The origin of water on the Earth can be dated to the period after the first billion years of the planet's existence when the Earth was hot and volcanoes were outgassing a mixture of carbon dioxide, vapor (H_2O), sulfur dioxide, hydrogen sulfide, nitrogen, and geothermally formed (abiotic) ammonia and methane (Windley, 2015). The presence of early carbonate sediments on the Earth is evidence of carbon dioxide in the atmosphere and its content was orders of magnitude greater than the amount in the present-day atmosphere.

Producing hydrogen by splitting water molecules into hydrogen and oxygen with energy from another source without emitting GHG would also falls in this category of hydrogen energy recovery that could be called "blue energy" as long as the energy source for splitting the water molecules is an excess of another blue energy (e.g., wind or hydroelectricity or stored photovoltaic energy). Produced hydrogen can be stored. Wind and hydropower are also blue energy sources related to the energy coming from the Sun.

Methane as a Source.   The third source of hydrogen energy is the reduced carbonaceous material contained in fossil natural gas, which is mostly methane (CH_4). Before the first photosynthetic microorganisms (mainly cyanobacteria) evolved, the early atmosphere was a composite of carbon dioxide with vapor, ammonia, gaseous sulfur, and some abiotic methane but no oxygen. The atmosphere was also very dense when compared to the present atmosphere. It was pointed out above that abiotic methane could be formed in subsurface geological layers and reach the atmosphere by volcanic eruptions and vents. The proof of the abiotic methane forming processes in the universe is evident in Saturn's moon Titan, which has clouds, streams, and lakes of liquefied methane and an atmosphere composed of carbon dioxide, nitrogen, methane (about 5%), and traces of many other more complex organic but allegedly abiotic compounds. However, biotic sources cannot be excluded even in Titan's very low surface temperatures of −179°C (Atrea, 2009). The GHG methane and carbon dioxide gases and clouds in Titan's atmosphere make its surface about 100°K warmer than it should have been but still extremely cold to allow life as we know it to develop. But in deeper geological layers there may be a "Goldilocks" (just right) temperature layer supporting some bacterial anaerobic forms capable of producing methane.

However, on the Earth, 95% of methane has been formed by intensive photosynthetic processes of microorganisms converting carbon dioxide to organic matter and its subsequent fermentation and methanogenesis in ocean and lake muck. This biotic process was initiated by cyanobacteria, sometimes erroneously called blue-green algae, about three billion year ago (Samson, 2016; Windley, 2015). When primeval anaerobic cyanobacteria appeared en masse in Earth oceans three billion years ago, they used solar radiation from the hydrogen fusion on the Sun and CO_2 in the atmosphere, water, and ammonium to form organic matter in the first photosynthesis, which released oxygen into water and the atmosphere, as illustrated by the basic equation of organic matter forming:

$$6 \, CO_2 + 6 \, H_2O + light \rightarrow C_6H_{12}O_6 + 6 \, O_2$$

However, for almost two billion years, the produced oxygen was mostly used in weathering and oxidation of minerals containing iron, calcium, magnesium, and other cations that combined with bicarbonate ions formed from dissolved CO_2 to form calcite and carbonate rocks limestone ($CaCO_3$) and dolomite ($CaMg(CO_3)_2$). Most carbonate deposits and rocks were formed biologically by corals, shellfish, algae, and other organisms that produced vast amounts of calcium carbonate skeletal deposits. The Dolomites (Figure 4.1) in northern Italy were originally carbonate bottom sediments in a shallow sea. A gargantuan sequestering of carbon dioxide–forming carbonic acid dissociated into bicarbonate alkalinity, interacting with calcium to form calcite. Calcite was also absorbed by small organisms to form shells in the sea. These processes formed the sedimentary rocks of limestone (calcium carbonate) and dolomite (calcium magnesium carbonate), which a massive tectonic uplift 280 million years ago raised to an elevation greater than 3,000 meters above sea level.

Organic Matter. The first organic matter fermenting biological processes were strictly anaerobic and so was the fermenting and methanogenic methane–producing decomposition processes approximately represented by this equation:

$$C_6H_{12}O_6 + 3 \, H_2O \rightarrow 3 \, CH_4 + 3 \, HCO_3^- + 3 \, H^+.$$

Because these processes were carried out in water, the biologically released carbon dioxide in the oceans was converted to bicarbonate (HCO_3^-) and carbonate ($CO_3^=$) alkalinity rather then released as carbon dioxide into the atmosphere. The four steps of anaerobic fermentation and methanogenesis are described in Chapter 7.

Figure 4.1. The Dolomites in Italy originated as gargantuan sequestering of carbon dioxide–forming carbonic acid that dissociated into bicarbonate alkalinity, interacting with calcium to form calcite, eventually rising 3,000 meters above sea level. *Source:* Pixabay image by Maricio.

Cyanobacteria (Figure 2.6) are very resilient and adaptable microorganisms that can use photosynthesis to create organic matter and use it as food in the fermentation process to grow and decompose organic matter. They can also fix inorganic ammonium and nitrogen gas as the source of nutrients for their growth. This fossil biotic methane is now found in shell geological layers formed from the organic deposits in the early oceans. Because of the very high concentrations of carbon dioxide in the atmosphere and its dissolution in the ocean water during these geological periods, the mass of bacteria and organic biodegradable mass was enormous. The above two equations have shown that in combining photosynthesis with anaerobic decay, fermentation of the produced biomass lowered the CO_2 content and eventually released oxygen into the atmosphere because O_2 has very low solubility in water. Chemical oxygenation reactions of alkalinity with calcium, magnesium, and other elements then formed calcium carbonate (limestone), calcium-magnesium carbonate (dolomite), and other minerals found today as sedimentary rocks uplifted by tectonic forces from the former oceans (as seen in Figure 4.1).

Hence, photosynthetic processes by early microorganisms were consuming carbon dioxide from the atmosphere and forming, to some degree, atmospheric and aquatic oxygen. Besides calcium and magnesium, alkalinity was also combining with other chemical compounds in water such as soluble reduced ferrous F^{++} iron, which was oxidized to mostly insoluble oxidized ferric F^{+++} complexes. Fossils of the early cyanobacteria formed stromatolites. Oxygen also reacted with ammonia to form nitrate gas and in the stratosphere ozone (O_3) was formed by ultraviolet radiation breaking the oxygen molecule (O_2) into two oxygen highly reactive atoms ($2O$), which then combined with the oxygen molecule to form ozone. However, for almost two billion years of photosynthesis by cyanobacteria the carbon dioxide content was decreasing, but the oxygen content of the atmospheres remained very low.

Mining fossil geologic methane (natural gas) and producing energy by combustion returns the carbon dioxide to the atmosphere. If all reachable methane in shell geological zones is mined by fracking and combusted, the CO_2 content of the atmosphere would greatly exceed the catastrophically dangerous levels. On top of this GHG contribution, GHG methane is released by global warming from melting permafrost in the tundra.

Landfill Sources of Methane and Carbon Dioxide. In the landfill (or a garbage dump), the biodegradable fractions decompose under anaerobic conditions and produce highly organically laden liquid called leachate containing acetates, organic acids, and emit landfill gases, mainly methane (50 to 75%), carbon dioxide, and hydrogen (see Chapter 8). Significant quantities of methane and some carbon dioxide are emitted from numerous garbage and solid waste landfills that are in developed countries engineered to reduce the emissions into the atmosphere and contain liquid leachate. In most developing countries, a landfill is a dump of sometimes gigantic proportions that attracts poor people, including children, for foraging and recycling. Solid waste deposited in landfills contains readily biodegradable food waste, slowly degradable paper (cellulosic) wastes, and biodegradable garden or commercial wastes. The decomposition in open dumps also may be partially aerobic, which decreases the methane fraction and increases carbon dioxide (Bogner and Mathews, 2003) and forms nitrous oxides. The global warming potential (GWP) of methane is 25, compared to that of carbon dioxide, which is 1 because CO_2 was used as a comparative base by the Intergovernmental Panel on Climatic Changes (IPCC). Because of this, many landfill operators in developed countries collect methane and mostly flare it but some convert it to energy. Gaseous nitrous oxides N_2O (not nitrogen gas, N_2, which is inert) have GWP around 300.

Insofar as atmospheric pollution is concerned, burning natural gas (methane) is cleaner than burning coal and the quantity of GHG carbon per unit of kW-hr produced (Table 4.1) is significantly less than that for oil and coal, yet, without CO_2 sequestration, using natural gas for energy and other purposes (e.g., making ammonium) is not sustainable and is designated as a "black" or, at best, a borderline "gray" source of energy if heating energy can be recuperated. Converting methane to hydrogen gas by methane reforming and sequestering CO_2 can resolve the problem (Chapter 8).

However, recent ($\sim$100 years or less antecedent period) methane biogas and liquid biofuels produced from carbonaceous sources such as wood chips, dead or harvested vegetation residues, municipal solid waste, manure, sludge from resource recovery facilities, or algae are generally considered as sustainable sources of energy and carbon dioxide emitted from energy production is not considered GHG by the IPCC. The reason is that the carbon dioxide withdrawn from the atmosphere or from alkalinity in water was recently photosynthesized into the carbonic energy generation matter and the amount of the carbon dioxide emitted in produced energy is generally less than that withdrawn from the atmosphere. The difference may be in ash and/or stable carbonic byproducts such as soil conditioners or man-made asphalt. This source of energy can be classified as "moderately green" or "neutral." "Energy crops" such as corn (alcohol), colza (biofuel), or even algae grown for energy (see Chapter 9) are suspect because the virtual energy needed for plowing the land, growth, harvesting, and preprocessing may be great, sometimes greater than the energy produced from the crop and the energy from a fossil fuel.

Coal and Oil.   Coal and oil derived from fossil organic carbon (Stach et al., 1982; Bouška, 1981) are the fourth source of energy. After land emerged from the oceans less than one billion years before the current time, the photosynthetic micro- and macroflora and -fauna developed on its surface in the pre-dinosaur (Cretaceous) period. When the atmospheric content of oxygen increased above 1% (the current oxygen content of the atmosphere is 21%), organisms relying on oxidation furnishing more energy developed in large masses, mainly in water because ultraviolet radiation on the land was still high. After settling on the bottom of the oceans, these small planktonic microorganisms represented the biomass that metamorphosed and formed oil during hundreds of millions of years of geological pressure and metamorphosis.

The origin of coal millions of years ago is the great density of small and large vegetation growing in vast swampy areas that formed peat after land emerged and the oxygen content of the atmosphere increased (Bouška, 1981). The Earth at that time was warm because atmospheric carbon dioxide was still high, but about 420 million years ago (Silurian period) oxygen levels in the atmosphere reached 10% of the present level, which produced enough ozone to protect the terrestrial organisms from harmful ultraviolet radiation. Consequently, terrestrial flora and fauna were developing in large masses and the energy source was safer solar radiation – light and photosynthesis. The macroflora consisted of large ferns and trees rich in resins that were different from the present. The dead plants and organic debris accumulated under anaerobic conditions in the muck of the swamps. Thick layers of plant debris were buried by sediments such as mud or sand, and these deposits were washed into swamps by flooding rivers. Coal was formed from the buried peat under the geological pressures for millions of years and has continued forming since the dinosaur (Jurassic) period.

The coal formed in coastal swamps (eastern US, Great Britain, central Europe, China) contains sulfur, while fresh water swamps contained less sulfur and produced better-quality coal (Wyoming). The burning of coal with higher sulfur content to produce energy and heating has caused rainfall containing damaging sulfuric acid that deforested vast areas in the eastern US (New Hampshire, Vermont), Scandinavia, and central Europe fifty years ago, and continues damaging nature in China and other developing countries today. Atmospheric pollution from coal burning also deposits mercury, which contaminates fish. Burning coal is not a clean process and requires expensive atmospheric pollution control scrubbers, which, however, do not remove CO_2. Because formation of coal from peat was anaerobic, coal deposits contain fossil methane that is dangerous, and many miners have been killed by explosions; therefore, methane must be ventilated from the mines. This represents unknown but large GHG emissions. Many miners also suffer from black lung disease that shortens their lives and leads to disability.

Both coal and oil are fossil fuels, and their combustion emits GHGs. Sequestering GHGs from these operations is complex and expensive. Energy from fossil fuels coal and oil is black energy, simply related to the color of the source and the fact that it contributes to global warming by emitting large quantities of GHGs. Market forces and the uneasiness of the population about global warming may make fossil fuels obsolete after 2050 (Jacobson et al., 2017). The cost of blue and green energy is decreasing even without subsidies.

Making cement from limestone. Often omitted as a fossil source of GHG, cement is made by heating fossil limestone ($CaCO_3$), with clay or shell in a kiln furnace at 1450°C, which releases CO_2 into the atmosphere and forms calcium oxide (CaO), known as quicklime. After mixing with water, quicklime converts to lime ($Ca(OH)_2$) and concrete and thus has been used in mortar for building since Roman times. Making quicklime is the fourth major fossil source of CO_2 in the atmosphere after burning fossil fuels coal, oil, and natural gas. CO_2 is also released during this process from burning the fuel to heat limestone. Recycling concrete is possible and has been practiced by some construction companies.

Hydrogen as a Source of Energy

H_2 is the most logical and most sustainable source of energy; it has a very high energy content per unit of mass, and when burned it produces clean water (steam) and no pollution. Hydrogen engines today are propelling rockets, sending people and scientific (and military) instruments into space, and are increasingly powering automobiles and even fishing boats in Iceland and trains and light rails in Europe and China. Hydrogen fuel cell engines are far more efficient than the gasoline or diesel oil combustion engines that emit GHGs and other pollutants that, along with coal-burning emissions, cause terrible smog pollution in Chinese and Indian cities by cheap mass-marketed cars and by industries and power plants. Hydrogen is also the key raw material in the industrial conversion of the atmospheric nitrogen into ammonium and subsequently into fertilizers that have saved humans from famine. It is also used in the pharmaceutical and cosmetic industries.

The idea of hydrogen as an energy source has become very popular in the automobile industry. Major Japanese, Korean, and even US car manufacturers have already developed automobiles using hydrogen-powered fuel cells that emit no pollution, just water. Hydrogen-powered trams (with no overhanging wires) are transporting people in some Chinese cities. Automobile manufacturers are developing hydrogen-powered vehicles not only because of the tightening on automobile emission limits but also because it is good for business and good news for the environment because of the lack of GHGs and toxic emissions. In 2017, Toyota reduced the price of the hydrogen automobile Mirai into the affordable range and planned on making tens of thousands of H_2 automobiles. In California hydrogen refueling in gas stations is becoming a reality. To top this development, the island country of Iceland located at the Arctic circle in the Atlantic

Ocean is planning to replace all fuel-powered vehicles and fishing ships with hydrogen manufactured by electrolysis splitting water using the island's abundant sources of blue geothermal energy and hydropower electricity sources (Woodard, 2009). On April 30, 2014, the US Department of Energy announced its renewed interest in hydrogen by launching a new project leveraging the capabilities of its National Laboratories in direct support of H2USA. The project is led by the National Renewable Energy Laboratory and Sandia National Laboratories and will tackle the technical challenges related to hydrogen fueling infrastructure.

The current drawback of hydrogen as an energy source is the fact that, while it is the most abundant element in the universe, there are almost no natural sources of hydrogen gas on the Earth (acidic wetlands and organic soils may be rare exceptions). Essentially, after hydrogen is unlocked from water and organic compounds, its molecules become both source and carrier of energy in pipelines, like electrons that carry electric energy in wires and natural gas delivered by pipelines.

The known processes of producing hydrogen are (US DOE, 2016a):

1. *Electrolysis* of water with electric current using an electrolyzer to split water into hydrogen at the cathode and oxygen at the anode. This process requires electric potential of at least 1.23 volts. However, the process used to break the hydrogen–oxygen bond is about 25% more that the energy gained from the produced hydrogen.

2. *Gasifying* coal and combustible biomass (sludge, combustible municipal solid waste, woodchips, plastics) into syngas (a mixture of carbon monoxide, hydrogen, and several other carbonic gases) and subsequent steam, reforming the produced CO in syngas into more hydrogen. The energy in the produced H_2 gas is greater than that to produce it.

3. *Steam reforming of gaseous organic carbon compounds* such as natural gas (most common) and biogas into hydrogen (see Chapter 8).

4. *By microorganisms* participating in the fermentation of sludge, fecal matter, and other biodegradable organic waste such as manure, algae, food waste, and wetland peat.

5. *Thermochemical and photo-electrochemical splitting* of water molecules.

6. *Photocatalytic conversion* processes of converting biomass to hydrogen in water with concurrent sequestering of produced CO_2 to carbonate.

Processes 2 and 3 emit carbon dioxide, which must be separated and sequestrated. The CO_2 in process 4 produced by microorganisms is sequestered by conversion into bicarbonate. It is also possible to use CO_2 for growing biomass that would be co-digested into methane and/or CO_2 can be used to neutralize high pH water (see Chapters 7 and 8). Processes 2, 3, and 4 are adaptable and very attractive for used water and waste solids (including agricultural waste) treatment and energy and resource recovery. Processes 1 and 5 are only feasible in areas with an abundance of blue and green energy, such as Washington State, Quebec Province in Canada, Iceland, France, Norway, Austria, or excess nuclear power production as France and Japan have had. However, after the Fukushima Daiichi 2011 nuclear plant catastrophe in Japan, building new nuclear power plants may be politically difficult and some countries (for example, Germany and France), are decommissioning their nuclear power plants and replacing them with green and blue energy but sometimes with natural gas. However, producing hydrogen by electrolysis (process 1) with excess power from nuclear, hydro, wind or geothermal power plants during off-peak hours

Table 4.2. Energy/heating characteristics of the gases.

Compound	Energy (Heating) Value		Molecular Weight gram/mole	Specific Density kg/m^3 at NTP
	kW-h/kg	kJ/mole		
Air				1.2
Hydrogen	39.4	283.4	2	0.09
Carbon monoxide	2.64	282	28	1.16
Methane	13.7	789	16	0.72
Carbon dioxide	0	0	44	1.8

Source: https://www.engineeringtoolbox.com/fuels-higher-calorific-values-d_169.html.
NTP = normal temperature (20˚C) and atmosphere at 0 sea water elevation.

can be efficient and highly sustainable and the produced hydrogen can be stored and used during times of energy shortage.

Process 6 is new (Wakerley et al., 2017, Kadam et al., 2014). In this process the light-driven photo reforming of cellulose, hemicellulose, lignin, and possibly other biomass to H_2 is performed by using semiconducting cadmium sulfide or nanostructured C, N, S-doped ZnO nanoquantum dots in alkaline aqueous solution with concurrent sequestering of produced CO_2 to carbonate.

Hydrogen (H_2) is eight times lighter than natural gas and, per unit of weight, has the highest energy content (Table 4.2). One kilogram of hydrogen has roughly the same energy as approximately 3 kg or 4 liters of gasoline; however, per unit of volume, at atmospheric pressure, hydrogen contains about 30% of the energy of the equivalent volume of natural gas (Lovins, 2005). Hydrogen can be compressed but reaching the energy compactness of the gasoline may be difficult.

In a comprehensive review of pros and cons of hydrogen energy, Lovins pointed out that in the automobile industry, a hydrogen fuel cell can convert hydrogen energy to motion about two to three times more efficiently than a normal gasoline engine. Hydrogen fuel cells with high efficiency convert hydrogen directly into electric energy and heat. Energy production by fuel cells is also an ultra-clean process in which the cell generates power by an electrochemical reaction between hydrogen, compressed in the tank on board a vehicle or supplied by a pipeline, and oxygen from air and the output of the process is steam that is emitted or condensed and reused. Hydrogen-powered automobiles can emit 25 liters (7 gallons) of water in a driving day cooled from steam to 80°C water by air. A good fuel cell can convert hydrogen energy to electricity with about 50–60% efficiency, while a typical gasoline combustion engine efficiency from gasoline input to output shaft is about 15–17%. Hydrogen fuel cell cost has been dropping and capacity increasing.

Hydrogen has an advantage over electric cars because the driving distance between refueling of hydrogen cars is about 480 km (300 miles) while the batteries of electric cars need to be recharged after about 150 km (94 miles). The time to refill a tank with hydrogen is about 10 minutes while it takes hours to recharge an electric car. Consequently, automobile executives and the environmentally conscious public are becoming excited with the hydrogen outlook, as pointed out in a 2015 *New York Times* article (Ulrich, 2015). However, the future of automobiles is neither hydrogen nor electric; the future is that there will be two zero-emission vehicles available on the market, and widely distributed as soon as the problem with hydrogen refueling is overcome. (The opportunities and advantages of the zero-emission hydrogen energy in urban used water municipal solid waste treatment, reuse and recycle, and energy recovery leading to net-zero GHG emissions are covered in Chapters 8 and 10.)

There have been some signs that the idea of hydrogen energy will also catch up in the waste and wastewater treatment and disposal domain. For one, hydrogen is an intermediate product of anaerobic digestion fermentation process before it is metamorphosed into methane by methanogenic bacteria. It is a co-product of gasification of coal and combustible organic waste that produces syngas, which is mainly a mix of carbon monoxide and hydrogen and it is relatively easy and efficient to convert methane and syngas into hydrogen and sequesterable CO_2 by well-known and tested methane/syngas steam reforming processes. New research is also changing the anaerobic decomposition processes from the traditional anaerobic digestion that produces methane to a process that produces mainly hydrogen, hence bypassing methanogenesis and methane reforming to hydrogen, which would otherwise reduce the energy yield (see Chapter 8). Also, the new development of hydrogen fuel cells will allow accepting syngas or even methane as a feed and reforming it internally to hydrogen and electricity.

There are several real or alleged drawbacks of hydrogen energy. After the early enthusiasm in the US government about hydrogen and green energy at the beginning of this century, articles in the general media and even in scientific publications raised doubts and put forth some condescending critiques to the point that the interest of the US Department of Energy in funding new hydrogen energy research dropped at the beginning of the second decade of this century. Lovin (2005) extensively addressed these concerns and hoaxes. However, the interest of private venture investors in the US and government and privately funded research in other countries (Japan, Germany, The Netherlands, China) is rising exponentially.

Based on US Department of Energy information, more than 95% of hydrogen produced today is by the steam methane (natural gas) reforming (SMR) process (covered in Chapter 8). Some abiotic hydrogen-producing processes require more energy to produce H_2 gas than that obtained in the produced hydrogen, which some critics deem a serious drawback considering that today most of the energy for hydrogen production is provided by fossil fuel–powered plants that emit GHGs. This is counterbalanced by the much higher efficiencies of hydrogen fuel cells and engines than combusting fossil fuel in making electricity. However, if society's goal is the reduction of GHGs to the point of zero adverse impact on the environment and life on Earth, hydrogen as the energy carrier as well as concurrently replacing most fossil fuels with renewable green energy sources or carbon sequestering makes a lot of sense. Furthermore, hydrogen can be produced by blue energy from the sun during times of excess energy production by photovoltaics or wind or hydro energies. Hydrogen then can be stored.

Producing energy by steam reforming methane (natural gas), carbon monoxide, and other organic carbonaceous compounds into hydrogen requires water. This is another argument raised against hydrogen energy alleged to become a problem in areas of water shortages. To reform one molecule of methane or carbon monoxide into hydrogen, two molecules of water are required for methane conversion and one molecule for CO conversion, respectively, or:

$$CH_4 + 2\,H_2O \rightarrow 4H_2 + CO_2$$

$$CO + H_2O \rightarrow H_2 + CO_2$$

Based on the information in Table 4.2 and stoichiometry of the reforming equations, per weight, one kilogram of methane would require 2.1 kg of water for reforming it into one half kilogram of hydrogen. The solution of this alleged problem is to balance the water budget of the entire process by combining the SMR process with energy production in the hydrogen fuel cell. In the latter energy-producing process, four molecules of H_2 produced from one molecule of methane combine in the hydrogen fuel cell with two molecules of O_2 from the atmosphere, hence producing four molecules of water, more than two that entered in the SMR process, or:

$$4\,H_2 + 2\,O_2 \rightarrow 4\,H_2O$$

Therefore, the two-step process of electric energy production by reforming methane (natural gas) into hydrogen and energy is producing water. Water recovery is not possible for hydrogen-fueled automobiles that produce water vapor, which is condensed and emitted as hot liquid, but it is possible in used water/solid waste energy recovery systems. The reward is also a more efficient conversion of methane into energy than that by combustion and the amount of water involved is not large. The water produced this way is ultra-clean.

Another argument against using green energy for producing hydrogen as a prime source of energy – for example, for running home appliances – is that there is no need to replace electricity as the energy carrier. This argument is valid and, apparently, in the future both electricity and hydrogen could be carriers of energy. Some or all natural gas pipelines can be converted to deliver hydrogen over long distances. Hydrogen for refueling can be produced locally by electrolysis driven by blue energy, which evidently is being done in Iceland.

Using renewable energy has also an alleged problem related to the availability of the energy produced by wind or hydropower during peak demand for electricity. The power companies relying on coal solved the problem of peak energy demand by building auxiliary peak power plants powered by oil or natural gas, or, in case of hydropower, by building large high-elevation storage basins into which water is pumped during the night for off-peak power by reversible hydro turbines or pumps, and then the accumulated water is used by turbines for generating energy during peak hours. Hydrogen provides the same opportunity for storing energy generated by solar or wind power or even by hydropower. Furthermore, large 100-MW batteries to store excess electric energy were installed by Tesla in 2017–2018 in Australia, North Korea, and the US. Duke Energy Co. plans to install 300-MW battery capacity in the next fifteen years.

The issues of safety have also been raised. Because of its lightness, hydrogen is safer as fuel or in transportation than natural gas or syngas. The density of carbon monoxide in syngas is greater than that of air (Table 4.2); therefore, if there is a leak or a pipe breakdown, carbon monoxide accumulates in lower places in higher concentrations and can spontaneously explode. Carbon monoxide in syngas is also a strong poison that kills in very small concentrations. Hydrogen, on the other hand, is not poisonous, it quickly rises and is diluted, and mixtures of hydrogen and oxygen are hard to explode without igniting (Lovins, 2005). This phenomenon makes hydrogen automobiles far safer from fires than gasoline cars. Hydrogen explodes only under specific containment conditions in higher concentrations. During the disaster of the hydrogen-filled airship *Hindenburg* in 1937, hydrogen caught on fire because of an electric spark but did not explode. Two-thirds of the passengers survived and most of the deaths were because passengers were not able to escape in time. It was a fiery spectacle and one would expect more deaths by burning, but there were only two deaths directly related to catching on fire. Natural gas (methane) and syngas explosions and poisoning have had proportionally far more victims. Hydrogen ignites easily to flame but the ease of ignition is not much different from natural gas and much less than ignition of gasoline vapors.

Vision of Hydrogen Role in the (Near) Future

The future to be reached in few decades is net-zero impact of energy production and use on the increase of the GHGs in the atmosphere in all economic sectors, including water, used water, and solid waste management. The reason for this promising outlook is the availability of several highly feasible alternatives for reaching this goal relatively quickly. The efforts can be divided into the following categories.

Category I: Sources of Energy.
Renewable sources. Availability of economic or even cheap blue energy from solar photovoltaic panels and wind renewable sources is now increasing rapidly. Availability of hydropower may remain steady because the suitable sites in the US have diminished and there is a strong resistance by an environmentally conscious public against building any new large hydropower plants. However, blue power of sea tides has not been tapped even though this is a known source of energy used in Boston two centuries ago to run mills.

Nuclear power. In the transition period that may last a few decades, nuclear power plants fueled by enriched uranium 235 fission into lighter elements and isotopes may operate till the end of their lifetime but in some countries (e.g., France, Germany, Japan) they may be gradually decommissioned at a faster rate and replaced by expanding renewable sources and by fossil natural gas. Several countries will be solely using blue and green energy in fewer than twenty years. Availability of hydrogen fusion reactors is still many years away and, realistically, it is unlikely after the experience with nuclear power. Nonetheless, because nuclear power plants cannot readily change their energy production, power from excess capacity times can be used to produce by electrolysis hydrogen that can be stored.

Carbonaceous green sources. The energy facilities producing energy from carbonic sources that have been classified as non-fossil fuel will be expanded but may be modified to minimize carbon dioxide emissions. Such sources include wood and organic matter (not coal or oil) gasifiers and biofuel production from woodchips and vegetation, including colza (rape) and corn. However, it should be noted that producing alcohol (fuel additive) from corn has relatively high virtual energy demand because of the use of the fuel for operating farm machinery and manufacturing industrial fertilizers. This can be minimized by using hydro-energy and/or biofuel for production of industrial fertilizers; however, nonpoint (diffuse) pollution from corn-producing farms using industrial fertilizers is relatively high. Another carbonaceous renewable green energy source is biogas from digestion of sludge, manure, food waste, airport deicing glycols, algal biomass, or methane captured from landfills and used as a source of energy and not flared. If CO_2 emitted from the electric or hydrogen energy production by reforming biogas, including syngas from gasification, is sequestered, it can be counted in the transition period as a GHG credit.

Natural gas. Natural gas is a fossil fuel; therefore, any carbon dioxide emissions from its conversion to energy count as GHG per IPCC. Instead of GHG emitting energy production by combustion, natural gas can be reformed into hydrogen, which emits concentrated and sequesterable CO_2. Sequestering CO_2 from this conversion can only be done in centralized and supervised H_2 production facilities and/or by hydrogen fuel cells and it cannot be made easily in local distributed refueling stations. Hydrogen fuel cells can concentrate CO_2 and enable more efficient sequestration while producing electricity.

Reforming methane and syngas to hydrogen.   Hydrogen production by fermentation is followed by microbial electrolysis cell that can recover more hydrogen than steam reforming methane from digestion. Gasification produces syngas that can also be reformed to hydrogen and CO_2 can be sequestered. Natural gas is mostly methane.

Coal and, to a lesser degree, oil are not acceptable future energy sources without sequestering emitted CO_2. Combusting coal is an air polluting process. In addition to high GHG emissions, low-quality coal-fired power plants were responsible for acid rain, smog, and regional air pollution by toxic metals in Europe. For example, mercury emitted into the atmosphere by coal-fired power plants is deposited into water and sediments, where it is converted by microorganisms to toxic methyl mercury (MeHg), which is absorbed and biomagnified by aquatic life and contaminates fish even in otherwise pristine distant areas such as the oligotrophic Everglades National Park in Florida, where it biomagnifies to very high MeHg toxic concentration in large fish, alligators, Florida panthers, and fish-eating fowl and exceeds toxicity standards (Julian, Redfield, and Weaver, 2016). The only measurable mercury source therein is atmospheric Hg deposition from regional coal-fired power plants. Air pollution controls are expensive and for some contaminants not adequate. Similarly, combusting oil is a GHG-emitting process.

Category II: Energy Carriers and Users.   *Electricity and hydrogen* are the most logical and clean energy carriers providing energy for cars, home appliances, and light, commercial, and industrial uses. It is not an "either/or" situation; both energy carriers will be transporting energy side by side. Automobile experts prefer hydrogen over electricity as fuel for automobiles or any vehicles, including boats (e.g., in Iceland) because of the problems with storing energy and recharging batteries that currently takes too long. Nevertheless, electricity outlets are already available in many parking garages and even parking lots, and hydrogen filling stations are already being built and operating. Hydrogen can easily replace natural gas in home appliances (stoves, heating) and commercial/industrial operations, probably with minimal or no conversions. Lights and machinery using electricity do not have to be replaced because most of the electric energy provided by the grid will be green or blue.

Which source of energy will be suitable for airlines that today solely rely on airline fuel derived from oil? It should be noted that hydrogen per weight contains three times more energy than gasoline and, hence, the same would be true for airline fuel. Since a significant portion of plane weight is in the fuel, planes could be lighter if they switch to hydrogen and/or could fly longer distances between refueling. Spaceships have already been powered by hydrogen for decades. Shell Global (2018) Sky Scenario envisions that airlines will switch to liquified biofuel and ultimately after 2050 to hydrogen (Jacobson et al., 2017).

Category III: Energy Storage.   Compressed hydrogen is superior to batteries for storing electricity because over 50% of the electricity used to charge a battery is lost when electric energy is transferred to the battery. Some energy is needed to pressurize the H_2 gas for a more efficient storage, but this energy can be partially recovered.

The scenarios of switching from fossil fuels to blue and green energy within a generation are realistic. If a major oil company, Shell Global (2018), can present a scenario by which the world should be net-zero carbon society and meet the Paris Agreement goal of limiting the warming to 1.5°C above the preindustrial period, then the Paris goal should be taken seriously. The Sky Scenario by Shell is based on the assumptions that the following will happen from now to 2070: (1) Consumer mindset will change, which means that people preferentially choose low- or zero-carbon emissions, high-efficiency options to meet their energy service needs; (2) There will be a step-change in the efficiency of energy use, which leads to gains above historical trends; (3) Governments will adopt carbon-pricing mechanisms globally over the 2020s, leading to a meaningful cost of CO_2 embedded within consumer goods and services; (4) While the rate of electrification of final energy will increase significantly over today's level; new green and blue energy sources will grow up to fifty-fold, with primary energy from renewables eclipsing fossil fuels in the 2050s; (5) About 10,000 large carbon capture and storage facilities will be built, compared to fewer than 50 in operation in 2020; (6) Net-zero deforestation will be achieved. In addition, an area the size of Brazil being reforested offers the possibility of limiting warming to 1.5°C, the ultimate goal of the Paris Agreement.

As oil and gas use falls over time even in the Shell Sky Scenario, redundant facilities will be repurposed for hydrogen gas storage and transport. Indeed, the growing liquified natural gas (LNG) supply in the early decades of the century has enabled hydrogen to gain a foothold and develop scale. An immense build-out of electricity networks and hydrogen pipelines will ensure secure and affordable electricity and hydrogen supply, which will stimulate switching from fossil fuels across sectors, particularly in transport and industry. In Shell Sky Scenario, hydrogen emerges as a material energy carrier after 2040, primarily for industry and transportation. Understandably, Shell Sky Scenario assumes that reduced oil production and consumption compensated by CO_2 sequestering and reforestation will continue after 2070. Other views and prediction on switching to blue energy will be presented in the subsequent section "Solar and Wind Blue Power."

4.3 CARBON DIOXIDE SEQUESTERING AND REUSE

Stopping the Atmospheric CO_2 Increase and Reversing the Trend

Worldwide, more than 35 billion tons of CO_2 are released annually into the atmosphere, almost all from the burning of coal, oil, and natural gas. Consequently, the Paris 2015 Agreement requires that to restrain global warming to the 1.5°C target, the CO_2 emissions into atmosphere must be dramatically reduced. However, reducing emissions may be enough. The National Academies Committee on Developing a Research Agenda for Carbon Dioxide Removal (NAC) (2018) outlined the global strategies for doing this before 2050. The actions to reverse the trend included *positive* and *negative* CO_{2eq} GHG emission. The NAC identified fossil fuel consumption, agriculture, land-use change, and cement production as the dominant positive anthropogenic sources of CO_2 to the atmosphere. The focus of climate mitigation is to reduce energy and cement-producing sectors emissions by 80–100%, requiring massive deployment of low-carbon technologies between now and 2050. Logically, all large to small CO_2 emissions should be reduced; however, the CO_2 emissions from combusting non-fossil fuels do not count as GHGs. The strategy for reducing positive emissions do not just include removal and sequestrating the GHGs from the emissions; they also count on converting production activities currently emitting GHGs from positively emitting CO_2 to sustainable net-zero carbon emission sources.

Negative Emissions Technologies (NETs) remove carbon from the atmosphere and sequester it. Removing CO_2 from the atmosphere and storing it has the same impact on the atmosphere and climate as simultaneously preventing an equal amount of CO_2

from being emitted. The NAC (2018) identified the following categories of actions to remove CO_2 from the atmosphere, with conclusions regarding the current prospects and feasibility:

- *Coastal blue carbon.* Land use and management practices that increase the carbon stored in living plants or sediments in mangroves, tidal marshlands, seagrass beds, and other tidal or salt-water wetlands. These approaches are sometimes called "blue carbon" even though they refer to coastal ecosystems instead of the open ocean.

  Coastal blue carbon approaches warrant continued exploration and support. The cost of carbon removal is low or zero because investments in many coastal blue carbon projects target other benefits such as ecosystem services and coastal adaptation. Understanding of the impacts of sea-level rise, coastal management, and other climate impacts on future uptake rates should be improved.

- *Terrestrial carbon removal and sequestration.* Land use and management practices such as afforestation/reforestation, changes in forest management, or changes in agricultural practices that enhance soil carbon storage ("agricultural soils").

  These practices and land management and biomass energy with carbon capture and sequestration (BECCS) can already be deployed at significant levels, but limited per-hectare rates of carbon uptake by agricultural soils and competition with food and biodiversity for land (for forestation/reforestation, forest management, and BECCS) will likely limit negative emissions but globally may have a limited effect.

- *Bioenergy with carbon capture and sequestration.* Energy production using plant biomass to produce electricity, liquid fuels, and/or heat combined with capture and sequestration of any CO_2 produced when using the bioenergy and any remaining biomass carbon that is not in the liquid fuels.

  These practices related to urban water and solids management are extensively covered in this handbook.

- *Direct air capture.* Chemical processes that capture CO_2 from ambient air and concentrate it, so that it can be injected into a storage reservoir.

  Direct air capture and carbon mineralization have high potential capacity for removing carbon, but direct air capture is currently limited by high cost and carbon mineralization by a lack of fundamental understanding.

- *Carbon mineralization.* Accelerated "weathering," in which CO_2 from the atmosphere forms a chemical bond with reactive minerals (particularly mantle peridotite, basaltic lava, and other reactive rocks), both at the surface (ex situ) where CO_2 in ambient air is mineralized on exposed rock and in the subsurface (in situ) where concentrated CO_2 streams are injected into ultramafic and basaltic rocks, where it mineralizes in the pores.

- *Geologic sequestration.* CO_2 captured through BECCS or direct air capture is injected into a geologic formation, such as a saline aquifer, where it remains in the pore space of the rock for a long time. This is not a NET, but rather an option for the sequestration component of BECCS or direct air capture.

Geological sequestration is presented in the subsequent section.

The NET actions are not fully covered in this book and the reader is referred to the NAC (2018) report. The NET actions include large-scale engineering geoengineering to reduce CO_2 from the atmosphere (National Research Council, 2015; Watts, 2013).

Sequestering CO_2

Strategies for reducing CO_2 emissions range from switching from coal burning in houses to greener methane gas since the end of the last century, and eventually in the future to hydrogen and heat pumps using green electricity as energy source, to large energy sources such as power plants, steel mills, and aluminum production. An integrated resources recovery facility (IRRF, formerly water and resources reclamation plant; see Chapter 8) with a power plant that processes both used water and biodegradable and combustible municipal solid waste could be a mid-size CO_2 emitter that, however, should not be counted as GHG source unless it emits methane and nitrous oxides.

It's possible to turn the black energy sources of oil and natural gas into neutral sources if the carbon dioxide emissions from the energy-producing processes are sequestrated. Sequestration will also be recommended for neutral energy producing carbon dioxide emitting CO_2 from gasification of sludge and municipal solid waste (see Chapter 8). The strategy of the Department of Energy to reduce large sources of GHG from burning fossil fuels is *carbon capture and storage* (CCS) that was recently extended and characterized as *carbon capture utilization and storage* (CCUS) (USDOE, 2010a, b; Eide, 2013). Oddly, utilization means using captured CO_2 for *enhanced oil recovery (EOR)* to increase oil production, which implies fracking, wherein CO_2 and other gases from coal and natural gas incineration in power plants are returned under pressure back into the geological zones and force the oil to the surface (Figure 4.2).

According to the US Energy Information Administration (2010b), at the end of the first decade of this century, 4% of US oil already came from using CO_2 for EOR, which may increase to 60% in the future, displaying a long-term growth trend that contrasts the long-term decline trend of the overall US oil production. EOR operators (large oil companies) are intensively looking for sources of anthropogenic high-pressure CO_2 sources that

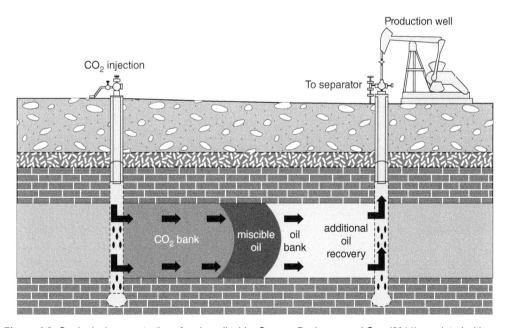

Figure 4.2. Geological sequestration of carbon dioxide *Source:* Buchanan and Carr (2011); reprinted with permission of Kansas Geological Survey.

could be also provided by the future IRRFs processing used water and MSW. Hence, there is money to be made on the high-quality, high-pressure CO_2. Blunt et al. (1993) wrote that for every kilogram of CO_2 injected, approximately one quarter of a kilogram of extra oil will be recovered, and for most projects at that time about as much carbon dioxide is disposed of in the oil zone reservoir as is generated when the oil is burnt. However, in general it is not clear what the ratio is between the carbon injected into EOR and the carbon emitted by burning the displaced oil, because companies using this technology to displace oil now recycle CO_2 at added cost. In either case, selling CO_2 to EOR operations is better than using large quantities of other gases or fluids (e.g., treated wastewater or fresh water) for EOR, which are very limited. Hence, carbon dioxide is temporarily becoming a valuable commodity to the point that fossil CO_2 is mined and sent by pipelines to the fracking operations in Texas (USDOE, 2010b). This obviously does not help to ease the problem with the global climatic change. Mining fossil CO_2 has grossly positive (increasing) GHG balance effect and does not make sense.

Quoting IPCC (2005), Eide (2013) and Herzog and Eide (2013) divided capturing CO_2 from power plants into the following categories:

(a) After combustion (post-combustion)
(b) By burning fossil fuels with oxygen such that the flue gas contains mostly CO_2 and water vapor (oxy-combustion)
(c) Before combustion (pre-combustion)

The first alternative (a) of sequestering CO_2 from a flue stack after combustion is unrealistic. The atmosphere contains about 78% of nitrogen and 21% of oxygen. Carbon dioxide is negligible (about 0.04%) by its weight but not by its effect on atmospheric warming and global climatic changes. Per weight, the proportion between N_2 and CO_2 in flue gases would be about the same as that between nitrogen and oxygen ratio before incineration because the atomic weights of nitrogen and oxygen are similar, 14 and 16, respectively. If all oxygen is converted to carbon dioxide, for example in a turbine, the proportion between N_2 and CO_2 would theoretically be about 75% N_2 and 25% CO_2, respectively, but because of inefficiency of burning, not all oxygen is used. Hence, the actual CO_2 concentrations in flue gases are about (Zeverenhoven and Kilpinen, 2002):

11% from pulverized coal combustion power plants
6–12% from waste incineration
3–4% from natural gas power plants

Because solid waste and methane (natural gas) contain molecules of hydrogen in their chemical composition, the CO_2 emissions are also diluted by water (steam). Therefore, the carbon CO_2 dioxide from the post-combustion flue gas emitted by smokestack of coal plants, natural gas plants, factories, and MSW (sludge) incinerators would need to be captured and condensed by passing the flue gases through an amine chemical bath, which binds the CO_2. The chemical is then heated to release the CO_2, which is then pumped into storage (Biello, 2014). The most common absorption chemical has historically been monoethanolamine (MEA) (MIT, 2007). The biggest challenge with post-combustion chemical absorption is that it requires significant amounts of heat to regenerate the solvent (Eide, 2013). Overall, the process is highly inefficient in terms of energy and GHG production. In 2017, Exxon Mobil–Fuel Cell consortium announced development of densification of CO_2 in flue gas from natural gas power plants by the molten carbonate fuel cell while producing more energy (see Chapter 8).

The efficiency of alternatives that use pure oxygen in combustion that would make flue gas mostly carbon dioxide appears at first look to be expensive. Manufacturing and transporting pure oxygen also use a lot of energy. However, as will be documented in Chapter 8, Section 8.7, in the sustainable hydrogen-driven system, pure oxygen can be produced by excess blue and green energy, stored and used for production of pure oxygen by electrolysis from extra clean water produced by the system that does not emit any GHG. Another promising method is the pre-combustion capture alternative, which is also the alternative proposed for the IRRF power plant outlined in detail in Chapter 8, Section 8.7. A similar CO_2 sequestration proposed for power plants is also known as Integrated Gasification Combined Cycle (IGCC) (Herzog and Eide, 2013; Rubin et al., 2012). In this process, coal or fuel consisting of a mixture of sludge and combustible MSW is gasified in the power plant to produce syngas, and then the syngas passes through a steam-gas shift reactor where CO reacts with steam to produce CO_2 and H_2. CO_2 is then captured and compressed for subsequent storage and reuse (or selling to EOR operators) and H_2 is used to generate electricity by a turbine or, as proposed in this manual, by the hydrogen fuel cell. The benefit of pre-combustion capture is that the flue gas is at high pressure and has a high CO_2 concentration. Even in coal power energy production, the pre-combustion capture process has a lower cost relative to post-combustion processes and may be a source of revenues.

Membranes that have a high CO_2-to-N_2 selectivity could separate excess CO_2 from N_2 in the cathode compartment of the molten carbonate fuel cell without the use of steam or chemicals. Membranes work well in the pre-combustion capture in IGCC (or IRRF power plants) where the flue gas has higher pressure and has a high CO_2 concentration (Rubin et al., 2012).

The captured concentrated carbon dioxide is then stored permanently underground or reused for beneficial purposes. There are four types of geological formations that have been proposed for long-term CO_2 storage (MIT, 2007):

1. *Oil and gas reservoirs.* In 2010 (USDOE, 2010a, b) 50 million tons of CO_2 per year was used for CO_2-EOR, of which 90% was obtained from natural reservoirs and pumped into the underground oil and gas fields (Tanner, 2010). These geological formations held oil and gas in place for millions of years and after the CO_2 is pumped therein to displace oil and gas to the surface to enhance recovery, the injected CO_2 gas must stay there for millennia, basically forever. Table 4.2 shows that CO_2 is about 1.5 times heavier than air. Hence, these reservoirs are good candidates for long-term storage. However, as stated above, pumping fossil CO_2 from natural underground geological zones where it had been stored millions or even billions of years and using it to push carbonaceous fossil fuel from deep underground to be burned is a GHG issue and has a negative effect on global warming.

2. *Deep saline formations and saline aquifers.* These formations consist of porous rock filled with brine. A key advantage is their abundance, allowing for large quantities of CO_2 to be stored. Both salt water aquifers and abandoned salt mines are good prospects for sustainable CO_2 storage.

3. *Basalt formation.* Basalt is a porous volcanic rock. When CO_2 is pushed into basalt rocks, it is turned into the carbonate mineral ankerite in less than two years, as found in a research project (McGrail et al., 2017). Because basalts are widely found in North

America and throughout the world, these volcanic rock formations could help permanently sequester carbon on a large scale. Wei et al. (2018) measured the carbon mineral trapping rate of 1.24 ± 0.52 kg of CO_2/m^3 of basalt per year and estimated that CO_2 in basalt formation would be completely saturated by carbonate in about 40 years, resulting in solid mineral trapping of ~47 kg of CO_2/m^3 of basalt.

4. *Unminable coal strata or abandoned mines.* This is becoming an option as coal is being phased out as a source of energy because of its high negative GHG emission consequences and regional air pollution when burned. While an abandoned mine might be a good place to be filled with waste CO_2 where it would replace mostly air, pushing CO_2 into underground coal seams would be replacing and pushing up methane in them, similarly to fracking. Methane gas that occurs in coal beds (seams) can explode if it concentrates in underground mines and the coal bed methane must be vented out of mines to make them safer places to work. Methane is 25 times more potent GHG than carbon dioxide; therefore, methane would have to be captured and used as a sustainable fuel, which only then would make the whole operation of CO_2 storage carbon neutral.

Sequestrated CO_2 must be safely stored for thousands or even millions of years in the geological storage reservoirs where it was injected. For the CO_2 emissions from IRRF to be disposed in this way, regional disposal systems for CO_2 must be developed. Currently, there are 5,400 km (3,600 miles) of pipelines that transport CO_2 from natural sources to EOR operations in the US (DOE, 2010; Tanner, 2010); however, using natural underground fossil CO_2 to enhance retrieval of oil and natural gas has no beneficial effect on GHG reduction.

To transport the gas in pipelines the CO_2 is pressurized to convert it into liquid, and the liquid is then pumped via pipeline to an appropriate storage site (Biello, 2014). This way of gas transporting is also attractive for the IRRF power plant CO_2 exhaust gas that is released under pressure. Herzog (2011) stated that *"transport of CO_2 over moderate distances (e.g., 500 km) is both technically and economically feasible."* Quoting the US Department of Energy, Biello also stated, "the US alone has an estimated 4 trillion metric tons of CO_2 storage capacity in the form of porous sandstones or saltwater aquifers and around the world, the potential storage resource is gargantuan." This estimate does not include sequestering in basalt rocks. This is a very good prospect that should lead to interconnected GHG CO_2 capture and storage (CCS) depositories and transfer in the US, China, or Europe.

Non-CCUS Reuse of Carbon Dioxide

Beside the CCS or more sustainable CCUS large-scale controls, pressurized and concentrated CO_2 released from the IRRF power plant (the only place where the CO_2 is released from the integrated resources recovery facility, IRRF) can be reused and after cleanup become a valuable commercial commodity. These CO_2 reuse opportunities include:

- *Commercial dry ice production.* Carbon dioxide turns from gas to an opaque white solid while under pressure and at a low temperature $-78.5°C$ $(-109°F)$. Dry ice is made from concentrated CO_2 by pressuring it at low temperature to liquid form and then reducing the pressure, causing a rapid lowering of temperature of the remaining liquid. Frozen CO_2 at the temperatures lower than its melting point can be found on Mars, Pluto, Saturn's moon Titan, and during winter in Antarctica, although the atmospheric content of CO_2 is very small there. Dry ice does not melt; instead it sublimates, meaning the solid ice turns directly into a gas (bypassing the liquid state) as the temperature rises and the solid begins to dissipate and forms smoke. Dry ice is not poisonous, but when the concentration of gaseous CO_2 in the air exceeds 5% (two orders of magnitude greater than its natural content), then it is toxic. Dry ice used for cooling or freezing foods must be very clean and considered "food grade" to ensure that food it may touch will not be contaminated. There are many uses of dry ice in food manufacturing and transporting for quick freezing of foods for future use at food processing plants, retarding the growth of active yeast at bakeries, and keeping foods chilled for airline industry catering. Other uses include slowing the growth of flower buds at nurseries to keep plants fresh for consumers, and flash freezing in the rubber industry during manufacture.

- *In seas and oceans, sequestering* occurs by dissolving gaseous CO_2 into dissolved carbon dioxide and forming carbonic acid. This will have an impact on coral reefs and algal blooms.

- *Growing algae in ponds and algae farms.* There is an uncertainty as to whether increased CO_2 input increases algae growth. Most studies have looked at the effects of CO_2 on very simple ecosystems and concluded that algae and plants are going to grow faster if CO_2 increases in the future. However, the growth is controlled by the limiting nutrients, which normally are phosphorus and/or nitrogen; CO_2 in the form or alkalinity is mostly in excess and is not a controlling nutrient. When the nutrient content is high or alkalinity very low (less than 20 mg $CaCO_{3-eq}$/L), then additional CO_2 input increases the algal growth rate. Adding CO_2 to a water body increases its dissolved and carbonic acid contents and reduces pH but generally does not increase bicarbonate alkalinity.

- *Use in beverage only after extensive cleaning.* Because of very high temperature the concentrated CO_2 from IRRF power plant flue gas is free of pathogens but may contain small amounts of other gases that would have to be purged.

- *Use for fire extinguishing.*

- *Possible innovative CO_2 reuse projects* listed by US DOE (2014) such as:
  - Conversion of CO_2 flue gas into soluble bicarbonate and carbonate, which can be sequestered as solid mineral carbonates after reacting with alkaline clay, a by product of aluminum refining. The carbonate product can be utilized as construction fill material, soil amendments, and green fertilizer.
  - Reforming CO_2 and waste fuel gas stream to produce synthetic natural gas. This is a highly endothermic reaction requiring more energy than in the produced fuel gas.
  - Converting CO_2 from an industrial waste stream into a plastic material that can be used in the manufacture of bottles, films, laminates, coatings on food and beverage cans, and in other wood and metal surface applications.

Recycling

All of the above carbon sequestering ideas are *a posteriori*, which means that society is reconciled with the facts of excessive GHG emissions and tries to remedy excessive emissions to avoid harm. Recycling and use of materials and energy that will not generate GHGs are *a priori* measures that include maintaining and increasing carbon storage in forests and soils and reducing use of fossil fuels. Recycling paper products made from trees that assimilates carbon dioxide saves the trees and keeps them assimilating (sequestering) CO_2. The same is true for reducing and recycling materials and products such as petroleum-based plastics made from the fossil fuels, which will keep the coal, oil, and natural gas underground and not cause GHG emissions or reduce use of aluminum, which requires a lot of energy to produce. These issues and methods are a part of the circular economy and life-cycle assessment (see Chapter 2).

4.4   SOLAR AND WIND BLUE POWER

Solar Power

Sun is the original primal power of our planetary system and the source of power on the Earth, and solar radiation is the ultimate blue energy that can be captured and converted into electricity, heat, and photosynthesis, turning GHG carbon dioxide to organic matter. The sun transmits its energy to the Earth as radiant energy emitted by the nuclear fusion reaction of hydrogen, which also creates electromagnetic energy. Specifically, solar cells capture certain wavelengths of solar radiation and convert them to electricity for homes, calculators, traffic signs, hot water heaters, or water treatment plants. In 2010, China unveiled the first solar-powered air conditioner and solar heat panels were common in many rural homes in China before 2010 (Novotny et al., 2010a). In some developing countries (e.g., Peru) in settlements without electricity, house solar photovoltaics panels run TVs and recharge phones. In Massachusetts, free solar panel installations are offered and are repaid by monthly fees that are less than the monthly payments to the electric utility before the installation. The cost of panels and installation can be repaid in less than five years by savings on payments for electricity. These already mass-produced devices are reducing fossil energy use in many countries.

Passive energy is another blue energy design methodology and energy standard that promotes super insulated, airtight homes or buildings that use 70–90% less energy for heating and cooling than a conventional new home or building. Passive energy measures such as insulation and heating with solar energy can be characterized per NAC (2018) as *negative emission technology*. In passive solar building design, windows, walls, and floors are made to collect, store, and distribute solar energy in the form of heat in the winter and to reject solar heat in the summer. This is called passive solar design (Figure 4.3) because, unlike active solar heating systems, it does not involve the use of mechanical and electrical devices. Biophilic building designs that use roof gardens and vertical forest to insulate and cool buildings are introduced in Chapter 6.

About half of solar radiation is in the visible short-wave part of the electromagnetic spectrum. The other half is mostly in the near-infrared part (heat), with some in the ultraviolet part of the spectrum. This radiation also warms the atmosphere that emits long-wave radiation, which can be sensed and measured as atmospheric heat. Surface water temperature is a result of water body absorption of both short- and long-wave radiations minus its own water body back radiation and reflection. A part of solar radiation reaching the Earth is returned back to the space. The ratio of returned and received radiation is called *albedo*. Albedo is related to the angle of solar radiation reaching the surface and to the color of the surface; white surfaces reflect radiation, while black surfaces absorb it.

Solar power harnesses the radiation energy of the sun to produce electricity and heat. Wind power turbines harness the power of atmospheric air movement caused by rotation of the Earth and different atmospheric pressures also caused by different temperatures of

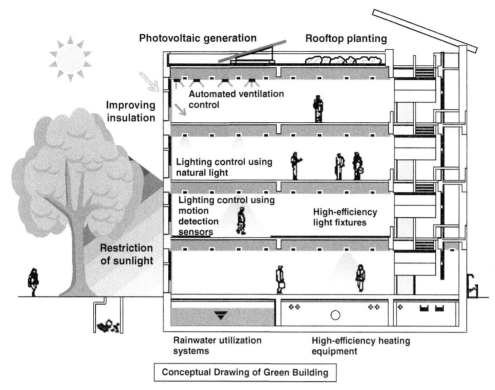

Photovoltaic generation — Rooftop planting

Improving insulation

Automated ventilation control

Lighting control using natural light

Lighting control using motion detection sensors

High-efficiency light fixtures

Restriction of sunlight

Rainwater utilization systems

High-efficiency heating equipment

Conceptual Drawing of Green Building

Figure 4.3. Passive energy savings features in high-performance buildings. *Source:* Ministry of Land, Infrastructure, Transport, and Tourism, Tokyo (2005).

water and soil. Solar and wind power are the cleanest, most sustainable, and most renewable energy resources.

Insolation is the power received by solar radiation on Earth per unit area of a horizontal surface. Albedo depends on the height of the Sun above the horizon, which means that the absorbed radiation varies with the time of the day. Because the angle (albedo) at which solar radiation reaches the Earth surface depends also on latitude, insolation is smallest in polar regions and highest near the equator. The unit of insolation and its measurements are Watts/m^2 or kW-hrs m^{-2}day^{-1}, reported over a horizontal surface. The high solar radiation of 6 kW-hrs m^{-2}day^{-1} is typical for deserts in Africa, Australia, and the Middle East; the middle range would be typical for the southwestern US, and lower incoming radiation would be typical for the rest of the densely inhabited regions of the world outside of the Artic (Figure 4.4).

According to the latest data from the US Energy Information Administration (2017), in 2016 the average annual electricity consumption for a residential utility customer was 10766 kW-h person^{-1}year^{-1}. If two comparably sized households in California and Massachusetts consume the average amount of electricity for an American household and the California insulation rate is 2000 kW-h m^{-2}year^{-1} and that in Massachusetts is 1500 kW-h m^{-2}year^{-1}, the one in California needs a photovoltaic system 25% smaller than a comparable household in Massachusetts. Homeowners in less sunny areas, like

GLOBAL HORIZONTAL IRRADIATION

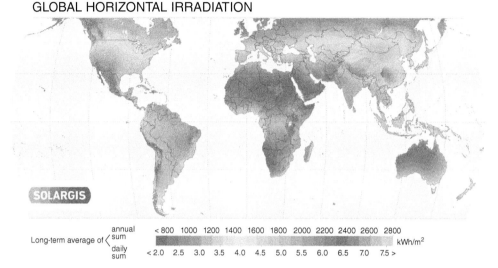

Figure 4.4. Solar insolation map in kW-h m^{-2} day^{-1}. *Source:* Global Solar Atlas, owned by the World Bank Group and provided by Solargis. © 2017 The World Bank.

Massachusetts, can compensate for this disparity by simply using more efficient panels or increasing the size of their solar energy system, resulting in more rooftop solar panels.

Solar power harnesses the natural energy of the sun to produce heat and electricity by:

- Concentrating solar radiation by computer-guided mirrors; heliostats focus the sunlight on a receiver containing a heat transferring fluid. The solar collectors concentrate sunlight to heat the heat transfer fluid to a high temperature. The hot heat transfer fluid is then used to generate steam that drives the power conversion subsystem, producing electricity. Thermal energy storage provides heat for operation during periods without adequate sunshine. This system works like a magnifying glass focusing the sun's energy onto a smaller area and creating high temperatures capable of igniting paper or wood. Figure 4.5 shows a detail of a concentrated solar panel array.

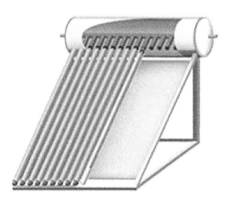

Figure 4.5. Energy Star evacuated tube solar water heat collector. *Source:* Courtesy United States Energy Association and DOE Energy Star.

Figure 4.6. Photovoltaic power plant installation in Newburyport, Massachusetts, with approximately 50 kW capacity. *Source:* Photo V. Novotny.

- Converting sunlight directly into electricity with photovoltaic panels made of semi-conductors (Figure 4.6). Photovoltaic cells in solar panels cause a reaction between photons and electrons on an atomic scale, which releases electrons that are then captured to produce electricity. Photovoltaic panels are very popular and provide wireless electricity for everything from pocket calculators to entire cities. Smart network electricity grids are generally capable of regulating power production during the day (i.e., reduce production when excess electricity would be provided by green developments and renewable sources) and activating more production (e.g., from hydropower storage power plants) during times when green energy would be insufficient. In 2017, more than 100-MW capacity storage batteries were already brought to the market in Australia by Tesla. Furthermore, Tesla is also realizing a giant virtual blue power plant by installing building 5 kW solar panels on 50,000 public housing buildings in Australia with their own batteries to store the excess energy and the entire system will be interconnected to create a giant virtual flexible blue power.

According to the International Energy Agency report (IEA, 2017), the total installed photovoltaic capacity at the end of 2016 globally amounted to at least 303 GW (gigawatts). The top 10 countries for total cumulative installed capacity in 2016 in gigawatts were:

(1) China	78.1 GW	(2) Japan	42.8 GW	(3) Germany	41.2 GW
(4) USA	40.3 GW	(5) Italy	19.3 GW	(6) UK	11.6 GW
(7) India	9 GW	(8) France	7.1 GW	(9) Australia	5.9 GW
(10) Spain 5.5 GW					

China alone in 2016 installed 34.45 GW and has become a leader in manufacturing and exporting photovoltaic panels and other blue energy (wind turbines). PV power now represents 2% of the total electricity power in the world. With rapidly declining prices in the last few years, PV power has appeared on the radar of policymakers in charge of energy policies in numerous countries and plans for PV development have increased rapidly all over the world.

Solar PV cells convert the sun's energy into electricity by converting photons into electrons. In 2018 the efficiency of commercial solar panels on the US market was about 15–22% and increased to 26.6% in Japan. Efficiency quantifies a solar panel's ability to convert sunlight into electricity. In theory, 30% energy-conversion efficiency is the upper limit for traditional single-junction solar cells, as most of the solar energy that strikes the cell passes through without being absorbed or becomes heat energy instead. A new solar cell design could raise the energy conversion efficiency to over 50% by absorbing the spectral components of longer wavelengths that are usually lost during transmission through the cell (Asahi et al., 2017). In December 2014, the highest solar cell efficiency of 46% was achieved by using multijunction concentrator solar cells, developed by collaboration efforts of Soitec, CEA-Leti, France, together with Fraunhofer ISE, Germany (Fraunhofer Institute, 2014).

The Solar Power Authority (2017) summarized the characteristics and advantages of photovoltaic solar power:

1. Most solar panels have 200- or 250-watt capacities and can be connected in series or parallel to provide more power.

2. Solar panels do not need direct sunlight to produce electricity. However, direct sunlight produces the most energy. Most modern solar units use net metering (a system hooked up to the city's power grid that measures the difference between the energy given back to the grid and the energy used from the grid). For controlling energy deficits, net metering is an easier and cheaper method than storing excess solar power in batteries. However, affordable 10- to 100-MW batteries are already available on the market for storing excess electric energy of sunny days on site. In the future, excess electric energy by photovoltaics can be used to produce by electrolysis hydrogen that can be stored as an energy carrier.

3. To power the entire Earth on renewable energy, solar panels would need to be installed on over 493,860 km^2 (the area of Wyoming). Considering there are over 147 million km^2 of land on Earth there is plenty of land surface to install solar power but it would take time. There are several other sources of blue power, such as wind and tide, that can even be installed in the oceans. Giant solar power plants are being built in China and Saudi Arabia.

4. A single house rooftop solar panel system can reduce pollution by 100 tons of carbon dioxide in its lifetime – and this includes the energy it took to manufacture the solar panels. Solar panels can improve future air quality for humans as well as nature.

5. Photovoltaic panels come in several shapes and forms. Of note are photovoltaic solar roof shingles, which are textured to resemble the granular look of roof shingles. Each shingle produced by OkSolar is 30 cm wide by 219 cm long and is nailed in place of roof decking. Electrical wires from the shingle photovoltaic panel pass through the roof deck, allowing interior roof or attic space connections.

6. As of 2016, the average cost of solar energy alone was about $0.12 per kilowatt, compatible with other energy sources. The cost has been dropping rapidly.

7. Solar power itself is a free source of energy. Once photovoltaic (PV) panels are installed on the home, maintenance is minimal, and returns are high.

The design of the solar power panel roof cover area is demonstrated herein:

Estimate the size of solar panels for a family of four in a house located in the northeastern US that would provide 100% of electricity need. Average residential electric energy use by one person is 11,000 kW-hrs/year requiring average power of 1.25 kW.

Assumption: 320 days that provide enough light energy

As seen in Figure 4.4, the average insolation in the northeastern US is 4 kW-h m^{-2} day^{-1}. The average efficiency of commercial solar panels in 2025 is expected to be about 25%. The solar panel area per one person is:

$$\text{Solar panel area} = \frac{11\,000\ (\text{kW-h capita}^{-1}\ \text{year}^{-1})}{320\ (\text{days}) * 4\ (\text{kW-h m}^{-2}\ \text{day}^{-1}) * 0.25} = 34.4\ \text{m}^2$$

or about 120 m^2 for the house. A set of panels covering area of 8 ∗ 15 meters would fit on one side of the roof of a larger size house if all electricity is to be provided by solar panels.

It is expected that the efficiency of the photovoltaic solar panels could double in the next 10–20 years, which would cut the size of the panels by half to 170 m^2/person. It is also expected that by installing more efficient appliances and heating systems, electricity use will decrease. Most likely, such panels would produce significantly more energy during the daytime than the family would need but less energy during dark peak energy use hour. The simplest solution is to sell the excess electricity to the power company and buy it during the hours of need. This requires a "smart" grid. It can be safely predicted that affordable house size kWatt or MWatt district size capacity batteries for storing excess electricity will become available on the solar energy market shortly after this book has been published.

Even though not extensively covered in this manual, power plants consisting of photovoltaics and wind turbines will be integral and possibly dominant parts of the future integrated water/energy and solids urban management that would lead to better than net-zero carbon emission cities. In 2017, Panda Green Energy (formerly known as United Photovoltaics) connected a 50-MW solar array to the grid in northwestern China, and solar power is rapidly becoming the cheapest source of blue energy. The Panda project is the first such installation the company under its collaborative agreement with the United Nations Development Program. China Merchants New Energy Group – Panda Green Energy's largest shareholder – signed an agreement with the UNDP in 2016 to build panda-shaped PV projects to raise awareness about sustainable development among young people in China. The plant was built in less than one year (Publicover, 2017).

Wind Power

People have used wind power and hydropower for centuries. Modern wind turbines fall into two basic groups: the *horizontal-axis* variety, like the traditional farm windmills used for pumping water, and the *vertical-axis* design, like the eggbeater-style Darrieus model, named after its French inventor. Most large modern wind turbines are horizontal-axis turbines (Figure 4.7) that are much larger, more sophisticated, and far more efficient than the traditional wind wheels. Wind turbines are often grouped together into a single wind power plant, also known as a *wind farm*, and generate bulk electrical power that can be built on land and on water. Electricity from these turbines is fed into a utility grid and distributed to customers, as with conventional power plants.

Figure 4.7. Three MW-size wind turbine installations in Gloucester, Massachusetts. *Source:* Fletcher, Wikimedia Commons, https://commons.wikimedia.org/wiki/File:Wind_Turbines_in_Gloucester.jpg.

Based on the US DOE (2016b) report the United States ranked second in annual wind additions in 2015 but was well behind the market leaders in wind energy penetration. Global wind additions again reached a new high in 2015, with roughly 63 GW of new capacity (mostly in Texas), 23% above the previous record of 51 GW added in 2014. Cumulative global capacity stood at approximately 434 GW at the end of the year (Navigant, 2016a). The United States ended 2015 with 17% of total global wind power capacity, a distant second to China by this metric. Based on wind power production, however, the United States remained the leading country globally in 2015 (AWEA 2016a). The US wind power market represented 14% of global installed capacity in 2015 but China is now leading by a great margin. Top 10 in wind power at the end of 2015 were:

(1) China 145 GW	(2) US 75 GW	(3) Germany 25 GW
(4) India 25.3 GW	(5) Spain 22.6 GW	(6) UK 13.4 GW
(7) Canada 11.2 GW	(8) France 10.2 GW	(9) Brazil 9.3 GW
(10) Italy 8.8 GW		

Wind power has comprised a sizable share of generation capacity additions in recent years – 41% in the US in 2015, up sharply from its 24% market share the year before and close to its all-time high. Since 2015, wind power has been the largest source of annual new generating capacity, well ahead of the next two leading sources, solar power and natural gas. Wind power is estimated to supply the equivalent of roughly 40% of Denmark's electricity demand, and 20–30% of demand in Portugal, Ireland, and Spain. In 2017 during his trip through northern Germany and Poland, the author saw the landscape dotted with perhaps a thousand MW-size wind turbines providing electric energy to almost one million people (per capita energy use in Poland is smaller than that in the US).

In the United States, the cumulative wind power capacity installed at the end of 2015, estimated in an average year, covers 5.6% of the nation's electricity demand. On a cumulative basis, Texas remained the clear leader among states, with 17 GW installed at the end of 2015, nearly three times as much as the next-highest state (Iowa, with 6.2 GW). Wind energy is a drought-resistant source of cash for farmers and ranchers who can rely on it to make a living and keep their land in the family. During 2016, US wind projects paid at least $245 million in lease payments to landowners and contributed taxes to rural communities.

A wind turbine's power is directly related to its "swept area" – the circular area covered by the blades' rotation and energy yield is obviously proportional to average wind speed (Figure 4.8). Wind turbine size varies, but in 2015 and earlier, typical wind farm towers were around 70 meters tall, with blades about 50 meters. The three turbines seen in Figure 4.7 in Gloucester, Massachusetts, located at the tip of the Cape Ann peninsula surrounded by Atlantic Ocean, take advantage of average wind of 26 km/hr (7 m/sec) and provide all the electric energy for the public offices, facilities, schools, and police for this community of 30,000 people. The largest of the three turbines, 2.5 MW and 165 m high, was the tallest in Massachusetts in 2017. Much taller wind turbines are now becoming a reality. In 2017, the upper end of power capacity of commercial wind turbines was 5 MW, which was enough to power about 1,100 homes, and 9-MW turbines with an overall height of 220 m and a diameter of 164 m were designed for the large Massachusetts offshore wind farm. Scientists are also changing the designs of the offshore "super turbines" expected to quadruple the current (2018) maximum power capacity. The biggest turbines in Europe, which now produce electricity that is cost-competitive with fossil fuels, are 240 m tall.

The most dominant material used for the blades in commercial wind turbines is fiberglass, which is reinforced thermoplastic plastic and is not currently recyclable. Resin technology has expanded to include both polyester and epoxy plastics on a broad scale. There is an unproven possibility that the plastic fraction could be gasified into hydrogen and sequesterable carbon dioxide (see Chapter 9) or some other recyclable materials will be invented. The

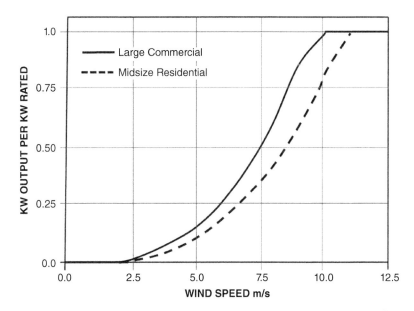

Figure 4.8. Estimated capacity factors for wind turbines related to wind speed. *Source:* US DOE Report (Lantz et al., 2016).

challenge to manufacturers of wind turbines blades is to develop recyclable resins. At the end of the second decade of this century, the fiberglass structures of wind turbine blades were expected to last for 25 years.

One of the largest wind farms to be built in the US – the Revolution Wind Farm, about 19 km off the shore of Martha's Vineyard in Massachusetts – would generate electricity and store some of it in large batteries built by Tesla. The project would have the capacity to generate 144 MW of wind power, or enough electricity to power 80,000 homes in Massachusetts.

However, the future of wind energy is not only in super large turbines. The US DOE National Renewable Energy Laboratory report (Lantz et al., 2016) finds that small distributed wind systems (less than 1 MW) are technically feasible for approximately 49.5 million residential, commercial, and industrial sites. The overall maximum resource potential for these small distributed wind turbines is 3 terawatts (TW) of capacity or 4,400 TW-hours (TWh) of energy generation, which is more electricity than the United States consumes in a year. Larger megawatt-scale distributed turbines could provide an additional 5.1 TW of capacity or 14,000 TW-h of annual energy generation, but in some cases this megawatt-scale resource potential overlaps with areas that would also be suitable for utility-scale (non-distributed) wind development.

Green and Blue Energy Storage

Energy demand varies throughout the day. It has morning and evening peaks and is the lowest during the night. However, the existing energy-producing systems are not very flexible. The backbones of the power systems are nuclear and coal-powered plants that in a boiler produce steam that runs turbines and generators. With this system it is very difficult to adjust electricity production to demand. Hydropower on large rivers is also not very flexible but it can adjust to the demand by simply shutting down and starting the turbines. Large river low head power plants are rarely used for peak power only. Currently, established peak energy producing is generally provided by high head hydropower with tall dams and oil and natural gas power plants that use combustion engines to run the turbines during peak times and stay idle or at reduced power during off-peak hours. Most of the hydropower is already developed; however, because of lower cost of fuel and flexibility, many power companies are today replacing coal-fired steam plants by combustion engines/turbine plants powered by natural gas. Many countries and US states are now undergoing this transition. The only energy storage available until now to power companies was pumped storage of water running a hydropower plant, which is very limited because it requires terrain with large elevation difference and storage volume is limited.

Until recently (before 2015), the main drawback with developing and using blue solar and wind energy was the inability to store excess energy because both sources are variable with the time of day and year and are dependent on weather. Hence producers and users had to rely on the electricity providers for buying the excess electricity produced during daytime and providing electricity when blue energy is less available.

A California study by Jacobson and Delucchi (2009) found that if the energy is provided by blue energy sources in a large system (a region, state) the problem is not as big as one might think. In a large renewable energy system, the sources will complement each other. For example, solar power produced during daytime may provide most of the day energy; wind power, especially in coastal areas, is strongest during evening peak hours, and if the sun is not shining the weather may be windy. Wind power of offshore or coastal wind farms is more constant but wind turbines can be turned on and off instantly as demand requires. This synergy can serve as one of safety factors but it is difficult to extrapolate universally to all areas. Biogas-powered plants that incinerate or gasify organic matter are not flexible, but geothermal energy is constantly available and can be turned on and off instantly. Until the end of the last century, use of rechargeable batteries was limited to small devices (from a phone to automobiles).

The following methods and technologies are available today for storage of energy.

Pumped Storage.   Until the end of the last century, only pumped storage hydropower system could store energy in the form of hydrostatic power. This system uses two water reservoirs with different elevations. The lower reservoir is connected to a power plant with turbines and pumps or reversible turbines. The upper reservoir is located at higher elevations. During the normal operations the power plant produces electricity for the system during the high use hours. During the hours of low electricity use the excess energy pumps water to the high elevation reservoir that provide hydrostatic energy to power turbines during the high use period. The pumping requires energy which is not fully recovered, nevertheless, the pumping is during the low-cost energy hours and energy is produced during the high cost peak hours.

Rechargeable Batteries.   Since the second decade of this century, the capacity of rechargeable batteries has been exponentially increasing. In the US water sector, the first 3.65-MW system was installed by Inland Empire Utilities Agency (IEUA) at the agency's six sites in Chico, California. The batteries, supplied by Tesla, reduce IEUA's demand for power during peak periods, saving the agency approximately $220,000 annually in electricity costs. The project charges the batteries at night when power costs are at their lowest and use the batteries during the day when grid demand is highest and costs are higher. The batteries are expected to reduce IEUA's peak energy demand by up to 15% and reduce annual greenhouse gas emissions by 5,436 tons. The agency installed 3.5 MW of solar power in 2008. Combined, these projects have provided more than 50% of peak energy demand agency-wide (https://www.ieua.org/about-us/programs/renewable-energy).

In 2017, a 100-MW battery was installed by Tesla in Australia and larger batteries installations followed in 2018 (Cutherson, 2018). South Korea was planning at the same time to install a battery with maximum power of 140 MW, and a 200-MW battery was being installed in California. These batteries were standard lithium-ion based (Cutherson, 2018). Meanwhile in China, Rongke Power Company in Dalien (population 6.3 million) in northeast China was manufacturing a super-large 200-MW/800-MW-hour vanadium flow battery (VFB) system, which is expected to peak-shave about 8% of Dalian's expected load when it comes online in 2020 (Ryan, 2017). In October 2018 Duke Energy Company announced a fifteen-year plan to install 300-MW battery storage in North and South Carolina. Apparently, there is no limit.

Special supercapacitors have been used for running electric buses in zones without overhead wires. Supercapacitors bridge the gap between conventional capacitors and rechargeable batteries.

Hydrogen Storage.   Excess produced hydrogen produced by the gasifiers and fermentation reactor can be stored in storage tanks similarly to natural gas (methane) and syngas produced by the historic gasification plants or in more modern water reclamation plants. Produced hydrogen can be also compressed. Excess blue and green electric energy can be also used to produce hydrogen by an electrolyzer during off-peak or excess energy hours that can be stored. The efficiency of an electrolyzer to convert electric energy into hydrogen energy and separated oxygen is about 80% and the reverse process of converting hydrogen into electric energy by hydrogen fuel cells is about 60% with the overall efficiency of hydrogen production, storage, and energy recovery of about 50% (Lovins, 2005). However, it is still worthwhile if the primary source is waste or unused renewable energy. Separated pure oxygen can be used in several applications in the IRRF, including production of ozone for disinfection.

4.5   FOOD/WATER/ENERGY/CLIMATE NEXUS

The *Water Energy Nexus in the IEA* (2016) report illustrated the interdependencies between energy and water that are set to intensify in the coming years, as the water needs of the energy sector and the energy needs of the water sector both rise. Water is essential for all phases of energy production; the energy sector is responsible for 10% of global water withdrawals, mainly for power plant operation and cooling as well as for production of fossil fuels and biofuels. These requirements will grow over the period to 2040, especially for water that is consumed (i.e., withdrawn but not returned to the source). In the power sector, there is a switch to advanced cooling technologies that withdraw less water but consume more. A rise in biofuels demand pushes up water use and greater deployment of nuclear power increases both withdrawal and consumption levels because efficiency of nuclear power plants (about 35%) is less than that of fossil coal and natural gas power plants (about 45%). Olsson (2012) documents that population increase, climate change, and an increasing competition between food and fuel production create enormous pressures on both water and energy availability. Since there is no replacement for water, water security is more crucial than energy security. This is true not only in developing countries but also in the most advanced countries, including the southwestern US. The water-energy issue is not only about technology. It is our attitudes and our lifestyle that can significantly influence the consumption of both water and energy, which are inseparable in the water–energy nexus.

Power plant energy efficiency is a ratio of electric energy produced divided by energy in fuel. Most common types of cooling systems used in thermoelectric power plants are (a) once-through flow cooling, which withdraws water from a large source (large river, lake or sea) and (b) closed recirculating system, located on smaller water sources, which withdraws water to replace the water lost by evaporative cooling and provide makeup water to flush out impurities and salts from the recycling water. Once-through cooling systems use about 20 times more water and have greater potential to negatively affect the aquatic life and ecosystems. The difference between the two is the heat dissipated by the plant cooling system, which is another important factor related to global warming affecting water availability for future fossil thermal electric (coal, oil, and natural gas) and nuclear power production. Such plants producing about 90% of the total energy in the US at the beginning of this century rely on wet-cooling, which requires a lot of water below prescribed design temperatures, both for cooling and operational efficiency. Such plants use 45% ($\sim$6.1 $*$ 10^8 m^3 per day) of total fresh water withdrawals. About 10% of heat is emitted with the flue gases. Water use of recycle systems is about 4% of once-through systems, but because recycle systems rely on evaporation for cooling, the consumption use is greater.

The analysis published in the IEA (2016) report provides a systematic global estimate of the amount of energy used to supply water to consumers. In 2014, some 4% of global electricity consumption was used to extract, distribute, and treat water and wastewater, along with 50 million tons of oil equivalent of thermal energy, mostly diesel fuel used for irrigation pumps and gas in desalination plants. US EPA estimate of energy use for water extraction, delivery, and wastewater treatment and disposal was 3% nationwide but increasing up to 7% in California. IEA estimated that by 2040, the amount of energy used in the water sector will more than double. Desalination capacity rises sharply in the Middle East and North Africa and demand for wastewater treatment (and higher levels of treatment) grows, especially in emerging economies. By 2040, 16% of electricity consumption and corresponding GHG emissions in the Middle East will be related to water.

Summarizing the wide nexus of impacts, the United Nation Report (UN Water, 2014) pointed out:

- Roughly 75% of all industrial water withdrawals are used for energy production.
- The food production and supply chain accounts for 30% of total global energy use.
- 90% of global power generation is water-intensive.
- Global water demand on water withdrawals is projected to increase by 55% by 2050, mainly because of growing demands from manufacturing (400% increase). More than 40% of the global population is projected to be living in areas of severe water stress by 2050.
- Power plant cooling is responsible for 43% of total fresh water withdrawals in Europe (more than 50% in several countries), nearly 50% in the United States of America, and more than 10% of the national water cap in China.
- By 2035, water withdrawals for energy production could increase by 20% and consumption by 85%, driven via a shift toward higher-efficiency power plants with more advanced cooling systems (that reduce water withdrawals but increase consumption) and increased production of biofuel.
- There is clear evidence that groundwater supplies are diminishing, with an estimated 20% of the world's aquifers being overexploited (mined), some critically. Deterioration of wetlands worldwide is reducing the capacity of ecosystems to purify water.

Ganguli, Kumar, and Ganguly (2017) analyzed the future water availability and temperature and found that power production in the US remains particularly vulnerable to water scarcity and rising stream temperatures. On one side, the efficiency of thermal power plants decreases with the increased temperature of the source of the cooling water, and on the other side, temperature of the outflow is regulated by water quality standards. During the periods of low flows, the plants are forced to operate at a reduced capacity and must be shut down temporarily if water temperature exceeds the standards. Their analysis of projected changes in low surface runoff and high stream temperature has predicted drying and warming patterns over most US regions between 2006 and 2035. Drought-induced water scarcity and heat waves driven by warm water have already impacted power productions in several parts of US, including Texas and California. According to Ganguli et al. (2017), assessments over the contiguous US predicted consistent increase in water stress under climate change and variability for power production, with about 27% of the production severely impacted by the 2030s and more thereafter. These factors will increase the demand for green and blue energy. "Managing energy-water-climate linkages (water-energy nexus) is pivotal to the prospects for successful realization of a range of development and climate goals" (IEA, 2016; Olsson, 2012).

Global trends of population growth, rising living standards, and the rapidly increasing urbanized world are increasing the demand on water, food, and energy. Added to this is the growing threat of climate change, which is expected to have a great impact on water and food availability. It is increasingly clear that there is no place in an interlinked world for isolated solutions aimed at just one sector. In recent years the "nexus" has emerged as a powerful concept to capture these interlinkages of resources and is now a key feature of policy making.

Higher temperatures increase the water requirements of crops and livestock. An increased annual variation in crop and livestock production is expected due to increased temperature and extreme weather events, exacerbating the productivity risks of rainfed production, particularly in semi-arid areas vulnerable to drought. Elevated CO_2 concentrations cause reduced nitrogen and protein content in most crops, which means a lower nutritious value and a need for greater fertilizer use to support crop growth. It also reduces the forage quality of grasses and thus the quality of livestock produce. By 2050, irrigated agriculture on 16% of the total cultivated area is expected to be responsible for 44% of total crop production (WBCSD, 2014). Dodds and Bartran (2016) brought together contributions by leading intergovernmental and governmental officials, industry specialists, scientists, and other stakeholder thinkers who are working to develop approaches to the nexus of water-food-energy and climate.

4.6   WORLD AND US ENERGY OUTLOOK

The Paris 188 Nations Agreement (Paris COF 21, 2015) on energy and limiting its impact on global climate change (warming) came into being in November 2016. However, in July 2017, the US president rescinded US participation and the US became the only one not committed to the Paris Agreement, the most important agreement of almost 200 countries to combat global warming. In this book we will follow the arguments and facts presented by Vandenbergh and Gulligan (2017), stating that the push for reducing GHG emissions and limiting worldwide temperature increases and, eventually, pressures for stopping and reversing the trend will be so strong that industries, cities, and the public – including in the US – will comply with the Paris Agreement simply to protect the life and health of humans and nature for future generations. Furthermore, it will also make economic sense because blue and green energy production and wide implementation will cost less than black energy, which is already happening. In 2018, the US government intended to subsidize coal mining because it could not compete in the energy sector with natural gas and renewable energy. What is needed is to mobilize the private sector, citizens, and local and state governments to adhere to the Paris Agreement. Hopefully, the US will rejoin the Paris Agreement; otherwise, some parts of the US (such as Arizona) may become uninhabitable and parts of Florida will be underwater, contaminated periodically by red tide flagellates.

The Organization for Economic Cooperation and Development (OECD) and its branch International Energy Agency issues annual Water Energy Outlook Reports. The following discussion of trends and the future of energy and compliance with the Paris Agreement is based on the 2016 report (IEA, 2016). Overall, the report was positive and pointed out that growth in energy-related CO_2 emissions stalled completely in 2015 and the changes already underway in the energy sector demonstrated the promise and potential of low-carbon energy, which will lead to meaningful action on climate change. The United Nation Environment Report (2017) highlighted and confirmed the facts that the accelerating irreversible trend toward sustainable energy and away from fossil fuel is based more on market forces than on governments. Worldwide, investment in renewables capacity in 2016 was roughly double that of fossil fuel generation; the corresponding new capacity from renewables was equivalent to 55% of all new power, the highest to date. The average dollar capital expenditure per megawatt for solar photovoltaics and wind dropped by over 10%. This prevented the emission of an estimated 1.7 gigatons of carbon dioxide. "Falling costs, expanding consumer demand, global concern over climate change, and profit-promising investments are making today's clean energy sector a classic example of a market-driven boom. The trend lines will continue to point upwards."

Although only a small factor in total power demand, the projected rise of electricity demand may be caused by the use of electric vehicles. IEA (2016) reported that the world-wide stock of electric cars reached 1.3 million in 2015, a near-doubling compared to 2014 levels. In their main scenario, this figure would rise to more than 30 million by 2025 and exceed 150 million in 2040 and significantly reduce the fossil oil fuel mining. In a more opti-mistic scenario, if tighter fuel-economy and emissions regulations combined with financial incentives become stronger and more widespread, the effect would be to have some 715 mil-lion electric cars on the road by 2040 and thereafter all vehicles produced will be electric or powered by hydrogen. It is therefore important that the portion of green and blue electric energy is increasing more rapidly. IEA did not evaluate the impact of hydrogen driven cars and technologies.

In the most optimistic scenario IEA projected that nearly 60% of the power generated in 2040 will come from renewables, almost half of this from wind and solar PV. The power sec-tor will be largely decarbonized in this scenario: the average emissions intensity of electricity generation could drop to 80 grams of CO_2/kW-h in 2040, compared to 515 g CO_2/kW-h today worldwide (620 g of CO_2 per kW-h in the US in 2000). In the four largest power markets (United States, China, the European Union, and India), variable renewables will become the largest source of generation, around 2030 in Europe and around 2035 in the other three countries. A 40% increase in generation from renewables, compared with our main scenario, comes with only a 15% increase in cumulative subsidies and at little extra cost to consumers. In some US states, solar power can be installed at no cost to homeowners and still result in reduced charges for electricity.

Shell Global (2018) Sky Scenario, which must be taken seriously but with caution, esti-mates the following proportions of energy sources in 2070:

Green/Blue: Solar 32%, Bioenergy 14%, Wind, 13% Nuclear, 10% Hydro 10%

GHG emitting: Oil 10% Natural gas 6% Coal 6%

By comparison, at the beginning of the twenty-first century, 70% of energy in the US was produced by GHG-emitting power generation (coal, oil, natural gas) with coal accounting for 51%. By 2050–2070 CO_2 sequestering technology will be greatly expanded.

In the Shell Global (2018) Sky Scenario, passenger electric vehicles reach cost parity with combustion engine cars by 2025. By 2035, 100% of new car sales will be electric in the EU, US, and China, with other countries and regions close behind. In the Sky forecast, nuclear power grows steadily, with 1,400 GW of capacity by 2070, up from 450 GW in 2020. This estimate is contrasted in Germany, France, or Japan, where nuclear power plants are expected to be decommissioned and replaced by other blue and green energy sources.

In contrast, in the most optimistic but realistic scenario, it may be expected that after 2050 the wind and solar power capacities in advanced countries (including China and India) but also today in developing countries (because of low energy use) will be so large and cheap that they will displace most fossil fuel and electric energy production, facilitate hydrogen production, and by providing blue energy to hydrogen and electric vehicles, trains, and boats the planet can be saved from global warming catastrophes. The Shell Sky Scenario estimate of 22% energy production anticipated to be provided after 2050 by fossil fuel represents a view of the oil industry giant that would be acceptable only if most of the GHG emissions were sequestered or, as envisioned by Shell, counterbalanced by massive reforestation, or both.

Despite governmental inaction in some countries, the future is already happening now. It is the market that drives the implementation of green and blue technologies on a large scale. Texas and Iowa are leaders in installing wind energy farms on a large scale, not because they were forced to do it by government regulation or subsidies but because it makes economic sense. Farmers still can farm their land in windswept Texas and have a significant additional – perhaps greater – source of income by letting wind power companies install dozens of wind turbines on their lands to provide clean blue energy that is economically cheaper than that from coal or oil. The same is even truer in Europe and China and elsewhere throughout the world.

Solar panels were much cheaper in 2018 than they were 10 years earlier and the cost is still going down, which means millions of homeowners who could not previously afford to do so are switching to solar. In Massachusetts, state rebates incentivize homeowners to use clean energy by reducing the cost of solar power projects by $0 payment for installations. With rebates being as high as they are, homeowners are able to drastically reduce their power bill without dealing with the upfront costs of installing solar panels. Before, solar companies would lose money if they were to rent or lease solar panels; in 2018 one of the US largest solar power providers claimed to be signing up a new solar panel customer every 5 minutes because of these incentivized programs and economic benefits. In Germany, there have already been times when all energy was provided by blue and green sources and wind turbine farms are ubiquitous.

NSTC (2008) estimated that by 2050 60–70% household electricity use will be achieved by more efficient appliances and lights, more efficient water heating, water savings through water conservation, and reduction of cooling energy by passive energy savings, insulation, and green roofs. GHG reduction from heating of buildings can be reduced by heat pumps powered by renewable energy sources deriving heating and cooling energy from air and geothermal sources. It is very likely that by 2050 all public transportation and personal automobiles will run on electric power or hydrogen. To top that prediction, many automobiles and subway (metro) trains will be driverless.

It is quite possible that electrification of shantytowns and remote communities in developing countries will not be accomplished by bringing energy through high-voltage power lines from distant black power electric power plants, but will be done by local blue energy by installing solar panels on the roofs with batteries to store energy locally. In 2017, Hurricane Maria destroyed the entire power system of Puerto Rico and more than a year later the energy and water supply systems were not fully restored. Instead of rebuilding the centralized power system distribution system, it would be far more resilient if the island communities had a local decentralized blue power system that could be restored in days and not in years. Electric energy use in these communities is significantly smaller than that in the US and other developed countries and the insolation is more intensive. It will be argued in Chapter 5 that a distributed system is generally far more resilient to catastrophes than centralized systems with long-distance transmissions of energy and water.

However, optimistic forecasts and expectations were cooled down during the 2018 conference of the International Panel on Climate Change in Korea. After intensive review of thousands of sources and reports, the highlights of panel findings are presented in the bullet points below, drawn from the IPCC (2018) report and the scientists' panel from the US, Canada, and Mexico (USGCRP, 2018b):

- Observed global mean surface temperature for the decade 2006–2015 was 0.87°C (likely between 0.75°C and 0.99°C) higher than the average over the 1850–1900 period. Human activities are estimated to have caused approximately 1.0°C of global warming above preindustrial levels, with a likely range of 0.8°C to 1.2°C. Global warming is likely to reach 1.5°C between 2030 and 2050 if it continues to increase at the current rate. Globally, atmospheric carbon dioxide (CO_2) has risen over 40%, from a preindustrial level of about 280 parts per million (ppm) to the current concentration of more than 400 ppm. Over the same time period, atmospheric methane (CH_4) has increased from about 700 parts per billion (ppb) to more than 1,850 ppb, an increase of over 160%.

- North American emissions from fossil fuel combustion have declined on average by 1% per year over the last decade, largely because of reduced reliance on coal, greater use of natural gas (a more efficient fossil fuel), and increased vehicle fuel efficiency standards. As a result, North America's share of global emissions decreased from 24% in 2004 to 17% in 2013 (USGCRP, 2018a). However, without additional efforts to reduce GHG emissions (business as usual) beyond those in place today, emissions growth is expected to persist, being driven by growth in global population and its economic activities. Without additional mitigation, this will result in global mean surface temperature increases in 2100 from 3.7°C to 4.8°C, compared to preindustrial levels the range is 2.5°C to 7.8°C when including climate uncertainty. Such increases would be disastrous. Warming greater than the global annual average is being experienced in many land regions and seasons, including two to three times higher warming in the Arctic (IPCC, 2018).

- Following devastating hurricanes in the US (Maria in 2017, Florence and Mathew in 2018), record droughts in Cape Town, forest fires in the US and even in the Arctic, and accelerated melting of the glaciers, the IPCC makes clear that climate change is already happening. Many land and ocean ecosystems and some of the services they provide have already changed due to global warming (IPCC, 2018).

- Reaching and sustaining net-zero global anthropogenic CO_2 emissions and declining net non-CO_2 radiative forcing would halt anthropogenic global warming on multi-decade timescales (high confidence). Decarbonizing electricity generation is a key component of cost-effective mitigation strategies in achieving low-stabilization levels.

- Several regional changes in climate are expected to occur with global warming up to 1.5°C, compared to preindustrial levels, including warming of extreme temperatures in many regions, increases in frequency, intensity, and/or amount of heavy precipitation in several regions, and an increase in intensity or frequency of droughts in some regions.

- Global warming effects will continue beyond 2100 even if global warming is limited to 1.5°C in the twenty-first century (high confidence). Marine ice sheet instability in Antarctica and/or irreversible loss of the Greenland ice sheet could result in a multi-meter rise in sea level over hundreds to thousands of years. These instabilities could be triggered around 1.5°C to 2°C of global warming by 2050.

- To prevent 1.5°C warming, greenhouse pollution must be reduced by 45% from 2010 levels by 2030, and 100% by 2050. The report also found that, by 2050, use of coal as an electricity source would have to drop from nearly 40% today to 1–7%. Renewable energy such as wind and solar, which make up about 20% of the electricity mix today, would have to increase to as much as 67%.
- The report concluded that in 2018 the world was already more than halfway to the 1.5°C mark. Human activities have caused warming of about 1°C since about the 1850s, the beginning of large-scale industrial coal burning.

The findings of the IPCC were accepted and further elaborated by a joint report of 13 US Government Scientific Agencies in *The National Climate Assessment* (US Global Change Research Program, USGCRP, 2018a and b). The reports noted that in the US (quoted from the reports):

(a) Annual average temperatures have increased by 1°C across the contiguous United States since the beginning of the twentieth century. Alaska is warming faster than any other state and has warmed twice as fast as the global average since the mid-twentieth century.

(b) The season length of heat waves in many US cities has increased by over 40 days since the 1960s.

(c) The relative amount of annual rainfall that comes from large, single-day precipitation events has changed over the past century; since 1910, a larger percentage of land area in the contiguous United States receives precipitation in the form of these intense single-day events.

(d) Large declines in snowpack in the western United States occurred from 1955 to 2016.

(e) Since the early 1980s, the annual minimum sea ice extent (observed in September each year) in the Arctic Ocean has decreased at a rate of 11%–16% per decade.

(f) Annual median sea level along the US coast (with land motion removed) has increased by about 23 cm since the early twentieth century as oceans have warmed and land ice has melted.

(g) Fish, shellfish, and other marine species along the Northeast coast and in the eastern Bering Sea have, on average, moved northward and to greater depths toward cooler waters since the early 1980s (records start in 1982).

(h) Oceans are also currently absorbing more than a quarter of the carbon dioxide emitted to the atmosphere annually by human activities, increasing their acidity measured by lower pH values.

Annual average temperatures in the United States are projected to continue to increase in the coming decades. Regardless of future scenarios, additional increases in temperatures across the contiguous United States of at least 1.1°C relative to 1986–2015 are expected by the middle of this century. As a result, recent record-setting hot years are expected to become common in the near future. By late this century, increases of 1.1°–3.7°C are expected under a lower scenario and 3°–6.1°C under a higher scenario relative to 1986–2015. Alaska has warmed twice as fast as the global average since the mid-twentieth century; this trend is expected to continue.

The carbon cycle is changing at a much faster pace than observed at any time in geological history These changes are primarily attributed to current energy and transportation dependencies on the burning of fossil fuels, which releases previously stable or sequestered carbon. Also contributing to rapid changes in the carbon cycle are cement production and gas flaring (by methane mining and from landfills), as well as net emissions from forestry, agriculture, and other land uses (USGCRP, 2018b).

High temperature extremes, heavy precipitation events, high-tide flooding events along the US coastline, ocean acidification and warming, and forest fires in the western United States and Alaska are all projected to continue to increase, while land and sea ice cover, snowpack, and surface soil moisture are expected to continue to decline in the coming decades. These and other changes are expected to increasingly impact water resources, air quality, human health, agriculture, natural ecosystems, energy and transportation infrastructure, and many other natural and human systems that support communities across the country. The severity of these projected impacts, and the risks they present to society, are greater under futures with higher greenhouse gas emissions, especially if limited or no adaptation occurs. These impacts are quantitatively assessed in the report. Intensifying droughts, heavier downpours, and reduced snowpack are combining with other stressors such as groundwater depletion to reduce the future reliability of water supplies in the region, with cascading impacts on energy production and other water-dependent sectors.

Many places are and will be even more subject in the future to more than one climate-related impact, such as extreme rainfall combined with coastal flooding, or drought coupled with extreme heat and wildfires. Damages of hurricanes in coastal communities are increasing followed by more severe and devastating flooding inland. The compounding effects of these impacts result in increased risks to people, infrastructure, and interconnected economic sector. Catastrophic wildfires followed by mudslides are becoming ubiquitous in California and several other western states. Wildfire trends in the western United States are influenced by rising temperatures and changing precipitation patterns, pest populations, and land management practices (Morello-Frosch et al., 2009). As humans have moved closer to forestlands, increased fire suppression practices have reduced natural fires and led to denser vegetation, resulting in fires that are larger and more damaging when they do occur.

US crops (such as corn, soybeans, wheat, rice, sorghum, and cotton) are expected to decline over this century as a consequence of increases in temperatures and possibly changes in water availability and disease and pest outbreaks. Increases in growing season temperatures in the Midwest are projected to be the largest contributing factor to declines in US agricultural productivity. Climate change is also expected to lead to large-scale shifts in the availability and prices of many agricultural products across the world, with corresponding impacts on US agricultural producers and the US economy.

Urban environments in North America are the primary sources of anthropogenic carbon emissions; hence, carbon monitoring and budgeting in these areas are extremely important. In addition to direct emissions, urban areas are responsible for indirect sources of carbon associated with goods and services produced outside city boundaries for consumption by urban dwellers. Careful accounting of direct and indirect emissions is necessary to avoid double counting of CO_2 fluxes measured in other sectors and to identify sources to inform management and policy.

Climate change threatens many benefits that the natural environment provides to society: safe and reliable water supplies, clean air, protection from flooding and erosion, and the use of natural resources for economic, recreational, and subsistence activities. Changes in temperature and precipitation can increase air quality risks from wildfire and ground-level ozone (smog). Projected increases in wildfire activity due to climate change would further degrade air quality, resulting in increased health risks and impacts on quality of life. Valued aspects of regional heritage and quality of life tied to the natural environment, wildlife, and outdoor recreation will change with the climate, and as a result, future generations can expect to experience and interact with natural systems in ways that are much different than today. Without significant reductions in greenhouse gas emissions, extinctions and transformative impacts on some ecosystems cannot be avoided, with varying impacts on the economic, recreational, and subsistence activities they support.

Carbon Management Opportunities in North America.   Analyses of social systems and their reliance on carbon demonstrate the relevance of carbon cycle changes to people's everyday lives and reveal feasible pathways to reduce greenhouse gas (GHG) emissions or increase carbon removals from the atmosphere. Such changes could include, for example, decreasing fossil fuel use (which has the largest reduction potential), expanding renewable energy use, and reducing CH_4 emissions from livestock and landfills. Increased afforestation and improved agricultural practices also could remove emitted CO_2 from the atmosphere (USGCRP, 2018b).

The USGCRP (2018a) report issued by the Office of the President of the US confirmed that if nothing or little is done rapidly, the climate changes and their very adverse impacts will be irreversible after 2050 without massive and expensive geoengineering.

DECENTRALIZED HIERARCHICAL URBAN WATER, USED WATER, SOLIDS, AND ENERGY MANAGEMENT SYSTEMS

5.1 ECONOMY OF SCALE DOGMA FORCED CENTRALIZED MANAGEMENT 45 YEARS AGO

The majority of local wastewater treatment plants built in some developed countries before 1960 had only primary treatment connected to anaerobic digestion of separated sludge (for example, Imhoff Tanks) or low-efficiency secondary trickling filter facilities or aerobic/anaerobic lagoons (Imhoff and Imhoff, 2007; Novotny et al., 1989). Few cities had activated sludge plants even though the first ones were built in the US and Europe in the first two decades of twentieth century – in Milwaukee, Wisconsin, in 1915; in the UK in 1914, and in Germany in 1924 (Wanner, 2014). However, in some developed countries at the end of the last century, large cities like Milan, Vienna, and Budapest still discharged untreated or partially treated wastewater into their receiving water bodies. Before 1964 (the year of the Tokyo Olympic Games) Tokyo and other large cities in Japan had no sewers. The majority of smaller cities had no treatment or operated local rudimentary lower-efficiency treatment plants and the most common treatment was a septic tank.

The passage of the Clean Water Act (CWA-PL 92-500) in 1972 by the US Congress – overriding the president's veto – gave an impetus to massive building of treatment facilities that were required to meet more stringent *best available technology economically achievable* (BATEA) effluent standards. These facilities also had to comply with regional pollution abatement plans (Section 208 the CWA) asking to select the lowest-cost alternatives. Concurrently, local potable water sources (wells) were also abandoned because of aquifer pollution and drawdown. Salinity and radioactivity of groundwater increase with the ground water table depth below the surface.

In the last century, traditional economic evaluation of the feasibility of urban infrastructure, including pollution cleanup, was to minimize the net cost. Typically, net cost would include the capital and operation and maintenance costs minus tangible (monetary) benefits of the project. Water supply and energy-providing facilities are typically regulated public utilities that can charge users fees for deliverables that would only marginally exceed the cost; essentially these facilities are and should be nonprofit. The same is true for the utilities that provided waste-water treatment or solid waste collection and disposal. Under these economic models, it was realized that the unit cost of environmental projects – for example, cost of treatment per unit of volume of treated water – in these economic evaluations decreases with the increased scope/size of the project. In "Section 208 planning" (Section of the 1972 US Clean Water Act requiring regional planning of water pollution control facilities), cost analysis was limited only to the present value of all tangible costs (capital and operation and maintenance discounted to time zero by a discount rate published by the Engineering News Record) over the bond amortization period or some arbitrary lifetime period. This accounting of only total infrastructure capital and OMR costs typically resulted in the lowest marginal cost for large centralized regional facilities connected with the communities by large water supply pipes and sewage interceptors and abandonment of small local treatment plants. In most cases, the intangible social and ecological costs, which are today included in the Triple Bottom Line Assessments, were considered only narratively in a separate Environmental Impact Assessment (EIA) (National Environmental Protection Act, 1970) prepared by a planning entity (consultant) other than the planner and only for projects where federal funds were used. Some states adopted their own EIA requirements. Typically, no resource or energy recovery or limiting carbon emissions were considered. The new large-scale activated sludge treatment facilities offered better efficiency capable of meeting the more stringent effluent BATEA standards and were managed by highly skilled professionals. Today, small (package) treatment plants can provide effluent quality that is comparable or better than that in the receiving waters and can be operated remotely by both public and private companies.

Planning and implementing large regional water supply facilities upstream from the urban areas or bringing water from large distances and situating the used water treatment plants (renamed often as "water reclamation plants") downstream from the city created sections of the urban water streams that were deprived of low flows and became effluent-dominated streams downstream from the effluent discharge from the urban area (Novotny, 2007; Novotny et al., 2010a). Under the economic benefit/cost analysis models of the second half of the last century it was impossible to develop systems that would capture and recycle water or change solids to resources instead of depositing them in the least costly but environmentally damaging landfills. Because capturing, treating, and using stormwater as a resource was somewhat costlier, stormwater was considered and declared as wastewater requiring capture and fast conveyance after some treatment into the receiving water bodies. In reality, captured rainwater and treated stormwater are the third best potable water sources after groundwater and clean freshwater from surface water bodies.

Triple bottom line assessment (or otherwise noted as TBL or 3BL) is an accounting framework with three parts: social, environmental (or ecological), and financial (Figure 5.1). Many organizations have adopted the TBL framework to evaluate their performance in a broader perspective to create greater business value. It is a method for assessing all sustainability benefits and costs of development. Methods today exist for valorization of formerly intangible benefits such as gaining (benefits) or losing (cost) recreation, fishing, clean versus polluted air and water, loss of natural assets by channelizing streams, and gains by their renaturalization. Such methods can be based on using scientific surveys of population to assess the *Willingness to Pay* for environmental originally intangible benefits and costs (Clark et al., 2001). Figure 5.1 shows the three (triple) components of the sustainability accounting.

Using triple bottom line accounting and enumerating social environmental benefits has dramatically changed the assessment of the total benefits and costs. The used water has become a resource that had significant local tangible and intangible benefits that couldn't be easily realized by pumping water from regional water reclamation facilities located great

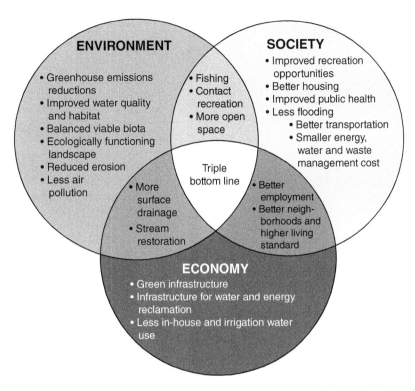

Figure 5.1. Triple bottom line sustainability assessment concept based on Elkington (1997).

distances from the points of reuse. This led to formulating and adopting the concepts of distributed water/stormwater/used water management, which is now being expanded to "circular economy."

5.2 DISTRIBUTED BUILDING AND CLUSTER LEVEL DESIGNS AND MANAGEMENT

The integration of a complete water management (urban water cycle) – which includes water conservation and reclamation, storage of reclaimed water and stormwater for reuse, used water (wastewater) treatment, and energy and resource recovery from waste – cannot be fully achieved in a linear system. Reuse and recycle is not considered in linear systems. However, not all management can be decentralized, and the reuse/recycle/recovery cycle cannot be fully closed. The differences between the centralized and decentralized management are outlined in Table 5.1. It will be documented herein that when water, energy and resource recovery is an issue, the systems perform best and most efficiently when the water/stormwater/used water systems are decentralized and are integrated with the urban solids management in a synergistic system (Anon, 2014). Thus, the concept of clustered distributed and decentralized water management with a regional resource recovery has been evolving (Lucey and Barraclough, 2007; Heaney, 2007; Daigger, 2009, Novotny et al., 2010a). Consequently, water and energy conservation and resource recovery, reuse, and recycle are hierarchical and are accomplished at three levels (Figure 5.2): (a) house or building level, (b) cluster/neighborhood (ecoblock) or district level, and (c) city/regional level.

Table 5.1. Centralized and decentralized components of urban water and energy management.

Component	Centralized	Distributed/Decentralized in Clusters
Stormwater rainwater management	None, stormwater management is local.	BMPs installation and management–pervious pavements, raingardens, green roofs, surface and subsurface storage, infiltration basins, and trenches.
Water conservation	Fixing leaking pipes, system wide education of citizens about water conservation, potable and nonpotable water distribution.	Wide variety of commercial water-saving plumbing fixtures and technologies for potable and nonpotable use; changing from lawns to xeriscape. Stormwater capture and reuse.
Treatment	Treatment for potable and some nonpotable reuse. Integrated resource recovery facility (IRRF) for recovering clean water, organic solids, methane, hydrogen, electricity, heat, and nutrients. Growing algae for more energy production.	Fit for reuse treatment for local potable use (from local wells and surface sources) and nonpotable reuse (from used water) in small cluster size water and energy reclamation units; stormwater treatment in biofilters, ponds and wetlands, effluent post treatment in ponds and wetlands. Gray water treatment and reuse. Possible source separation into black water, gray water, and urine flows.
Energy recovery	Anaerobic treatment, digestion of residual organic solids and MSW, microbial electrolysis cells. Gasification of MSW and solids to produce syngas. Electricity from methane, syngas, and hydrogen by hydrogen fuel cells.	Capture and distribution of heat and cooling energy (heat pumps); geothermal, wind, and solar energy production. Small scale biogas production by digestion (outdoor in developing countries). Battery energy storage for district peak use and emergency. Municipal solid waste (MSW) collection and recycling
Nutrient recovery	Land application of biosolids, struvite (ammonium magnesium phosphate) precipitation and recovery. Anammox removal of N.	Irrigation with reclaimed water with nutrients left in it; reclaimed irrigation water distribution to parks, golf courses, and homes; urine separation and recovery.
Source separation	Treatment of concentrated black wastewater and organic solids with energy (biogas) production. Recycle of MSW and recover resources from gasification ash.	Supply of potable and nonpotable water; treatment of black, gray (laundry and kitchen), and yellow water for nonpotable reuse (irrigation, toilet flushing), concentration of residual used water flow with removed solids for further processing in the IRRF. Sorting MSW for recycling and gasification.
Landscape management	Daylighting and habitat restoration; fish management and restocking, wildlife management in ecotones, floodplain restoration.	Stream and ecotones maintenance, including ponds and wetlands; on and off water recreation, incorporating flood storage and extreme weather resiliency into landscape.

Source: Adapted from Daigger (2009) and Novotny et al. (2010a).
BMP = Best Management Practices

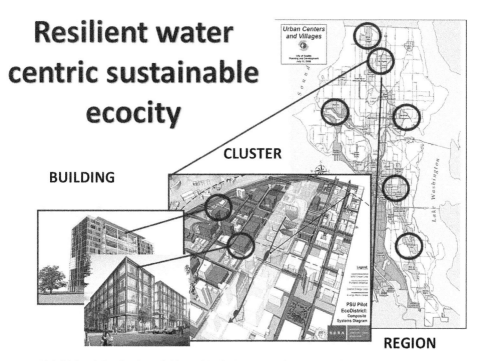

Figure 5.2. Hybrid (partially distributed) hierarchical urban water/storm water/used water and energy system. *Source:* Adapted from Moddemeyer (2010); courtesy CollinsWoerman, Seattle.

Cluster or Neighborhood Level Water and Energy Recovery

House or Building Level. At the house level (an individual house, apartment, or commercial building), water- and energy-saving devices are installed along with outdoor sustainable landscape – xeriscape with minimum, mostly natural rainfall irrigation.

With current advancing green technologies for houses and buildings, most of the energy and water neutrality can be accomplished in buildings. The fundamentals were already discussed in Chapter 2. Houses (buildings) have a choice to apply some or all of the following measures:

(a) Passive energy conservation by insulation, shade, and "smart" blinds in the windows
(b) Installation of green roofs and vertical forests (Chapter 6)
(c) Installation of photovoltaics and thermal recovery solar panels
(d) Potable water conservation
(e) Used water separation into black and gray water flows
(f) Geothermal heat, sensible heat from water, air and cooling energy reclamation by heat pump

Energy. Scientific research quoted in the National Science and Technology Council (NSTC, 2008) report indicates 60–70% of energy reductions in buildings in cities can be achieved with passive heating (Figures 5.3 and 4.3) and cooling incorporated in the architecture of the building, more efficient appliances such as better water and space heaters and heat pumps, significant reduction of water demand by water conservation and use of rain and stormwater, organic solids management for energy and resources recovery, and other improvements. NSTC also estimated that 30–40% of energy can be produced by renewable sources, including heat recovery from used water and/or extracted from the ground and groundwater. The estimate of renewable energy recovery today is more optimistic, more than double, as explained below.

Figure 5.3 shows the Platinum LEED-certified Bullitt Building in Seattle, which outperforms two triple net-zero goals (water and energy) and by MSW separation it contributes to the no-landfill goal. The Bullitt building in early 2000s was an experiment and to some degree an overkill, but it has received worldwide positive attention. The British newspaper the *Guardian* in a feature article (Nelson, 2015) reported that in 2014 the Bullitt's rooftop photovoltaic panels produced roughly 244,000 kW-hours of energy while the building used only 153,000 kW-hours, resulting in enough surplus solar energy to power eight additional homes. The building energy demand is also less than 40% of a LEED Platinum building and less than 60% of a typical net-zero energy building. Consequently, the roof solar panel area, which gives the building its unique futuristic look, could have been smaller but, on the other hand, could be a part of the district net-zero carbon program. Significant energy savings on building level could have been also achieved by green roofs helping to manage stormwater flows and providing insulation and cooling, passive heating (Figures 4.3 and 5.3) and cooling energy savings, insulation, and so on. Based on the recent experience with the net-zero buildings, net-zero energy use and GHG emissions goals can be met at the building level. Bullitt Building produces excess energy that is delivered to the grid and provides a district-wide credit.

Solar panels are becoming a common feature of houses; as of 2017 the US government and the governments of most of EC countries had provided subsidies and incentives for installing photovoltaic panels on individual buildings. Currently, even without subsidies, installation of solar panels is economically attractive and, in many cases, achieving the net-zero carbon goal at the building level. Homeowners in the US today even have an option to rent the solar system from solar installation companies, pay no upfront cost,

Figure 5.3. Bullitt Center in Seattle. The practices incorporated in the building include energy conservation and geothermal energy, roof solar panels, rainwater harvesting, waterless toilets, black water aerobically composted, and gray water treated by wetland and reused. *Source:* Nic Lehoux; reprinted with permission of Bullitt Center, Seattle.

and purchase electricity at less cost for five to ten years and thereafter to own the system. Small and large electricity-producing wind turbines are already commercially available and producing energy on a large scale in Texas, Iowa, Europe, and China.

Water.   Table 5.2 shows the results of an older study by the American Water Works Association in 12 North American communities of domestic water use, divided between indoor and outdoor (primarily irrigation) uses. The study found that indoor water use was consistent throughout the US. In this millennium, it has become obvious that water demand without conservation at those levels is clearly unsustainable. This has been magnified lately by severe droughts and water shortages, such as in California in 2013–2016, long-term water deficiencies in the southwestern US, droughts in Europe and China, and severe water shortages in Cape Town, South Africa, starting 2017. The California state government imposed mandatory cuts in domestic and commercial water use in 2014 and Cape Town reduced water use to 50 L capita^{-1}day^{-1} in 2017. Even with cuts, California's per capita water use is greater than an average use typical for Europe.

District Level.   Water and heat energy can be efficiently recovered in water management/recovery clusters/neighborhoods. A cluster or district in this context is a semiautonomous consortium of buildings (neighborhoods) managing its stormwater (using LID concepts), preprocessing used water and recycling gray water if it practices black and gray water separation. (See Table 5.3 for definition of colors and various water conservation and recycle schemes.) A cluster (water/energy management district) ranges in size from a large building (for example, a large hotel or apartment complex with hundred or more units, a high-rise) to a subdivision or neighborhood (Figure 5.2) or a large government or

Table 5.2. Domestic water use by AWWA study.

Water Use	Without Water Conservation*		With Water Conservation		Water Use (L cap^{-1}day^{-1})
	Liter Cap^{-1} day^{-1}	%	Liter Cap^{-1} day^{-1}	%	**Black water before conservation** $- 17.5 +$ $63 + 3.6 = 84.1$
Faucets	**35***	**14.7**	**35***	**25.0**	
Drinking water and cooling	**3.6**	**1.2**	**2.0**	**1.5**	**After** $= 17.5 + 15 + 2 =$ **34.5**
Showers	42	17.8	21	15.3	
Bath and hot tubs	6.8	2.0	6.0	4.5	**Gray water before**
Laundry	54	22.6	40	29.2	**conservation** $= 17.5 +$
Dishwashers	3.0	1.4	3.0	2.2	$42 + 6.8 + 54 + 3 =$
Toilets	**63**	**26.4**	**15**	**10.9**	**123,3**
Leaks	30	12.6	15	10.9	
Total indoor	238	100	137	100	**After** $= 17.5 + 21 + 6 +$
Outdoor	112	47	0**	0	$40 + 3 = 88.5$
Total	350	147	137**	100	
Energy to deliver and treat water kW-h cap^{-1}day^{-1}	0.8		0.3		**Conservation water saving** $= 350$-$137 = 213$

Source: AWWA RF (1999); Heaney, Wright, and Sample (2000); and Asano et al. (2007).
*Faucet water ½ kitchen sink (BW) ½ bathroom sink (GW). Drinking water is converted to urine (black water).
**Converting from lawn irrigation to xeriscape, irrigation by recycled treated gray water and captured rain and stormwater. No potable water use for irrigation.

Table 5.3. Definitions of water and energy reuse terms.

Reused/reclaimed water classification	
Blue	Water quality suitable for potable use from springs and wells, from high-quality surface water bodies, unpolluted rainwater, and water treated to potable quality
White	Mildly polluted surface runoff
Gray	Untreated household wastewater that has not come in contact with fecal and urine sewage (black water); includes medium polluted water from laundries, and bathrooms
Dark	More polluted drain water from kitchen sinks with garbage disposal units and dishwashers; can be counted as gray water (without disposal unit) or as black water
Black	Highly polluted wastewater from toilets and urinals containing urine and excreta
Yellow	Separated urine (about 1% of the total flow) rich in nutrient content
Green	Water used for irrigation
Processing	
Water reuse	Use of the used water effluents for the same or another beneficial use with treatment before the reuse
Water reclamation	Capturing, treating, or processing of used water and/or stormwater to make it reusable
Water recycle	Repeated recovery of water from a specific use and redirection of the water back either to the original use or another use
Product	
Reclaimed water	The end-product of used water reclamation that meets water quality requirements for biodegradable materials, suspended matter, and pathogens for intended reuse
Direct reuse	The direct reuse of reclaimed water for applications such as agricultural and urban landscape irrigation, cooling water and other industrial uses, urban applications for washing streets or firefighting, and in dual water systems for flushing toilets. Direct potable reuse is still rare.
Indirect reuse	Mixing, dilution, and dispersion of treated used water by discharge into an impoundment, receiving water, or groundwater aquifer prior reuse, such as in groundwater recharge with a significant lag time between the use and reuse
Potable water reuse	Use of highly treated use water to augment drinking water supplies
Nonpotable reuse	Includes all water reuse applications other than direct or indirect use for drinking water supply

Source: Novotny et al. (2010a); Allen et al. (2010); Levine and Asano, (2004).

commercial building. A subdivision or a smaller community connected to a citywide or regional utility can be managed as a semiautonomous water/energy cluster (Figure 5.4).

Instead of bringing recycled water from a distant regional utility water reclamation plant, *fit for reuse* reclaimed treated gray water from the cluster gray and stormwater water reclamation units can be used locally in applications currently using potable water, such as:

(a) Irrigation of public green areas and community gardens
(b) Street flushing and pavement cooling

Figure 5.4. Grow community (subdivision) buildings on Bainbridge Island, Washington. Note the solar panels on the roof. The development received OPL certification. *Source:* Photo by Steve Moddemeyer, CollinsWoerman, Seattle.

(c) Public fountains and architectural water features

(d) Protecting aquatic ecosystems by providing ecological base flow

(e) Return reclaimed water to buildings as a nonpotable water supply for toilet flushing, laundry, and bathroom supplements

(f) After extended retention and dilution in surface lakes or ponds or groundwater aquifers reclaimed gray water after additional treatment can be returned to buildings as potable water.

(g) Recharging groundwater to enhance available water and/or preventing subsidence of historic buildings built on wood piles that need to be surrounded by groundwater to prevent rotting

By starting with building satellite treatment in upstream portions of the urban drainage area, used water (wastewater from the local collection system) can be intercepted and treated to a high degree required for various above-mentioned reuse (Asano et al., 2007). This concept was implemented in the Solaire Battery Park residential complex in New York City where reclaimed water is used for toilet flushing, irrigation, and cooling. The remaining effluent is then conveyed to a central (regional) treatment plant (Novotny et al., 2010a).

Fit for reuse treatment at the cluster level implies that the reclaimed water products may have different water quality. If reclaimed water in the cluster is used for landscape irrigation and/or toilet flushing, separation into black and gray water flows may not be needed. Most of the nutrients are contained in black water and removing nutrients does not make sense because the nutrients eliminated from reclaimed water would have to be replaced by industrial fertilizers, which would defy the purpose of reclamation and reuse. The needed water can also be obtained by capturing the sewage flow from the cluster for treating and reusing or from the sanitary (or combined) sewer and treating it on-site in a packaged automated treatment plant. Toilet flushing may require reduction of turbidity, disinfection (primarily to control bacterial growth in the toilets and urinals), and adding some color, if needed. For providing ecological flow to lakes or streams, the effluent from the cluster package treatment plant must comply with the effluent and receiving water quality standards, organic solids, and turbidity should be removed by enhanced primary treatment and microfiltration; nutrients should be recovered (e.g. by recovering struvite or urine separation) and not just removed (e.g., in sludge deposited in a landfill). Residual solids and reject water from the cluster treatment should be conveyed to the regional integrated resource recovery facility. Note that this reuse will result in higher concentrations of COD and other compounds sent to the regional resource recovery (water reclamation) plant.

The enabling technologies contributing toward closing the cycle and approaching the net-zero goals include recent developments in membrane treatment technologies and small automated treatment plants, rapidly dropping cost of photovoltaic panels that are effective even in areas with less direct solar radiation, and advances in urban planning. Closing the water cycle can never be 100% because of other factors described later in this and subsequent chapters. The most important factor is the accumulation of conservative or difficult-to-remove harmful pollutants in the cycle. That may include endocrine disruptors and pharmaceutical compounds. For example, in Qingdao, China, proposed by the University of California and Masdar (UAE) reclamation/reuse systems, only about 25% of the water demand is brought from the municipal potable water grid. This would imply some water in the cycle may be reused three to four times with a very short lag time. The amount of import of clean potable water is a key decision parameter (Novotny et al., 2010a).

5.3 FLOW SEPARATION: GRAY WATER RECLAMATION AND REUSE

Gray water (from dishwashers, showers/baths, and laundry) and black water (from toilets and kitchen sinks with food waste) is the most common and realistic separation to be considered for the triple net-zero cities (Cities of the Future, COF). However, it should be recognized that a greater portion of the nutrient mass will remain in black water and, hence, it will not be recovered at the house or cluster level. Henze and Comeau (2008) estimated that about 85% of total nitrogen and 84% of total phosphorus (total N and P), respectively, is in black water (Table 5.4) which, based on Table 5.2, represents only about 10% of the total used water flow. This reduces the value of reclaimed and treated gray water flow for irrigation. Furthermore, if separated, most of the nitrogen and about half of the phosphorus is in urine, which is only about 1% of the total flow. Using reclaimed gray and storm (white) water as a supplement for nonpotable water in bathrooms and laundry needs more vigorous treatment to remove bacteria, viruses, and harmful trace-accumulating chemicals, which may require black and gray water separation, microfiltration/nanofiltration, and possibly reverse osmosis. On the local cluster/ecoblock scale, aquifer recharge is accomplished mostly by infiltration of captured (harvested) rainwater and LID treated stormwater.

Gray water accounts for more than 75% of the total flow in houses without water conservation and about 90% in houses with water conservation (Table 5.2); therefore, it is an attractive source of water for reuse, along with rain and stormwater. However, its quality is still poor, which does not allow direct reuse without proper, functioning, approved, and

Table 5.4. Unit loads of key pollutants in total and separated used water flows in g/capita-day.

Parameter	Type of Used Water			
	Traditional Combined	Gray Water with Ground Food Waste In	Gray Water Without Food Waste[1]	Black Water
COD	130	55	32	75
BOD	60	35	20	25
N	13	2	1.5	11
P[2]	2.5	0.5	0.4	2

Source: Henze and Comeau (2008).

[1] Food waste collected and reused with municipal solid waste.

[2] The difference between the phosphorus content of gray water and combined used water reflects the ban on use of phosphate detergents.

supervised treatment. Gray water reuse for toilet flushing can reduce the in-house net water consumption by 40–60 L capita^{-1}day^{-1}, leading to 10–20% reduction of the urban water consumption (Friedler, 2004). Gray water contains suspended solids (dirt, lint), organic material, oil and grease, sodium, nitrates and phosphates (in countries where phosphate detergents are still sold), increased salinity and pH, chlorine bleach from laundry, hair, organic material, and suspended solids (skin, particles, lint), oil and grease, and soap and detergent residue from bathtubs and showers (Allen et al., 2010). Friedler (2004) reported that gray water was found to contribute as much as 55–70% of the specific daily load of total suspended solids (TSS) and BOD$_{total}$ in municipal sewage. Kitchen sinks with garbage disposals, common in the US, were major contributors of volatile suspended solids (VSS), COD$_t$, and BOD$_t$ with 58%, 42%, and 48%, of their total daily load, respectively. Washing machine was recognized as a significant contributor of sodium, phosphate, and COD$_t$ (40%, 37%, and 22% of the total load). The dishwasher, although contributing only 5% of the flow, was found to be a significant source of phosphate and boron. However, the phosphate loads reported by Friedler represent countries that in the early 2000s allowed use of phosphate detergents, which have been banned in the US and other developed countries. Gray water also has high concentrations of pathogenic bacteria and viruses that multiply in warm water (Table 5.5); therefore, household reuse using home treatment units is not safe and may not be allowed if the recovered gray water is used to irrigate lawns where children play or in vegetable gardens (Maimon et al., 2010). An extensive survey on gray water characteristics in Europe was published by Eriksson et al. (2002). Tables 5.5 and 5.6 present the bacterial and chemical content of gray water in three geographical locations.

Table 5.5. Fecal coliform densities of gray water components.

Source of Water	Number of Fecal Coliforms cfu/100 ml
Bath, shower, and washing machine with cloth diapers	10^4–10^7
Bathing and shower	10–10^6
Washing machine, bathroom sink and shower, kitchen sink	3.44×10^6
Washing machine with children	10^4–10^6
Washing machine without children	10^2–10^4

Source: WHO (2006), Allen et al. (2010), Friedler (2004), Eriksson et al. (2002).

Table 5.6. Gray water chemical characteristics.

Parameter:	Massachusetts Study[1]		China[2]		Europe[3]
	Mean	Range	Mean	Range	Range
COD mg/L			432	202–639	13–361
BOD$_5$ mg/L	129	22–358	186	68–271	90–290
TSS mg/L	53	8–200	810	335–2656	
TKN mg/L	12	3–32			2–31.5
NO$_3^-$ mg/L	1.5	<1.0–17.5	0.9	0.13–2.2	0–5
Total N mg/L			14	7–25	0.6–5.2
PO$_4^{3-}$ mg/L	0.9	<0.5–3.7	1.2	0.4–2.9	4–35[4]
pH	7	5.3–10.8	6.3	5.7–6.6	

[1] Veneman and Stewart (2002).
[2] Chen et al. (2010).
[3] Eriksson et al. (2002).
[4] All European PO$_4^{3-}$ data reflect use of phosphate containing detergents and soaps.

The reuse of gray water for toilet flushing and garden irrigation has an estimated potential to reduce domestic water consumption by up to 50% (Maimon et al., 2010; Allen et al., 2010), most of it being used for toilet flushing. Garden (lawn) irrigation reuse potential is the second potential reuse, but the demand will be vastly different between densely populated urban centers and suburbs. In densely populated urban areas in Israel the demand for recycled (treated) gray water for irrigation was at most 8–10 L person^{-1}day^{-1} (Friedler, 2004), which, as seen in Table 5.2, would be less than 10% of the irrigation demand of the average US house at the end of the last century. Recovered gray water contains only a fraction of the nutrients contained in used water in countries that banned use of phosphate detergents, otherwise the phosphate load in gray water would be significant. Gray water obviously contains emulsified soaps and detergents and could be alkaline. Consequently, the chemical properties of soils are important. Calcium and magnesium in soils may precipitate the emulsified soaps and contribute to clogging of soils. Besides soaps, gray water also contains various household chemicals, cosmetic products, toothpaste, shower creams, mouthwash, bleaches, dyes, and even potentially toxic biocides and fungicides (Eriksson et al., 2002).

Recycling of gray water obviously requires dual plumbing and sewers. There are significant health and other risks connected with the reuse of gray water, especially at the household level. Because of the relatively high contents of microorganisms in gray water there is a risk of exposure to pathogenic microorganisms if the water is used for irrigation or toilet flushing where the risk of bacterial contamination could be in the form of aerosols generated by irrigation sprinklers and during toilet flushing (Eriksson et al., 2002). Far greater risk may occur if the recycled water is used as a nonpotable water supplement in baths. California health standards for use of recycled water for toilet flushing limit fecal coliforms to less than 2.2/100 ml. Untreated gray water turns septic, devoid of oxygen in a few days (48 hours) during warm weather; therefore, gray water codes do not allow gray water to be stored (e.g., Queensland, Australia; California).

The gray water treatment systems can be categorized as follows (Allen et al., 2010; Friedler et al., 2005; Meda et al, 2012):

- Simple home filtration units (unreliable and not allowed in many communities because of health and safety concerns)

- Cluster treatment by submerged flow wetlands and some filtration post treatment that may also include disinfection for irrigation and nonpotable uses
- Microfiltration (membrane) for toilet flushing, followed by reverse osmosis and UV disinfection if the recovered gray water is intended for human nonpotable use such as laundry or supplement for showers and baths
- Biological aeration system with membrane filtration – membrane bioreactors (MBR) or biological aerated filter (BAF) or sequencing batch bioreactors (SBR) in areas of concentrated gray water flow, such as in China (Meda et al., 2012). These units could be followed by ozonation. Due to a large variability in the quality of gray water influent, an equalization/buffer tank with a minimum of 24 hours detention time is recommended and sometimes required by the authorities.

After extensive investigation of gray water treatability in China, Chen et al. (2010) pointed out that the selection of the treatment process depends on the presence or absence of kitchen water (with food residues) in the gray water flow. The biodegradability improves if kitchen water is present; therefore, biological processes are preferred for combined gray water. For physical treatment requiring chemical coagulant additions such as microfiltration, membrane separation should be selected for gray water without kitchen water because of lower concentrations of pollutants, namely of biodegradable fractions, low nutrient content, and less demand for chemicals.

A gray water treatment filtration unit for toilet flushing and irrigation may contain a short hydraulic residence time (HRT ~ one day) equalization tank, a filter (sand, diatomaceous earth), UV disinfection, and distribution pipes with a pump. Because emulsified detergents are not easily removable by sand or diatomaceous earth filtration and leave the effluent hazy, a crushed limestone filter may be included. Limestone (calcium carbonate) and dolomite (calcium-magnesium carbonate) increase hardness, which could break the soap and some detergent emulsions and result in clearer effluent. Using membrane biological reactors (MBRs) may be difficult because of great variations in gray water organic content (from 10 to 400 mg/l of COD), insufficient nutrient content, and use of chlorine bleach chemicals in the laundry. A manual by Allen et al. (2010) includes an extensive discussion and listing of available gray water treatment systems.

Meda et al. (2012) reported good results with the biological aerated filter (BAF) that after a two months of startup period resulted in relatively clean effluent with BOD_5 of less than 5 mg/L. The recommended volumetric COD loading of the BAF unit ranged from 1 to 4 kg m^{-3}day^{-1}, which is significantly smaller than the 8 kg m^{-3}day^{-1} recommended for conventional biofilters treating municipal used water (Tchobanoglous et al., 2003). The biologically treated effluent should be polished by membrane microfiltration, possibly followed by reverse osmosis if it is to be used as a nonpotable supplement for showers/baths and laundry

In the areas with great water shortages or even no fresh water availability (e.g., Masdar in the United Arab Emirates, Northwest China, Saudi Arabia), the focus of the water/energy management is on minimizing fresh water demand from outside sources yet maintaining comfortable water availability in households. Such a system has been proposed by the University of California for the Chinese city of Qingdao, which, in an adapted variant form, is presented on Figure 5.5. Unlike the original UC system that allowed some reuse of the reclaimed water for potable purposes, the gray water loop on Figure 5.5 provides enough fresh water from the grid (clean) water supply to cover all potable water demands for direct potable use.

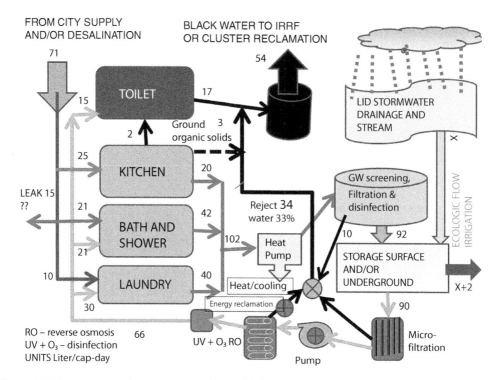

Figure 5.5. Cluster water and energy reclamation double-loop system. The numbers represent the water flow in liters capita^{-1} day^{-1} within the loop and the water demand is based on AWWA-RF (1999) data in Table 5.2 assuming water conservation. Black water plus reject water is sent for resource and energy recovery to the IRRF (Integrated Resource Recovery Facility). *Source:* Novotny et al. (2010a).

This system would implement black and gray water separation in the cluster, recycling gray water supplemented by captured and stored rainwater and sending black water to a regional integrated water-resources reclamation facility. Gray water would first be treated by enhanced filtration or biofiltration and after disinfection stored in an architecturally designed surface pond or underground basin along with collected rainwater. During times of excess this water would be fit for outdoor irrigation or ecological stream flow. After storage, the recycled water would be treated to be fit for nonpotable reuse in the building. Heat energy in water could be reclaimed in the cluster by a heat pump on the gray water outflow line, and some static energy from the high-pressure reject flow from RO can be recovered by a turbine and generator.

Because it is an absolute minimum, this system does not provide water for irrigation and requires a portion of the recovered gray water to be used for personal (nonpotable) hygiene. Avoiding this problem and providing water for irrigation and ecological flow can be achieved by increasing the input water flow and reducing leaks. Such a system will be presented in Chapter 10 (Figure 10.6).

For this dual loop cluster system, data from Henze and Comeau (2008) indicate that if the used water flow is reduced to black water flow of 80 L capita^{-1} day^{-1} that carries most of the pollutant load from toilets and kitchen and the pollutants in reject flow from the treatment of separated gray water, the maximum high concentrations of pollutants in the reduced flow could be as follows:

Flow 80 L/cap-day	COD	2 750 mg/L	BOD	1 125 mg/L
	Tot. N	184 mg/L	Tot. P	35 mg/L

In this cluster system, backwash water from filter and reject water from reverse osmosis (if installed) along with the separated black water would be sent to the regional resource recovery facility (IRRF) for water, energy, nutrients, and other resources recovery. The water demand on the outside freshwater source (in Masdar, in many other Arabian Peninsula communities, and even in Florida requires expensive desalinization) would be reduced to about 70 L capita^{-1} day^{-1}, or less if leaks are controlled, yet would maintain a comfortable 120–130 L capita^{-1} day^{-1} water use in buildings. The reject water from filtration and reverse osmosis (RO) directed for processing with black water may represent up to 30% of the recovered gray water. The low water input from the grid would be at the limit for water reuse and needs to control dissolved solids and nonremovable pollutant buildup within the cycle. The black and reject water sent to the regional resource recovery facility and the leak losses could make the loop water deficient during dry weather conditions. If there is a deficit it should be covered by captured stormwater and rainwater preprocessed in LID treatment and stored in surface and subsurface basins or aquifers dedicated for this purpose. Use of rain and treated stormwater should be maximized.

Most of the irrigation and ecological flow demand in these frugal water systems should be covered by captured stormwater; therefore, long-term storage is crucial. This could revive the old concept of communities having a central or nearby multipurpose wet pond/lake for storing water and providing recreation and enjoyment. Masdar envisioned a similar type of management but there is no used water separation in what will be a high-density city of about 50,000 permanent inhabitants (when completed) into semiautonomous water/energy districts. However, in 2017, only a fraction of the town had been built – less than 5% of the original planned six square km "green" city – and the completion date has been pushed to 2030 (*Guardian*, February 2016) but the pace of building seems to be accelerating. The dream of a fully net-zero city is not lost.

The reject water from RO is under high pressure and this hydrostatic energy potential can be converted to energy. Supplementary heat energy can be provided to each house by heat pumps retrieving energy from the neighborhood reclaimed integrated nonpotable water-energy loop (gray water, stormwater, cooling water) and/or ground and geothermal sources. In many warm, dry areas such as India or the Middle East, surface storage would be subjected to very high evaporation losses. In Southeast Asia on one side of the world, or in the southwestern US on the other side, most of the rain occurs during the months of the monsoon seasons. Therein, underground aquifer storage of stormwater may be the only feasible option that would provide water for irrigation between the rain seasons. Perhaps a return to the Roman (Pompeii) or Minoan (Crete, in the Mediterranean Sea) civilizations where thousands of years ago captured rainwater was collected from the roof and stored under the house (see Novotny et al., 2010a) or under a town square in the cluster is not a bad idea. It may be better and/or cheaper than using energy-demanding desalinated seawater for irrigation and it is a great example of building- and/or district-level water management.

A concept integrating house water management by combining with cluster-level recycling and landscaping is also illustrated on Figure 5.6. In this system, water treated in the potable water treatment plant (PWTP) is directed to kitchens and bathrooms. Black water generated by toilets is sent to a (cluster or regional) biological black water treatment plant (BWTP) that recovers biosolids as fertilizers, energy, and potentially other resources

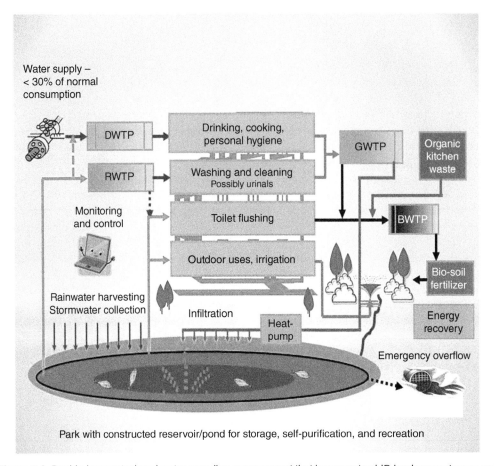

Figure 5.6. Double-loop water/used water recycling management that incorporates LID landscape storage and treatment proposed by Prof. Hallward Ødegaard. DWTP = potable (drinking) water treatment plant, RWTP = recycled water treatment plan, GWTP = gray water treatment, BWTP = black water treatment. *Source:* Prof. Hallward Ødegaard, Scandinavian Env. Technology, Hellemsweiden, Norway.

(e.g., hydrogen). It is not specified whether the BWTP is aerobic or anaerobic. Captured stormwater in the LID system is recycled back into the recovered water loop, where the water for indoor use is treated in the recycled water treatment unit (RWTP). This system may cut the water demand from the grid by more than 70% or even more in areas that previously wasted water.

Tap a Sewer, Keep the Liquid, and Sell the Solids

The previous system of district water management requires dual sewer systems (black and gray) and dual water supply (potable and nonpotable). Black water is sent to the regional IRRF for processing and energy and resources recovery and gray water is treated by micro-filtration for in-house reuse. This system may overtreat the water for toilet flushing and may undertreat the water for bathing. Furthermore, it sends major part of nutrients to the IRRF that could have been used to enhance nutrient level in the recycled irrigation water in the district.

The "tap the sewer" concept was included in Asano et al. (2007). Because current package (mostly aerobic) treatment plants with membrane reactors have become so versatile, they can be put in a small place such as a basement or garage of a large building. The original idea presented in the Asano et al. manual was to tap into the sanitary or combined sewer passing through the district. The district water and energy recovery unit would be automated and operated remotely with periodic visits of supervising and maintenance engineers. These district systems have been installed at the end of the last century and the beginning of this one in many places, including the Battery Park City development in New York City or the New England Patriots stadium complex in Foxborough, Massachusetts. In the Dockside Green sustainable development district in Victoria, British Columbia, the district facility treats 100% of its sewage (Figure 5.7) and storm water on-site, and treated water is reused for flushing toilets, landscape irrigation, and water features.

Reuse of combined used water is a sensitive issue in most communities. The advancement of used water treatment at the end of the last century has already certainly reached the level that would guarantee safe potable reuse but the resistance of the population to "toilet to tap" reuse has been very strong when it was proposed in Southern California, where the time between water reclamation, recovery, and reuse was short. There seems to be little or no resistance to potable reuse if treated used water is safely discharged into a natural surface or groundwater aquifer storage that provides longer-term retention and storage of several months and dilution with fresh water that had not been contaminated by sewage. For example, the relatively small Trinity River in Texas provides potable water for the 6 million people in the Dallas–Fort Worth metroplex. The treated effluent is discharged into the same river at a distance downstream, making the river effluent dominated. After about 6 months of flow and retention in a large downstream reservoir, the Trinity River water

Figure 5.7. Water reclamation plant in the Dockside Green sustainable urban community in Victoria, British Columbia, located in the basement of a residential/commercial building. The reclaimed water is used for ecological flow and other nonpotable uses in the district. *Source:* Photo courtesy Dockside Green, Victoria, British Columbia, Canada.

is used after treatment for the potable water supply of another metropolis, Texas's largest city of Houston. Nevertheless, on a district level, direct reuse of combined used water for potable and human hygiene should be avoided. This leaves the potential use of reclaimed and fit for reuse water for (1) irrigation, (2) toilet flushing, and (3) partially for laundry. For laundry, nonpotable water use could be limited to hot water while the final rinse could be done with potable water.

Again using the water demand data for a household with water conservation, Figure 5.8 presents an idea for a recycle loop for combined used water district water and energy recovery (DWER) unit that provides reclaimed and treated water fit for reuse for irrigation and in-house nonpotable water demand. This recovery provides the total water flow to the loop building connections of 120 L capita^{-1} day^{-1}, of which ~100 L capita^{-1} day^{-1} is provided from the treated fresh water supply grid. This amount of fresh water flow satisfies all potable water needs, and 35 L capita^{-1} day^{-1} is recycled water for the nonpotable uses of laundry and toilet flushing. Outputs from the loop include 62 L capita^{-1} day^{-1} for irrigation and assume 13 L capita^{-1} day^{-1} to be losses. The sanitary sewer interceptor to the regional integrated resource recovery facility (IRRF) will receive 23 L capita^{-1} day^{-1} of solid and concentrated liquid flow into sanitary sewer interceptors, conveying the flow to a regional integrated resource recovery facility (IRRF) for processing that will yield water, energy, and resources. This 75% sanitary sewer interceptor flow reduction is significant because in many communities with flat terrain, interceptors are deep and must be pumped by lift stations. The treatment at the district level should be anaerobic membrane reactors (AnMR; see

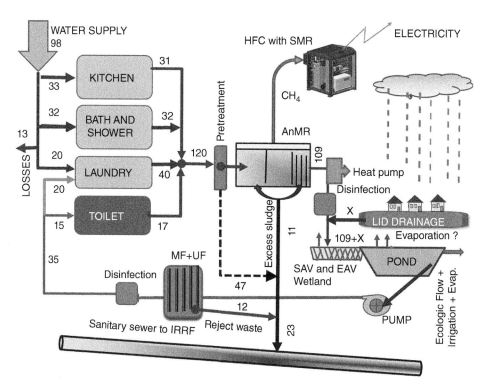

Figure 5.8. District water and energy recovery loop that processes combined used water. MF+UF is the return flow treatment unit consisting of a microfilter followed either by an activated carbon filter or reverse osmosis. Hydrogen fuel cell (HFC) with steam methane reforming (SMR) unit emits non-GHG carbon dioxide.

Chapters 8 and 9 for possible alternatives). The best candidates are upflow sludge blankets and fluidized bed membrane reactors that, in addition to clean reusable effluent, will emit methane that could be converted on site to electricity in a smaller-megawatt-capacity hydrogen fuel cell (HFC) combined with a steam methane reforming (SMR) component that converts methane to syngas (see Chapter 6). Both syngas and hydrogen can fuel the HFC. Figure 5.9 is a small, district-size HFC installed in the Hammarby Sjőstad in Stockholm, one of the first ecological triple net-zero adverse impact communities where the average fresh water use is about 100 (L capita^{-1} day^{-1}). The anaerobic fluidized bed treatment units perform well at temperatures above 20°C. These hermetically enclosed reactors should not have odor problems. Typically, the small DWER recovery units would be located indoors in heated basements or outdoor heated shelters, which would solve the problem with processing used water in colder climatic conditions. Heat can be provided by the hydrogen fuel cell and a heat pump extracting heat from the effluent.

The pretreatment unit may consist of sieves with a solids comminutor. The anaerobic treatment can decompose biodegradable organic solids. The post-treatment, after disinfection (with ozone or UV radiation), could be in a wetland that has submerged and emerged aquatic vegetation (SAC and EAV) followed by a pond. The SAV and EAV wetlands provide the best treatment efficiency for removing phosphorus (see Chapter 9, "Phosphorus Flow in the Distributed Urban System") and the vegetation could be harvested and processed in the IRRF for energy. More than half of the effluent flow from the DWER unit is available for irrigation and other outdoor uses such as waterscapes and cooling. Combined with the rainwater and stormwater captured by the low-impact development (LID) drainage, the district will have enough water for indoor and outdoor needs at the average fresh water input of 100 L persom^{-1} day^{-1}. Reducing the fresh water flow input to less than 100 L capita^{-1} day^{-1} would require providing high-level advanced treatment and acceptance of the district tenants to use some of the highly purified recycled water for non-potable hygienic uses such as baths, showers, and laundry hot water supplements, as shown in Figures 5.5 and 5.6. Significant reduction of water use will lead to oversized systems and water stagnation that may result in microbial growth and pathogenic contamination.

Figure 5.9. Small hydrogen fuel cell producing DC current from biogas in the Hammarby Sjőstad community in Sweden. *Source:* GlashusEtt, Hammarby, Stockholm, Sweden.

Designers of interceptor sewers must get accustomed to much smaller residual outflow from the district that will be concentrated with solids and reject water. The flow reduction saves energy because in flat terrain the depth of interceptor sewers can become large and sewage must be pumped by lift pumping stations.

Integrated District Water and Energy Providing Loop

Another integrated cluster (district) water and energy management system that could also incorporate solar heat panels, geothermal heat exchange, LID drainage, and surface and subsurface storage is shown on Figure 5.10. This water-energy recovery loop has originally been proposed for the Seattle Yesler Terrace sustainable neighborhood, which may become one of the first triple net-zero neighborhoods in the US (Moddemeyer, 2010). This economically viable development, which is close to triple net-zero, could achieve up to 45% reduction in potable water use and 70% reduction in wastewater flows. Over 90% of heat and cooling energy will be derived from on-site renewable sources and 25% reduction of energy from the grid (Puget Sound Energy) will make Seattle a net-zero GHG energy utility. The Yesler is undergoing a dramatic urban renewal that would also include public transportation by electric street cars, solid waste separation and recycling, green architecture, and other aspects of sustainable urban development.

As stated before, the systems presented on Figures 5.5, 5.6, and 5.10 bring back to forefront the importance of the multipurpose roles the village or city ponds and lakes used to play since medieval times. The permanent ponds that stored water over an extended time were often near the center of the community. They provided some recreation during summer, skating in winter, and some recreational fishing, and they were sources of limited nonpotable water. These functions could be enhanced in the future.

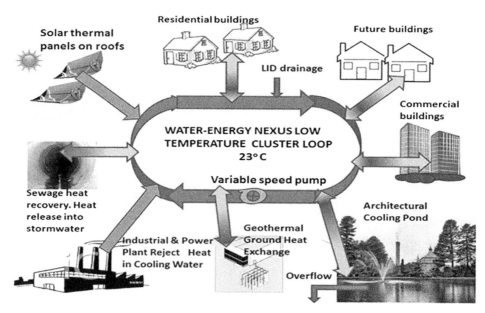

Figure 5.10. Nonpotable water/energy nexus and reclamation loop in a cluster. *Source:* Inspired, adapted and modified from Moddemeyer (2016), CollinsWoerman, Seattle.

Energy Savings and GHG Reduction by Gray Water Reuse in Clusters

Figure 5.11 schematically shows the water-energy nexus of reducing residential water use by water conservation and reuse and its effect on energy use. It should be emphasized that complete 100% reuse is not feasible, even on the International Space Station. Figures 5.5 and 5.11 also suggest that reducing fresh water input from the grid to 50 to 70 L capita^{-1} day^{-1}, with some additional water provided by captured clean or pretreated storm runoff and recycling used water at the cluster, may be the reuse limit.

The author in Novotny (2010) hypothesized that there is a minimum inflection point beyond which further reduction of water use will increase energy demand and urban water metabolisms because of increased use of chemicals, energy, and infrastructure (materials). A relationship can be developed for relating the cost of providing water to the magnitude of the water demand. The water–energy nexus has three phases (Figure 5.11) (Novotny et al, 2010a):

1. *The water conservation phase,* in which energy and GHG emission reduction is proportional to the reduction of the high-water use
2. *The inflection phase,* in which additional and substitute sources of water demanding more energy are brought in, treated, and used
3. *The rising energy (cost) phase,* in which energy use is increasing while water demand of the development is reduced by used water reclamation and multiple reuses

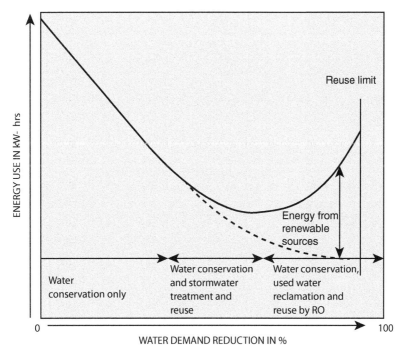

Figure 5.11. Water use, water conservation, and reuse relationships to energy. *Source:* Novotny et al. (2010a), Novotny (2011a).

Water Conservation. The best energy use/GHG emission reduction benefits come from water conservation. In the preceding chapter it was estimated that the average energy use for extracting, treating, delivering, and discharging water and used water in the US is about 1.03 kW-h capita^{-1} day^{-1}, which is 1.87 kW-h/m^{-3}. Thus, reducing water use through conservation from 550 L/cap-day to the conservation use of 137 L capita^{-1} day^{-1} in a linear system (no recycling) would represent energy reduction of:

$$E_c = (0.55 - 0.137)[m^3 capita^{-1} day^{-1}] * 1.87 \ [kW\text{-}h \ m^{-3}] = 0.77 \ kW\text{-}h \ capita^{-1} \ day^{-1}$$

Following the concept of reuse presented in Figures 5.5 and 5.6, and introducing additional treatment and reuse of gray water would bring further benefits in recovering sensible heat from gray water but would also result in additional energy demand for microfiltration, reverse osmosis, and ozonation (Novotny et al., 2010a; Novotny, 2011a). Specific energy uses for microfiltration, ozonation, and reverse osmosis for 90 L/capita-day of recycled used gray water and captured stormwater is given in Table 5.7 using data on energy use in Table 5.8.

Meda et al. (2008) estimated energy use for gray water biological treatment with filtration ranging from 0.1 to 0.7 Kw-h m^{-3}, depending on the method.

Sensible heat recovery by heat pump from 1m^3, assuming temperature difference of 12°C and heat pump coefficient of performance COP = 4, is 10.47 kW-h m^{-3} (Novotny et al.,

Table 5.7. Energy use for filtration of used water and stormwater.

Process	Specific Energy Use (kW-h m^{-3})	Gray Water Recycle Use kW-h capita^{-1} day^{-1}
Microfiltration	0.6	0.05
Reverse osmosis	2.0	0.18
Ozonation	0.1	0.01
Total E$_t$		0.24

Table 5.8. Parameters of membrane treatment processes.

Process	Pore Size. μm	Operating Pressure Bar	Energy Use kW-hr/m^3	Compounds Removed
Microfiltration	0.008–2.0	0.07–1	0.4–0.8	Fine particles down to clay size, bacterial, some colloids
Ultrafiltration	0.005–0.2	0.7–7	Wide range	All of the above plus viruses and large molecules
Nanofiltration	0.001–0.01	3.5–5.5	0.6–1.2	All of the above and about 50% ions
Reverse osmosis	0.0001–0.001	12–18	1.5–2.5	Small molecules, 98% to > 99% ions
Used water desalination			1.5–5	

Source: Asano et al. (2007).
1 bar = 100 kPascals = 100,000 N/m^2 = 1.03 atmospheres = 14.5 psi (at 0 elevation)

2010a). From Figure 5.5 the recycled gray water is 0.102 m^3 capita^{-1} day^{-1}, the additional sensible heat (cooling) energy recovery is then:

$$\Delta E_s = 0.102 \, (\text{m}^3 \text{capita}^{-1} \, \text{day}^{-1}) \, * \, 10.47 \, (\text{kW-h m}^{-3}) = 1.7 \, \text{kW-h capita}^{-1} \, \text{day}^{-1}$$

The total net energy saving by full recycle of gray water is then:

$$\Delta E_T = E_c + E_s - E_t = 0.77 + 1.7 - 0.24 = 2.23 \, \text{Kw-h capita}^{-1} \, \text{day}^{-1}$$

Using the average US equivalent of GHG emissions for energy from fossil fuel power of 0.62 kg of CO_2/KW-h, the daily GHG emission reduction by cluster gray water reuse based on Figure 5.5 schematics could be:

$$\text{GHG reduction} = 0.62 \, [\text{kg CO}_2/\text{kW-h}] \, * \, 2.23 \, [\text{kW-h cap}^{-1} \, \text{day}^{-1}]$$
$$= 1.38 \, \text{kg CO}_2 \, \text{capita}^{-1} \, \text{day}^{-1}$$

or about 0.5 ton CO_2 capita^{-1} year^{-1}.

Instead of capturing the heat energy from gray water by heat pump, treated warm gray water can be recycled as a supplement to hot water for laundry and possibly for bathing to reduce heating energy.

Water Scarcity and Crises. Cape Town is the most desirable historic large tourist desig-nation city in the Republic of South Africa. According to the South African Weather Ser-vice, two of the driest seasons ever recorded for the city happened in the three-year period from 2015, when 549 mm (21 inches) of rain fell, to 2017 – the driest year on record – when annual rainfall totaled 499 mm (19½ inches). Consequently, the South Africa government has declared a national disaster over the drought-afflicted southern and western regions, including Cape Town, and has restricted water use to 50 L capita^{-1} year^{-1} (City of Cape Town, 2018) starting July 2018 (i.e., the time when the key water supply reservoirs dried up). Note that the dry year half-meter annual rainfall in Cape Town is more than twice as large as the average annual rain in Phoenix, Arizona; however, Phoenix receives large vol-umes of water from the Colorado River by the Central Arizona Canal, which is not available to Cape Town.

The United Nations (2015a) declared that around 700 million people in 43 countries suffered from water scarcity. By 2025, 1.8 billion people will be living in countries or regions with absolute water scarcity, and two-thirds of the world's population could be living under water-stressed conditions. With the existing climate change scenario, almost half the world's population will be living in areas of high water stress by 2030, including between 75 million and 250 million people in Africa. In addition, water scarcity in some arid and semi-arid places will displace between 240 million and 700 million people.

The worst situation of water scarcity magnified by global warming has emerged in the second decade of this century in India, as reported in MIT Technology Review. In some major cities water is being delivered to millions of shantytown poor urbanites by trucks from which water is siphoned chaotically to pales and plastic jars. During the 2019 dry period the trucks arrived once in ten days and on average provided only 60 liters of water per family and day in New Delhi and Bangalore (not per person like in Cape Town). Hundreds of thousands of people are dying now in India because of water shortage. These shortages are magnified because the dry period before monsoon period is longer and hotter due to global warming. More than 600 million Indians face "acute water shortages and seventy percent of the nation's water supply is contaminated, causing an estimated 200,000 deaths a year. Some 21 cities could run out of groundwater as early as 2020 year, including Bangalore and New Delhi. Forty percent of the India population, or more than 500 million people, will have "no access to drinking water" by 2030. (https://www.technologyreview.com/s/613344/indias-water-crisis-is-already-here-climate-change-will-compound-it/)

Hence, population increases and global climate change create a permanent water crisis, which for urban areas is magnified by water pollution and mining of groundwater resources. Adding desalination to cover the water shortage of growing urban areas increases the price of water and energy and is economically available only in affluent coastal cities.

BIOPHILIC SUSTAINABLE LANDSCAPE AND LOW IMPACT DEVELOPMENT

6.1 URBAN NATURE AND BIOPHILIC DESIGNS

Biophilic or *close to nature* designs have been included in European cities since medieval defense walls were removed in the nineteenth century. These designs are different from the manicured architectural gardens and parks surrounding nobility and royal castles since the Middle Ages. In the nineteenth century and before, walled European (and Asian and African) cities were filthy and disease was rampant. After removing the walls, most historic cities used the vacant spaces for creating parks that were supposed to provide health benefits to the urban population. US cities did not have walls, but the situation was not much different. In nineteenth century, because of heavy pollution, smaller and medium urban streams were converted to combined sewers and buried underground.

In the US, thanks to Frederick Law Olmsted, the father of American urban biophilic landscape, and his followers, parks that mimicked nature were built in many US cities during the second half on the nineteenth century. The design of Central Park in New York City embodies Olmsted's social consciousness and commitment to egalitarian ideals. Olmsted believed that nature must always be accessible to all citizens, in contrast to Europe and China, where the best parks and gardens were reserved for royalty and aristocracy (Versailles in France, Beihai Park in Beijing). Olmsted's best-known park projects are nature-mimicking ecosystems accessible to the public and not the manicured garden parks common in Europe. Olmsted's designs include the Muddy River environmental corridor (see Novotny et al., 2010a) and arboretum in Brookline, Massachusetts; Lake Park in Milwaukee, Wisconsin; and Jackson Park in Chicago.

The concept of modern multipurpose urban biophilic designs is more recent and more comprehensive than the islands/parks of nature placed in otherwise polluted city environments of the last century. These new designs incorporate the concepts of natural cleansing of city lands, waters, and air. Professor Beatley (2011, 2017) of the University of Virginia defined biophilic nature mimicking city this way:

> *A city abundant with nature, a city that looks for opportunities to repair and restore and creatively insert nature wherever it can.*

Biophilic Designs

Biophilic concepts can be implemented in many places and scales throughout the city, from rooftops to entire regions. Beatley also emphasized that "biophilic cities understand urban environments as spaces shared with many other forms of life and recognize an ethical duty to work towards co-existence." The biophilic city concept also loosely coincides with "green city." In the present context, urban nature is also supposed to ameliorate injuries and environmental stresses caused by pollution and urban living and make city environments more resilient to global warming such as urban "heat island" effects and increasing occurrences of extreme weather. Other benefits of greening the city include lower stress and better health of citizens, as well as economic benefits. It also enables wildlife to enter the urban landscape, such as bald eagles nesting in the Twin Cities in Minnesota or falcons in Milwaukee, Wisconsin; the appearance of wild turkeys or deer in subdivisions, and even eagle and falcon nests, but also undesirable invasions from scavenging bears and coyotes in many northern states and mountain lions in California.

This treatise focuses mainly on the triple net-zero benefits of the integrated urban management of water, energy, and resources. The new biophilic designs covered in works by Beatley (2011 and 2017), Wheeler and Beatley, (2014), and Ahern (2007, 2010) emphasize symbiosis and the adherence to the current best management practices (green roofs, rain gardens, biofilters), urban lakes, urban wetlands, urban environmental corridors, as well as river and floodplain restoration to the biophilic design concepts introduced in the last decades. This movement broadened the greening and naturalizing urban landscape concepts from two-dimensional horizontal designs to three-dimensional. The direction of low-impact development (LID) practices has changed from engineering designs of drainage and stormwater storage facilities to mimicking and preserving nature in the urban landscape and, at the same time, achieving key urban landscape sustainability benefits.

The biophilic designs emphasize biodiversity as well as social, physical, and economical diversities. Maintaining or enhancing biodiversity is an essential component of urban landscape resilience (Ahern, 2010). Urban horizontal ecosystems support a suite of functions, including hydrological drainage and flood controls, nutrient and biochemical processes, biomass production, climate regulation, waste disposal, wildlife habitat, and more.

Expanding the tree canopy is an example of a multifaceted beneficial ecosystem, which has the following biotic and abiotic functions:

Abiotic:	Shading and cooling to reduce global warming and urban heat island effects
	Hydrologic interception storage; up to 5 cm of rainfall or 15%–25% of the rainfall intercepted by foliage followed by evaporative cooling
	Increased infiltration and transpiration
	Absorbing CO_2 from the atmosphere by photosynthesis
	Absorbing air pollutants
	Fallen leaves and woodchips from pruning becoming a source of biomass for green energy production
Biotic:	Habitat for nesting and sources of food for birds and other organisms
	Nutrient cycling
	Maintaining healthy microbial population and organic composition of soils by creating humus.]
	Pollination by bees and flying pollen

Similar benefits can be attributed to other biophilic components of the urban ecosystems introduced in this chapter, such as wetlands, ponds, xeriscape, rain gardens, or floodplain ecosystems that also provide pollution removal and stormwater treatment and flood storage. Urban greenery and greenways provide not only stormwater conveyance but also enjoyment and recreation for the population.

The biophilic environmental ecotone corridor concept shown in Figure 6.1 was conceived in Toronto in the last decade of the previous century. Recently, Google selected Toronto as a place to develop the most advanced sustainable digitalized smart community on the 3.25 km² (800 acres) of undeveloped Toronto lakefront, which used to be a port. This design will adopt elements of the biophilic horizontal and vertical flora and forest concepts (Mitchel, 2017).

Connectivity of urban ecological terrestrial systems and waterways is needed to provide conditions for sustainability of aquatic and terrestrial ecology (see also Chapter 3). If the biota is disturbed or lethally impacted by a stressor (e.g., toxic spill), biotic systems can be repopulated by migration from neighboring unaffected ecotones. Fragmentation of the urban ecosystem (i.e., separation into isolated landscape elements separated built-in or impervious zones) is still a common but undesirable element of the urban ecosystems. The concept of connectivity also applies to water flow and aquatic components of the urban landscape. Disruption of the water system by lining and channelization, water drops, and dams are major concerns that should be replaced by green landscape planning that includes dam removal or fish ladders (Novotny, 2003a; Ahern, 2010).

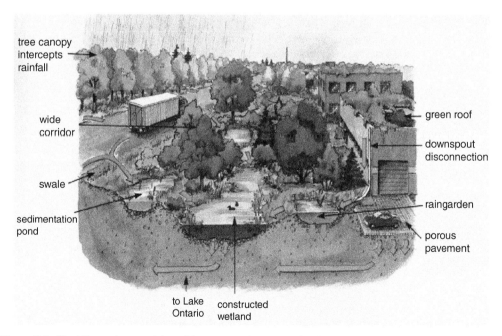

Figure 6.1. Biophilic environmental corridor concept for separating Lake Erie and the urban Toronto zone. *Source:* Illustration by Hether Collins in *Greening the Toronto Port Lands,* by Michael Hough, Jeff Evenson, Beth Benson, Waterfront Regeneration Trust, Toronto, 1997; courtesy Toronto Waterfront Regeneration Trust.

Biophilic landscape designs bring a dimension of nature into urban planning to incorporate all "green" LID/SUDS components (Beatley, 2011, 2017) and add more nature. They consist of horizontal interconnected ecological green and forested zones and vertical forests. These designs can also alleviate the impacts of global warming. (See Section 6.3.)

6.2   LOW-IMPACT DEVELOPMENT

The concept of low-impact development (LID) is a key component of biophilic urban landscape design that deals predominantly with urban runoff and, originally, provided control and even elimination of combined sewer overflows. In the US, the LID approach to urban stormwater management selects "integrated management practices," which are distributed small- and medium-scale controls that aim to replicate the predevelopment hydrology. Similar sustainable urban drainage controls and management programs are known in Europe under the acronym SUDS (sustainable urban drainage systems). They are also known in the US as GI (Green Infrastructure). Their goal is to achieve the highest hydrological efficiency of the urban and suburban open land by approximating predevelopment conditions. In December 2013 China's president, Xi Jinping, announced a variant of the LID/SUDS concept called "Sponge Cities" (*Guardian,* 2016). Originally, LID/SUDS goals were aimed primarily at controlling urban runoff and water conservation; other aspects of "green" development are not *a priori* considered. LID developments are sometimes situated in rural/suburban settings with very high open/built space ratios, which could imply long-distance travel and urban sprawl. Extensive descriptions and design parameters of urban runoff pollution controls, including erosion and LID designs, have been presented in the previous publications by the author (Novotny, 2003a; Novotny et al., 2010a). However, LID concepts can also easily be implemented in medium-density urban zones (population 100–200 people/ha), which is the optimum density for the sustainable COF (see Figure 1.6) and many practices can also be implemented in high-density cities. Green LID/SUDS practices are part of biophilic designs.

Figure 6.2 shows traditional urban stormwater drainage and high-flow runoff mitigation. It becomes immediately obvious that such drainage designed 50–100 years ago, using design flows based on "pre-global warming" meteorological data, will not be capable of handling more extreme flows expected to occur in this century. The simple reason behind this is the fact that the 100-year flow estimated in 1970 may be a design flow of 25 years or less in 2020 and less than a 10-year flow in the second half of this century under the business-as-usual scenario.

Pollution of Urban Runoff.   Urban storm runoff in fast conveyance drainage systems has been classified in the US as a point source that is to be captured and treated before discharge into receiving waters. Enlarging the sewers and more treatment would be very costly and would not bring other benefits beyond partially alleviating the drainage and pollution dilemmas. LID commonsense concepts provide multiple-benefit solutions to urban drainage problems. LID urban drainage concepts have been developed from the engineered urban stormwater management practices of the previous century.

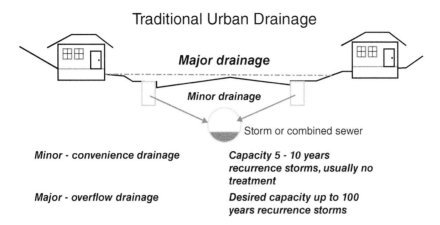

Figure 6.2. Twentieth-century urban drainage schematic.

The sources of pollution in traditional fast conveyance systems were extensively covered in Novotny and Olem (1994) and Novotny (2003a) and numerous reports and articles referenced and synthetized in these monographs. The sources of urban runoff pollution are:

- *Pollution contained in precipitation*, both rain and snow. This pollution originates from and is correlated to air pollution.
- *Acidity of precipitation*. Rainfall is acidic. Natural pH is about 5.6 and decreasing because of increasing CO_2 content in the atmosphere. Anthropogenic acidity (pH < 5.6) is caused by burning sulphur-containing coal (acid rain) and by traffic emissions of nitrates. Acid rain dissolves toxic metals (zinc, copper) from metal roofs and gutters, and PAHs (carcinogenic polycyclic aromatic hydrocarbons) from asphalt shingles and pavements. PAHs are also generated by incomplete combustion of organic materials (e.g., coal, oil, trash) and subsequently absorbed by precipitation.
- *Erosion of pervious lands, mainly construction sites*. Construction erosion is several orders of magnitude larger than natural erosion.
- *Dry atmospheric deposition*. Atmospheric pollution by fine particles contains fine ash particles from burning and traffic, dust from erosion, infrastructure deterioration, desert and dry land dust sometimes coming from long distances (for example, dust from the Sahara in Central Europe or from the Gobi Desert in Beijing).
- *Street refuse accumulation near the curb*. These deposits come from vegetation, street deterioration, traffic, animal feces, and litter.
- *Traffic emissions*. Vehicles deteriorate and rust.
- *Urban lawns clipping and fertilizer applications.*
- *Street deicing chemicals.*
- *Oil and fuel drips on parking lots.*

LID Best Management Practices.   Making the transition from runoff being a menace and hazard classified previously as a point source of pollution to an asset, a source of good-quality water, and a social and ecological benefit requires a change from:

- *Fast conveyance drainage systems* (Figure 6.2) characterized by high imperviousness, curb and gutter runoff collection, and conveyance by underground sewers, culverts, and concrete-lined channels on the surface carrying highly polluted white water containing toxic metals, pathogenic microorganisms, high salinity (from deicing salts), and medium levels of BOD, COD, and nutrients (see Novotny, 2003a)

to:

- *Storage-oriented slow-release systems* with biophilic design (Figure 6.3) characterized by *storage* through vegetation, landscape, nature-mimicking ponds, green flat roofs, underground cisterns, preserved or restored lakes, and the like; *infiltration* into shallow aquifers; *soft treatment* through rain garden biofilters, earth filters, wetlands, and ponds; *slow conveyance* in grassed swales (rain gardens); and natural or nature-mimicking surface channels.

LID drainage, in addition to collecting and conveying stormwater, includes treatment of rainwater and stormwater in enhanced rainwater infiltration (rain gardens), pervious pavements, infiltration ponds, and wetlands. The term "rainwater" in technical practice describes captured precipitation that is not polluted during overland flow. An important part of LID landscape is the switch from thirsty lawns with high evapotranspiration to xeriscape that uses native grasses, flowers, and mulch.

Lawn irrigation represents the largest water use of an average US home located in a subdivision (see Table 5.2). LID-designed landscape systems capture, treat, and store stormwater in the neighborhood and represent an additional important source of quality reclaimed water and could also provide cooling (see Figure 5.10). Some LID treatment unit processes such as wetlands can also be used for polishing the effluent from the cluster treatment plant,

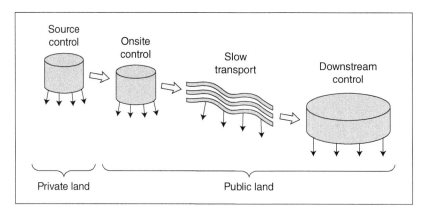

Figure 6.3. LID components. *Source:* P. Stahre (2008).

and the biomass grown in the treatment wetland can be harvested and transported to the regional IRRF for energy production and nutrient recovery. In general, captured rainwater and clean stormwater is the third best source of fresh water for a community after groundwater and clean fresh water from clean surface water resources.

The LID concept was first developed and demonstrated in Maryland (Prince George's County, 1999) in the US with the objective of trying to maintain the predevelopment hydrology of an area to control runoff and the transport of pollutants. Traditional pre-LID stormwater management practices concentrated only on reducing peak flow rates to prevent flooding, whereas LID techniques also try to reduce runoff volumes by utilizing storage and infiltration systems to more effectively mimic the predevelopment hydrology (Dietz, 2007). The LID/SUDS practices focus on reducing or eliminating runoff volume and pollution through onsite infiltration, evaporation, and/or reuse of rainwater (Prince George County, 1999, 2014; Novotny at al., 2010a).

The key components of LID are:

- Reduction of imperviousness
- Providing storage to reduce peak stormwater flows
- Infiltrating excess flow into aquifers, from which water can be retrieved during the time of shortage
- Reducing pollutant loads in stormwater by sedimentation in ponds and biofiltration

LID practices also reduce or eliminate stream channel degradation by attempting to mimic natural flow and reduce stream erosion by bioengineering. These measures reduce the frequency and severity of combined sewer overflows by limiting the volume and peak flows in areas where disconnection of stormwater inputs into sewers is not possible. Integration of stormwater management into landscape by incorporating ecological corridor increases visual and recreational quality of urban areas and provide habitat to wildlife. The fundamental characteristics and components of LID best management practices-BMPs are:

- Eco-mimicry for the landscape
  - Developing urban landscape that would mimic but not necessarily reproduce the predevelopment natural system
  - Aesthetically pleasing xeriscape
  - Minimizing imperviousness
  - Drainage service mostly on the surface
  - Interconnected riparian buffers and wetlands
- Green ecologically functioning space for recreation along the rivers and throughout the city
- Urban brownfield remediation and development
- Ecologically sound stream restoration and daylighting, including restoring base flow

Water and waterscapes are great assets to a city and to the new water-centric architecture (Dreiseitl and Grau, 2009), and clean urban runoff, stored and reused, is a good source of water for potable and nonpotable uses analogous to raw water brought from distant rivers or from deep saline groundwater aquifers or from salty seas desalinated at a great cost and high energy use. The rediscovered values of rainfall and urban runoff from the LID watersheds are (Novotny et al., 2010a):

- A valuable and easy to treat source of potable water
- A source of recharge of depleted groundwater aquifers
- A good traditional source of water for irrigation
- A source of base flow of restored/daylighted urban streams, ponds, and lakes
- A source of water for decorative purposes of urban waterscapes such as water arts and fountains in plazas of cities and water festivals and inspiration to painters
- For cooling of homes (e.g., by green roofs) or evaporation cooling systems
- Enjoyment of people and especially children to play in waterscapes of urban parks and recreational areas
- Source water in private courtyards and swimming pools

LID/SUDS stormwater management practices and management are implemented on the cluster/neighborhood level, although many components (e.g., rainwater harvesting, green roofs, rain gardens, reuse of reclaimed water) require homeowners' participation and are implemented on individual building or city block scale. LID was created to decrease the impact of urbanization and increased impervious surfaces. Similar drainage practices with incorporated best management practices (BMPs) are also used to control runoff from medium- to high-density highways, freeways, and commercial/industrial sites. Runoff from roads with high-density traffic contains high concentrations of toxic metals (copper, nickel) and asbestos. Salt for road deicing and snow removal magnifies the toxicity effects of toxic metals and adds complex cyanides to the melt that under sunlight can be converted to toxic hydrogen cyanide (Novotny et al., 1998).

The hydrologic designs of the current LID drainage practices have the following goals:

(a) Prevent flooding caused by increased imperviousness of the watershed and providing storage to reduce peak flows.
(b) Reduce runoff pollutant loads in runoff
(c) Control degradation and erosion of stream channels receiving urban runoff by reducing frequency and severity of flows from stormwater drainage sewers and surface channels and significantly reduce or even eliminate combined sewer overflows (CSOs).

LID drainage designs should be holistic and incorporate biophilic designs such as ecological corridors and riparian areas surrounding streams, especially floodplains and ephemeral surface channels, urban forests ecotones. This increases visual and recreational quality of urban areas and provides habitat. Streams buried by development during the last century should be uncovered and converted to natural channels and some wetlands could be restored. Uncovering and revitalization of streams buried by urbanization is called "daylighting."

The "Sponge City" program in China is broader than more traditional LID/SUDS programs. It involves water governance, financing structures, and technical measures. The *Guardian* (2016) reported that the program and its technical aspects, represents an integrated urban water system that incorporates low-impact-development practices, upgrades of the traditional urban drainage system, and provides solutions for excessive run-off discharge. The sponge cities program is designed to address flooding but also to offer a solution to water shortages. China has much less freshwater per capita than most developed countries and many fresh water resources are polluted. Initially 16 cities were selected as pilots for China's "sponge cities" program, but eventually it will be rolled out nationally. The practices will adopt LID to a great degree but will also include watershed-wide measures such as connecting rivers and streams to enhance in-stream storage and flood conveyance.

Classification of LID (SUDS) Practices

Urban best management practices (BMP) based on LID concepts are an integral part of the COF type sustainable communities. The engineering and natural designs of urban BMPS were covered extensively in Novotny (2003a) and Novotny et al. (2010a). Figure 6.3 presented the holistic concept of LID practices. In the original engineering and science definitions they were divided into:

- Source control measures that mainly involve rainwater harvesting, infiltration in rain gardens, pervious driveway, construction erosion controls at individual building or construction sites
- On-site controls and hydrologic modification focusing on infiltration implemented in the city block or subdivision on both public and private lands
- Reduction of delivery in transport runoff (slow transport) and pollutants from the source area to the "end of the pipe" point that is accomplished mainly by biofiltration and reducing the velocity in the channel
- Downstream storage and treatment in ponds, wetlands and by infiltration into groundwater aquifers

Source Controls.   Source controls on the building or a city block scale include:

- Rain harvesting and onsite storage
- Green roofs and roof storage
- Rain gardens
- Biophilic building designs, including vertical forests

Rain harvesting.   As stated previously, use of captured rain water is nothing new and was discussed previously (see the section "Water Footprint" in Chapter 3). Rain was harvested thousands of years ago in water-short areas around the Mediterranean Sea by Greek, Minoan, and Roman civilizations and in the southwestern US by Anastasi Native Americans and today in the British colony of Gibraltar. Native Americans' rainwater capture can still be seen in the Sky City Acoma Pueblo Historic Native American settlement in New Mexico that is located on top of a mesa with no sources of fresh water other than rain.

By collecting runoff from a roof or other impervious surfaces and using it after the rainfall event for irrigation or even drinking water, the peak flows from a development can be reduced. The simplest but not very efficient rainwater harvesting technology consists of a barrel with water and overflow outlet to which the building downspout is connected. The best storage of collected water is in underground cisterns or tanks, which in some historic Mediterranean and Byzantine cities have been used for more than 3,000 years. Underground basilica cisterns capable of storing 80,000 m³ (21.1 mg) of water were built at the beginning of the seventh century in Constantinople (today Istanbul), the capital of the Byzantine Empire and is still in a working shape (see Novotny et al., 2010a). Archeological excavation on Crete found Minoan cisterns built 2500 BCE. Underground storage reduces evaporation and keeps water colder during hot summer days. However, once rainwater falls on the street surface, it becomes urban or highway runoff, which is polluted and must be treated before reuse.

Green roofs.   Green roofs are LID biophilic systems designed to limit the area of impervious surfaces of the city. They are the first steps for collecting rainwater and using it for building cooling and insulation and providing water for rainwater harvesting. Green roofs are designed to help mimic the predevelopment hydrology of an urban development by incorporating vegetation and soils on top of the impervious rooftop. The systems consist of a vegetation layer, a substrate layer used to retain water and anchor the vegetation, a drainage layer to transport excess water from the roof, and specialized waterproofing and root-resistant material between the green roof system and the building structural support (Mentens et al., 2003).

Figure 6.4 presents an award-winning roof garden on the top of Chicago City Hall and shows that in addition to the main benefit of reducing runoff from the roofs' surfaces, there are other benefits of green roof installation. Roof gardens provide excellent heat and cooling effects that contribute to mitigating the urban heat island effect, improve building insulation and energy efficiency, increase biodiversity and aesthetic appeal, and reduce runoff water temperature. The bottom part of Figure 6.4 shows that roof areas without vegetation had a hot sunny day temperature of 66°C, while the temperature on vegetated surfaces ranged from 23°C to 48°C (US EPA, 2003). They also insulate the buildings and reduce heating and cooling energy demand by as much as 25–30%. By restricting the downspout flow capacity, flat roofs can temporarily store rainfall and help to reduce peak flows causing combined sewer overflows (CSOs).

Rain gardens.   A rain garden (Figure 6.5) is a shallow depression containing spongy soil and a variety of plants that collect stormwater, rainwater, and runoff from impervious surfaces, such as roofs, streets, and parking lots, and filter pollutants out of the runoff water. There are two basic types of rain gardens: (1) Rain gardens with no underdrain allow captured runoff water to infiltrate into the underlying soil. They are therefore effective at removing water and pollutants. (2) Rain gardens that have an underdrain to convey accumulated water when subsoils have poor permeability or are frozen during snow melting (Minnesota Pollution Control Agency, 2019). Biophilic rain gardens are also places beautified by flowers on public and private lands where stormwater can infiltrate (Figure 6.6) and receive some treatment by biofiltration. Infiltrating water through a soil media decreases surface runoff, increases groundwater recharge, and achieves removal of the total mass of some pollutants. As noted, in Seattle, the Milwaukee suburbs, and many other northern cities promoting rain gardens and other natural stormwater BMPs, rain gardens improve with age. As the plants grow larger and microbes in the soil multiply, rain gardens will absorb and filter stormwater more effectively, but they require maintenance practices (mulching, weeding, and watering) for the first year or two. However, because fertilizers and even pesticides are commonly used by homeowners and cities for supporting vegetation growth, it

Figure 6.4. Chicago City Hall roof garden and its cooling effect. *Source:* Courtesy Chicago Climate Action Plan, Chicago Municipal government, http://www.chicagoclimateaction.org/filebin/pdf/finalreport/CCAPREPORTFINALv2.pdf.

was observed that concentrations of nutrients and sometimes pesticides in the outflow from rain gardens may be greater than those in the inflow, but the flow is much smaller, and often none for smaller storms.

Figure 6.5 shows a rain garden with an overflow to a traditional curb and gutter drainage. In many communities, roadside ditches (swales) are converted by homeowners and communities into rain gardens but without significant infiltration and biofiltration benefits over grassed waterways. Rain gardens/swales can be engineered to increase infiltration and filtration, as shown in Figure 6.5 from the Seattle Street Edge Alternative development. Figure 6.7 suggests an obvious fact that the roadside swale may not be just a decorative rain garden with flow only during rainfalls. These aesthetic rain gardens can replace simple grassed swales and channels that are hydrologically and hydraulically designed to collect stormwater runoff and other clean water inputs from a larger area and directing it to another LID stormwater management system or conveyance. The design must account for increased hydraulic resistance by plants growing in the swale. The swale in Figure 6.7 indicates more permanent flow that, in addition to roof downspouts and road runoff, can originate from basement sump pump drainage, upstream creeks, construction site dewatering, overflows from storage ponds, wetlands, and other sources. The Street Edge Alternative LID system retains 98% of the water of a two-year-design storm. Because most pollution

Figure 6.5. Rain gardens in the Street Edge Alternative (SEA) of Seattle. Note connection to the street runoff. *Source:* Courtesy public domain pictures of the Seattle Public Utilities.

Figure 6.6. Street Edge Area swale with infiltration. *Source:* Courtesy public domain pictures of the Seattle Public Utilities.

Figure 6.7. Improved side street swale serving as a more permanent nature mimicking conduit. *Source:* Courtesy public domain pictures of the Seattle Public Utilities.

is carried by storms of lesser magnitude, the pollution load is also significantly reduced, which in Seattle reduces the impacts of urban environments on receiving streams inhabited by salmon. The swale design procedure has been presented in a book by Novotny and Olem (1994).

On-Site Controls.

Porous pavements.   There are several types of porous or pervious pavements (Figure 6.8). Old historic pavements made from stone (granite) pavement bricks and cobblestones in streets or flagstones laid over a sand subbase are partially pervious, and the subbase provides some storage for runoff permeate. Today, pervious pavements use asphalt or concrete mixtures from which fine fractions are missing. The subbase is made from gravel and is thicker than that in conventional pavements (Figure 6.9). The main benefits of installing pervious pavements are: (1) a significant reduction or even complete elimination of surface runoff rate and volume from an otherwise impervious area. If the pavement is designed properly, most or all of the runoff can be stored and subsequently allowed to infiltrate into the natural ground; (2) aquifer recharge by infiltrated water; (3) reduced need for storm drainage or even its elimination; (4) removal of contaminants from the runoff both through the filtration of the water through the soils and in the reduction of the amount of salt needed for deicing purposes.

Typical construction of a porous asphalt system includes several layers (Figure 6.9). The top layer is the pervious pavement layer, which can range between 10 and 15 cm thick. In this layer the fine sand particles are removed from the pavement mix, creating a void space in the pavement of 18–20%. Below the surface layer is a choker course layer created with 2 cm crushed stone followed by the filter course and filter blanket layers consisting of finer filter materials such as sand and gravel. Below the filter blanket material is a reservoir course layer, which can include a subdrain if needed. Below this layer would be the native materials. An optional impervious layer could be installed between the reservoir course layer and the

Figure 6.8. Porous pavement in Street Edge Seattle development. *Source:* Courtesy public domain pictures of the Seattle Public Utilities.

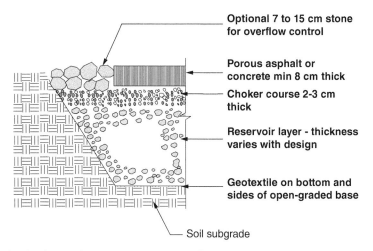

Figure 6.9. Design fundamental of pervious pavements. *Source:* Minnesota Stormwater Manual; photo courtesy Minnesota Pollution Control Agency (2019), Copyright Minnesota PCA, https://stormwater.pca.state.mn.us/index.php?title=Design_criteria_for_permeable_pavement.

native materials if infiltration is undesirable in the area, allowing for the system to become a reservoir with the runoff exiting through the drainage pipe. Regulations and designs of permeable pavement can be found in ASCE manual (Eisenberg et al., 2015).

Field (1986) summarized the results of several early US EPA studies on the experimental applications of porous pavements. Results from a study in Rochester, New York, indicated that peak runoff rates were reduced by as much as 83% where porous pavement was used. In a more recent study at the University of New Hampshire, no surface runoff was observed during the 4 years of operating a parking lot constructed with pervious pavements even for two 100-year storm events (New Hampshire Porous Asphalt fact sheet). Typically, hydraulic conductivity (permeability) of porous pavements is much greater than rainfall/runoff rates.

Biofilters. Any time urban runoff flows over a grass surface it is cleansed and attenuated. This includes rain gardens, swales, environmental corridors and buffer zones, grass filter strips, and, to an extent, green roofs. If designed properly, these practices minimize or even eliminate the need for street sweeping. A grass area can be designed as a biofilter and an infiltration area (Figure 6.10).

Several processes contribute to effective attenuation: (1) Infiltration of runoff into soil that reduces runoff volume and provides absorption and adsorption of many pollutants on the soil; (2) slowing down the runoff flow that reduces peak flows; (3) provision of surface storage that enhances settling; and (4) filtering and adsorption effect of grasses.

According the University New Hampshire Stormwater Center, the efficiency of vegetated swales to remove pollutants on an average annual basis was:

Total suspended solids	60%
Zinc (toxic metals)	88%

Most of the removal was due to infiltration and grass filtration of solids. Some biofiltration systems simply consist of an excavated trench or basin containing vegetated filter media. Below the filter media is a perforated pipe that is used to collect the treated water and deliver it to a stormwater drainage network or directly to a waterway or can be reclaimed for reuse.

Figure 6.10. Example of biofilter connected with retention and infiltration basin located on the University of New Hampshire campus. *Source:* Photo V. Novotny.

Figure 6.11. Perennial urban drainage with stepping stones and public access in Hammarby Sjöstad. *Source:* Courtesy Malena Karlsson, GlashusEtt, Hammarby Sjöstad.

Slow Transport: Reduction of Delivery of Pollutants. In the LID (SUDS) communities the COF biophilic natural perennial surface drainage replaces storm and combined sewers, leaving only sanitary sewers in place, carrying smaller flows, as it has been done in Seattle (Figure 6.7); in Hammarby Sjöstad, Stockholm (Figure 6.11); in Berlin, Germany (Ahern, 2010); in Malmö, Sweden (Stahre, 2008); and in many other "green" cities.

Implementation of natural storm drainage is important in the new and retrofitted sustainable communities and is a part of LID. It eliminates the need for storm sewers and makes existing combined sewers oversized with no CSOs. The pollution control benefits are significant (see subsequent sections); measurements in Wisconsin have documented that pollution loads from urban and suburban low- and medium-density zones using surface (natural) drainage even without best management practices are 10–30% of the loads if the same area had storm sewers.

The solution to natural drainage today is not ditches and concrete-lined urban streams. Surface drainage conduits can be ephemeral, with occasional flows, or close to perennial, carrying some permanent flows that may originate in urban areas from high groundwater table contributions, basement sum pumps and de-watering, cooling water flows, or from upstream natural inflows and wet weather flows from rain gardens and grass swales. Natural, restored, or created perennial streams can also form an eco-corridor that supports wildlife and represents the urban ecotone interconnecting urban biological community areas such as urban forests, parks, lakes, ponds, and the like, and/or convey the flow for natural treatment in landscape features such as ponds and wetlands (Figures 6.12 and 6.13).

Eco-corridors are natural or nature-mimicking biophilic landscapes that treat (attenuate delivery) and slowly convey urban runoff from source areas to the final place of treatment (large pond or wetland) and discharge into receiving waters. Eco-corridors also include walking and bike paths and other recreational opportunities.

Figure 6.12. Biophilic Eco-corridor with a perennial channel in Philadelphia. *Source:* Photo courtesy Mark Maimone, CDM-Smith.

Figure 6.13. Wet stormwater pond. Minnesota stormwater Manual. *Source:* Photo courtesy Minnesota Pollution Control Agency (2019), Copyright Minnesota PCA, https://stormwater.pca.state.mn.us/index.php?title=Stormwater_ponds_combined.

Downstream Storage and Treatment.

Storage basins and ponds. Storage basins have the following roles in LID/SUDS: (1) Hydrologically, they attenuate the peak flow and if they have permeable bottoms, they reduce the volume of the runoff by infiltration; (2) they store water for subsequent reuse; (3) in cold regions they store polluted high-salinity chemical melt for subsequent dilution by less polluted spring runoff; (4) they provide treatment by settling and biochemical self-purification in the basin.

Ponds can be either permanently wet (Figure 6.13) or dry (Figure 6.14), collecting water during storm events and releasing it shortly after the event. In high-density built-up areas typical for historic European cities, modular storage basins can be built under plazas or parking lots. Biophilic wet ponds, sometimes combined with a pre- or post-treatment wetlands, are more desirable.

Storage basins in LID/SUDS watersheds designed by landscape architects are natural and architectural assets that can even increase the value of the homes that may surround them. Designed engineered wet pond (Figure 6.15) consists of (1) a permanent water pool, (2) an overlying zone in which the design runoff volume is temporarily stored by increasing the pool depth and released at the allowed peak discharge rate, and (3) a shallow littoral zone acting as a biological filter.

If the main purpose of the pond is to store rainwater and snowmelt water for reuse, a permeable bottom is not desirable. Capping and sealing the bottom can be done by synthetic liners or by clay or bentonite. While the main function of liners is to prevent seepage and hold water, they would also prevent seepage of phosphorus and other chemicals from the nutrient-rich bottom muck to the water above. Pond liner longevity is very important because repairs and replacements require dewatering and refilling the pond.

However, impermeable (e.g., vinyl) liners alone should not be installed over muck wherein microbial decomposition is occurring and the high organic matter soil (muck) tends to give off methane gas. This and an occasional high groundwater table could cause the liner to float from the bottom to the surface. Using a proper pond commercial *underliner* venting will help prevent this. An underliner is made of perforated tubes wrapped by nonwoven geotextile for more protection and function. The lifetime of a vinyl liner is about 20 years. Clay and bentonite liner can last a century.

Figure 6.14. Dry pond outlet detail.

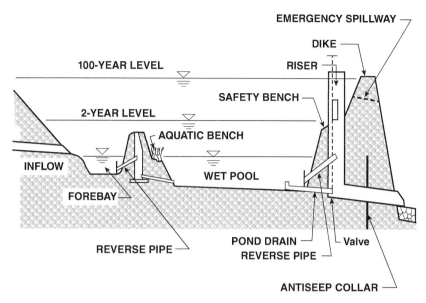

Figure 6.15. Engineered wet pond modified for winter operation. *Source:* Adapted from Center for Watershed Protection and Novotny (2003a).

Dry pond is a stormwater detention facility that is designed to temporarily hold stormwater during high peak flow runoff events. The outlet is restricted to activate the storage during flows that exceed the outlet capacity (Figure 6.14). A grass dry pond is usually a part of the landscape and can serve for other purposes; e.g., as a playing field. An alternative to a grassed dry pond is an underground modular of concrete storage basin constructed under parking lots or public squares in densely populated areas such as historic centers of the cities where open land is not available, and demolitions of historic sites are not allowed. These storage basins provide solely hydrological runoff peak flow attenuation or reduction of surface runoff inflow into combined sewers.

The objective of the storage in the sustainable (COF) triple net-zero cities is not only to attenuate the peak flows and volume to reduce flooding, which historically was the main and sometimes the only goal of urban watershed planning and pollution abatement, but also to include water conservation done by storing harvested rainwater (urban runoff) during dry periods as it is being done in the Potsdamer Platz commercial/residential development in Berlin (see the section on waterscapes in this chapter). Hence, the design parameters of the storage basins must be changed from the current designs requiring stored water above the permanent pool to be released in a period of 24–48 hours after the precipitation event to much longer retention periods that should cover the entire dry period between the precipitation events. Consequently, the storage capacity should be determined by long-term simulations and not just by sizing storage for a certain design storm. As a matter of fact, the concept of single-design storms may not be appropriate in the LID/SUDS or COF designs.

In some cases, ponds and wetlands may provide ample surface storage to accomplish the goals of the development as a source of water. Underground storage basins are needed and have been used in densely populated urban zones since Roman and Minoan civilizations thousands of years ago. Today and in the future, wet ponds and retention areas are designed by or in close cooperation with landscape architects. These new designs are treated by the public as urban lakes that increase the value of riparian land and provide recreation and enjoyment, which contrasts the past engineered, often lined rectangular retention basins with signs to keep the public away. Well-designed biophilic ponds can be stocked with fish. Nutrient control by educating homeowners about the eclogical harm to the water body by the overuse of fertilizers is necessary to prevent eutrophication or hyper-eutrofication with harmful algal blooms and vegetation.

Wetlands.   Both natural and manmade wetlands have been used for runoff pollution control and flood mititgaton. However, natural wetlands have been designated in the US as receiving waters and are subjected to water quality standards and restrictions. Constructed wetlands could be considered as treatment facilities and the standards mostly apply to the outlet from the wetland (Kadlec and Wallace, 2008). Only constructed or restored wetlands may be used for treatment of runoff and CSOs and for polishing the secondary effluents (see Chapter 9).

Wetlands combine both sedimentation and biological utilization effects to remove pollutants from runoff. Wetland construction for runoff and wastewater pollution control is different from wetland restoration. Wetland restoration efforts are aimed at restoring the nature's cleansing capability and creating habitat in places where former wetlands were drained and lost. Such created wetlands are located mostly in riparian zones of water bodies and serve as a buffer against pollution. Constructed wetlands have been used for (Mitsch and Goselink, 2015, Kadlec and Wallace, 2008):

Flood control

Wastewater treatment

Stormwater or nonpoint source pollution control

Ambient water quality improvement (e.g., riparian and instream systems)

Wildlife enhancement

Fisheries enhancement

Replacement of similar habitat (wetland loss mitigation)

Research wetland

Providing supplemental biomass for energy production

There are two types of constructed wetlands (1) free water surface (FWS) systems and (2) subsurface flow systems (SFS) shown in Figure 6.16 (Kadlec and Wallace, 2008; Vymazal, 2005; Vymazal and Kröpfelová, 2008). These wetlands were specifically designed to treat combined sewer overflows and urban runoff. A stormwater wetland and pond manual was prepared by the Center for Watershed Protection and published by the US EPA (2009). Table 6.1 presents removal efficiencies for the stormwater control wetland featured in Figure 6.17.

Macroflora in the FWS systems (highly recommended) can be classified as submerged aquatic vegetation (SAV) and emerged aquatic vegetation (EAV). Maintaining and managing healthy vegetation greatly enhances the treatment efficiency of wetlands treating permanent flows. The role of vegetation is important when wetlands are designed to treat permanent flow from water reclamation plant effluents, to remove nutrients from irrigated agriculture and subdivisions with lawns. As it will be documented in Chapter 9, vegetation (SAV and EAV) in constructed FWS wetlands, if properly designed, compartmentalized, and continuously wetted, provide higher phosphorus removal efficiencies (80–90%), far more than that reported in Table 6.1 for intermittent surface runoff flow treating wetlands.

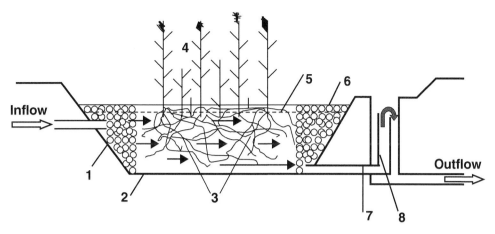

Figure 6.16. Schematics of a constructed wetland with subsurface horizontal flow: 1. distribution zone; 2. impermeable liner; 3. filtration medium (gravel, crushed rock); 4. vegetation; 5. water level in the bed; 6. collection zone filled with large stones; 7. collection drainage pipe; 8. outlet structure maintaining water level in the bed. *Source:* Vymazal (2005), copyright Elesevier.

Table 6.1. Average annual removal efficiencies of the gravel wetland on the campus of the University of New Hampshire treating runoff.

Total suspended solids	100%
Dissolved inorganic nitrogen	100%
Zinc	100%
Total phosphorus	55%

Source: UNH Stormwater Center.

SAV also provides oxygen to water and prevents anoxia. The wetland plants represent a source of biomass that can be harvested and codigested with sludge from the used water reclamation and resource recovery plants, along with food waste and other biodegradable organic solids to produce energy and fertilizers.

However, while wetlands cleanse and detoxify urban and highway runoff, concerns about global warming and GHG emissions should be considered. Wetlands sequester carbon dioxide that is converted by photosynthesis into wetland flora and organic peat soils, but the decay of organic sediments deposited by the dead wetland vegetation could emit methane, which is a potent GHG.

The wetlands should be compartmentalized (see the section "Phytoseparation of Nutrients" in Chapter 9). Logically, the first compartment removes the settleable solids that also contain adsorbed pollutants (toxic metals, phosphorus, organic solids). The important biological filtration and biochemical attenuation continues in the second compartment in which growth of wetland plants also absorb and remove nutrients. The third compartment continues biochemical processes and provides polishing of the effluent.

If the wetland is not compartmentalized, a pond-wetland combination has been proposed as the best design for stormwater storage and pollution control (Figure 6.18). In the integrated water management of the sustainable cities, in addition to capturing and storing runoff, the wetland-pond system can accept treated gray water, cooling water, and clean

Figure 6.17. Experimental research gravel wetland treating runoff from a parking lot at the Stormwater Center of the Campus of the University of New Hampshire. The compartmentalized arrangement works best for urban stormwater. *Source:* Photo by V. Novotny.

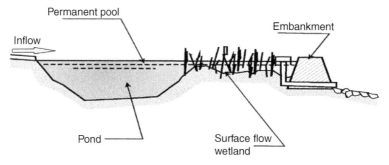

Figure 6.18. Schematics of a pond with emerged flow wetland combination that enhances treatment. *Source:* US EPA.

water from drainage of basements and other sources and, after treatment, subsequently can provide clean nonpotable water in the cluster (Figures 5.5–5.7).

The efficiency of wetland treatment is related to the hydraulic loading:

$$q = 100 * Q/(A)$$

where Q is the design flow in m³/day, A is the surface area in m², and the unit for q is centimeter/day. Q for free surface wetlands treating surface runoff and combined sewer overflows was recommended as 2.5–4 cm/day and for submerged bed flow wetlands as

6–8 cm/day (Vymazal, 2005). The efficiency also depends on the type of vegetation that can be either submerged aquatic vegetation or emerged (e.g., cattail) or both. Cattails generally do not increase the treatment efficiency of FSWs and are irrelevant in SFW.

Xeriscape. Unlike European single-family homes that are typically built on a smaller lot, the majority of typical suburban US homes have a larger lot with a lawn. This is a tradition of the last 50–60 years after lawns replaced vegetable and flower gardens in many US suburbs. Because of using drinking water for lawn irrigation grass, water use in the US is very high, especially in the dry and hot US states, where it could become astronomical because homeowners transplanted from more humid and colder northeast cities try to grow lush grass in areas that used to be desert. Table 5.2 shows that the average irrigation water use is much greater than in-house domestic use. In order for a lawn to stay green in dry weather it needs about 25 mm (1 inch) of rainfall and/or irrigtion every week during growing season. If a lawn covers 0.2 hectares (0.5 acre), the daily water use during a dry week would be 7,143 liters/day, which is five times more than the domestic daily use of a four-member family.

Accordingly, during summer many suburban communities impose lawn irrigation bans periodically and with increasing frequency. Global warming represents an increasing problem to the water utilities. The severe five-year drought (2011–2016) in California led to irrigation water restrictions, which are now becoming frequent during increasingly parched and dry summers over most of the US, as well as Europe and Asia. Droughts are becoming more common, as global warming models have predicted. In 2015 and 2016, the entire US Midwest and Northeast suffered from very hot temperatures and drought and irrigation restrictions were ubiquitous. California and New England were the epicenters. It was already stated that 10 of the hottest years on record have occurred in this millennium and 2016 was the hottest (and driest) at the time of this writing. The 2017–2018 period was also very dry and hot in Europe.

The consequences of global warming are obvious. Water utilities just cannot supply potable water for irrigating lawns; therefore, irrigation bans and brown and yellow lawns during summer will be increasingly common during in summertime unless alternate sources of water are found (unlikely) or people realize that their landscape must be changed to something less thirsty than a lawn of imported grass. It should be pointed out, as was documented in Chapter 5, that gray water recycling alone can only provide a fraction of the summer irrigation need for a typical urban lawn. The final answer to reducing outdoor water use, pollution, and GHG emissions in the US and elsewhere is switching from growing imported water-demanding non-native grasses to *xeriscape* that uses native plants and decorative surface covers not requiring irrigation in the more arid areas, to grasses not requiring irrigation and natural flowers in the prairies and northeastern US zones. For example, in more humid areas, native prairie flowers and grasses can be grown, while in the southwest, desert plants and natural mineral ground covers result in far more pleasing urban landscapes with almost no irrigation water requirement, as shown in Figures 6.19 and 6.20. Xeriscape landscaping essentially refers to creating a landscape design that has been carefully tailored to withstand drought conditions, which could also be more aesthetically pleasing than lawn. The term *xeriscape* means "dry scape" but xeriscape concepts can be used anywhere.

Under a paradigm in the United States in the previous century, all water delivered to a home was potable – treated to the highest drinking water standards to maintain human health free of water-related diseases, which is a waste if used for irrigating lawns. Exceptions can be found in some arid states like Utah, where many homeowners in Salt Lake City and other urban areas have a dual water supply, one treated for home use and the other mostly

Figure 6.19. Xeriscape landscaping of a home in Flagstaff, Arizona. *Source:* Courtesy public domain pictures, City of Flagstaff.

Figure 6.20. Sustainable landscaping with native prairie flowers and grasses.

untreated natural flow from mountains for irrigation. Selecting xeriscape landscape may run into opposition in some "upscale" suburbs requiring uniform, mostly non-native lawns to be laid during constructing a home. In some communities, prairie grasses and flowers would violate the height and maintenance requirements embedded in archaic local codes.

Control of irrigation water use by pricing water is not effective. For one, those living in large mansions with large lawns will be able to pay the increased cost, while small homeowners with small irrigation water use will not be able to pay. Mandatory restrictions will lead to violations and/or to unsafe use of poorly treated or untreated contaminated gray water. Hence, public education and subsidies for xeriscape may be the beginning of the process that would bring about the reduction of the extremely high US use of potable water. New "green" triple net-zero developments should be built based on the COF concepts.

Good references about US LID programs are available from the following websites:

Seattle and Portland http://bluegreenuk.com/references/case_studies/YW%20USA%20SuDS%20Report%202012.pdf

Philadelphia http://bluegreenuk.com/references/case_studies/Maimone_Lords_2012.pdf

Minnesota https://stormwater.pca.state.mn.us/index.php?title=Main_Page

Software for LID design: https://stormwater.pca.state.mn.us/index.php?title=Calculator

6.3 RESTORING, DAYLIGHTING, AND CREATING URBAN WATER BODIES

Stream Restoration

Rivers have always been a lifeline for cities. In coastal areas river mouths offered safe harbors and a location for a commercial and historical center. They provided water for various uses, transportation, fishing, recreation, and enjoyment. In the post-industrial period many urban streams were damaged, lined with concrete or masonry and, in extreme cases, disappeared from the surface to become underground conduits and oversized sewers. Restored or daylighted urban surface water bodies are today serving multiple purposes and attract people again. After 'being one of the most polluted urban rivers in the nation, the Charles River in Boston has become swimmable again and the city has developed water and entertainment parks and an environmental corridor along its banks. Today, Boston Harbor claims to be the cleanest harbor in the US. The Emscher River in the industrial Ruhr area of Germany, converted to an open sewer at the beginning of the twentieth century, has been restored to a renaturalized urban stream serving multiple purposes. However, while efforts are being made to revitalize the urban streams and bring them to a state that mimics natural rivers, urbanization has changed the natural hydrology to the point that in most cases they cannot be fully restored to their pre-urbanization conditions. For one, restored urban streams may have insufficient low (base) flow, often to the point that they have become ephemeral or effluent dominated because the groundwater level dropped as a result of the developments that installed sump pumps to keep basements dry and deep sewers that suffer from infiltration-inflow clean water inputs. Without best management practices and control of imperviousness, high peak flows have become 4 to 10 times larger than predevelopment flows. LID/SUDS practices can to some degree remedy the hydrology but not quite enough. Furthermore, documented cases in Philadelphia, Boston, Tokyo, and many other larger cities show that often more than 75% of the original streams in the urban areas were lost, buried, and became sewers. Stony Brook in Boston was in the nineteenth century a lifeline

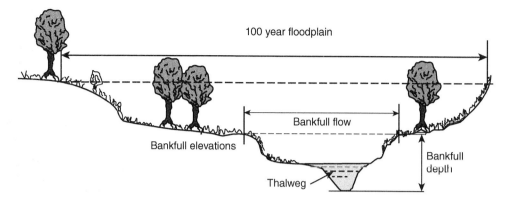

Figure 6.21. Typical cross section of a natural channel.

river of the city, but today only names of subway stations and streets remind people the lost river buried in culvers and until 2000 serving as an oversized combined sewer.

A restored stream should be biologically and morphologically functional (Figure 6.21) and should contain a low-flow channel (thalweg) that defines the centerline of the channel, the main channel between the stream banks and a floodplain. The bankfull flow in a stable main channel of natural streams usually has a probability of being exceeded once in two years (Leopold, Wolman, and Miller, 1992). If the bankfull flow occurs more frequently than once in two years, the channel is unstable unless the banks are reinforced by erosion-resistant vegetation or engineered natural-looking stone lining. Urbanization and global warming increase the frequency of bankfull flows, which causes bank erosion and channel enlargement until a new balance between the flows and channel cross-section is achieved (See Figure 2.5).

Based on the geomorphological/hydrological classification, streams are categorized according to their orders. *First-order streams* have no tributaries and may not have perrenial flow. *Second-order streams* have first-order streams as tributaries; *third-order streams* have second- and first-order tributaries, and so forth. Almost all first-order streams and many second-order streams might have been lost to urbanization and, in most cases, new created natural channels with rain gardens (Figures 6.7 and 6.11) could take on the role of the original first-order streams that were lost. Restored or daylighted small and medium urban streams would be second-order creeks and small rivers. The term *daylighting* refers to uncovering the buried streams, restoring and cleaning up their flows, bringing them back to the surface, and, if possible, restoring their riparian flood zones (e.g., as a park) (Novotny et al., 2010a). Figure 6.22 shows the concept and classfication of urban streams that mimic natural geomorphology. (IRRF stands for Integrated Resource Recovery Facility; see Chapter 8.)

Many stream restoration projects have been carried out by communities striving to become sustainable and green. An example featured in this treatise is Lincoln Creek in Milwaukee, Wisconsin (Figures 6.23 and 6.24). A highly publicized award-winning finished restoration of an urban stream previously lined with concrete is the Kallang River in Singapore. Ghent, Belgium, and Seoul, South Korea, have examples of daylighting of streams and canals that were covered and buried to make room for parking lots and some greenery and now are tourist attractions. Daylighting is now common in many European water-centered cities. (For extensive descriptions of urban stream restoration and daylighting, see Novotny et al., 2010a.)

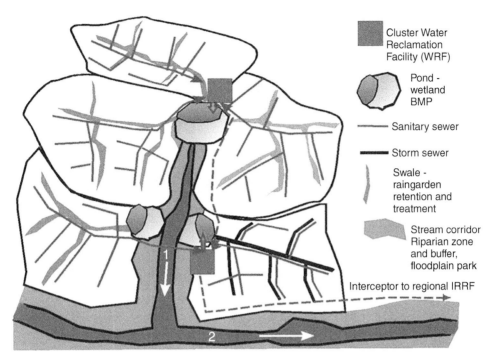

Figure 6.22. Concept of the restored urban hydrological/morphological stream system with cluster used water infrastructure and LID BMPs.

Figure 6.23. Fragmentation by culverts and habitat destruction by concrete lining of the Lincoln Creek in Milwaukee. Small depth channels with no pools are uninhabitable by aquatic biota except by some attached growth slimes. *Source:* Photo V. Novotny.

Figure 6.24. Renaturalized middle section of Lincoln Creek in Milwaukee. The stone used for protecting the channel banks was excavated during construction of the deep tunnel/interceptor for storage of sanitary and combined sewer overflows. *Source:* Photo MMSD Milwaukee, 2019, https://www.mmsd.com/what-we-do/flood-management).

Lincoln Creek was one of the first restored streams in the US that in the past had been fully converted into a concrete channel. It drains a 54 km^2 (21-square-mile) urban watershed, which is mostly located in the city of Milwaukee. Lincoln Creek watershed had undergone a period of urbanization after World War II until 1980, when it became completely urbanized. The lower part of the watershed contains architecturally interesting, desirable prairie- and bungalow-style single-family houses built in the post-1930 Depression period. Because of the increased imperviousness of the watershed, the hydrology and frequency of high flows in the creek changed (Figure 2.5) and many homes and entire sections of the city found themselves in a floodplain and subjected to frequent flooding. Between 1960 and 1997, more than 4,000 serious house flooding problems were reported along the creek. The problem was magnified by frequent combined sewer overflows in the creek. The city answer to the increased flows and frequent floods was to begin lining the creek with concrete (Figure 6.22) and straightening the channel.

However, conversion of this creek into a concrete channel met with resistance from the homeowners in the watershed. Supported by the initiatives after the Clean Water Act period, funds were obtained to embark on renaturalization of the entire creek (Figure 6.23). Although the project focused on reducing the risk of flooding for homes and businesses to the 1% probability storm event commonly referred to as the 100-year storm, it also included measures to enhance the attractiveness of the corridor; improve water quality; restore, stabilize, and protect eroding banks; and provide a suitable habitat for fish, birds, and other wildlife.

In some sections, the new floodplain storage was excavated and converted into riparian wetland. The project also built two large off-line detention basins (Milwaukee Metropolitan Sewerage District, 2012). Because the combined sewer overflows (COSs) in the Milwaukee Lincoln Creek watershed were eliminated as a part of the Milwaukee Pollution Control Program, the water quality is now good, and salmon runs from Lake Michigan are now occurring each year in fall. However, it took several years for the creek biota to reach a reasonably healthy state supporting balanced aquatic life. Obviously, the creek is not a replica of the previous water body before urbanization because the hydrology and morphology changed. There is not enough base flow in the creek, which is partially mitigated by storing water during high flows in the off-line storage basins and releasing during low flow.

Waterscapes

While the original intent of the LID/SUDS drainage was to use natural best management practice and mimic the natural hydrologic cycle, both natural and architectural man-made waterscapes can be used in medium- to high-density areas and effectively convey, treat, and provide for reuse of urban stormwater. Waterscapes, water parks, and ecological corridors provide opportunities for recreation and enjoyment as well as aesthetic enhancement of the community.

Waterscapes could be but generally are not designed to support viable balanced aquatic biota. However, as documented in Figures 5.5, 5.6, and 5.10, in addition to their aesthetic and recreational values, they may be a part of the cluster of water reuse and energy recovery loop.

The Potsdamer Platz (Figure 6.25) urban complex in Berlin encompasses a combination of green and non-green roofs harvesting 0.55 m of annual rainfall, which entirely feeds the system of water recycling needs. Captured rain is stored in underground cisterns and then used for toilet flushing, irrigation, and fire-extinguishing systems. Excess water flows into

Figure 6.25. A part of Potsdamer Platz waterscape in Berlin for conveyance of captured treated rainwater. *Source:* Photo V. Novotny.

the pools and canals of the outdoor waterscape and this water is fed into a succession of narrow pools and a larger main wetland/pond on the south side. Planted purification biotopes are integrated as a part of the overland landscape and serve to filter and circulate the water that runs along streets and walkways (Dreiseitl and Grau, 2009); the water therein, as the picture attests, is clean. From the wetland/pond, the effluent discharges into a canal.

Vertical Forests and Systems

Vertical biophilic components begin with green roofs but also include vertical green walls and forest and foliage on large buildings (Figure 6.26) (Horton 2017). The idea is not new. Vines have provided cooling and aesthetics on smaller and even medium-size buildings in hot regions of the Mediterranean and elsewhere for centuries. But the idea of "vertical forests and gardens" was born only a decade or so ago as a contrast to traditional skyscraper architecture that uses excessive amounts of glass with poor thermal isolation. The beneficial impact of green roofs on heating and cooling has been already presented in this chapter. Vertical forests and gardens envelop the entire building with designed greenery of trees, flowers, and other vegetation, as shown in Figure 6.26.

The idea of vertical forest application to large and tall buildings is attributed to Italian architect Stefano Boeri. In Boeri's concept, an equal number of plants grow on the building as they would on the ground comprising the building's footprint (Horton, 2017). The idea advanced by Boeri is "to include not only green surfaces but also trees that could reduce the amount of CO_2 and dust particles in the atmosphere." Bosco Verticale helps to build a microclimate, filters dust particles present in the urban environment, produces oxygen, and protects people and houses from the sun and from noise pollution. Boeri defined the structure as "a model for a sustainable residential building, a project for metropolitan reforestation that contributes to the regeneration of

Figure 6.26. Bosco Verticale (Vertical Forest), a building in Milan designed by Stefano Boeri. These designs are expanding in China and other countries. *Source:* Copyright Stefano Boeri Architects, Milan, Italy.

the environment and urban biodiversity without the implication of expanding the city upon the territory. They help to set up an urban ecosystem where different kinds of vegetation create a vertical environment that can also be colonized by birds and insects, and thus becomes both a magnet for and a symbol of the spontaneous recolonization of the city by vegetation and by animal life." The first vertical forest high-rise buildings were built in Milan and the idea spread to China and Singapore, including the magical Gardens by the Sea. Singapore has the highest biodiversity of any city in the world (Kenny, 2016). In China, entire vertical forest cities are planned and built (https://www.stefanoboeriarchitetti.net/en/project/forest-city/).

Vertical forests obviously need water, and using potable water from the grid would violate the water sustainability concepts advanced in this book. Because of the atypical nature of the plants that cannot have a great soil water storage, irrigation could be more frequent. There is, however, an opportunity to use treated gray water (without nutrient removal) drip irrigation cascading on the building wherein the vegetation may provide gray water post-treatment and nutrient removal. The plants must be resilient to winds.

The tenants and architects of vertical forests and gardens should also realize that vertical forests will become their own full ecosystems. Tenants must live with the birds and animals (such as squirrels) the plants will attract and must be committed to the maintenance of the ecosystem.

6.4 BIOPHILIC URBAN BIOMASS MANAGEMENT AND CARBON SEQUESTERING

Biophilic green cities full of flowers, lawns, parks, and gardens could also become large *negative emission* systems. This means that urban vegetation removes carbon dioxide from the atmosphere and in the form of alkalinity from urban waters and converts it into a lot of biomass that can be transformed into blue and green energy and the CO_2 can be sequestered. Based on the Negative Emission Management categories defined by the National Academies Committee (NAC) (2018) and reported in Chapter 4, trees, grass, flowers, roof gardens, rain gardens, aquatic flora in wetlands and ponds, and vertical forests can absorb enormous amounts of CO_2 and other chemicals like nitrogen oxides from traffic and other atmospheric pollution. These negative emissions have been characterized by the NA Committee as *terrestrial carbon removal and sequestration*, which includes the land use and management practices such as urban afforestation, changes in forest and nature management, and changes in land use practices that enhance soil carbon storage. These practices and land management and biomass energy with carbon capture and sequestration (BECCS) into soils and aquatic system (as alkalinity) are negative emissions.

The second category defined by the NA Committee applicable to the greenery of biophilic cities is *bioenergy with carbon capture and sequestration,* wherein produced urban biomass is collected and instead of being wasted or burned it is used for energy to produce electricity, liquid fuels, and/or heat, combined with capture and sequestration of any CO_2 produced when using the bioenergy and any remaining biomass carbon that is not in the green liquid and blue gaseous fuels.

Diverse urban vegetation practices related to drainage and insulation of buildings (roof gardens and urban forests) are extensively covered in this chapter. This includes grass clippings from home and public lawns, garden vegetation residues, fallen leaves, vegetation residues from urban and suburban agriculture, wood chips from trees and branches, and aquatic vegetation (e.g., cattails in wetlands). There are several methods for managing vegetation residues in biophilic cities.

Lawns and Grass Clippings

As pointed out earlier in this chapter, thirsty urban lawns are not a preferred ground cover in sustainable urban communities and, wherever possible, these should be replaced by the xeriscape or natural grasses that do not require excessive irrigation. While growing grass does absorb carbon dioxide from the atmosphere, combusting gasoline from lawn mowers releases emissions that counterbalance the CO_2 absorption. Hence, grass cutting should be reduced. If grass is still desired, it should be species that require minimum or no watering, and the grass clippings should be mulched and left in the lawn to decompose and be incorporated in the soil. In this way the nutrients in the clipping will be recycled, which will first reduce and later eliminate the need for the industrial fertilizers that themselves require a lot of fossil energy to be produced. As a matter of fact, the nitrogen and phosphorus in urban rainfall will be used by grass and accumulate in the soil if grass clippings stay in the lawn. Phosphorus and nitrogen from rainwater absorbed by grass and soil do not evaporate. Growing grass and retaining grass clippings will recycle nutrients and will have a small effect on retaining and accumulating carbon in the soils.

The third factor that could lead urban lawns to becoming negative CO_2 emitters is reducing the high use of water for watering. On average, an urban lawn needs 2.5 cm of a combination of irrigation water and rain each week during the growing season, which may last on average 4 months in temperate zones and more in warm zones. In most cases fully treated potable water is used for irrigation by homeowners. Irrigation water use is highest in the southwestern United States. The energy use to deliver and treat water to potable quality is 1.8 kW-hour/m^3, which currently is provided mostly by electricity from fossil fuel. The corresponding average GHG CO_2 emissions have been 0.62 kg CO_2/kW-h (see Chapter 5). If 50% of irrigation water is provided by potable water, the positive annual CO_2 emission per 1 m^2 of the lawn is:

$$0.5 * 16 \text{ [weeks]} * 0.025 \text{ [m of water/weel]} * 1 \text{ [m}^2\text{]}$$
$$* 1.8 \text{ [kW-h/m}^3\text{]} * 0.62 \text{ [kg CO}_2\text{/kW-h]}$$
$$= 0.22 \text{ kg CO}_2\text{/m}^2 \text{ of the lawn}$$

Adding positive irrigation water uses CO_2 emissions with emissions from lawnmowers and manufacturing fertilizer, diminishing the negative CO_2 emission by photosynthesis.

Other Vegetation

Vegetation residues such as fallen leaves, flowers, decorative plants, and other vegetation residues can be either composted or codigested with other organic solids in the proposed Integrated Resource Recovery Facility (see Chapter 8).

Composting. This is an old and popular process that can be done at the individual or communal level (University of Illinois Extension, 2018). In the composting process, organic vegetation residues are decomposed by organic bacteria that operate in *psychrophilic* (less than 25°C), *mesophilic* (25–40°C), and *thermophilic* (optimally between 50°C and 60°C and up to 70°C) ranges. The composting bacteria are strictly aerobic, which implies that composting produces heat and releases difficult-to-sequester CO_2. This is not the same nor even a similar decomposition process as anaerobic decomposition (covered in Chapter 8). Nevertheless, the emitted carbon dioxide is less than that incorporated into humus (composting product).

Composting is done in piles that should be periodically mixed. In the composting process, slow-acting psychrophilic bacteria heat the pile to the mesophilic temperature. The piles can be small, static, large, aerated, or turned windrow composting, which is suitable for large volumes such as that generated by entire communities and collected by local governments, and high-volume food-processing businesses (e.g., restaurants, cafeterias, packing plants). This type of composting involves forming organic waste into rows of long piles called *windrows* and aerating them periodically by either manually or mechanically turning the piles. The ideal pile height is about 1.25–2.5 meters, with a width of 4–5 meters. This size pile is large enough to generate enough heat and maintain temperatures, but small enough to allow oxygen flow to the windrow's core.

In the process, psychrophilic bacteria first warm the temperature of the pile to the mesophilic range. At that point mesophilic aerobic microorganisms take over and rapidly decompose the organic matter, producing acids, more carbon dioxide, and heat. When the pile temperature rises above 25°C, the mesophilic bacteria begin to die off or move to the outer part of the heap and are replaced by heat-loving thermophilic bacteria, which continue the decomposition process and the temperature will become thermophilic. As the thermophilic bacteria decline and the temperature of the pile gradually cools off, the mesophilic bacteria again become dominant. The mesophilic bacteria consume the remaining organic material with the help of other organisms. The entire process takes several months to complete. There are no pathogenic microorganisms in the humus, even when some waste materials with pathogens are composted. These pathogenic microorganisms are killed by the high temperature and other microorganisms. Several research projects have shown that soil amended with compost can help fight fungal infestations (University of Illinois Extension, 2018). Composting will yield significant amounts of compost, which might require assistance to market the end product. Local governments may want to make the compost available to residents for low or no cost (US EPA, 2010).

No biogas production.   While composting is very popular with gardeners and communities, it is not an ideal process when reduction of CO_2 emissions is necessary to control global warming. Composting emits CO_2 that cannot be sequestered but also incorporates some organic carbon from photosynthesis into humus and soils. In this sense it would be a mild negative emission process, per National Academies Committee classification. The emitted CO_2 is not classified as GHG by the IPCC. Composting also produces heat that is not currently recovered. An alternative is to send the vegetation residues after some preprocessing by compressing to the IRRF, where it would be codigested with other organic solids and produce methane and hydrogen (see Chapters 8 and 9).

Waste Woodchips.   Waste from tree cutting and trimming and scrap wood should not be burnt *en mass*. The best sustainable process is indirect gasification (Chapter 8) with other combustible municipal solid waste and organic solids from the IRRF that produces syngas (carbon monoxide and hydrogen) for making clean blue energy and biofuel.

BUILDING BLOCKS OF THE REGIONAL INTEGRATED RESOURCES RECOVERY FACILITY (IRRF)

7.1 TRADITIONAL AEROBIC TREATMENT

Typically, influent into a traditional water reclamation plant is a combined flow of black and gray water diluted by infiltration-inflow clean water additions in separated sanitary sewers. In combined sewer systems the flow is further diluted by wet weather urban runoff and winter snowmelt. Thus in the US, the influent flow has relatively low COD concentrations on the order of 300–400 mg/L; it is higher in developed countries in Europe or Asia because of markedly smaller water use and about 10% smaller COD load because houses and apartments there do not have in-sink garbage disposals.

A typical water reclamation plant consists of primary treatment, aerobic secondary treatment, waste sludge separation, sludge handling, and post-treatment that includes effluent polishing and disinfection. In the traditional activated sludge process, nutrient removal is limited to the primary sludge and uptake by bacteria in the waste sludge biomass (sludge yield). Figure 7.1 shows late-twentieth-century alternatives for biological treatment. System A presents an aerobic activated sludge membrane bioreactor (MBR). Introduction of membrane filters and converting the more traditional aeration systems followed by secondary clarifiers into MBR was a breakthrough at that time and brought more efficient treatment of used water, both total mixed flow and black water. In these systems, high-quality effluents can be reliably produced in compact units that can be installed close to the points of reuse or even in the basement of large houses (e.g., the Solaire apartment complex in New York City or Dockside Greens Development in Victoria, British Columbia, Canada; see Figure 5.7 in Chapter 5). The filtration capability of membranes is very high due to small pore sizes (0.04 to 0.4 μm) in the most common membranes capable to remove colloidal particles and microorganisms. Developing membrane solids and even smaller particles and colloid separation systems revolutionized the treatment process and eliminated the need for secondary clarifiers. The effluent quality from these systems is very high, typically exhibiting BODs and total suspended solids (TSS) concentrations of less than 10 mg/L.

Alternative B is a classic trickling filter system that uses significantly less energy than the classic activated sludge system employing aerators. Treatment efficiency of a traditional trickling filter is less than that of the traditional activated sludge aeration system; therefore,

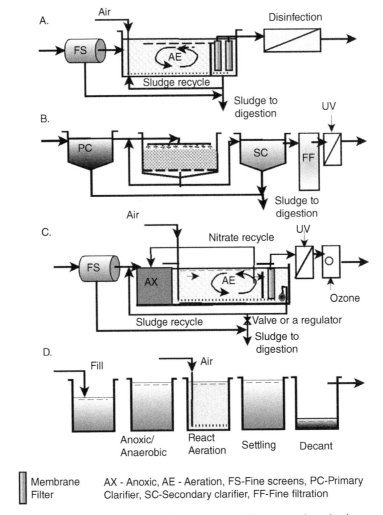

Figure 7.1. Alternative biological processes for water treatment and reuse, treating mixed sewage or separated black water, dating from the end of the previous century. (A) Aerobic membrane bioreactor, a modification of the conventional activated sludge process; (B) trickling filter; (C) Bardenpho process for BOD, TSS, nitrogen, and phosphorus removal; and (D) sequencing batch reactor. *Source:* Adapted from Asano et al. (2007) and Novotny et al. (2010a).

these systems were abandoned 45 years ago, but may have a resurgence as a post-treatment following anaerobic treatment and energy-producing systems. A more modern and popular Bardenpho treatment process (shown in Figure 7.1 as Alternative C), named after its inventor, James Barnard, begins with an anoxic process unit, followed by aerobic unit with a high recycle. The latter process oxidizes Total Kjeldahl Nitrogen (TKN = organic nitrogen plus ammonium) into nitrate (NO_3^-) in the aerobic units that is recycled back into the first anoxic unit, wherein nitrate is converted mostly to nitrogen gas (N_2). Adding an anaerobic unit up front increases phosphorus uptake by bacteria as luxury uptake that is then removed with waste sludge.

Alternative D is a sequencing batch reactor (SBR). The picture shows the phases of the operation in a single tank. For continuous flow, at least two SBR reactors will be needed.

For BOD removal only, aeration begins once the reactor is filled, followed by settling and decanting the clean supernatant. Sludge wasting occurs during the aeration period. For nitrogen removal, the filling period is followed by several periods of aeration and no aeration so that nitrification and denitrification can occur. SBR treatment can be automated. These treatment systems remove pollutants, sometimes at a great expenditure of energy, and in many cases require industrial additives (e.g., coagulating and pH-adjusting chemicals, activated carbon), and they do not recover the resources (e.g., nutrients) for reuse, with exception of dried sludge solids.

Two types of sludge are produced: (1) more biodegradable primary sludge containing high densities of pathologic microorganisms, and (2) secondary sludge or sludge, excess which is less biodegradable but contains densities of pathogens that are orders of magnitude smaller, hence it is more amenable to agricultural land disposal. Direct disposal of sludge on agricultural land producing food crops should not be allowed. Bacterial indicators such as *Escherichia coli* or fecal coliform densities cannot predict the presence and removal of all pathogens, considering that bacteria, viruses, and other resistant enteric pathogens have different survival capacities in the environment and different resistances to treatment processes and disinfecting chemicals (Gianico et al., 2015; Harwood et al., 2005). In the follow-up process of handling the residual solids, sludge is dewatered and digested, or heat treated. These processes require electric energy to run the centrifuges and belt press filters for sludge dewatering and heat energy for sludge drying and thermal disinfection and for heating mesophilic or thermophilic digesters. Sludge digestion produces methane biogas. With the increased interest in energy production and becoming self-sufficient (carbon neutral), the biogas containing methane and some hydrogen is combusted to produce energy that supplements the energy needs of the utility. In this century, some utilities have been using produced purified biogas as a fuel for their vehicle fleet. However, it will be subsequently pointed out (Table 7.1) that the energy potential of used water solids is insufficient for making the utility net-zero carbon.

Constructed wetlands (Figures 6.20–6.23; see also Chapter 9) are special biological treatment units that involve several physical and biological processes. For more detailed study and design of traditional and more modern treatment processes, the reader is referred to standard wastewater treatment texts such as those published by Tchobanoglous, Burton, and Steusel (2003); Imhoff and Imhoff (2007); Asano et al. (2007); or Kadlec and Wallace (2008). Constructed wetlands represent a promising technology because of

Table 7.1. Theoretical per capita methane energy production potential from used water.

COD per capita unit load (US, including in-sink grinders)[1]	130 g/cap-day
minus COD in the effluent (5%)	123 g/cap-day
Maximum methane produced 0.3 L/g of COD removed[2]	39 L of CH_4/day
Methane energy 9.8 W-h/L of CH_4[2]	0.38 kW-h/day
TOTAL ENERGY POTENTIAL (All COD to biogas)	140 kW-h/cap-year
Only about 33% of energy potential can be converted by combustion to electric energy (rest is heat into cooling)[3]	42 kW-h/cap-year
TOTAL AVEGAGE ENERGY USE RELATED TO MUNICIPAL WATER USE (WITHOUT HEATING)[4]	375 kW-hr/capita-year

[1] Henze and Comeau (2008).
[2] Metcalf and Eddy (2003), Van Lier et al. (2008).
[3] Eurelectric (2003).
[4] US EPA (2010).

their zero-energy use and potential for recovery of vegetation for energy production and nutrient recovery. They are especially suited for conditions in developing countries and are popular in many developed countries for used water treatment in smaller communities. However, if improperly designed they could emit methane (Kadlec and Wallace, 2008; Mitsch and Gosselink, 2015), but this can be mitigated (see Chapter 9).

Maximum theoretical energy yield from the traditional systems is very small and does not even cover the energy demand of the treatment plant (Table 7.1). The calculation in this table presumes that all COD will be converted to biogas, which is not the case even in the best-functioning current anaerobic treatment plants. A large part – 30 to 60% of influent COD in aerobic treatment plants – is converted in the aeration basin into CO_2 and not to CH_4.

GHG Emissions from Traditional Regional Water/Resources Recovery Facilities

Traditional wastewater treatment and water reclamation plants do not produce excess energy; as a matter of fact, they are relatively high energy users. In addition to electric energy use that is responsible for GHG emissions of CO_2 emitted from the fossil fuel power plant, a typical traditional aerobic wastewater treatment facility could emit GHG nitric oxides and some methane, which are byproducts of the removal of the organic compounds in the treatment process and sludge digestion. Apparently, IPCC (2007) panel justifiably did not include the CO_2 emitted from degradation of organics in the traditional wastewater treatment plants into its greenhouse emissions balance. Theoretically, in a well-functioning plant, CO_2 emissions are less than CO_2 assimilated by the crops from which consumed food was processed and used water formed in a very short antecedent period of few months or years. The difference between the CO_2 assimilated and emitted is due possible incorporation of solid residues into soil or landfill where degradation may last centuries and turns the solids into peat. Greenhouse gas (GHG) designation is for emissions from household stoves, furnaces, combustion engines, boilers for electric power generation, and thermal plants burning and combusting fossil fuels formed millions of years ago.

However, organic matter decomposition in traditional aerobic treatment processes also produces methane in digestion of sludge and nitric oxides that are byproducts of nitrification and denitrification. The global warming potential (GWP) of methane is 25 over 100 years, that of nitrogen oxides is more than 300, while that of carbon dioxide is 1 (IPCC, 2007). Methane is produced by anaerobic processes, which are an integral part of treatment, specifically in anaerobic digesters but also in dead zones of primary clarifiers and sludge handling, and small amounts of gaseous nitric oxides are intermediate by-products in nitrification and denitrifications processes. Older treatment plants simply let methane and nitrogen oxides escape into atmosphere. If the digester gas is flared or, better, converted to energy, the emitted CO_2 may not be considered a GHG. Besides CO_2, Gupta and Singh (2012) also estimated methane and nitric oxides emissions from aerobic sequencing batch reactor (SBR) treatment plants in India and found that $CO_{2\text{-eq}}$ of emitted methane and NO_x is approximately 5% of that attributed to the energy use by the water reclamation plant provided by a fossil fuel power plant.

Monteith et al. (2005) investigated 16 Canadian wastewater treatment plants and found that the principal GHG type emission was carbon dioxide with a small fraction of methane. Emission rates ranged from $0.005\,\text{kg}\,CO_2/\text{m}^3$ of treated wastewater at primary-treatment-only facilities, $0.26\,\text{kg}\,CO_2/\text{m}^3$ for conventional activated sludge with anaerobic sludge digester, to $0.8\,\text{kg}\,CO_2/\text{m}^3$ for extended aeration with anaerobic sludge digestion. While IPCC (2007) did not include these emissions in their GHG estimates, reducing them would provide the utility with a credit that could counterbalance the CO_2 emissions due to use of fossil energy that are considered GHG. In the simplest way, this is being achieved by converting methane from digestion of biodegradable organic matter in the incoming organic compounds into energy used to run the treatment plant (Keller and Hartley, 2003). Keller and Hartley estimated that in a fully aerobic process up to 1.4 kg CO_2 per kg of removed COD originates from power generation and all this energy is derived from fossil fuel. As shown in Table 7.1, the energy recovery from combusting digestion methane is not sufficient to cover the energy needs of a typical aerobic treatment plant.

7.2 ENERGY-PRODUCING TREATMENT

Because of water conservation savings at the home (building) and cluster level of recycling (Figures 5.5–5.6) in future communities, black water plus reject water flowing into the new integrated resource recovery facility (IRRF) will be concentrated with COD, most likely exceeding 1000 mg/L (Henze and Comeau, 2008). Accordingly, the core of the future IRRF is anaerobic treatment and resource recovery. Conventional aerobic treatment processes are incapable to handle high concentrations of biodegradable COD; on the other hand, anaerobic treatment processes are inefficient for low concentration diluted wastewater entering current "water reclamation plants." The effluent BOD (COD) from anaerobic treatment may be elevated because it may contain the residual dissolved methane, which is measured in chemical analyses as BOD and COD. Because of smaller volumes of concentrated used water entering into existing sanitary or combined sewers, there may be problems with solids deposition and methane formation in sewers and in some cases sewers may have to be modified; for example, this has been done by inserting plastic pipes into existing sewers and renting the remaining cross-sectional area to other utilities (cable or phone companies) in Tokyo (personal communication by Professor Shoichi Fujita, former director of Tokyo Sewerage Bureau). Vacuum sewers have also been installed in numerous locations in Europe, Middle East, Alaska, and some African countries. They have also been considered, but as of 2017 not yet installed, in Venice, Italy (en.wikipedia.org/wiki/Vacuum_sewer). Distributed used water management and reuse is considered herein.

Anaerobic Digestion and Decomposition

Anaerobic digestion is a process in which concentrated biodegradable solids and liquids are degraded in the absence of oxygen by anaerobic and facultative bacteria. These processes take place in the anaerobic treatment reactors such as upflow anaerobic sludge blanket (UASB) or anaerobic fluidized bed membrane (AFBM) reactors or in sludge digesters and to some degree also in microbial fuel cells, in wetlands and even in septic tanks. The anaerobic digestion process (Figure 7.2) progresses in four stages (Van Lier et al., 2008):

1. Hydrolysis of proteins, polysaccharides (cellulose), fats and solids into organic acids, sugars, and alcohols
2. Acidogenesis/fermentation and oxidation of amino acids, sugars, alcohols, and fatty acids into hydrogen and carbon dioxide, and intermediate products

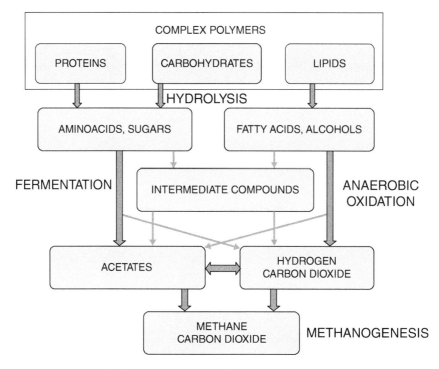

Figure 7.2. Anaerobic digestion schematics. *Source:* Adapted from Van Lier et al. (2008).

3. Acetogenesis formation of acetic acid acids into acetates, acetic acid, and hydrogen with an intermediate product of volatile fatty acids
4. Methanogenesis in which acetates are broken down to methane and hydrogen and carbon dioxide are scavenged by methanogenic microorganisms to form methane

The fourth step, methanogenesis, is endothermic and requires heating, pH above 7, and longer hydraulic retention time HRT, otherwise acetates and hydrogen are the "dead-end" products of fermentation. The conversion of organics to acetic acid, acetates and hydrogen is sometimes called *dark fermentation*. Van Lier et al. (2008) and other authors point out that COD is not removed in the first three digestions steps because hydrogen and carbon dioxide produced by fermentation are used by methanogens to form methane. The removed COD is in the methane because it is a gas that has a relatively low solubility. At pH above 7, CO_2, which has a high solubility, is converted to the bicarbonate ion (HCO_3^-), which is not removed by volatilization.

Methanogenesis is performed by a specific group of bacteria, methanogens, which use acetates and hydrogen for their growth, converting the final products of dark fermentation into methane. Their growth rate is very slow, resulting in doubling times to days or weeks, which requires long start-up times for digesters. In contrast, the fermenting and acidifying bacteria have growth rates that are five times faster than methanogens. Also, methanogens require higher pH, while acidifying bacteria are active even at pH down to 4 to 5 (Van Lier et al., 2008). The apparent problem with the methanogenesis is a subject of new research that suggests bypassing the methane formation and focusing on converting anaerobically COD to hydrogen (and CO_2) as it will be discussed in the subsequent section on microbial

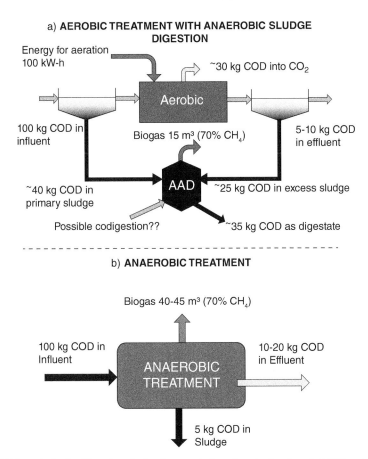

Figure 7.3. COD flows in the traditional centralized concept of wastewater treatment; AAD is anaerobic digester. *Source:* Bartáček and Novotny (2015) and Van Lier et al. (2008).

fuel cells. The anaerobic dark fermentation cannot easily break down complex carbohydrates such as cellulose and proteins but can decompose algal biomass by co-digestion with sludge. Figures 7.3a and b compare the COD mass balance in the aerobic and anaerobic treatment processes. The comparison of the two concepts shows the advantages, along with a few drawbacks, of making a paradigm shift from aerobic treatment to anaerobic.

Chemical Oxygen Demand of Organics. The organic content of the used water and industrials wastewater has been measured and expressed as Chemical Oxygen Demand (COD), Biochemical Oxygen Demand (BOD), or Total Organic Carbon (TOC). Van Lier et al. and other authors argue that for anaerobic processes (and the same arguments can be used also for aerobic processes), COD has been accepted over the last 25 years as the most adequate parameter to quantify the concentration of organic materials, better than BOD_5 or TOC. If the chemical composition of the organic waste or other materials is known in the most common form as $C_mH_nO_p$, COD can be calculated by the following equation, assuming that the organic matter is fully oxidized:

$$C_mH_nO_p + (m + n/4 - p/2)\, O_{2-} \rightarrow m\, CO_2 + n/2\, H_2O$$

$$\leftarrow \quad COD \quad \rightarrow$$

The equation above is expressed in moles. Atomic weight of carbon = 12, that of hydrogen = 1, and oxygen = 16, respectively. All are in grams. For example, the weight of one mole of hexose $(C_6H_{12}O_6)$ is:

$$MW = 6 * 12 + 12 * 1 + 6 * 16 = 180 \text{ grams}$$

The COD of one mole of hexose then will be:

COD = $(m + n/4 - p/2) * 32 = (6 + 12/4 - 6/2) * 32 = 192$ grams, or:
COD of one gram of hexose is COD = $1 * 192/180 = 1.07$ g O_2.

Using the $COD/C_mH_nO_p$ molar ratio, COD of 1 gram of methane(CH_4) would have $COD_{CH4} = (1 + 4/4) * 32/16 = 4$ g of O_2. $COD/C_mH_nO_p$ ratio for acetate $(C_2H_4O_2)$ is also 1.07.

Comparison of Aerobic and Anaerobic Treatment and Energy Recovery (Use) Processes

Because the secondary excess sludge separated from the mixed liquor of the traditional classic aerobic activated and anoxic/oxic Bardenpho processes (Figure 7.1C) is significantly less biodegradable than primary sludge, it makes sense to maximize in the traditional system the COD removal in the primary settling units and process the primary sludge separately along with the sludge yield in the secondary units (Figure 7.3a). The separation of primary sludge could be enhanced by precipitation with ferric chloride, alum, polymers, or other coagulants but adding these inorganic coagulants increases inert mass of the sludge. Jeníček et al. (2012) have shown that enhanced primary treatment can remove 55–60% of the total suspended solids and up to 30% of the COD load could be then transformed into methane; however, virtual energy (energy used to produce and deliver the chemicals to the facility) and related GHG emissions must also be considered in the life-cycle assessment. Primary settling may not be needed if the treatment reactor(s) is anaerobic and heavier inert inorganic particles (for example, sand) are removed by pretreatment. The specific gravity of organic particles is about 1.1 g/cm^3, while that of sand particle is about 2.65 g/cm^3.

The digestate in Figure 7.3a includes COD and nutrients in the highly concentrated liquid supernatant that could be returned to the influent, and in the residual waste solids that must be dewatered and disposed. ANAMMOX unit and/or struvite recovery units could be installed on the supernatant return line. As of the second decade of this century, this is currently the most common system in the more advanced water reclamation plants.

The COD balance in Figure 7.3a indicates that in the current aerobic (activated sludge) or anoxic/oxic (denitrification/nitrification) unit processes, only about 30% of the incoming COD in the used water could be converted to biogas for energy production, of which about 70% is methane. The efficiency of the biogas conversion into electric energy in the traditional process is about 30%; hence the energy recovery in the current water reclamation plants is around 10%. Based on the calculations in Table 7.1 the recoverable energy would be only about 14 kW-hr capita^{-1}year^{-1}, which is minor when compared to the energy used for aeration, which is about 1 kW-hr/1kg of incoming COD or 48 kW-hr capita^{-1}year^{-1} (Van Lier et al., 2008). Table 7.1, based on US EPA (2010) data, shows the total average energy use related to municipal water use is 375 kW-hr capita^{-1}year^{-1}. In contrast, the biogas yield from anaerobic processes is (Figure 7.3b) about 80% of the incoming

COD. The gas yield is then 0.4 to 0.45 m^3/kg of COD or 0.3 kg of CH$_4$ per kg of COD removed.

The full anaerobic treatment recovers far more energy than the traditional aerobic treatment with sludge digestion (see Chapter 7). The generic anaerobic treatment system – represented in Figure 7.3b by a red box, based on the estimates in Van Lier et al. (2008) – produces three times more biogas without using appreciable energy from fossil fuel than the best aerobic treatment process with the digester. The other advantages of anaerobic treatment and resource/energy recovery outlined by the above authors are:

- Reduction of excess sludge production by about 90%
- Smaller volumes of treatment reactors with a corresponding reduction of space for the facility
- Producing about 13.5 MJoule CH$_4$ energy/kg COD removed, which may yield 1.2W-h electricity in the conventional turbine gas engine/generator energy-producing system
- No use of chemicals
- Excess sludge has a market value
- High rate of anaerobic system facilitates water reclamation and recovery and implementing hybrid loops of recycle

A disadvantage of the conventional anaerobic water/energy reclamation system is that it operates best in the mesophilic temperature range (between 25 and 45°C; the optimal temperature is 37°C). However, producing electric energy provides enough heat to keep the reactors mesophilic even under cold ambient temperatures. Furthermore, a significant portion of the added heat to keep the reactors mesophilic is recoverable from the effluent by a heat pump or heat exchanger.

Another problem is the residual dissolved methane in the effluent (Figure 7.4). As calculated previously, the COD of 1 mg of CH$_4$ = 4 mg of O$_2$. Methane can be oxidized by the aerobic and facultative sewage microorganisms, which increases BOD and COD of the effluent. The higher COD and BOD in the effluent caused by dissolved methane concentrations warrant aerobic post-treatment such as trickling filter and/or using activated granular

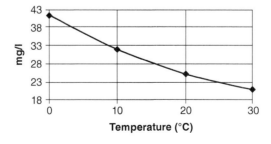

Figure 7.4. Methane solubility in pure water. *Source:* Yamamoto et al. (1976).

carbon with membrane filters, or simply stripping it of methane and sending it to gas storage (McCarty, Bao, and Kim, 2011). Stripped methane cannot be released; at worst it could be flared but this would be a waste of a resource and money. Methane solubility is strongly affected by temperature (Figure 7.4).

Another obvious drawback of anaerobic resource recovery facilities is the smell, which requires all reactors to be enclosed. Having odor controls may be optional because the emitted gas is collected for energy production in closed systems and cleaned. For maintaining good conditions for methanogenesis, alkalinity may need to be added.

Acid Fermentation and Its Hydrogen Production

With the increasing emphasis on hydrogen, several research studies focused on maximizing production of hydrogen in the anaerobic decomposition processes instead of methane. Theoretically, anaerobic breakdown of one mole of hexose could yield 12 moles of H_2 (Logan, 2008; Logan et al., 2008; Van Lier et al., 2008), or stoichiometrically:

$$C_6H_{12}O_6 + 6\,H_2O \rightarrow 12\,H_2 + 6\,CO_2$$

This partially explains the interest of scientists and energy agencies in hydrogen. Most of the hydrogen on the market is produced by the processes fundamentally described in the next sections (i.e., gasification of coal and oil to syngas and steam reforming syngas and natural gas to hydrogen). Electrolysis of water to hydrogen and oxygen is less energy efficient but is also acceptable and often more convenient if the energy source for electrolysis is an excess blue or green energy source.

Hydrogen is produced in the anaerobic digestion process in the second *acidogenesis* and third *acetogenesis* steps of the anaerobic digestion process (Figure 7.2). Acidogenesis and acetogenesis processes further break down the organic compound from hydrolysis and produce acetates, volatile fatty acids (FVA), bicarbonates, and hydrogen (H_2) gas and hydrogen ion (H^+), which implies a reduction of pH. This is the fastest process in the anaerobic digestion food chain. Acidogenic microorganisms could function at pH as low as 4 but their optimum is in pH range between 5 and 6. Acetogenic microorganisms convert short-chain fatty acids, alcohols, and lactate into acetate, hydrogen gas, and carbon dioxide. The following are examples of hydrogen-producing fermentation reactions (Van Lier et al., 2008).

Acetic acid (acetate) production from hexose:

$$C_6H_{12}O_6 + 3H_2O \rightarrow 2CH_3COO^- + 2HCO_3^- + 4\,H^+ + 3H_2$$

Butyric acid formation and conversion to acetate:

$$C_6H_{12}O_6 \rightarrow CH_3CH_2CH_2COO^-H + 2CO_2 + 2H_2$$

$$CH_3CH_2CH_2COOH \rightarrow CH_3COO^- + H^+ + 2CO_2 + 2H_2$$

Acetate and hydrogen from ethanol:

$$CH_3CH_2OH + H_2O \rightarrow CH_3COO^- + H^+ + 2CO_2 + 2H_2$$

Sucrose to acetic acid and hydrogen:

$$C_{12}H_{22}O_{11} + 9H_2O \rightarrow 4\,CH_3COO^- + 4HCO_3^- + 8H^+ + 8H_2$$

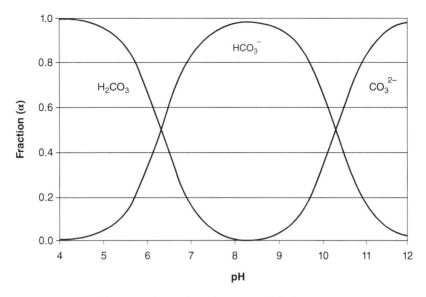

Figure 7.5. Inorganic carbonate equilibria in water.

Figure 7.5 shows the carbonic acid/carbonate equilibria. The fermentation phase reactor with pH at or below 6.5 is separate from the methanogenic reactor that has pH above 7. Then the molar content of dissolved carbonic acid and bicarbonate in the fermentation reactor is about the same. However, in the methanogenesis with pH > 7 inorganic carbon is mostly in the form of bicarbonate alkalinity.

Generation of other acids in the fermentation process (propionic, lactic) and alcohol does not produce hydrogen. Therefore, theoretically, the barrier for fermentation of glucose (hexose) is either 2 or 4 moles of H_2 to be produced by fermentation of one mole of hexose (glucose). However, the real hydrogen yield by fermentation of glucose is less than the stoichiometric yield, about 2–3 moles of H_2/mole of glucose (Cheng and Logan, 2007) or even less for mixed wastewater, as confirmed by the Paudel et al. (2015), who recovered only 1.01–1.025 moles of H_2/mole of glucose from mixed organic wastewater in an anaerobic reactor operated at pH 5 to 5.5.

Figure 7.6 presents carbon dioxide solubility under the normal carbon dioxide partial pressure. It indicates that the CO_2 formed by fermentation may be in dissolved gas and carbonate forms and not be emitted, unless the fermentation process is very intensive, as happens in fermentation-producing alcoholic beverages by yeast wherein glucose and fructose concentrations typically range from 160 to 300 g/liter. Added dissolved CO_2 by fermentation reduces pH.

Acidogenesis reactions are the fastest reactions with the highest free enthalpy (ΔG) of all anaerobic decomposition reactions (Van Lier et al., 2008). Therefore, the acid fermentation reactors in a two-step digestion process have much shorter HRT (in hours) than the methanogenesis reactor (days or weeks). Also, these former reactors work satisfactorily in psychrophilic (less than 25°C) temperature range and may require less or no heating, while methanogenesis works best in mesophilic (25°C to 40°C) or thermophilic (optimal range 50°C to 60°C) ranges and requires heating.

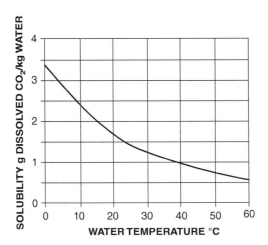

Figure 7.6. Solubility of carbon dioxide in water at different temperatures. *Source: Engineering tool box*, http://www.engineeringtoolbox.com/gases-solubility-water-d_1148.html.

In the fermentation and oxygenation phases where acetates, hydrogen, and carbon dioxide are formed, acetates cannot be converted by microorganisms into hydrogen. Hydrogen and acetates are the final (or dead-end) product of the first three phases and methane is produced in the final methanogenesis step. The fermentation reaction equations presented above document that, theoretically, two to four molecules of hydrogen can be produced per mole of substrate entering the fermentation digestion process, which is far from the hydrogen potential of 12 molecules. Hence, acetate formation is a barrier to hydrogen formation (Cheng and Logan, 2007 and 2008).

Antonopoulou et al. (2008) postulated that hydrogen production by fermenting bacteria is strongly dependent on the conditions of the process, such as pH, hydraulic retention time (HRT), and gas partial pressure, which affect the microbial metabolic balance and subsequently the fermentation end products. The dominant metabolism in a mixed acidogenic culture depends strongly on the pH of the microbial culture. Quoting literature sources, they also confirmed that maximum hydrogen yields in the fermentation process are achieved when the pH of the culture medium is between 5 and 6. Therefore, researchers speculated that preventing methanogenesis by lowering pH to less than pH < 6 would maximize the hydrogen yield but these attempts were not fully successful. Antonopoulou et al. concluded that digestion processes producing hydrogen do not significantly reduce the organic carbon content of the feed. Usually, chemical oxygen demand (COD) removal is below 20% during the hydrogen production process, which corresponds to a mean hydrogen production of 2.5 mole H_2/mole glucose. The low H_2 yield was confirmed by Paudel et al. (2015), who, after 20 days of acclimation period in the digester processing low- and high-strength synthetic wastewater with decreasing pH from 7 to 5.5, obtained 1.2 moles of hydrogen per one mole of glucose removed. The COD removal in the fermentation digester was also low, about 20–30% for the high load process; hence, the waste (sludge and other organic solids) would require additional treatment by full digestion, and both COD removal and hydrogen production would not be efficien t or economical.

Thompson (2008) studied hydrogen production by anaerobic fermentation of synthetic wastewater (SW), dairy manure, and cheese whey. These three constituents were combined under anaerobic conditions in batch reactors to determine the optimal H_2 production and COD removal. Thomson also found that the optimal pH for hydrogen production is pH = 5.5, which confirms the previous results mentioned herein. Anaerobic acidic fermentation of manure only did not produce any hydrogen, but when manure was mixed with the synthetic wastewater containing glucose or with cheese, hydrogen was produced. The hydrogen yield for mixes of SW and manure and SW cheese ranged between 0.29 to 0.41 L/g of COD removed. However, the COD removals in the acidic fermentation were low. COD removals for manure and SW mixtures were between 2.8 to 9.8% and those for cheese whey and SW mixtures were ranging between 4.4 to 12.5%. It was proven again that this step does not remove COD but it preprocesses the solids for the next step, which could be the traditional higher pH methanogenesis reactor or microbial fuel cell (see next the chapter).

The low H_2 yield by dark fermentation was attributed to the "fermentation barrier" forcing formation of organic acids along with some hydrogen as the end products of fermentation (Rabaey et al., 2005; Pham et al., 2006; Chang and Logan, 2007; Logan et al., 2008; Cotterill et al., 2016). However, these authors pointed out that the acetate barrier can be removed by abiotically manipulating electron flow, which is the foundation of microbial electrolysis cell presented in the next chapter, Section 8.4.

Methanogenesis is the fourth step in the anaerobic digestion and decomposition of biodegradable organic compounds. There is no sharp division between the methanogenesis and acidogenesis if the process is contained to one reactor; however, the methanogenesis reactions are endothermic and require heating in most cases. The methanogenic microorganisms are also sensitive to pH and alkalinity and need pH > 7 to develop and grow. In this phase, methanogenic microorganisms both reduce carbon dioxide using hydrogen as electron donor and decarboxylate acetate to form methane (Metcalf and Eddy, 2003; Van Lier et al., 2008). The final products of digestion are then methane (CH_4), carbon dioxide (CO_2) that could dissociate to bicarbonate alkalinity, ammonium/ammonia (NH_3), hydrogen sulfide (H_2S), and water. Methanogens first scavenge carbon dioxide and hydrogen and after carbon dioxide reduction to bicarbonate alkalinity methanogens form acetates (Van Lier et al., 2008) and this affects pH:

$$CO_2 + H_2O \leftrightarrow H^+ + HCO_3^-$$

$$2\,HCO_3^- + 4\,H_2 \rightarrow CH_3COO^- + 4\,H_2O$$

Produced acetate, along with the acetates transferred from the fermentation step, is then converted by methanogens to methane, which affects the bicarbonate alkalinity pool:

$$CH_3COO^- + H_2O \rightarrow CH_4 + HCO_3^-$$

If fermentation and methanogenesis are compartmentalized, CH_4 produced in the methanogenesis compartment result in COD removal. Since only about 3 molecules, maximum 4, and not 12 molecules of H_2 can be produced in the fermentation compartment due to the acetate formation barrier, the maximum theoretical COD removal in the fermentation compartment is about 25%. Consequently, if COD removal is required, fermentation must be followed by another COD removing process, which would be methanogenesis, or microbial electrolysis cell, both combined with membranes. If the entire digestion process takes place in a single reactor, methanogens scavenge CO_2 and hydrogen to produce methane. However, per molar basis, reducing 1 mole of CO_2 to methane would use 4 moles of hydrogen, wherein only 1 mole of H_2

may be available. This leaves excess of 3 moles of dissolved CO_2 in the effluent that would eventually be converted to carbonic acid and reduce pH, but this does not affect alkalinity.

Anaerobic Treatment

In their seminal work, Professor McCarty with collaborators from Inha University, South Korea (McCarty, Bao, and Kim (2011), compared traditional aerobic activated sludge treatment with anaerobic energy recovery systems, shown in Figure 7.7. They outlined the potential of both systems to produce energy and the treatment capacity. In the traditional aerobic system, energy loses and uses are significant, beginning with the conversion of higher energy carbohydrates into methane, which is the main energy output and carrier. This loss represents about 8% of the initial energy potential. Another 7% of energy is lost in converting COD into microorganisms necessary to carry out the reactions, and another 5% is lost because the of efficiency of the treatment plant is not 100%. That implies that the produced CH_4 contains only about 80% of the original potential. Considering that the electric energy yield by combustion from the produced methane is about 30% (Figure 7.8), the rest being heat, the overall electric energy yield even from a well-functioning traditional water reclamation plant would be less than 25%.

The suggested hypothetical but very realistic and more energy-efficient anaerobic treatment and energy recovery plant was shown in Figure 7.7. All components included in the schematic are based on well-known and tested technologies. In the comparison, the anaerobic treatment plant produces twice as much methane, less than half the sludge solids, and more energy than it uses.

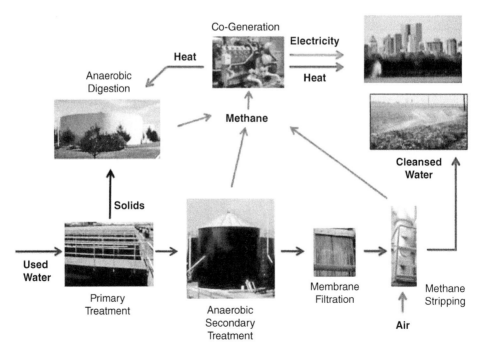

Figure 7.7. Anaerobic fluidized bed pilot treatment and energy recuperation treatment plant. *Source:* Reprinted with permission from McCarty, Bao, and Kim (2011); copyright © American Chemical Society.

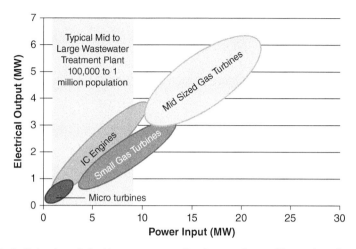

Figure 7.8. Typical efficiencies of electric power generation from methane of internal combustion engines and turbines. *Source:* Knight et al. (2015).

7.3 TRIPLE NET-ZERO: COF FUTURE DIRECTION AND INTEGRATED RESOURCE RECOVERY FACILITIES

Integrated Resource Recovery Facility (IRRF), Resource Recovery Plant, and similar terms are new names for new or reconstructed regional wastewater treatment and water reclamation plants that may also include processing of liquids other than municipal and industrial used water and resource recovery from waste solids by pyrolysis and gasification. Including incinerating solids in the integration is not considered. First the engineering departments and later whole societies have begun to realize that used water and municipal solid waste (MSW) are not "waste," but are sources of energy and resources. This is the foundation of the paradigm shift from "use, waste, and dispose" to fifth paradigm of "conserve, recover, and reuse."

The integrated resource recovery facility (or integrated facilities) will provide economic recovery of water, energy, nutrient fertilizer, and other potential resources (char, biofuel, soil conditioners, possibly purified trace metals, and others) not only from the concentrated used water and sludge produced by the municipality but also from organic solids such as vegetation residues, and highly concentrated liquids such as food waste, glycols from airport deicing, and possibly manure from animal operations. Other waste materials that are valuable resources include difficult-to-degrade combustible organic waste such as wood, paper, plastic, or tires that today may be processed and disposed separately (in many instances) but the final resource recovery can be integrated with the used water and biodegradable waste solids. Practically, while less polluted gray water processing and recycling is possible on the cluster level in small packaged automated treatment and recovery plants, a more sophisticated regional facility is needed because co-processing residual high-concentration solids (black water and sludge) and biogas recovery may not be advisable in the simpler building or cluster reclamation facilities and may be objectionable to the citizens living nearby in the cluster. Certain reuse and reclamation processes are not suitable for house or cluster applications, such as generating large volumes of explosive methane and poisonous carbon monoxide. Also, resource recovery is best accomplished on a larger scale with skilled labor operation and sophisticated smart controls. Interestingly, once the organics from used water and MSW solids are converted to biogases (methane and syngas), conversion of the produced gases into hydrogen, sensible heat and electric energy are possible on the cluster level in compact and clean hydrogen fuel cells (Whitlock, 2018). However, reforming the biogas into hydrogen in smaller operations would emit CO_2, which would be difficult to sequester on building or district scale but could be sequestered and even become a resource in more sophisticated resource recovery operations. (See Chapter 8.)

Goals of the Future IRRFs and Enabling Technologies

The main currently achievable objectives of the future IRRF are:

1. Treating and reclaiming used water for:
   a. Ecological flow of the receiving water body
   b. Providing clean water to downstream uses for irrigation, water supply from alluvial deposits, and recreation
   c. Groundwater aquifer recharge after additional treatment
   d. Nonpotable water reuse in buildings for toilet flushing, shower, and laundry (after mixing with fresh water)
   e. Street flushing and car wash facilities
   f. After high-efficiency treatment and longer underground and surface storage with some dilution and additional treatment, recovered used water can be used for potable uses.
2. Recovering fertilizer phosphorus and nitrogen:
   a. Chemically as struvite, or precipitated calcium phosphate by adding iron salts or calcium or magnesium oxides (hydroxides), and considering recovery of ammonium as ammonium sulfate fertilizer by reaction with sulfuric acid
   b. As nutrient-rich solids from digesters and/or from pyrolysis/gasification processes
3. Providing water, nutrients, and carbon dioxide (alkalinity) to algal aquaculture producing biomass for additional biofuel and energy production and additional carbon dioxide sequestering
4. Recovering sensible heat and producing energy for heating the anaerobic treatment and fermentation units as well as the facility and buildings in surrounding urban areas
5. Producing biogas that may include methane and/or syngas (a mixture of carbon monoxide and hydrogen) and/or hydrogen
6. Converting biogas and hydrogen into electricity in the IRRF power plant component
7. Deriving all energy needs from on-site energy recovery and additional renewable sources (solar and wind) in the IRRF power plant, sequestering carbon and selling the excess energy in the form of biogas, hydrogen, biofuel, and electricity
8. Additional biodegradable waste separated from the municipal solid waste (MSW) and commercial sources (food production, glycols from airports) or agricultural solids can be co-processed and/or co-digested with the residual sewage solids to produce more biogas. hydrogen and energy.
9. Unlike the CO_2 in flue gas from incineration, which is highly diluted by nitrogen and atmospheric pollutants and contains toxins, CO_2 released from syngas, methane reforming, and hydrogen fuel cells is concentrated and may be considered as a raw material for production of dry ice or sequestered regionally in underground CO_2 sequestering formations or commercially used. CO_2 is not released from fermentation and anaerobic treatment processes.

In the US, Water Environment Research Foundation (WERF) sponsored research that in 2012 revealed only about 25% of the municipal water reclamation plants (wastewater treatment plants) with a treatment capacity greater than 3,800 m^3/day ($\approx$1 mgd = megagallons per day) employed anaerobic digestion of sludge (Beecher and Qi, 2012) and only a fraction thereof of recovered energy in a form of electricity and biogas. In the US, almost no municipal WWTPs in the last century used anaerobic treatment for municipal used water; nevertheless, anaerobic treatment has been common as a pretreatment for high-strength industrial wastewater (brewery, food processing, stockyards, etc.). In contrast, in the first decade of this millennium, Holland (population 17 million) constructed 2,500 full-scale registered anaerobic treatment operations; the number was exponentially increasing, and many such facilities were used for symbiotic treatment of domestic and agricultural wastes (Van Lier et al., 2008). A similar situation can be seen in Denmark (population 5.7 million). In these two countries, with large dairy cattle operations, soils are becoming overloaded by manure and losing their capacity to adsorb nitrogen and phosphorus, which leads to severe groundwater and surface water pollution. Anaerobic treatment of used water has been used in some tropical countries, such as in Brazil or Indonesia.

The new IRRF linked to the decentralized used water management and to symbiotic processing of MSW solids and some industrial and agricultural organic waste, should not release pollution (GHG and other emissions) and should produce excess electric and heat energy. These facilities could be a net sequester of carbon GHG and recover resources such as nutrients or even commercial carbon dioxide. Good reviews of the state of the art and future outlooks have been presented by McCarty et al. (2011), Verstraete et al. (2009, 2010), Metcalf and Eddy (Asano et al., 2007), Novotny et al. (2010a), and Holmgren et al. (2016). Laboratory and field-tested technologies that propose this revolutionary resource recovery system include:

- New developments of the more than century-old anaerobic treatment and digestion of organic solids in upflow anaerobic sludge blanket (UASB), extended granular sludge blanket (EGSB) reactors (Lettinga and Hulshoff-Pol, 1991; Verstraete et al., 2009; Van Lier et al., 2008), anaerobic membrane bioreactors, anaerobic fluidized bed AFBR (Stuckey, 2012; van Lier et al., 2008), and fluidized bed membrane bioreactors (AFBMR) (Kim et al., 2011; McCarty et al., 2011).

- Recently discovered Anammox (anaerobic ammonium oxidation) microbiological process by which bacteria transform ammonium (NH_4^+) by oxidation with nitrite (NO_2^-) into nitrogen gas (N_2) and water (H_2O) without adding organic carbonaceous sources. Since full nitrification to nitrate (NO_3^-) followed by denitrification is avoided, Anammox needs less energy for aeration than the energy needed for the traditional nitrification-denitrification processes. Also, the reaction has a smaller sludge yield (Van Niftrik and Jetten, 2012).

- Co-digestion of used water excess sludge with other organic solids and high concentration liquids (food wastes, organic industrial wastewater, beverage production, vegetation residues, manure, stillage, aircraft deicing glycols) (Zitomer et al., 2001, 2008; Zitomer and Adhikari, 2005; Hamilton, 2012).

- Focus on and development of acid (dark) fermentation-producing acetates and hydrogen as a precursor to more energy recovery in the form of hydrogen in microbial electrolysis cells (Benemann, 1996; Hallenbeck and Benenmann, 2002; Kim, Han, and Shin, 2003; Nandi and Sengupta, 1998; Chang and Logan, 2007; Logan, 2008, Heaney, Wright, and Sample, 2000; Antonopoulou et al., 2008).

- Promising development of microbial electrolysis cells (MEC) and bio-electrically assisted microbial reactors (BEAMR) converting biodegradable organic matter to hydrogen (Ditzig, Liu, and Logan, 2007; Logan, 2008; Logan et al., 2008; Mohan Venkata et al., 2008; Arends and Verstraete, 2012).
- Fast advances in developing and implementing hydrogen fuel cells with steam methane and syngas reforming, which convert biogas (methane and syngas) to hydrogen and electricity (US DOE, 2016a; Dowaki, 2011; Fuel Cell Energy, 2013).
- Reemergence of pyrolysis and indirect gasification producing heat, hydrogen-enriched syngas (mixture of hydrogen and carbon monoxide), and biofuel as an alternative to less efficient and polluting incineration for processing organic solids and sludge, and making hydrogen (Dowaki, 2011; Panigrahi et al., 2002; Ahmed and Gupta, 2010; He et al., 2009; Lee et al., 2014, Callegari et al., 2018).
- Heat/cooling energy recovery from water by heat pumps and other sensible heat and cooling energy reclamation devices.
- Production of struvite (ammonium magnesium phosphate) and calcium phosphate fertilizers from used water effluents and digester supernatants (Barnard, 2007; Negrea et al., 2010; Wang et al., 2005; Holmgren et al., 2016).
- Possible recovery of ammonium as ammonium sulfate fertilizer and or soil conditioner (Jiang et al., 2010; Pacific Environmental Services, 1996; UNIDO, 1996; Holmgren et al., 2016).
- Improved production of nutrient rich solids from sludge (Verstraete et al., 2010)
- Commercial production of algal biomass along with harvesting biomass from nutrient removal reactors (wetlands, algal ponds) (Randrianarison and Ashraf, 2017; Ajjawi et al., 2017; Exxon Mobil, 2018) that can be co-digested to produce methane and subsequently hydrogen.
- New and more efficient and affordable capture of renewable solar energy by concentrated solar panels and photovoltaics and more efficient and smaller (<100 kW) and large (>4 MW) wind turbines and power plants ("Solar and Wind Blue Power" in Chapter 4).

Knight et al. (2015) in their synthesis and analysis of the current energy from waste conversion processes noted that in Western Europe, because of significantly higher energy cost than in the US, the energy recovery process selection tends to favor technologies that could achieve high electrical efficiency. This led to a dominance of systems with internal combustion engines with generators converting methane to energy. In the US, due to lower energy prices in many areas, including low cost of methane from fracking natural gas from shale (a polluting process), the incentives to implement energy recovery from waste are less favorable. Figure 7.8 showed the typical efficiencies of energy recovery from biogas (digester methane). The plot shows that conversion of methane (syngas) to energy by internal combustion (IC) engines is about 33% and by turbines is about 28%. This chart does not include recovery of sensible heat.

Energy Recovery in a Centralized Concept with Anaerobic Treatment and Digestion as the Core Technology

As shown in the preceding section, anaerobic biotechnology is the key process for recovery of chemical energy from wastewater. Various modifications of reactor arrangements have been used for this purpose. Most often, reactors of the upflow anaerobic sludge blanket (UASB) type are currently used in full-scale applications (Mahmoud et al., 2004) and they can be combined with a membrane unit (Ozgun et al., 2015; Kim et al., 2011; McCarty et al., 2011; Chang, 2014). The anaerobic complete mix system is similar to a conventional aerobic activated sludge system and has been used as pretreatment of wastewater with high concentrations of COD (BOD).

Without membrane solids separation technology, the efficiency of anaerobic treatment units treating municipal used water (wastewater) is relatively low in cold and temperate climatic zones. The performance of this technology reported in literature varies significantly. For example, without membranes typical COD removal efficiencies of UASB reactors (75 to 85%) are smaller than those of a typical aerobic activated sludge unit ($\geq$85% with secondary settler) but it should be realized that the influent COD and TSS concentrations in the distributed system will be much higher, more than 1000 mg/L, and a typical aerobic activated sludge process even with pure oxygen aeration cannot handle such high concentrations of organic matter. Majority of authors obtained effluent COD concentrations of anaerobic units without membrane separation (depending on process temperature) ranging from 50 to 200 mg/L (Gouveia et al., 2015; Ozgun et al., 2015) with completely mixed reactors exhibiting the lowest COD removals. For example, one of the largest completely mixed anaerobic treatment plants treating wastewater and river flows from a part of Belo Horizonte (population 2.5 million in 2015) in Brazil at the beginning of this century had BOD removal efficiency $\approx$ 50% and all methane generated by the treatment was flared (personal observation of the author in 2006). This large plant operated in the psychrophilic to low mesophilic digestion temperature range in a subtropical zone of Brazil. Shin et al. (2014) obtained in a classic one reactor digester about 40% transformation of influent COD into biogas. Similarly, Smith et al. (2013) obtained approximately 35% transformation of COD into biogas using a complete-mix AnMB (anaerobic membrane bioreactor) with submerged flat-sheet membrane.

Because complete anaerobic digestion of organic biodegradable matter is very slow at low temperatures, it is advantageous to thicken and transfer the solids accumulated in the psychrophilic (temperatures of bacterial growth 25°C or less) anaerobic reactor into a second, smaller, mesophilic (heated) reactor where solids can be further degraded. The rates of hydrolysis and fermentation with acctates and hydrogen as final products in the fermentation processes in the pre-digester are much faster than those for methane generation; therefore, the particle breakdown and formation of acetates and hydrogen can occur in much shorter hydraulic retention time (HRT) than in a typical digester and some hydrogen. About 25% of the total hydrogen potential can be recovered from the first fermentation digester. After a shorter HRT, the pre-processed mixture full of acetates can be sent into the UASB influent or AnFMBR (anaerobic fluidized bed membrane bioreactor) for methane production or to a follow-up microbial electrolysis cell (MEC) reactor to produce more hydrogen. Moreover, co-digestion with other biodegradable solid materials (wastes) collected centrally is possible in this predigesting fermentation reactor. (The use of MEC is further elaborated in Chapter 8.)

Bioflocculation (Akanyeti et al., 2010; Zhang et al., 2015) or biosorptive activated sludge process (Diamantis et al., 2014) can be used for the separation of up to 70–80% of total COD load. Sludge separated in this process is highly degradable (up to 70%), thus 50% of total COD load can be transformed into biogas. Alternatively, forward osmosis can be used for pre-concentrating used water (Nasr and Sewilam, 2015). The pre-concentration of municipal used water avoids the problems with dissolved methane at low temperatures and minimizes the production of sulfide.

Residual digested sludge still contains a high mass of difficult-to-digest COD that is typically disposed on land with a low agricultural value or, more likely, into landfills or incinerated. The availability of these disposal alternatives is diminishing; furthermore, poorly managed landfills emit methane, which is a strong GHG. In the future, residual digested sludge after dewatering can be co-processed with the organic combustible municipal solid waste that is difficult to decompose, such as waste wood, plastics, rubber, etc., by pyrolysis or gasification, as discussed in subsequent sections.

Anaerobic Energy Production and Recovery Units and Processes

Traditional sludge digesters are slowly mixed or even unmixed tanks that do not have good efficiency of COD removal. Essentially, their function is to break complex organic solids and solids, stabilize the sludge, and, in doing so, convert about 30% to maximum 50% of COD into methane. Anaerobic completely and slowly mixed activated sludge units treating liquid wastewater have been built in the last century to treat or pretreat high strength organic wastewater from food and chemical industries or from animal operations and slaughterhouses. With the emphasis on energy recovery, attempts have been made to use similar system operating under anaerobic conditions also for low-strength municipal wastewater. However, such systems like the aforementioned large anaerobic treatment plant in Belo Horizonte (Brazil), despite operating in a subtropical region, exhibited relatively low COD removal efficiency and the produced methane was flared. Under psychrophilic temperatures these systems are suitable for sludge digestion or for the small-scale wastewater treatment (Chang, 2014). In temperate zones, complete mix anaerobic systems need to be heated to maintain mesophilic operating conditions. Solid and liquid separation is also a problem since the traditional secondary settling tanks must be covered and emitted methane and hydrogen sulfide gasses somehow removed.

Anaerobic Trickling Filters. Unlike the traditional downflow aerobic trickling filters that have been used for almost one hundred years (Figure 7.1C), the anaerobic biological filters are upflow fixed-bed biological reactors with two or more filtration chambers in series (Figure 7.9). As wastewater flows through the filter, particles are trapped, and organic matter is degraded by the active biomass that is attached to the surface of the filter material (Tilley et al., 2014). The upflow mode is preferred because there is less risk that the fixed biomass will be washed out. The water level should cover the filter media by at least 0.3 m to guarantee an even flow regime.

The filter has no moving parts. Such filters are used as a low-cost treatment mainly in developing countries. Biogas should be collected from the vent and used for energy. The recommended hydraulic retention time (HRT) is 12 to 36 hours and the organic load up to 10 kg COD/(m^3-day).

This filter is similar to constructed submerged treatment wetlands. The first compartment is essentially a septic tank that must be periodically pumped out and the sludge transported for another disposal such as a co-digestion in a regional resource recovery facility. These filters are not recommended for high strength organic wastewater and are acceptable mainly for soluble types of wastewater. They also have a problem with clogging (Van Lier et al., 2008). Final solids separation can be done by a membrane compartment or a less efficient settling tank.

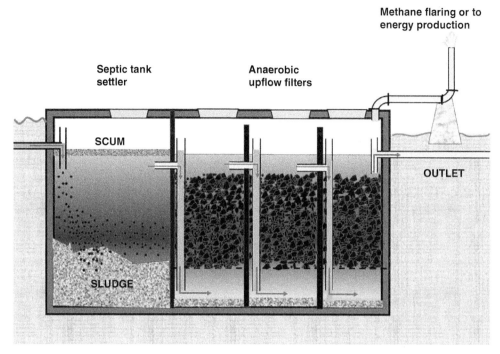

Figure 7.9. Upflow anaerobic trickling filter. *Source:* Reprinted from Tilley et al. (2014), under license from Swiss Federal Institute of Aquatic Science and Technology; copyright © EAWAG.

High Rate Anaerobic Treatment Systems

Van Lier et al. (2008) presented the development of high-rate anaerobic wastewater treatment (AnWT) systems that can treat cold and/or low-strength wastewater (used water). In addition to low-strength municipal used water, many industrial high-strength wasters are discharged at low temperatures, such as beer and malt industry waters. This shows the robustness of the modern anaerobic treatment and new improvements can be expected. The separated and distributed municipal water/energy collection systems covered in this text for black water with high COD concentration is highly acquiescent for the anaerobic treatment. The influent can be mixed with liquid and even wet solid high-organic-strength municipal and agricultural wastes and industrial discharges. Separated low-strength gray water can and should be treated locally.

However, the high-rate anaerobic systems defined as *anaerobic contact process* that copied the aerobic activated sludge systems with external primary and secondary solids separation by settling and sludge recycle used for medium-strength wastewater were not successful. The reason was that the bioreactor required rather intensive agitation that was detrimental to the sludge structure, flocs formation, and settleability (Van Lier et al., 2008).

Anaerobic Membrane Bioreactors (AnMBR). Development of AnMBRs integrates organic (COD) removal in one tank is shown in Figure 7.10b. An AnMBR can be defined as a biological treatment process operated without oxygen that uses membranes to provide

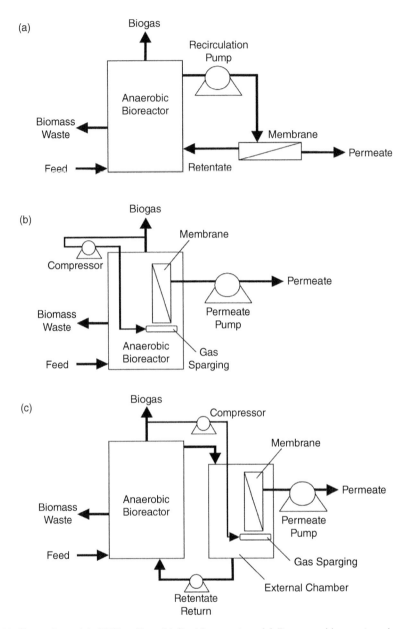

Figure 7.10. Alternatives of AnMBRs with solids/liquid separators. (a) Pressure-driven external cross-M membrane. (b) Vacuum-driven submerged membrane with the membrane immersed directly into the reactor. (c) Vacuum-driven submerged membrane with the membrane immersed in an external chamber. *Source:* Liao, Kraemer, and Bagley (2006); copyright © Taylor and Francis; reprinted with permission.

complete solid-liquid separation. As a matter of fact, the notion of effluent suspended solids may become meaningless because the pore sizes of the membranes are typically less than the pore size of the filters used in determining the suspended solids (SS) concentrations. These reactors were originally designed on the same principles as the complete mixed aerobic membrane activated sludge tanks except mixing is provided by turning paddles or similar slow devices and not by aeration turbines or air diffusers. In these reactors, sludge settleability is not an issue.

However, adding low-pressure or vacuum microfiltration membranes directly into tank (Figure 7.10b) result in problems with membrane fouling (Chang, 2014; Ramesh et al., 2007; Stuckey, 2012; Dvořák et al., 2016). It can be classified under three categories: pore clogging, sludge cake formation, and adsorption of extracellular polymeric substances (EPS). The problem can be resolved by connecting a separate fluid-solids membrane separator on the effluent line (alternatives a and c in Figure 7.10). A portion of the produced biogas can be diverted to the separator to prevent fouling of the membranes. Alternative c is the most realistic alternative for the design of the AnMBR. The membranes can operate under pressure or vacuum (Liao et al., 2006).

In later designs of AnMBR similar to the activated sludge aeration tanks, the solids residence time (SRT) is separated from the hydraulic residence time (HRT) because majority of the solids separated from the effluent permeate by the membrane filter either stays in the bioreactor (alternative b in Figure 7.10) or is returned from a separate filtration unit (alternatives a and c in Figure 7.10). Consequently, the HRT is much smaller than the STR. Excess sludge can be released from the reactor or from the sludge recycle line continuously or daily.

Membrane materials can be divided into three basic categories: polymer, ceramic, and metallic. Polymer membranes have the advantage of lower cost compared to ceramic or metallic membranes. From an operational point of view, ceramic membranes are a more suitable option for AnMBRs than polymer membranes. Hollow-fiber membrane modules tend to be the most commonly used in current AnMBRs, followed by flat-sheet membrane modules, which show good stability and are easily cleaned or replaced when defective (Dvořák et al., 2016).

Hydraulic retention times (HRT) reported by Chang (2014) in AnMBR treating municipal wastewater at temperatures ranging from 20°C to 30°C were 0.25 to 1.5 days and resulted in in COD removal efficiencies of 84–94%, and methane production of 0.24 m^3 CH_4/kg of COD removed, respectively. The energy recovered from the produced methane was about the same as energy needed to provide turbulent mixing to keep the solids in suspension and prevent fouling of membranes, which remains the major factors limiting the efficiency of the complete-mix AnMBR. Dissolved methane in the effluent must be recovered.

Upflow Anaerobic Sludge Blanket (Fluidized Bed) and Extended Granule Sludge Blanket (EGSB) Reactors.
The upflow anaerobic sludge blanket (UASB) or, in a similar arrangement, anaerobic fluidized bed reactor (Figure 7.11) was developed by Lettinga and colleagues (Lettinga et al., 1980; Lettinga and Hulshoff-Pol, 1991) 40 years ago. Applications of UASB reactors has more promise than using complete-mix AnMBRs. USAB reactors are widely used for many application and types of wastewater, from moderately low organic content to very high COD concentrations.

Influent into the UASB reactor enters through the bottom inlet, creating upflow conditions that keep the formed solids in suspension as long as the upflow hydraulic velocity is near or slightly below the settling rate of the sludge granules formed in the tank. Correspondingly, a UASB reactor combines a high-rate bioreactor with the settling space on the top. A typical UASB reactor consists of four parts (Figure 7.11): (1) sludge bed, (2) sludge blanket, (3) liquid/solids separator, and (4) gas/solids separator. Hence the UASB is a suspended growth reactor wherein the primary mechanism of COD removal is by absorption, biological action of the anaerobic microorganisms, fermentation of organic dissolved and suspended solids in the sludge blanket, and methanogenesis of the

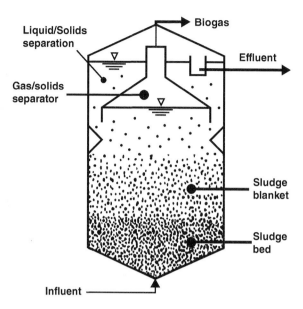

Figure 7.11. Simple schematics of the UASB reactor.

fermentation products. The products of anaerobic decomposition and fermentation are excess biomass of the sludge, methane and carbon dioxide, and bicarbonates. The bubbles of the biogas rise through the rector and keep the sludge flocs in suspension. Methane is captured and is a source of energy. The gas collection dome above the reactor creates more quiescent conditions that prevent the washout of the sludge flocs.

The height/surface area ratio is important, especially for treating low-strength wastewater (Van Lier et al., 2008) and reactors reaching heights up to 20–25 meters have been installed. Therein, the smaller surface area facilitates the feeding of the influent into the reactor and the height concentrates the formation of the biogas that creates turbulent conditions needed to keep the sludge flocs in suspension. The increased height also provides better contact time between the sludge flocs and pollutants in the treated wastewater. Pumping energy is not expensive and could be provided by an internal source.

EGSB reactors.   A variant of the upflow anaerobic sludge blanket (UASB) digestion concept for anaerobic wastewater treatment is the EGSB reactor. Granular sludge exhibits high settling velocities and activity rates that reduce required reactor volumes and increase allowable organic loading rates. The distinguishing characteristic of EGSB reactors is a faster rate of upward-flow velocity and shorter HRT. Typically, these systems retain granular sludge by employing specially designed, often proprietary gas-liquid-solids (GLS for short) separation devices restricting the solids from entering the liquid effluent compartment.

Minor modifications of the UASB reactor and combining it with the side stream membrane solids/liquid separator resulted in the development of the *anaerobic fluidized bed membrane reactor (AnFBMR)*. In this reactor, the solids are kept in suspension as a sludge blanket by the upflow velocity that should be about the same as the gravity-settling velocity of the flocks. Hence the operation of the reactor needs to be closely monitored and controlled. These reactors maintain solids inside the unit in very high concentrations that enable shortening the HRT and maintain large solids residence time (SRT).

UASB (and EGSB) reactors work at temperatures in the psychrophilic range (10–25°C) but function best in mesophilic temperatures. They are popular in subtropical regions of

Latin America. However, in the context of the IRRF, heat produced by the subsequent energy production units or waste heat from nearby thermal power plant and, soon by hydrogen fuel cells, could solve the problem of USAB/EGSB reactors in temperate or even colder regions. These reactors produce sludge in much smaller quantities than the sludge production in the aerobic activated sludge processes. The key design parameters for the UASB design are given in Table 7.2.

A more modern and efficient two-tank anaerobic fluidized bed membrane reactor (AFBMR) was developed in North Korea by combining two-step fluidized bed process with recycling and with a membrane filter in the second reactor (Kim et al., 2011; McCarty et al., 2011). The pilot set-up of the units is shown on Figure 7.12. The AFBMR unit treating

Table 7.2. Design parameters of UASB and AnFB reactors.

Input CCOD concentrations, g/L	0.5–15
Maximal input of total suspended solids (TSS), g/L	6–8
Reactor temperature, °C	>20
Concentration of solids (VSS) in the sludge blanket, g/L	30–50
Hydraulic residence time, HRT, hrs	
Temperature 20–26 °C	7–10
>26 °C	6
Upflow velocity UV= reactor height/HRT m/hour	0.6–0.9
Solids residence time (SRT=sludge mass/sludge yield), days	30–50
Organic loading kg COD/(m³-day)	
Influent COD 1–2 g/L	8–14
2–6 g/L	12–24
COD removal percent with a settler	75–95
with a membrane	95–99
Sludge yield kg VSS/kg of COD	0.08
Methane production m³/kg of COD removed	0.2–0.4
Energy used to run AnMB reactor kW-hr/m³ of flow	0.06
Energy content of methane at 25°C kW-hr/m³ of CH$_4$	0.2

Sources: Lettinga and Hulshoff-Pol (1991); Metcalf and Eddy (2005); and McCarty et al. (2011). Table updated and reprinted from Novotny et al. (2010a).

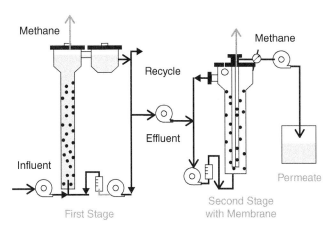

Figure 7.12. Two-stage anaerobic fluidized bed reactor, followed by anaerobic fluidized bed membrane reactor pilot arrangement with granulated activate carbon addition in the second reactor. *Source:* Adapted from Kim et al. (2011); copyright © American Chemical Society 2011.

typical municipal used water (COD $\geq$ 500 mg/L) can achieve 99% COD and suspended solids removals with energy expenditure of less than 1/10 of a typical aerobic membrane bioreactor. The electric energy runs the recycle and creates membrane pressure differential. Activated carbon granules are added to enhance the COD removal. Nutrients removal in UASB and AFBM reactors is limited to the nutrient uptake into sludge. HRT is also much shorter (7–12 hours) than that in a typical digester, which is more than 10 days, or even that in typical completely mixed AnMBR, which is about one day. Heating is required when ambient temperatures drop below 20°C.

Table 7.2 indicates that these reactors work best when the inflow COD concentration is between 0.5 and 15 g/L. For larger COD concentration, Lettinga and Hulshoff-Pol (1981) suggested using primary a clarifier or pre-digester. Sludge can be removed daily or continuously. Without adding granular activated carbon, effluents from the anaerobic reactors may include dissolved methane that is measured as COD (up to 30 mg/L) (Kim et al., 2011).

Liao et al. (2006) reported that approximately 67% of the anaerobic membrane reactors in use have been completely mixed-tank reactors (CMTR), 15% anaerobic filters, 10% upflow anaerobic sludge bed (UASB) or UASB hybrid, 7% fluidized bed, and 2% septic tank. In part, this may be due to the ease of use and construction of a CSTR reactor, but it is also likely due to the almost exclusive use of external cross-flow of external membranes. However, after 2010 by switching from CMTR to fluidized bed reactors (AnFBR), better performances have been reported with the AnFBRs, which with membranes achieved more than 95% COD removals (Kim et al., 2011).

In the second decade of 2000s, the research on anaerobic processing of organic wastes started shifting from methane production to maximizing hydrogen yield. In these arrangements, the first phase reactor could be CMTR with acid (dark) fermentation that produces some hydrogen and acetic acids at pH between 5 and 6. A CMTR in which the solids residence time (SRT) is the same as the hydraulic residence time (HRT) requires long HRT. Fluidized bed or UASB reactor could also be used. A membrane-coupled acid reactor can separate the SRT from the HRT. This dark fermentation acid reactor is then followed by microbial electrolysis cell (MEC) in which more hydrogen at relatively short HRT is produced by adding some DC current. This novel technology will be described in the next chapter. However, Lu et al. (2009, 2010) pointed out that dark fermentation is efficient for decomposition of carbohydrates such as monosaccharides (glucose, fructose), oligosaccharides (sucrose, lactose), or polysaccharides (cellulose, starch, hemicellulose) and less efficient for decomposition of proteins.

Dark fermentation products (organic acids) can be utilized by another type of bacteria under illuminated conditions and nitrogen-deficient conditions to produce more hydrogen by the photo-fermentation process (Hallenbeck and Benemann, 2002). However, this process cannot use proteins to produce hydrogen.

Efficiency of Membranes to Remove Pollutants. Introduction of membrane filters and membrane reactors in the second half of the last century revolutionized the solids and colloid removal and, to a great degree, replaced conventional secondary settling and tertiary filters. In membrane filtration processes pretreated or treated water (by biological or physical/chemical treatment) passes under pressure or vacuum through thin membranes that retain particulate materials, organic matter, nutrients, large molecules and ions from the permeate and remove them into reject flow. Membrane processes include *microfiltration (MF), ultrafiltration (UF), nanofiltration (NF)* and *reverse osmosis (RO)* (Asano et al., 2007). The membrane pore size and operation characteristics are presented in Table 7.3. Membrane units can be used in several configurations, placed directly into biological treatment units or into external membrane chambers, as described in the preceding section. After a thorough pretreatment, fine pore membranes (with nanometer or smaller pore sizes) can be used in nanofiltration or reverse osmosis for separation of molecule size ions and salts.

Most membrane microfiltration filters consist of a bundle of long hollow fibers with the pore diameter ranging from 0.04 to 0.4 µm (Asano et al., 2007), hence such filters can remove colloidal particles and microorganisms. Plate and frame/flat sheet membranes and tubular membranes are used in the range of micro- and ultrafiltration. The filter unit can be installed in the aeration tank or in a separate tank (Figure 7.9) and filtration is accomplished by creating a small negative (suction) or positive larger pressure differential (Asano et al., 2007). Hence, force or energy is needed to generate the pressure (before the membrane) or vacuum differential (after the membrane). The difference between these processes is in the size of millipores in the membrane. Table 7.3 presents the typical characteristics of the processes. In nanofiltration, membrane reject multivalent ions (>99%), monovalent ions (<about 70%), and organic compounds with molecular weight above the molecular weight of membrane (>90%). Nanofiltration systems typically operate at lower pressures than RO, which requires very high pressures and energy.

The most common types of membrane used by MBRs are hollow-fiber membranes, plate-and-frame/flat sheet membranes, and tubular membranes in the range of micro- and ultrafiltration (Judd and Judd, 2011). Nanofiltration (NF) and reverse osmosis (RO) (Figure 7.13) systems can remove molecules and ions. These systems are used in water reuse wherein the product water could be used for nonpotable reuse. Nanofiltration is used mainly for treatment of used water. Reverse osmosis is used for both (a) final purification of the reclaimed water to remove monovalent ions before reuse, and (b) for desalination of sea water for input fresh water in the water supply system.

Typically, the permeate from the RO is injected into groundwater aquifer with residence time of several months and then, after mixing with some groundwater, can be brought back for potable and non-potable uses. The NF and RO units should not be used for removal of particulates; hence, MF or UF with or without sand filter can serve as pretreatment before NF and RO. RO may also be required if reclaimed and treated water is used in-house as a supplement for bath or laundry (Figure 5.5).

As the permeate water is produced, a concentrated stream containing all impurities rejected by the membranes is also produced. This concentrated stream, sometimes referred

Table 7.3. Parameters of membrane treatment processes.

Process	Pore Size. µm	Operating Pressure Bar	Energy Use kW-hr/m^3	Compounds Removed
Microfiltration	0.008–2.0	0.07–1	0.4 – 0.8	Fine particles down to clay size, bacterial, some colloids
Ultrafiltration	0.005–0.2	0.7–7	Wide range	All of the above plus viruses and large molecules
Nanofiltration	0.001–0.01	3.5–5.5	0.6–1.2	All of the above and about 50% ions
Reverse osmosis	0.0001–0.001	12–18	1.5–2.5	Small molecules, 98 to >99%
Used water desalination			1.5–5	ions

Source: Asano et al. (2007).
1 bar = 100 kPascals = 100,000 N/m^2 = 1.03 atmospheres = 14.5 psi (at 0 elevation)

to as brine (in desalination) or reject, is a polluted flow that cannot be directly reused but could be treated and some compounds (e.g., nitrate) removed by NF-RO units can be handled by the anoxic unit of the biological treatment. Reject water ranges from less than 5% of flow for MF but may represent up to 25% of water treated by RO.

Sharrer, Rishel, and Summerfelt (2010) conducted experimental research to determine the removal efficiencies of key pollutants, nutrients, and toxic metals from a high-concentration wastewater from fish culture–producing facilities treated by sequential anaerobic and aerobic completely mixed reactors (CMTR) with membranes immersed in the aerobic tank. This configuration may resemble (with or without the aerobic tank) a high-inflow concentration treatment of the concentrated used water (black water) in the IRRF. According to the manufacturer of the membrane rack used in the research, the membrane filter had a nominal 0.4 μm pore size, which after biofilm formation was reduced to 0.04 μm. Table 7.4 presents the treatment of mixed liquor particulate and dissolved solids concentrations of the compounds after the solids build up in the anaerobic and aerobic reactors and of the permeate along with the treatment efficiencies in percent. The removal of TP in the aerobic reactor was tied to particulate solids either as a part of the sludge cells or adsorbed on organic matter of the cells and other organic matter.

Membranes Removing and Separating Dissolved Gases (CO_2, Methane).

The information in Tables 7.3 and 7.4 refers to design parameters and removals of particulates and dissolved ions (salts). Both aerobic and anaerobic treatment processes and syngas and methane conversion (reforming) to hydrogen produce carbon dioxide. Furthermore, residual dissolved methane in the effluents from anaerobic treatment is measured as COD (BOD) and this decreases the treatment efficiency.

Methane, hydrogen, and carbon dioxide (if needed) can be separated from water by wetted micro/nanoporous hydrophilic membranes that can block gas due to a large water

Figure 7.13. Reverse osmosis treatment plant in Orange County, California. *Source:* Courtesy CDM-Smith.

Table 7.4. Performance of anaerobic and aerobic CMTR with microfiltration/ultrafiltration membrane treating high-concentration fish-processing wastewater.

Compound	Inlet Concentration mg/L	Anaerobic Mixed Liquor Concentration mg/L	Aerobic Mixed Liquor Concentration mg/L	Permeate Concentration mg/L	Treatment Efficiency%
TSS	1827 ± 272	16988 ± 767	20007 ± 682	0.3 ± 0.3	99.9
VSS	1488 ± 224	12499 ± 564	14983 ± 493	0.2	99.9
BOD$_5$	1073 ± 182	–	–	0	100
Total N	185	995 ± 51	1168 ± 57	3.1	97.5
Total P	55	677 ± 41	845 ± 51	0.1 ± 0.01	99.8

Source: Sharrer, Rishel, and Summerfelt (2010).

surface tension in the entrance of micro/nano pores to withstand the gas pressure, while it allows water to permeate with overcoming the flow resistance of pores. In contrast, hydrophobic membranes permit gas to penetrate but prevent water flow (Zhu, 2009).

7.4 CO-DIGESTION OF SLUDGE WITH OTHER ORGANIC MATTER

Figure 3.2 documented that about 30–40% of the municipal solid wastes (MSW) are decomposable components that, currently, are sources of methane produced in landfills that is either flared to CO_2 or released as a GHG. Methane greenhouse gas potential is 25, while that of carbon dioxide is one. On top of the biodegradable MSW fractions a city produces other high organic liquid and solid compounds, such as food production wastes (brewery, cheese and yeast production wastes, fish production wastes, stockyards), tannery wastes, landfill leachate, and glycols from winter deicing operation of airports. These, if untreated, are not normally acceptable for discharging into aerobic municipal sewerage systems because of overloading the water reclamation plants. In suburban rural areas, agricultural operations such as cattle feedlots, dairy, poultry, and pig operations produce very large quantities of organic waste, which today is generally untreated and mostly stored in putrescible objectionable lagoons. Land disposals of sludge and manure onto soil are questionable because they may lead to soil overloads by nutrients and cause contamination of groundwater, not counting the bad smell from such operations, especially in proximity to suburban and urban areas. All are candidates for co-digestion with the sludge produced from the treatment of municipal used water. In many cases, because the chemical energy in the organic content (COD) of used water is relatively small, the mass and COD content of the other biodegradable digestible waste will exceed the mass of the municipal used water residual solids. Figure 7.14 shows the large digesters at the Changi used water and energy reclamation facility in Singapore (population 5 million).

Unlike the UASB and AnFBM reactors that work at temperatures as low as 20°C, anaerobic digestion of sludge and organic solids that produces methane is a process that progresses either in mesophilic (about 37°C) or thermophilic (about 55°C) conditions. Obviously, the heat energy requirement is greater for the latter thermal conditions, but this can be compensated by better and faster COD removal rates, smaller hydraulic retention times, and greater methane production (Perez, 2013). Most of methane-producing digesters in the current water reclamation (resource recovery) plants are mesophilic.

Figure 7.14. Sludge digesters in Singapore Changi used water treatment and recovery plant designed by CH2M-Hill (today CH2M). *Source:* Photo. V. Novotny.

Municipal water reclamation utilities with excess digestion capacity are looking for other degradable solid and liquid sources that could be co-digested with the sludge from the treatment operation and produce more methane. Including the solid waste and liquids with high biodegradable COD concentration into the energy balance will clearly make a net positive gain. In the new anaerobic IRRF organic waste liquids and used water can be more efficiently treated by the AnFBM reactors; co-digestion could accept wet biodegradable solids and the sludge produced by the liquid treatment reactors. The benefits of co-digestion can be illustrated with the following example:

Assume net methane yield from digested vegetation residues and food waste is about 0.25 m³ CH₄/kg of solids (Hamilton, 2012) and the per capita organic solids production in this category is 0.5 kg/c-day. Then the additional methane energy production is

$$\textit{Energy from codigestion} = 0.5 \ (\text{kg/day}) \times 0.25 \ (\text{m}^3 \ \text{CH}_4/\text{kg of solids})$$

$$\times \ 9.8 \ (\text{KW-h/m}^3\text{CH}_4) = 1.22 \ \text{kw-h/capita-day}$$

$$= 0.45 \ \text{MW-h/cap-year}$$

which could yield about 130 kW-h $\text{cap}^{-1}\text{year}^{-1}$ electricity and 317 KW-h $\text{cap}^{-1}\text{year}^{-1}$ of heat energy. This additional energy yield could now make the utility energy self-sufficient (net-zero). About 10% of heat would be lost through the stack in a traditional incinerator boiler-turbine electricity-producing system.

In the IRRF the produced sludge can be co-digested with many of the highly biodegradable, methane- and hydrogen-producing organic waste compounds. In such cases the digester may become the process unit that is receiving a higher organic load than the upflow anaerobic fluidized bed reactors receiving the liquid influents. In the future, a two-step dark fermentation digester followed by Microbial Electrolysis Cell (MEC-reactor) process could become even more efficient and economical (see the section on microbial fuel cells in the next chapter).

Biodegradability and biogas production potential of other solids co-digested with used water treatment sludge are not uniform. This has led to the classification of the co-digested waste compounds as being *synergistic, antagonistic*, and *neutral* (Zitomer and Ahhikari, 2005, Zitomer et al., 2008; Navaneethan et al., 2012), as listed in Table 7.5, which presents an approximate classification of co-digestate compounds digested with the municipal used water sludge. A neutral outcome occurs when the rate of methane production from the mixture is equal (plus or minus few percent) to the sum of weighed averages of the production rates observed when the compounds are digested separately. A synergistic effect is observed when the overall methane production rate during co-digestion is greater than the

Table 7.5. Effect of co-digestates on methane production in municipal (conventional) anaerobic sludge digesters.

Added Co-digestate	Effect	Source	Comment
Airport deicer (propylene glycol)	Synergistic	Zitomer et al. (2001, 2008)	Synergisticity decreases with higher loading of digestate
Yeast waste	Highly synergistic	Zitomer et al. (2008)	
Organic fraction of municipal solid waste (OFMSW)	Synergistic	Ara et al. (2015)	Adding grinded OFMSW to primary and secondary sludge increases biogas production
Landfill leachate	Slightly below neutral (5–15%)	Montusiewicz (2014)	Methane production decreases with higher loads of co-digestate
Restaurant waste	Neutral	Zitomer et al. (2008)	Digestate added after primary clarification
Slaughter facility sludge	Highly synergistic	Hamilton (2012)	Very high methane yield, almost all COD converted to methane
Food flavoring waste	Highly synergistic	Zitomer et al. (2008)	Better than average methane production attributed to trace nutrients
Potato peels and other food scraps	Highly synergistic	Hamilton (2012)	
Boiler cleaning waste	Antagonistic	Navaneethan et al. (2010)	Higher concentrations of toxic metals in the digestate
Animal manure	Somewhat antagonistic	Jepsen, (2005), Perez (2013)	Digestion of manure alone produces less methane. Production can be improved by co-digestion with other organic waste and sludge and by thermophilic conditions.
Algae (*Chlorella*)	Synergistic	Wang et al, (2013); Olsson (2018)	Adding algae improved dewaterability of the digested residual solids

weighted sum of the separate digestions of the compounds. This would typically occur when an original compound does not have enough nutrients that are provided by the co-digested compounds or when some toxic compounds present in the original compound are diluted below the toxicity threshold by the other compounds. Waste containing potential toxicants in larger amounts can result in antagonistic effect that would reduce the methane production of the mixture. Professor Zitomer and his coworkers, all from Marquette University (Milwaukee, WI) Water Quality Center, used in their studies Biochemical Methane Potential test (Owen et al., 1979) as a short-term screening tool of treatability and methane production under anaerobic conditions (Zitomer et al., 2008).

Sources of co-digestion energy tested by several authors include brewery and dairy waste solids and liquids, waste glycols from airport plane deicing (a liquid high COD waste that cannot be processed by a AFBM reactor), manure from dairy and pig farms, fish-processing plants and stockyards, and biodegradable municipal solid wastes (McMahon et al., 2001; Zitomer et al., 2008; Welch et al., 2013, Sands, 2014; Angelidaki and Ellegaard, 2003; Perez, 2013, and others).

Denmark and Holland, countries known for their milk and cheese production, have large cattle operations. Thus, in many areas of the two countries, water reclamation plants used to be overloaded with dairy wastewater and soils overloaded with manure were oversaturated by phosphate and caused nitrate and phosphorus pollution of groundwater and surface waters.

Consequently, at the beginning of this millennium Denmark already had 22 centralized plants that co-digested manure with sewage and other organic waste solids (Angelidaki and Ellegaard, 2003; Jepsen, 2005). Denmark expects that by 2030 a great part of the green energy will be produced by co-digestion of manure and sewage sludge. This will help to achieve a target of 33% of the national energy consumption being provided by renewable energy by the year 2030. A major part of this increase is expected to come from new centralized biogas-producing plants (Veolia, 2017). The annual potential for biogas production from biomass resources available in Denmark was estimated to be approximately 30 petajoules (PJ = 10^{15} Joules) and manure comprises about 80% of this potential.

In 2010, 190 methane energy production facilities using biogas from digestion of manure were installed in the US (US EPA, 2010). The greatest number of such facilities were in Wisconsin (25) and the largest 1.4 MW manure digestion energy plant was installed in 2013 near Oshkosh. The US EPA AgStar report assessed the manure digestion and co-digestion potential in the US. The study estimated that there were total of 8,241 dairy and swine operations in the US with the total power and energy potential of 15,667 MW and 13,536,288 MW-h/year, respectively. Without digestion and energy production these facilities were responsible annually for 2×10^6 tons of CH_4 emissions, of which more than 90% were recoverable as energy. The urgency for this implementation is magnified by the fact that agricultural soils in Denmark and in some places in the US are already overloaded with manure, resulting in severe impact on surface and groundwater quality. Soil overloads with manure and fertilizers are suspected to be a cause of obnoxious harmful algal growths/blooms (HAB) along the southern coast of Florida.

In 2017, the co-digestion concepts had been implemented at the Metropolitan Milwaukee Sewerage District (MMSD), which has a goal to be a net energy producer within a few years. Dairy and swine manure is co-digested with municipal sludge in Madison, Wisconsin (Welch et al., 2013). Utilities in Oakland, California, Philadelphia, and other communities worldwide also practice co-digestion and are becoming net-zero carbon utilities by accepting other biodegradable compounds.

Putting swine waste in lagoons in North Carolina backfired during Hurricane Matthew in October 2016 when catastrophic rains and flooding flushed large amounts of liquid swine manure with the nutrients from the lagoons into the Neuse River, which feeds one of the nation's largest and productive coastal estuaries. A similar situation occurred in 2018 with Hurricane Florence. Co-digestion can prevent these animal waste lagoon overflows and overloads of important water resources.

Models for estimating methane and energy yield use both volatile suspended solids and COD as the design input parameters for anaerobic digestion processes, the traditional textbook relationship between the two parameters for food waste, manure and digestate is COD = 1.4 ∗ VSS (Metcalf and Eddy, 2003; Hamadek, Guilford, and Edwards, 2015; Gűngőr-Demirci and Demirer, 2004). It is higher for paper and cardboard (COD = 1.5 ∗ VSS) and less for woodchips (COD = 1.3 ∗ VSS), respectively.

7.5 CONVERSION OF CHEMICAL AND SENSIBLE ENERGY IN USED WATER INTO ELECTRICITY AND HEAT

Energy in used water is contained in reduced organic compounds (organic solids, dissolved organics,) and as sensible heat. Organic compounds can be decomposed anaerobically by microorganisms into energy-producing methane and hydrogen. In a simpler but less efficient way, dry organic solids can be incinerated and produce heat. This highly polluting burning is practiced in less developed countries even for cooking, and burned solids sometimes include dry manure. Composting organic solids is also an exothermic process but the produced heat is almost never recovered, and composting emits carbon dioxide (not counted as GHG). Sensible heat is related to the change of water temperature by heating and cooling and energy can be recovered in a heat exchanger or by a heat pump (Novotny et al., 2010a). Solar (light) energy can be added to treated used water that can stimulate production of additional chemical energy containing biomass (algae) by photosynthesis and by adding more sensible heat from concentrated heat solar panels. The energy recovery can be realized by:

Incineration of combustible organic solids (dewatered residual solids from IRRF) and combustible fraction of MSW that produces steam for heating and low-efficiency electricity generation by a turbine and generator. Because the water content of dewatered sludge is high (more than 70%), this traditional and persisting process is very inefficient, often requires external fuel (Takaoka et al., 2015), and is generally polluting by emitting GHGs and toxics. Nevertheless, in Japan, more than 70% of residual sludge is incinerated and incineration of MSW is now also becoming widespread in China, with severe adverse environmental consequence. Because of these problems, as pointed out in Chapter 4, no new MSW and/or sludge incinerators were installed in the US after 1995, but they are being installed at an accelerated pace in China and many have been installed in Sweden and Denmark.

In Hartford, Connecticut, an existing sludge incinerator was converted into an energy-producing facility by recovering heat from the incinerator exhaust flue and running it through boilers to produce steam to power a turbine and produce electricity (Tyler, 2016). In the first two years of operation, this power plant generated 7,600 and 9,600 MW-hours electricity, respectively, which covered about 40% of the energy demand of the plant. Incineration of sludge and municipal solid waste may not be an option under the new paradigm of the energy and resource recovery from municipal used water and waste.

*Combusting digester gas*es (approximately 70% methane) produced by anaerobic decomposition of primary and secondary treatment residual solids and co-digested biodegradable organic waste solids and liquids is used as a fuel in internal combustion engine or turbine running generator for producing electricity. Other high organic content municipal, industrial, and agricultural wastes can be co-digested with residual solids to produce more biogas from the IRRF. Based on IPCC classification, emissions from combustion of the biogas produced by the IRRF may not be counted as GHG but no credit will be given for reducing global warming. Both heat and electric power from internal combustion (IC) engines or turbines powering generators for production of electricity and low- and high-temperature heat can be recovered (Knight et al., 2015).

Pyrolysis/gasification of dewatered removed digested primary and secondary digester solids with combustible municipal solid waste (MSW) organic solids produces syngas, which powers a turbine and generator. Both electricity and heat can be recovered. Pyrolysis and direct gasification of residual sludge solids alone is not efficient. Note that in indirect gasification water is one of the feeds; therefore, dewatering may be reduced or unnecessary (Chapter 8).

Microbial fuel cells (MFC) producing electricity.

Microbial electrolysis cells (MEC) (bioelectrically assisted microbial reactors) producing hydrogen from digestible organic solids and liquids (Logan, 2008; Cheng and Logan, 2007; Pham, et al., 2006; Li, Yu, and He, 2014) followed by hydrogen fuel cell.

Methane and syngas reforming to hydrogen, followed by hydrogen fuel cell.

Hydrogen fuel cells, in which hydrogen reacts with air oxygen to produce energy and water.

Electrolysis of water. Hydrogen can also be produced efficiently by electrolysis of water by excess green or blue energy, stored, and used to produce energy by hydrogen fuel cells.

Methane, carbon monoxide, and hydrogen and to some degree heat are the endpoints of energy recovery that can be converted to electricity in the IRRF either by combustion and turbine/generator or, in the very near future (the future is already now), by hydrogen fuel cells. The option based on the traditional digestion with methane as end product is to produce biogas (mostly methane) and then reform it to hydrogen and produce electricity by hydrogen fuel cell. This system is energy positive regarding the difference between produced green energy minus energy used in producing it and should not emit waste greenhouse gases, with the exception of concentrated hot CO_2 flue gas, which would not be counted as GHG and could be beneficially reused (e.g., for growing algae or making dry ice) or sequestered. Sensible heat energy can also be recovered from the flue gas and from the process units and the excess energy can be commercialized.

A better option emerging in the laboratories is to bypass methanogenesis in the digestion process and, with the help of electrons supplied from an auxiliary source of DC electricity or produced by the hydrogen fuel cell (about 10% of the total energy produced) or "free" solar photovoltaics power, convert the dead-end products of anaerobic fermentation (mostly acetates and proteins) into hydrogen in an electrochemically assisted anaerobic

Table 7.6. Composition of digester biogas from biodegradable municipal solid waste (MSW), used water sludge, and agricultural wastes.

Biogas Component	Household Waste (Biodegradable MSW)	Used Water Treatment Sludge	Agricultural Waste	Agricultural Industry Waste	Natural Gas, Methane
CH_4 % vol.	50–60	60–75	60–75	68	97
CO_2 % vol.	34–38	19–33	19–33	26	2.2
N_2 % vol.	0–5	0–7	10		0.4
H_2O % vol.*	6	6	5	6	
H_2S % vol.	100–900	1,000–4,000	3,000–10,000	400	
NH_3 mg/m^3			50–100		
PCS** kW/m^3	6.6			7.5	11.3

Source: Naskeo Environment, copyright © 2009; http://www.biogas-renewable-energy.info/biogas _composition.html.
*At 40°C.
**Upper calorific power (including latent heat of water).

bioreactor (see the next section). Table 7.6 presents approximate compositions and heating values of digester gas from digesting the most common feed stock. These feedstocks are synergistic and could be digested together. Methane/natural gas values in the last column of Table 7.6 were added as reference. The values for syngas produced by pyrolysis and gasification are presented in the subsequent section. The digester gas (methane) and syngas from pyrolysis and gasification can be directly combusted to produce energy by combustion engines or turbines and generators (heat and electricity). The efficiency of these processes to produce electricity is relatively low, around ≤30%.

Digester methane gas can be purified and reformed to syngas containing carbon monoxide and hydrogen mixed with the syngas produced by pyrolysis and gasification, and then internally in the hydrogen or solid oxide fuel cell converted to hydrogen to produce electricity and high-quality heat (see the section on fuel cells and IRRF power plants). The latter process is considerably more efficient and cleaner than combustion; MCFC produces small quantities of ultraclean water (more than used in reforming methane and carbon monoxide) and emits no GHGs. This is important if hydrogen becomes the energy supplying carrier, as discussed in Chapter 4.

At the end of the previous century, most of nonbiodegradable combustible solids had been incinerated or deposited in landfills. It should also be noted that the mass of sludge solids from resource recovery unit processes handling used water is 20–30 times smaller than that from MSW and other organic solid and liquid waste. Since high-efficiency gasification to hydrogen requires water (steam), the need for dewatering of residual used water solids may be reduced or even eliminated and ultra-clean hot water (steam) – more than that needed for reforming the input gases – can be recovered from the hydrogen fuel cell.

INTEGRATING GASIFICATION AND DEVELOPING AN INTEGRATED "WASTE TO ENERGY" POWER PLANT

8.1 TRADITIONAL WASTE-TO-ENERGY SYSTEMS

When focusing on treatment and disposal of used water, 50% of the cost of traditional aerobic activated sludge plants is for sludge management and disposal processes that used to require sludge dewatering and aerobic or anaerobic digestion followed by drying residual solids. In an International Water Association-sponsored study, Holmgren et al. (2016) identified two major routes to recover energy from used water in a traditional treatment plant. One is to convert sludge on-site into biogas through anaerobic digestion, thus recovering electricity with a turbine and using the heat to heat up the reactor. The other major route is to concentrate the sludge and transport the digested sludge product to central incineration. The IWA report by Holmgren et al. did not directly address integrated resource recovery from used water and MSW.

In the aftermath of the passage of the Clean Water Act (PL 52-900) in the early 1970s there was a scant interest among scientists and some utility managers to use already known technologies of pyrolysis and gasification of organic solids (e.g., low-quality coal used during WWII for producing fuel and heating/cooking gas) to process waste sludge and even municipal solid waste but these efforts were abandoned because the cost was not favorable when compared to the cost of thickening, drying the solids, and disposing them in landfills or onto land. Today, efficacy and environmental impact of pyrolysis and gasification have significantly improved and with the increasing interest in producing sustainable "green" biofuel and biogas, these processes are receiving a second look as the utilities search for other sources of organic carbon for co-processing with the sludge to biogas.

In an integrated process where the goal is economical production of green energy from sludge (net-zero carbon goal) and eliminating the need for landfilling (no new landfills), co-processing of residual sludge from co-digestion with the municipal and agricultural waste solids would make sense and would be efficient. Figure 3.2 has shown that in addition to processing the dried solids (ds) sludge residues (about 0.07 kg ds capita^{-1} day^{-1}) and other biodegradable MSW solids and liquids (about 0.5 kg capita^{-1} day^{-1}), plus unaccounted glycol from deicing airport operations, liquid food waste, and manure, after co-digestion there is an opportunity to process these residues by pyrolysis/gasification with additional 1 to 1.5 kg capita^{-1} day^{-1} of dry combustible but nonbiodegradable or difficult to degrade organic solids to produce fuel oil and syngas. Calorific value of sorted organic MSW (without concrete, metals, glass, etc.) is high, around 7 kW-h/kg of dry MSW (Valkenburg et al., 2008) generated in developed countries. Adu and Lohmueller (2012) found that solid waste from dumps in Ghana had a wet heating value of 17 MJ/kg (4.72 kW-h/kg) with a biodegradable fraction of 60% and a moisture content of 50%. About 0.5 kg of solid waste is generated daily by an average Ghanaian (less than 25% of the US per capita generation).

Table 8.1 presents the fundamental reactions and parameters of incineration, pyrolysis, and gasification by which the organic feed compounds are converted to energy.

Incineration

Before introducing pyrolysis and gasification it is necessary to discuss incineration, which has been used for burning MSW since nineteenth century. Incineration also has been used to produce some energy and heat. MSW combustion is a rapid exothermic process in which organic carbon and some other reduced chemicals in the MSW are oxidized by oxygen (O_2) from air. It is the same process as burning coal, wood, natural gas, hazardous waste, etc. The pre-dominant reactions are between organic carbon and oxygen, producing carbon dioxide (CO_2), and between hydrogen (H) in the feed and oxygen, producing water vapor (H_2O). Incomplete combustion of organic compounds in the waste feed stream produces some carbon monoxide (CO) and carbon-containing particles. Hydrogen also reacts with organically bound chlorine to produce hydrogen chloride (HCl). In addition, many other reactions occur, producing sulfur oxides (SO_x) from sulfur compounds, nitrogen oxides (NO_x) from nitrogen compounds and from the nitrogen in the air, metal oxides and vapors from some metallic compounds, and other gases; therefore, the pollutants of primary concern are compounds that contain sulfur, nitrogen, halogens (such as chlorine), and toxic metals. Specific compounds of concern include CO, NO_x, SO_x, HCl, cadmium, lead, mercury, chromium, arsenic, beryllium, dioxins and furans, PCBs, and polycyclic aromatic hydrocarbons (PAHs) (Committee on Health Effects of Waste Incineration, 2000). Many pollutants released into the environment by incinerators are in the category of persistent organic pollutants (POPs). Dyke and Foan (1997) identified MSW incinerator residues as the largest dioxin release to land in the UK. Therefore, the 2001 Stockholm Convention intended to protect human health and the environment by reducing and eliminating POPs lists waste incinerators among *"source categories that have the potential for comparatively high formation and release of these chemicals to the environment"* (Petrlik and Ryden, 2005).

Residues generated by incinerators include bottom ash, fly ash, scrubber water, and various miscellaneous waste streams. Until the first decade of this century, most of the attention has been paid to releases into the air, whereas the content of POPs in wastes and waste waters has been left aside for a long time. However, bottom ash is also toxic and generally it must be safely disposed in landfills. Up to 25% of the quantity of municipal solid waste (MSW) fed to the grate incinerator furnaces ends up as bottom ash after the combustion process. Bottom ash is also known as *slag* and must be disposed onto landfill. Fly ash is composed of small dust particles in flue gases that are captured by electrostatic precipitators (ESP filters) after the flue gases leave the boiler. Approximately 1–5% of the municipal solid waste fed to grate furnaces end up as fly ash after the combustion process and is added to toxic solid waste disposal after removal (Petrlik and Ryden, 2005).

The air pollution control methods for MSW incineration, because of environmental regulations, are complex, costly, and use energy. That is reflected in the low energy yield. Table 8.2 lists the unit processes for removing air pollutants (Quina et al., 2011).

Table 8.1. Conversion reactions and energy of one mole of methane and char (organic carbon) into syngas and hydrogen (one mole of CH_4 is 16 grams).

Conversion	Formula	Temperature °C	Energy Input kJ/mole	Energy Gained (Used)[+] kJ/mole	Energy in Produced Hydrogen[*] kJ/mole	Net Energy Output H_2 + CO ± Heat kJ/mole
Combustion of organic carbon (OrgC) (wood, coal, char, MSW, dry sludge)	$OrgC + O_2 \rightarrow CO_2$			394	0	<394** in heat
Methane combustion	$CH_4 + 2O_2 \rightarrow CO_2 + 2H_2O$		790	790	0	<790 (part in heat)
Methane reforming in low O_2 atmosphere to syngas	$CH_4 + \frac{1}{2}O_2 \rightarrow CO + 2H_2$	>400	790	38	$2*284 = 568$	$568 + 282 = 850$ Heat = 38
Methane steam reforming to hydrogen enriched syngas	$CH_4 + H_2O \rightarrow CO + 3H_2$	650-800 (High temp. shift)	790	(206)	$3*284 = 852$	$852 + 282 = 1134$ Heat = −206
CO steam lower temperature reforming to H_2 and	$CO + H_2O \rightarrow CO_2 + H_2$	230-400 (Low temp. shift)	282	41	284	284 Heat = 41
The above two reactions combined	$CH_4 + 2H_2O \rightarrow CO_2 + 4H_2$		790	(165)	$4 * 284 = 1136$	1136 Heat = −124
Direct gasification of MSW to syngas per mole of Org C***	$CH_{5/3}O_{2/3} + 1/6\,O_2 \rightarrow CO + 0.83\,H_2$	>900°C		111	$284 * 0.83 = 237$	$282 + 237 += 519$ Heat = 111
Char, tar, combustible OrgC indirect conversion to syngas	$OrC + H_2O \rightarrow CO + H_2$	>850°C		(131)	239	$282 + 239 = 527$ Heat = −131
Indirect gasification of MSW to syngas per mole of Org. C	$CH_{5/3}O_{2/3} + (0.33H_2O \rightarrow CO + 1.17\,H_2$	>850°C		(131)	$284 * 1.17 = 332$	$282 + 332 = 614$ Heat = −131
Total MSW energy yield in indirect gasification after converting syngas to CO_2	$CO + 1.17H_2 + H_2O = CO_2 + 2.17\,H_2$		$282 + 33 = 2 = 614$	41	$284 * 2.17 = 616$	$616 + 41 = 657$ Heat = −90

Source: Halmann and Steinberg (1998) and Richardson et al. (2015).

[+] Gibbs reaction energy.

* Energy value of CH_4, H_2, and CO from Table 4.2.

** At least 10% of heat is lost with flue gases (CO_2) in incineration and direct gasification.

*** Formula for MSW $C_6H_{10}O_4 \rightarrow CH_{5/3}O_{2/3}$.

**** Syngas diluted by nitrogen gas.

213

Table 8.2. Processes for removing air pollutants.

Pollutant	Removal Process	Reduction %
Fly ash	Electrostatic precipitators and fabric filter	95–99
SO_x and HCl	Wet scrubbers or multicyclones	50–90
NO_x	Selective catalytic reduction	10–60
Heavy metals	Dry scrubbers and electrostatic precipitation	70–95
Dioxin and furans	Activated carbon and fabric hose filters	50–99.9

The only internationally licensed incinerator in the Czech Republic (Ostrava Region) described in the report by Petrlik and Ryden has a capacity of processing 10,000 tons per year of municipal and industrial solids waste and, for air pollution control, consumes 1,134 tons of calcium hydroxide and 140 tons of activated carbon. It transforms the feed solids into hazardous solid waste that must go to a landfill. The contaminated activated carbon is incinerated in the incinerator. A notable caveat related to deposition of the very hot ash waste from incinerator into landfill is formation of syngas from residual organic carbon in the landfill and its spontaneous reforming into hydrogen under anaerobic conditions in the landfill. Because of the air pollution concerns and stringent air pollution control regulations, no incinerators were put into operation in the US in this century and the regulations in the developed countries of Europe are also becoming more stringent. However, in China today implementing incineration of MSW with lax air pollution controls has been progressing at a feverish pace, causing a serious air pollution problem.

In the US, most burning of MSW is known as *waste-to-energy* (WtE) or *energy from waste* (EfW). It involves complete combustion of MSW by an incinerator with or without preprocessing. Incineration with refuse-derived fuel (RDF) differs from mass burn because the incoming waste is processed before combustion to improve fuel performance. These facilities located mainly in the US Northeast states process 28 million tons of waste per year, which is about 13 percent of the total mass of MSW minus the recycled fraction. They have the capacity to produce 2,720 megawatts of power (US EPA, 2016), which corresponds to the energy yield of 0.85 kW-hr/kg of combusted MS, about 12% of the total energy value of MSW heating value estimated by Valkenburg et al. (2008) as 7 kW-hr/kg of combusted MS. However, the energy produced by WTE systems is less than 1% of the total electric energy produced in the US. Ash from MSW incineration, because of its toxicity, must be put in engineered safe landfills. The electricity yield of 12% is less than estimates made in the last century that did not include the electricity use for air pollution control such as electrostatic precipitators, pumping water to run scrubbers, and pushing air through filters.

The process schematic is presented in Figure 8.1. First, recyclable and noncombustible materials are removed from the MSW and the refuse material is shredded, dried, and/or compacted into pellets or cubes, to produce a more homogenous refuse-derived fuel (RDF). RDF can be used as a fuel in either a dedicated or an existing boiler alone or with other fuels such as natural gas. Heat generated by incineration heats water (high-quality heat) in the boiler to produce steam than runs a turbine and generator. The process produces fly ash and bottom ash and flue gas that must be filtered and scrubbed of undesirable and toxic pollutants (Gershman, Brickner, and Bratton, Inc., 2013; Setyan et al., 2017). Figure 8.1 presents a schematic of an incinerator in Switzerland with the flue gas exhaust treatment system that has a very high efficiency to remove airborne particles. In addition to high-quality heat generating steam for electricity production low-quality heat is recovered by a heat exchanger.

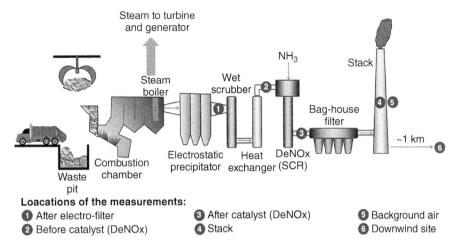

Figure 8.1. Incineration process schematics of a Swiss incinerator. *Source:* Setyan et al. (2017*), Atmospheric Environment* 66: 99–109, Copyright © Elsevier (2017).

Only a fraction of about 13% of the total MSW is processed in the US by incineration, most of MSW still goes to landfills. Expanding incineration Sweden has now reached about 50% and, because of extensive recycling, Sweden has almost reached the goal of zero land-filling (Fredén, 2018).

Heat Energy to Dry the Solids

For effective incineration and for converting the residual digested sludge into a fertilizer for commercial distribution, such as the popular Milorganite distributed by the Milwaukee Metropolitan Sewerage District, the residual sludge solids must be dried and disinfested by heat. If the solids are to be burned along with dry MSW solids as a fuel in incineration, the solids must be dried and moisture vaporized, which has also a disinfecting effect. Drying and even more so evaporating water requires heating energy. In early small wastewater treatment plants of the middle of the last century, sludge was dried by natural evaporation in sludge drying beds, which allowed water to drain by gravity and the rest to evaporate by solar and atmospheric heat. The sludge beds were not efficient during winter; consequently, they had to provide ample storage for winter. At the end of the last century and currently, heat is provided by sludge incineration, which takes away a significant heat released from incineration and often additional heat may have to be supplemented by natural gas during winter. Gasification does not require drying but, as it will be documented in the next session, moisture in the process is vaporized to steam, which is the feed of the indirect gasification reaction. Indirect gasification is endothermic, but it produces a lot of hydrogen-rich syngas.

Heat necessary to evaporate water from MSW and residual sludge fuel consist of two processes (http://www.engineeringtoolbox.com):

$$Heat\ to\ increase\ water\ temperature\ to\ boiling = \Delta T * W * c_p$$

where ΔT is the temperature difference to the boiling point in degree °K or °C , W is water mass in kg, and specific heat capacity of water is $c_p = 4.187$ kJ°C^{-1} kg^{-1} which is the heat required for increasing temperature of 1 kg of water by 1°C. Temperature difference of 1 °K (Kelvin degree) is the same as that of 1°C (Celsius).

Heat to evaporate water to change to gaseous (vapor) state

Heat of vaporization is 2257 kJ/kg.

Note that the heat of vaporization is about five times the heat required to bring 0°C cold water to the boiling point of 100°C. The importance of including the heat of drying solids in the overall heat balance can be shown by the following example:

Estimate the heat needed to dry MSW solids that moisture content of 40% and express the heat in kJoules and kW-hours per kg of dry solids. 1 kg of wet MSW contains 0.4 kg of water and 0.6 kg of solids., which, expressed per kg of dry solids, would be: 0.66 kg H_2O/kg of ds MSW. 1 kW-hour = 3600 kJoules.

$Heat_1$ to increase water temperature to boiling

Starting with the temperature of 20°C of the solids delivered for incineration:

$$Heat_1 = \Delta T * W * c_{p\text{-}w}$$
$$= 80 \ (°C) * 0.66 \ (H_2O/\text{kg of ds MSW}) * 4.187 \ \text{kJ} \ °C^{-1} \ kg^{-1}$$
$$= 221 \ (\text{kJoule/kg ds}) = 0.061 \ \text{kW-h/kg ds}$$

$Heat_2$ to evaporate water to change to gaseous (vapor) state

Heat of vaporization is 2257 kJ/kg, then:

$$Heat_2 = 0.66 \ (\text{kg}) * 2257 \ (\text{kJ/kg}) = 1490 \ \text{kJ} = 0.41 \ \text{kW-h/kg ds}$$
$$\text{Total heat required} = H_1 + H_2 = 0.066 + 0.41 = 0.471 \ \text{kW-h/kg ds}$$

This heating value can be compared with the heat of combustion, which is 394 kJ/mole for organic carbon (see Table 8.1 in the preceding section). One mole of carbon has 12 grams. Dry combustible MSW are approximately 50% organic carbon (see the next section on MSW gasification). Then the heat generated by incineration of 0.5 kg of dry solids (1 kg of wet MSW) is:

$$\text{Heat of combustion} = 0.5 \ (\text{kg OC/kg ds MSW}) * 1000 \ (\text{g/kg})$$
$$* 394(\text{kJ/mole})/[12 \ (\text{g OC/mole})]$$
$$= 16 \ 416 \ \text{kJ/kg ds MSW} = 4.56 \ \text{kW-h/kg ds MSW}$$

OC stands for organic carbon. The efficiency of heat recovery from incineration of MSW is about 50–60%; hence, under normal circumstances and average conditions, enough heat should be available from combustion of MSW. The efficiency of incineration to produce electricity is 25–28%, which would represent electricity production of about 1.2 kW-h/kg ds MSW, or about 500 kW-h/ton of MSW before drying.

8.2 PYROLYSIS AND GASIFICATION

These processes are different from incineration. In this technology, organic carbon in MSW and other solid fuels (wood chips, coal, vegetation residues, nonrecycled plastics, tires) are thermally converted to biogas (syngas) and biofuel with limited oxygen (direct gasification) or without oxygen (indirect gasification with added hot steam). This technology potentially offers feedstock flexibility and customization for generating a range of desirable products. Gasification's main product is synthesis gas (syngas) that can be further processed into electricity, ethanol, diesel, or other organic chemicals. As it is subsequently documented, syngas can be thermally reformed with steam to hydrogen and carbon dioxide, which provides new market opportunities for gasification. Hence, these processes are not just for disposing MSW. Therein, MSW and other formerly waste materials are turned into commercial resources. There are 147 companies offering gasification technologies in different stages of development worldwide, most of which is marketed in the US through licensees. In the US, in the second decade of the twenty-first century, 21 companies had more than 21 total pilot and demonstration facilities and 17 commercial-scale facilities have been under development and/or under construction (Gershman, Brickner, and Bratton, 2013). The report by Gerschman et al. also provides a comprehensive review of gasification.

Figure 8.2 shows the pyrolysis and gasification as two sequential processes, but they do not have to be separated. If both biofuel and biogas are the desired outputs, lower-temperature pyrolysis as the first step is the process of choice. If high-quality biogas containing mostly hydrogen is the desired product, higher-temperature gasification will be selected. Hence the proportion between syngas, biofuels, and char can be regulated by temperature or separated into two steps. Small amounts of ash produced in gasification contains nutrients (mostly phosphorus) that could be extracted and used as soil conditioner and long-term degradation fertilizer. Ash does not have to go to a landfill. Because of high gasification temperature and lack of O_2 dioxins are not produced.

Processes of pyrolysis and gasification of organic solids (e.g., coal and woodchips) have been known for more than 150 years and were used extensively during and after WWII in the European countries that did not have access to oil. They provide charcoal, gas, and fuel to millions of apartments, houses, and buses and, during war, they powered war machine vehicles. As said above, pyrolysis received a scant interest in 1970s for processing residual solids from treatment plants, but these attempts were abandoned because of the economics based on minimizing cost only. Callegari, Hlavinek, and Capodaglio (2018) attempted in

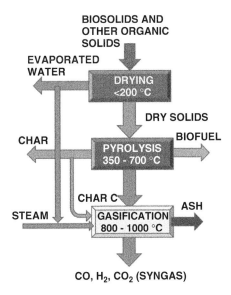

Figure 8.2. Concept of pyrolysis and gasification to obtain biofuel and a high-quality gas for energy production.

laboratory tests to enhance microwave-assisted pyrolysis and gasification of waste sludge to recover biofuel, syngas, and biochar.

Pyrolysis and gasification are different from and more efficient than energy generation by combustion and incineration. The difference between "pyrolysis" and "gasification," as shown in Figure 8.2, is the internal temperature of the process that also determines the output from the process. Pyrolysis at temperatures between 300°C to 700°C thermally decomposes biomass (MSW and organic sludge residues) into gaseous, liquid, and solid (char) products in low oxygen or no oxygen environments. Gasification requires temperatures greater than 900°C up to 1500°C (Richardson et al., 2015; Belgiorno et al., 2003; He et al., 2009; Arena, 2012; Lee et al., 2014). The products of pyrolysis include syngas $(CO + n\,H_2)$, some methane and other low-carbon combustible gases (e.g., ethane, or propane), liquid biofuel, char, tar, and possibly nitrous oxides if air is used as the agent for converting organic carbon to syngas in direct gasification. Because indirect gasification (see the next section) is not oxidation, air is not used and oxygen is not present, dioxins and furans are not formed by indirect gasification and nitrogen is not present. In direct gasification at lower temperatures there is a remote possibility of forming these dangerous pollutants. Destruction of chlorinated dioxins and furans present in the waste feed stream can take place at temperatures as low as 730°C and safely accomplished when feed is gasified at temperatures greater than 900°C (Committee on Health Effects of Waste Incineration, 2000), which is the optimum temperature of gasification.

Major oil energy companies are researching making energy from garbage and biomass. Of note are experiments of BP to develop production of jet fuel from garbage and Exxon-Mobil making fuel from algae.

Gasification of Digested Residual Used Water Solids with MSW

Gasification is vastly different from incineration. On average, conventional waste-to-energy plants that use mass-burn incineration can convert 1 ton of MSW to about 550 kilowatt-hours of electricity. With gasification technology, 1 ton of MSW can produce more than 1000 kilowatt-hours of electricity by a more cleanly and efficiently. These estimates by the Gasification Technology Council (GTC, 2014) still assume that the produced syngas is combusted to produce energy and not converted to hydrogen, which provides more efficient conversion of syngas to energy. The next section of this chapter will document that if hydrogen is the main product, the energy yield is potentially substantially greater that the GTC estimates. In the gasification process, MSW is not a fuel; it is a feedstock for a high-temperature chemical conversion process. There are no burning and GHG emissions. Gasification can convert MSW and sludge that typically would be incinerated into a useful "synthetic gas," syngas $(CO + n\,H_2)$ with little ash. At lower gasification temperatures $(\sim 750°C)$, syngas will contain other gases, such as CH_4, C_2H_y. Syngas can then be used to produce energy and valuable products, such as chemicals, transportation fuels, hydrogen, fertilizers, and electricity.

Typically, the mass of the residual sewage sludge in traditional activated sludge water reclamation plants is about 50% of the raw sludge solids but this ratio can be increased up to 70%. Consequently, integrated co-processing of residual sludge with MSW and green fuel production instead of landfilling makes sense. However, the separate biofuel-producing facilities sprouting throughout the world processing woodchips, corn, or colza, and those planned to replace municipal incinerators produce waste liquids that are overloading the aerobic municipal treatment plants. Therefore, the time has come to combine the two fuel-producing technologies into one synergistic system, although not necessarily in one facility. For example, sludge can be pumped to the MSW gasification plant and polluting liquids from gasification, if there are any, pumped back to the Integrated Resource Recovery Facility (IRRF).

It should also be pointed out that the amount of secondary excess sludge from anaerobic processes is much smaller than that produced by aerobic treatment units, including also Bardenpho denitrification plants, shown in Figure 7.1a. Also, Lettinga and Hulshoff-Pol (1991) suggested that primary settlers are not necessary when USAB reactors are used in the anaerobic treatment of used water because both COD removal and sludge/organic solids digestion can be performed simultaneously in the anaerobic reactor. About 50–60% of the influent COD can be converted to sludge mass in aerobic treatment processes while only 5% of COD becomes sludge in anaerobic treatment (Van Lier, Mahmoud, and Zeeman, 2008).

Gasification process can be direct or indirect (Figure 8.3). In direct gasification processes, the heat is provided by low oxygen burning in the same reactor. In indirect gasification processes, the heat is externally or internally supplied through a heat-transfer medium or via the physical separation of the combustion and gasification zones. Because air is used in the direct gasification the produced gas (syngas) is greatly diluted by N_2 from air and has much less hydrogen. Because steam instead air is used in indirect gasification, the produced gas is not diluted and no expensive inert N_2 separation is needed (Grotjes et al., 2016). MSW is a far greater source of recoverable energy than residual used water solids that are not disposed in agriculture. The combustible fraction of MSW (paper and cardboard, wood and woodchips, plastics and tires) represent about 50% of the total mass of MSW, or about $1\,\mathrm{kg\,cap^{-1}\,day^{-1}}$, compared to about $0.07\,\mathrm{kg\,cap^{-1}\,day^{-1}}$ of total residual municipal used water sludge and $0.5\,\mathrm{kg\,cap^{-1}\,day^{-1}}$ of biodegradable and co-digestible MSW. Gasification

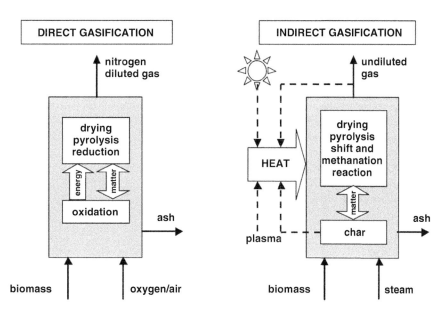

Figure 8.3. Direct (left) and indirect (right) gasification schematics. *Source:* Replotted from Belgiorno et al. (2003), *Waste Management* 23: 1–15; Copyright © Elsevier (2003).

of the residual co-digested sludge (about 50%) with the combustible MSW solids is attractive and makes sense because:

- Gasification is a well-known and tested energy producing process dating back to the nineteenth century when syngas produced from coal gasification was providing "town gas" to illuminate the cities before electric lights were invented. Town gas was also used in cooking. Gasified woodchips and low-quality brown coal provided gas (syngas) on a large scale to city apartments and fuel to vehicles before, during, and after WWII when many European communities had problems with fly ash and other pollutants emitted from smelly gasification plants, sometimes located near the center of the city and smog from burning. These plants used direct gasification concepts. The author recalls such a smelly eye-burning low-quality (direct) coal gasification plant in the post-WWII years located in the center of the second largest city of the former Czechoslovakia (Czech Republic).

- Efficient indirect gasification requires moisture that is converted to more hydrogen (see Table 8.1); hence, energy demanding drying of residual used water solids and MSW before processing may be limited or even unnecessary. Gravity and/or belt press dewatering of wet diluted solids would suffice especially when sludge is gasified with combustible MSW and the sludge could be pumped to the gasifying plant for gasification to syngas and further steam reforming to hydrogen. In the gasification of organic carbon to syngas one mole of water is needed to form one mole of CO plus one mole of hydrogen in syngas and another mole of water is needed to reform CO to CO_2 and more hydrogen. Per unit weight of organic carbon in the processed biomass, this represents roughly two kilograms of water per one kilogram of organic carbon. Under traditional circumstances this "high" water demand would be an impediment in water-deficient areas; however, plenty of water in the form of hot steam is recovered in the subsequent hydrogen fuel cell and in the moisture of MSW and sludge. Indirect gasification plants do not emit air pollution except small amounts of CO_2 if a small portion of the produced syngas is used to provide endothermic heat to the gasifier. Heat can be provided by blue/green electricity produced by the IRRF, which does not emit GHG.

- Energy yield of combustible MSW solids by gasification is relatively high, up to 85% of the MSW heating value (Ahmed and Gupta, 2010), which is 7 kW-hour/kg dry MSW and is much greater than that from MSW incineration. The technique combining gasification and pyrolysis is more energy efficient and environmentally friendly than any other process used for energy recovery from coal and municipal solid wastes (Malkow, 2004; Farzad et al., 2016).

The gasification reactions expressed in moles of gases, water, and steam are shown in Table 8.3.

Syngas from direct gasification will be diluted by nitrogen from air supplying oxygen to the direct gasification process in Figure 8.3. A two-stage gasification process – direct gasification followed by or combined with indirect – can be considered (Materazzi et al., 2016) or the conversion of syngas to hydrogen can occur in the anode of the molten carbonate fuel cell (MCFC). MILENA biogass gasifier and gasification technology (https://www.milenatechnology.com/) being developed in The Netherlands is an example of such combined gasification.

At temperatures greater than 900°C, gasification also converts char and oils from pyrolysis to more syngas, small amounts of other combustible hydrocarbon gases, carbon

Table 8.3. Gasification reactions.

Energy from	Reaction	Comment
Combustion	Org $C + O_2 \rightarrow CO_2$ $C_mH_nO_p + [m + 0.5 * (n - p)] O_2$ $\rightarrow mCO_2 + 0.5 n H_2O$	Flame is present, CO_2 is emitted.
Direct gasification	Org $C + 0.5 (O_2 + 3.7 N_2)$ $\rightarrow CO + 1.85 N_2$ $C_mH_nO_p + 0.5(m - p) (O_2 + 3.7N_2)$ $\rightarrow mCO + 0.5n H_2 + 1.85 (m - p) N_2$	N_2 diluted syngas is produced. Syngas is the terminal product of direct gasification.
Indirect gasification	Org $C + H_2O \rightarrow CO + H_2$ $C_mH_nO_p + (m - p) H_2O$ $\rightarrow m CO + (m + 0.5n - p) H_2$	H_2 enriched syngas is produced.
Conversion of CO to CO_2 and H_2	$CO + H_2O \rightarrow CO_2 + H_2$	Could occur in indirect gasification or in anode of MCFC to produce more H_2.

dioxide, and ash (He et al., 2009). The Gasification Technology Council (2014) brochure also specified that unlike incineration that emits toxins, dioxins, and furans – which need sufficient oxygen to form or reform – the oxygen deficient or oxygen-absent atmosphere in a gasifier does not provide the environment needed for dioxins and furans to form or reform and they are not present in the produced syngas or in the ash. However, direct gasification produces syngas diluted by nitrogen from the atmosphere.

Valkenburg et al. (2008) and Belgiorno et al. (2003) pointed out that MSW is a heterogeneous feedstock containing materials with widely varying sizes and shapes and composition. Raw MSW can be difficult to feed into many gasifiers and can lead to variable gasification behavior if used in as-received condition. A possible solution is the production of a refuse-derived fuel (RDF) with homogeneous and controlled characteristics. In any case, gasification is particularly suitable for many homogeneous agricultural and industrial wastes (rubber, paper and cardboard wastes, wood wastes, food wastes, plastics, manure, etc.). It is logical that recyclable inert materials (metals, glass), recyclable paper, cardboard, and textiles must be recovered and sent to reprocessing and inert nonrecyclable materials must be removed. RDF is a preprocessed form of MSW where significant size reduction, screening, sorting, and, in some cases, palletization is performed to improve the handling characteristics and composition of the material to be fed to a gasifier. The inert particles that are not removed from feed increase the mass of ash. Residual sludge from IRRF treatment of used water is more homogenous and is an asset to the RDF in the gasification process. Preprocessing and gasification of RDF and sludge residues may also produce small amounts of reject water containing sulfur, phenols, and others organic compounds (Bridgewater, 1994). This wastewater can be conveyed to the liquid influent to IRRF.

Gasification of Municipal Solid Wastes (MSW)

MSW represent the largest contribution of the biomass that can be sustainably converted to energy and resources. Obviously, this waste mass is highly heterogeneous. Figure 3.2 classified MSW into biodegradable, combustible, and recyclable fractions. Tchobanoglous and Kreith (2003) and Themelis, Kim, and Brady (2002) and other New York City (NYC) researchers classified MSW on the basis of their moisture into "dry" and "wet" materials (Table 8.4).

Table 8.4. Types of MSW in New York City.

	% by Weigh	Moisture	Volatile Matter %	Noncombustible
Dry combustibles				
Paper	26.6	10	76	5.4
Cardboard	4.7	5.2	77.5	5
Mixed plastics	8.9	2	96	2
Textiles and leather	4.8	10	66	6.5
Wood	2.9	20	68.5	0.6
Wet biodegradable and combustible waste				
Yard wastes	0.7	60	30	0.5
Food waste	12.7	70	21.4	5
Recyclable noncombustible (glass bottles and other glass, metals, aluminum containers)	9.8			
Hazardous waste	0.7			
Bulk items (appliances, furniture, etc., can be made recyclable)	9.9			

Source: Tchobanoglous et al. (1993) and Tchobanoglous and Kreith (2003).

This division is important when there is a plan to incinerate MSW by combustion because "wet" components have a low calorific value and incineration may require auxiliary drying by heat provided by natural or digester gas. This attribute becomes less important, almost mute, when the combustible wet and dry MSW are to be processed by indirect gasification, which requires water (steam). The third category is "dry noncombustibles" – metal (bulk to small items), glass, concrete and mortar from demolitions, and other inorganic compounds that have no heating value recoverable energy value and must be removed. In the US, highway concrete and asphalt are recycled.

MSW are not pure carbon. Generally, MSW components also contain hydrogen and oxygen in significant amounts. Other elements such as nitrogen, phosphorus, and sulfur are also present but in insignificant quantities. The organic C, H, and O contents of MSW are in Table 8.5. A lot of MSW is cellulose and hemicellulose.

The formula for the COD content of organic combustible compounds is theoretically the same as that for incineration presented previously, or:

$$C_m H_n O_p + [m + 0.5\,(n - p)]\,O_2 \rightarrow m\,CO_2 + 0.5\,n\,H_2O$$

$$\underbrace{}_{\text{COD}}$$

Themelis, Kim, and Brady (2002) estimated the molecular formulae corresponding to each of the combustible components of MSW:

Mixed paper, textiles, diapers	$C_6 H_{9.6} O_{4.6} N_{0.036} S_{0.01}$
Mixed plastics, rubber, and leather	$C_6 H_{8.6} O_{1.7}$
Mixed food wastes	$C_6 H_{9.6} O_{3.5} N_{0.28} S_{0.2}$
Yard wastes and wood	$C_6 H_{9.2} O_{3.8} N_{0.01} S_{0.04}$
Miscellaneous	$C_6 H_{9.6} O_{3.8}$

Table 8.5. Composition of combustible municipal solid waste in New York City.

Component of Waste Stream	% Dry Weight in NYC	% by Weight				
		Carbon	**Hydrogen**	**Oxygen**	**Nitrogen**	**Sulphur**
Paper	26.6	43.5	6.0	44.0	0.3	0.2
Cardboard	4.7	44.0	5.9	44.6	0.3	0.2
Plastics	8.9	60.0	7.2	22.8	—	—
Textiles	4.7	55.0	6.6	31.2	4.6	0.2
Rubber and leather	0.2	69.0	9.0	5.8	6.0	0.2
Wood	2.2	49.5	6.0	42.7	0.2	0.1
Glass	5.0	0.5	0.1	0.4	<0.1	—
Metals	4.8	4.5	0.6	4.3	<0.1	—
Other	4.6	26.3	3.0	2.0	0.5	0.2
Atomic weight	g/mole	12	1	16	14	32.1

Molar formula for combustible New York MSW is about $C_6H_{9.6}O_{3.6}$, rounded to $C_6H_{10}O_4$
Combustible (volatile) MSW content is about 47% and moisture content is 40% of the total weight of MSW. Specific weight of MSW = 1 600 kg/m³

Source: Tchobanoglous et al. (1993) and Tchobanoglous and Kreith (2003).

They also showed that the hydrocarbon formula $C_6H_{10}O_4$ closely approximated the mix of elements in the combustible organic wastes in MSW.

Direct or Indirect Gasification?   Direct gasification or low temperature pyrolysis of wood have been practiced probably for centuries. Gasification of wood and coal to produce syngas dates back to the nineteenth century when the produced gas was used in street lamps and later for cooking and as a substitute for gasoline for running vehicles (Siedlecki, de Jong, and Verkooijen, 2011). Consequently, it is more known and used than indirect gasification. When air is the gasification agent, then syngas from direct gasification contains large proportions of nitrogen and nitric oxides and other gases from air, which significantly reduces in direct gasification its calorific value. On the other hand, the direct gasification of MSW to syngas is exothermic while indirect gasification is endothermic.

The reaction of syngas forming direct gasification low oxygen provided by air is:

$$C_mH_nO_pN_xS_y + 0.5*(m - p + 2x + 2y)(O_2 + 3.7N_2) \rightarrow mCO + 0.5\,n\,H_2 + 1.85$$
$$(m - p + 2x + 2y)N_2 + xNO_2 + y\,SO_2$$

Dry air contains by volume 78.09% nitrogen, 20.95% oxygen, 0.93% argon, 0.04% carbon dioxide, and the rest are methane and other gases and aerosols. If N and S content in the gasified compounds are small when compared to the three main elements (C, H, and O) of the gasified organic compound, the syngas for each mole of carbon will contain one mole of carbon monoxide, one or less molecules of hydrogen, and almost two molecules of nitrogen gas. Char and tar are also produced. Hydrogen-poor syngas is the terminal product of direct gasification. Adding more oxygen would make more carbon dioxide and reduce the energy value of the gas. The process would turn into incineration.

Indirect gasification with steam produces hydrogen-enriched synthetic gas (syngas) that, depending on the temperature of the reactor, can contain carbon monoxide and dioxide, methane and other low-carbon gases, char (organic carbon), and tar. The overall equations is (Lee, Chung and Ingley, 2014):

$$C_mH_nO_p + z\,H_2O \;\rightarrow\; \underbrace{m_1\,CO + m_2CO_2 + m_3\,CH_4 + x\,H_2}_{\text{Syngas}} + \underbrace{m_5\,C_s}_{\text{Char/tar}}$$

where $m = m_1 + m_2 + m_3 + m_4 + m_5$ and z is the molar quantity of steam needed to form syngas and char that come from the moisture of the feedstock and from steam produced by the system – for example, by the hydrogen fuel cell. Indirect gasification is the most frequent type covered in the recent gasification literature. The equation for indirect gasification in a more specific form is:

$$C_mH_nO_pN_xS_y + (m - p)\,H_2O \rightarrow m\,CO + x\,NH_3 + y\,H_2S$$
$$+[m + 0.5 * n - p - (1.5x + y)]H_2$$

wherein for each atom of carbon of the gasified compound two molecules of hydrogen are included in the produced syngas and there is no dilution by nitrogen gas.

Table 8.1 documented that (theoretically) indirect gasification process retrieving syngas from MSW and many other organic compounds produces more energy because it generates more hydrogen-enriched gas and contains no diluting nitrogen. However, it requires heat and a source of water (steam) that generally is available in the MSW moisture and/or both heat and steam can be provided by the hydrogen fuel cell. The volume of needed water is very small (in liters cap^{-1} day^{-1}) when compared to treated used water flow (hundreds of liters cap^{-1} day^{-1}).

If the temperature inside the indirect gasifiers is increased to 850–950°C and correspondingly more steam is added, methane and lower carbonate gases are reformed to carbon monoxide and hydrogen, and ultimately by adding another mole of steam, carbon monoxide in an exothermic reaction is converted internally or externally in a reforming unit to carbon dioxide, and hydrogen and char/tar are converted to mostly inorganic ash and more hydrogen. The steam-reforming process converting syngas to carbon dioxide and hydrogen is exothermic.

The intermediate gasification processes in the indirect gasification are (Lee, Chung, and Ingley, 2014):

Org. solid $C + H_2O \rightarrow CO + H_2$ *Char and coal gasification.* Temperature >850°C, low pressure

Org. solid $C + 2\,H_2 \rightarrow CH_4$ *Methane formation from char.* Low temperature, high pressure

$CH_4 + 2\,H_2O \rightarrow CO + 4\,H_2$ *Methane reforming* endothermic reaction, temp. >700°C

$CO + H_2O \rightarrow CO_2 + H_2$ *CO reforming* exothermic reaction, temperature >250°C

Gasified organic matter may – but should not – contain other elements like chlorine, nitrogen, and sulfur. Organic N and S combine with hydrogen and are released into syngas as ammonia and sulfite in indirect gasification and in direct gasification as nitrogen oxides and sulfate (SO_x). Sulfate upon release into the atmosphere combines with the atmospheric moisture and forms acid rain. Chlorine is a part of polyvinyl chloride plastic (see Section 8.6) and nitrogen and sulfur are in sludge and in biomass. Other possible elements are mostly negligible.

The equation to reform carbon monoxide to carbon dioxide and hydrogen in indirect gasification is (Table 8.1):

$$CO + H_2O \rightarrow CO_2 + H_2$$

and the overall hydrogen yield equation in the indirect gasification after reforming syngas CO to carbon dioxide is:

$$C_mH_nO_pN_xS_y + z\,H_2O \rightarrow m\,CO_2 + x\,NH_3 + y\,H_2S$$
$$+ [2m - p + 0.5 * n - (1.5x + y)]H_2$$

where $z = 2m - p$ is molar quantity of water needed to generate syngas (hydrogen) from the feed, which is provided by the feedstock moisture and steam generated by the system. Unlike in direct gasification where syngas is the terminal product, adding one molecule of steam to carbon monoxide in the indirect gasification is an exothermic process that also generates more hydrogen.

Based on MSW composition in Table 8.5, it appears that textiles, rubber, and plastics are a good source of syngas and ultimately of hydrogen. The total dry weight percentage of the NYC MSW is 62%. Lower heating value (LHV) of nonrecycled plastics (NRP, about 80% of the 14 million tons of plastic containers and packaging in the US) is about 32 MJ/kilogram. The calculated heating value of the NRP is higher than that of the average grades of coal and petroleum coke available on the US market (Gershman, Brickner, and Bratton, 2013).

Differences in heating balance of direct and indirect gasification are also important in the selection of the process. Heat in direct gasification is generated internally by exothermic heat of the gasification process and by burning/gasifying char and tar. However, the feed into the gasifier must by fully dried before entering the gasifier. In contrast, in indirect gasification, moisture in the feed biomass is a source of more hydrogen and energy. Furthermore, air entering the direct gasification reactor is also heated to high temperature while indirect gasification is airless. This partially counterbalances the endothermic heat losses of indirect gasification, which will be illustrated by the subsequent example.

At the time of writing this manual it was not clear which type of gasifiers will prevail 15 to 30 years in the future. Indirect gasification, which avoids external full drying of solids and produces syngas rich with hydrogen and undiluted with nitrogen, appears to be a logical choice. Hybrid gasifiers, which combine advantages of both gasification concepts, have been investigated in Holland and elsewhere. These systems can employ two reactors or compartments, one direct gasification or combustion reactor and the other one indirect gasification or a combination of direct exothermic gasification with air providing the necessary heat followed by fluidized endothermic bed compartment with steam (Richardson et al., 2015). The direct exothermic gasification section accepts char and tar and portion of solids and generates some syngas and endothermic heat.

Example of Gasification of Combustible MSW. Using the formula for the NYC MS of $C_6H_{10}O_4$, the CO_2 yield from indirect gasification of 1 kg of dry solids (ds) after all of CO was internally reformed in the gasifier and/or in the anode of MCFC is:

$$CO_2(Content) = [6 * (12 + 32)/(6 * 12 + 10 * 1 + 4 * 16)] * 1$$
$$= 1.8 \text{ kg of } CO_2/\text{kg ds MSW}$$

and hydrogen yield potential is:

$$\text{Hydrogen yield} = [(2 * 6 + 0.5 * 10 - 4)] * 2/(6 * 12 + 10 * 1 + 4 * 16) * 1$$
$$= 0.18 \text{ kg of } H_2/\text{kg MSW}$$

Water needed as steam for converting 1 kg (dw) of MSW to hydrogen is then:

$$H_2O \text{ need} = [(2 * 6 - 4)(2 * 1 + 16)/(6 * 12 + 10 * 1 + 4 * 16)] * 1$$
$$\approx 1.0 \text{ kg } H_2O/\text{kg of dry combustible MSW}$$

However, in the synergistic use of a gasifier making fuel for a molten carbonate fuel cell, the indirect gasifier would produce syngas [$mCO + (m + 0.5 * n - p)H_2$] and CO reforming to hydrogen could be completed in the anode compartment of the MCFC, which has a lot of hot steam. In this case, the water demand for gasification would be:

$$H_2O \text{ need} = [(6 - 4)(2 * 1 + 16)/(6 * 12 + 10 * 1 + 4 * 16)] * 1$$
$$\approx 0.25 \text{ kg } H_2O/\text{kg of dry combustible MSW}$$

which could be totally provided by the MSW and sludge moisture content, and less heat would be needed to vaporize the moisture water and bring the steam to 900°C. The hydrogen yield would be the same.

The indirect gasification water need is important because the moisture content of the NYC MSW is about 40%; meaning that 1 kg of MSW contains 0.4 kg of water and 0.6 kg of solids, which, expressed per kg of dry solids, would be 0.66 kg H_2O/kg of ds MSW. This water content is more than the steam needed to gasify the MSW and solids biomass to syngas. This implies that there is no reason for excessive drying MSW, just measuring the moisture content. Lee, Chung, and Ingley (2014) documented that providing excess steam improves hydrogen productivity in the indirect gasification reactors. Additional steam can be provided in the MCFC. (See Section 8.7, "The IRRF Power Plant," which proposes the synergistic system of gasification and fuel cells.) The amount of internal heat needed to evaporate the moisture content and bring the steam to a reactor temperature greater than 850°C must be accounted in the thermal energy balance. The moisture content of dry combustible NYC MSW is less than 10%. If only dry combustibles are gasified and wet (moisture content [60–70%] biodegradable MSW are digested, then the heat energy needed to elevate moisture to 900°C would be less by a fraction of kW-h/kg ds.

Based on Table 4.2, the energy value of hydrogen obtained by gasification of 1 kg of combustible MSW and sludge from IRRF energy yield by gasification of MSW would be:

$$E_{MSW} = 0.18 \text{ (kg of } H_2) \times 39.4 \text{ (Kw-hr/kg } H_2) = 7.1 \text{ Kw-hr/kg of MSW}$$

This theoretical value of the energy yield is almost identical to the experimental value of 7 kW-hr/kg MSW, measured by Valkenberg et al. (2008). If this paper is used as a reference, then the energy efficiency of indirect gasification would be about 98%, which far exceeds incineration. It is reasonable to assume that the measured yield will be less than the theoretical maximum. Using a similar methodology and data of NYC MSW, Themelis, Kim, and Brady (2002) estimated the heating value of combustible fraction of MSW as 18,490 kJ/kg of ds MSW, or 5.23 kW-h/kg. The heating value is linked to less efficient incineration/combustion and not to indirect gasification that produces H_2 enriched syngas.

From Table 4.2, the specific density of hydrogen is 0.09 kg H_2/m^3 and that of carbon dioxide is 1.8 kg CO_2/m^3, respectively. Then the volume of hydrogen-enriched syngas from 1 kg combustible ds MWS is:

$$V_s = V_{H_2} + V_{CO_2} = 0.18/0.09 + 1.8/1.8 = 3 \text{ m}^3$$

The volume of syngas from direct gasification would be significantly greater and energy value smaller because of dilution by atmospheric nitrogen.

Finally, the heat for indirect gasification on the last line of Table 8.1 is negative $(-131 + 41) = -90$ kJ per mole of organic carbon, hence the process of indirect gasification is endothermic and needs auxiliary heat. Using the information in Table 8.5 reporting composition on NYC MSW for illustration of the energy need, the weighted average of the organic C content of combustibles (without metals, glass, construction waste, etc.) is 47% of the dry weight. Hence 1 kg of dry weight of combustible dry weight (dw) of MSW contains the following number of moles (AW, atomic weights) of organic carbon:

$$AW_{Org\ C} = (470 \text{ (g of C in MSW)}/(1 \text{ (kg dw MSW)} * 12)$$

$$= 39.2 \text{ } (AW_{Org\ C})/\text{kg dw MSW}$$

The theoretical heat energy used in gasification to convert Org C to syngas is then:

$$\text{Gasification endothermic heat H}_e = 39.2 \text{ } (AW_{Org\ C}/\text{kg}) \times 91 \text{ } (\text{kJ}/AW_{Org\ C})$$

$$= 3567 \text{ (kJ/kg)}/3600 \text{ (kJ/kW-hour)}$$

$$= 0.99 \text{ (kW-h)}/(\text{kg dw combustible MSW})$$

$$\approx 1 \text{ kW-h}/(\text{kg})$$

To complete the heat balance, the heat necessary to evaporate the solids moisture and bring the steam temperature inside the gasifier to 900°C is estimated. The heat demand is calculated only for the solids moisture (0.66 kg H_2O/kg of ds MSW) because any steam delivered from the molten carbonate fuel cell is already hot and under pressure. The heat of bringing the water to boiling ($Heat_1$) and vaporizing heat ($Heat_2$) is the same as the heat for drying the MSW and other solids before incineration, or:

$$Heat_1 + Heat_2 = 0.061 + 0.41 = 0.471 \text{ kW-h/kg ds}$$

Unlike in incineration where $Heat_1$ and $Heat_2$ are drying the solids, in the indirect gasification they change liquid water to steam that is used to make syngas and energy. During gasification, steam is heated to ~900°C, which is $Heat_3$, which is calculated as follows:

Select mid-range isochoric specific heat capacity, which for steam is $c_{p-s} \approx 1.9$ kJ. Then the heat to increase steam temperature from 100°C to 900°C is (www.engineeringtoolbox.com)

$$Heat_3 = \Delta T * W * c_{p-s} = (900 - 100) * 0.66 * 1.9 = 1003 \text{ kJ} = 0.28 \text{ kW-h°C}^{-1} \text{ kg}^{-1}$$

Total heat to warm moisture to 900°C is:

$$Heat_t = Heat_1 + Heat_2 + Heat_3 = 0.061 + 0.41 + 0.28 = 0.74 \text{ kW-h}$$

Isochoric specific heat capacity assumes change of temperature but not much of pressure. After adding the gasification endothermic heat, the heating of the gasifier will require:

$$\text{Gasifier heat} = H_t + H_g = 0.74 + 1 = 1.74 \ (\text{kW-h})/(\text{kg dw combustible MSW})$$

This heat will be provided by an internal or external burner, and transfer of heat from the fuel cell power and in the overall energy balance will be subtracted from the energy produced by the MCFC in the power plant. (The IRRF power plant section concludes this chapter.)

The same amount of heat energy (0.74 kW-h/kg ds) is used for auxiliary drying of feed solids for direct gasification. However, unlike indirect gasification, which is airless, cold air enters the direct gasification reactors and is heated to the reactor temperature. The isochoric heat capacity of air ranges from 0.72 kJ $^{\circ}$C^{-1} kg of air^{-1} at 18°C to 1.17 kJ $^{\circ}$C^{-1} kg^{-1} at 900°C, the mid range being 0.95 kJ/kg (www.engineeringtoolbox.com).

$$\text{Air use} = 0.5 * (m - p)(O_2 + 3.7N_2)/(C_mH_nO_p)$$
$$= 0.5 * (6 - 3.6) * (32 + 3.7 * 14)/(6 * 12 + 9.6 - 3.6 * 16)$$
$$= 0.72 \ \text{kg of air/kg of dry MSW solids}$$

$$\text{Heat to warm air} = 0.72 \ (\text{kg air/kg ds MSW}) * 1.09 \ (\text{kJ } ^{\circ}\text{C}^{-1} \ \text{kg}^{-1}) * 872 \ (^{\circ}\text{C})$$
$$= 684 \ \text{kJ/kg ds}$$
$$= 0.19 \ \text{kW-h/kg ds}$$

Gasification Schematics and Reactors   A simplified schematic of the indirect gasification reactor is seen in Figure 8.4. Indirect gasification to syngas is an endothermic process but it produces higher-quality syngas with more energy in hydrogen and in carbon monoxide without being diluted and polluted by nitrogen gases from the air. Furthermore, because water is required, the feedstock does not have to be excessively dry, which represents significant energy and clean water savings in preprocessing the feed solids. The produced carbon

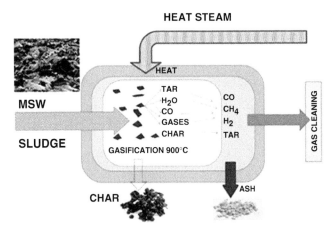

Figure 8.4. Indirect gasifier schematic. *Source:* Adapted from Arena (2012), Richardson et al. (2015), and Gómez-Barea (2010).

monoxide or methane could be used to provide heat in indirect gasification, but in this heating CO_2 and possibly other gases are emitted, and the exhaust gas may require air pollution control. However, if the burning of hydrogen is used for heating, there is no waste gas exhaust, only steam and heat that can be directed into the gasifier. Produced syngas may require some treatment to remove sulphur before using. Based on the circular economy paradigm, it is better, more logical, and more efficient to recycle textiles, paper, and cardboard to make paper and cardboard products and other recycled materials than to make syngas (ICF, 2015).

Gasifier types. No detailed gasifier designs are featured herein. This treatise is about the future and it is not known what designs will be used 20–30 years from the present. Hence, the schematics in Figures 8.3 and 8.4 are generic and not a design. Belgiorno et al. (2003), Valkenburg et al. (2008), Arena (2012), Richardson et al. (2015), and Farzad (2016) have described the following general types of gasifiers:

1. *Fixed bed direct gasifiers* in which the biomass moves slowly downward through a fixed zone supported by a grate at the bottom of the gasifier. Fixed bed gasifiers are relatively small and are mostly heated directly by heated air. These gasifiers have limited capacity because of the relatively long residence times of the biomass.

2. *Fluidized bed indirect gasifiers* (FBIG) use sand or another medium made to become essentially a buoyant weightless quicksand by introducing the gas stream below the sand bed at sufficient velocity to barely lift (fluidize) the sand or any other heat carrying particles. In the second decade of this century the Energy Research Centre of The Netherlands has developed a 1-MW prototype FBIG that uses olivine as the heat carrier medium because of its catalytic activity, attrition resistance, and cost. Olivine is a silicate mineral in which magnesium and some iron cations are embedded in the silicate tetrahedral, which resulted in high H_2 yields (Aranda et al., 2014. Grootjes et al., 2016).

3. *An indirectly heated fluidized bed gasifier* withdraws a portion of the fluidized bed contents, including the char in the bed, and transfers it to a second fluidized bed where air is introduced into the bottom of the second vessel to combust the char, thereby heating the heat carrier to a high temperature. The hot carrier is then transferred back to the gasifier, where it is used to heat the biomass. Only steam is introduced to the bottom of the gasifier in this configuration to fluidize the bed. The efficiency of a fluidized bed indirect (hybrid) gasifier is about five times of that of a fixed bed direct gasifier (Belgiorno et al., 2003).

4. *Entrainment flow indirect gasifier* is like the fluidized flow gasifier described above except the gases are introduced into the bottom of the gasifier vessel at much higher velocity causing the fluidized medium and biomass to become entrained and carried out from the top of the gasifier. The medium solids are then removed from the product gas, reheated in the hydrogen or syngas burner, and returned into the gasifier. This gasifier is a possible candidate for the IRRF power plant.

5. *Plasma gasifiers* use plasma, which is a low-density, high-temperature ionized vapor, to heat the feed biomass stream. Air, nitrogen, argon, carbon dioxide, and steam may be used as the plasma gas. Feedstock materials are treated using extremely high temperatures that convert solid or liquid fuel streams into syngas (likely a mixture of H_2 and CO_2 if the temperature is very high) and vitrified slag. The temperature of the plasma jet may reach up to 6000°C. Thermal plasmas may be generated by either an electric arc or even by a radio frequency induction discharge (Zhang et al., 2011). This may be the future.

A commercial gasifier with steam reforming unit for processing of organic biomass solids and converting them to hydrogen has been developed in Japan and India and is promoted by Japanese Intech Corporation under the name Blue Tower™ (Dowaki, 2011). A similar system was announced and commercially implemented in Arizona by Concord Blue Energy in collaboration with Lockheed Martin Corporation in 2017 (http://www.concordblueenergy.com/). In the former system the final product is hydrogen while the latter gasifier's end product is syngas. Both can be processed by the MCFC to produce electricity and steam. Germany is also actively researching and rapidly implementing medium size (>10 MW) gasifiers. Other gasification and plasma applications throughout the world were listed in Valkenburg et al. (2008). Undoubtedly, the research on developing less expensive gasifiers will continue.

Tanthapanichakoon and Jian (2012) described two alternative direct gasification processes used in producing biofuel in which biomass is used to produce syngas. The first syngas production is through partial direct combustion that was also published by Roddy and Manson-Whitton (2012). Syngas is generated according to the following equations:

$$C + H_2O \rightarrow CO + H_2 \tag{1}$$
$$C + O_2 \rightarrow CO_2 \tag{2}$$
$$CO_2 + C \rightarrow 2CO \tag{3}$$

In this process biomass is first reduced to carbon (char) by pyrolysis, followed by alternating blasts of steam and air. To initiate the first reaction, hot steam is passed through the mixture. As the reaction proceeds, temperature falls to a point when air must be passed through the mixture to initiate reactions 2 and 3. Reaction 2 is exothermic, while reaction 3 is endothermic. The two combined reactions to produce carbon monoxide are exothermic. Subsequently, the temperature rises to a point where reaction 1 can be restarted. This cycle continues until all carbon is consumed.

Syngas can also be produced through *catalytic partial oxidation*. The University of Minnesota team (O'Connor, Klein, and Schmidt, 2000) developed a metal catalyst that reduces the biomass gasification reaction time by up to a factor of 100. The entire process is autothermic; therefore, heating is not required.

Both pyrolysis and gasification require catalysts. Adding dolomite as a catalyst in gasification significantly increases hydrogen content of the produced syngas (Figure 8.5). To maximize hydrogen fraction, steam is added. He et al. (2009) investigated using a calcined dolomite (calcium-magnesium hydroxides) as a catalyst and found magnesites as active chemical solids for cleaning raw hot gas from biomass gasifiers with steam. These additives are derived from natural dolomite, which is calcium-magnesium carbonate (Figures 4.1 and 8.5). Dolomite is an inexpensive and abundant mineral that significantly reduces the tar content of the product gas. He et al. pointed out that with the cheap dolomite catalyst at temperatures higher than 800°C tar production can be eliminated, resulting in more purified gas. Other catalysts have also been investigated (Luo et al., 2012). Magnesium hydroxide and oxide are also used in nutrient recovery to produce struvite (ammonium magnesium phosphate) from phosphates and ammonium in used water and supernatants. (See Chapter 9, "Nutrient Recovery.")

In a simple traditional energy production by a utility, syngas from pyrolysis/gasification and methane from digestion can be mixed and by combustion the mix produces steam in a boiler to run a turbine and generate electricity. This traditional energy-producing process is not efficient. Most of the energy produced in the traditional combustion followed by turbine running a generator is heat, about 65%, which is released by the cooling system and flue gas

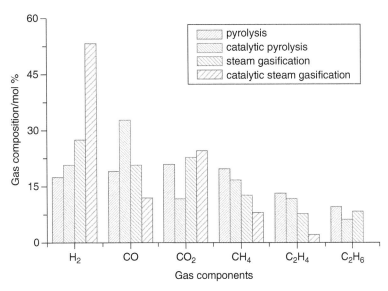

Figure 8.5. Effect of dolomite catalyst on product syngas composition from gasification. *Source:* Reprinted with permission from He et al. (2009), *International. Journal of Hydrogen Energy* 34: 195–203; Copyright © Elsevier.

into atmosphere or into water resources. Consequently, efficiency of electricity production is about 30–35% (Table 8.6, Eurelectric, 2003). Reported values of syngas yield from wood-chips by direct gasification were 2.2–2.5 m^3/kg of woodchips and the heating value of syngas was 1.42 to 1.53 kW-hr/m^3. Wan et al. (2013) quoted that on average 1 kg of biomass processed by pyrolysis produced 0.75 kg of biofuel (oil), 0.1 kg of char, and 0.15 kg of gas. As stated above, char and biofuel can be gasified at higher temperatures into more syngas.

In studying indirect gasification, He et al. (2009) found that at the reactor temperature of 950°C, the product gas contained 70 mole % of syngas (53% H_2 and 17% CO), 22% CO_2, 6% CH_4, and 2% of low carbohydrates. Lee, Chug, and Ingley (2014) added more steam over the stoichiometric steam dosage and measured the composition of the gas produced, which by reforming CO and CH_4 in the reactor increased the molar hydrogen and CO_2 yields to 65% and 35%, respectively. This indicates that CO and CH_4 in syngas can be reformed to CO_2 and H_2 in the gasification reactor. (The section "Hydrogen Fuel Cells" discussed how carbon monoxide reforming can also take place in the anode compartment of the cell.)

Smaller unit values of syngas production should be expected from gasification of unsorted MSW. Luo and Yi (2012) reported that production of 1.75 m^3 of syngas/kg of MSW solids was achieved by gasification at 900°C, while He et al. (2009) by adding steam under similar conditions achieved 1.65 m^3 of (dry) syngas per kilogram MSW, respectively. These values are about 50% to 60% of theoretical calculations that estimated the theoretical yield of 3.2 m^3 of syngas/kg of ds MSW that was sorted. The syngas and hydrogen yield obviously depend on composition of the MSW. Hence, these lower values represent MSW that most likely include nocombustible waste and/or waste that have high moisture content. Lower heating value (LHV) of syngas is approximately 3.2 kW-h/m^3 (He et al., 2009), which is more than that reported by Wan et al. (2013) and Panigrahi et al. (2002). Belgiorno et al. (2003) and Arena (2012) reported the produced heating value of gas from indirect gasification of MSW ranging from 3 to 5.5 kW-hr/m^3, which is three times greater than that from direct gasification with air because in the latter process significant portion of produced gas is inert nitrogen and polluting nitrogen oxides. Note that the term heating value implies incineration, which is not efficient. Besides energy, syngas is a valuable raw material for many other products, such as biofuel, alcohols, plastics, and fertilizers (Gershman, Brickner, and Bratton, 2013; Arena, 2012, Grootjes et al., 2016). Grootjes et al. also pointed out that the key feature of the indirect gasification process is that the produced gas keeps 70–80% of the chemical energy initially contained in the initial solid fuel.

Gasification ash. The amount of ash from gasification is much smaller than that from incineration, less than 10% of the MSW mass and volume. Arena (2012) and Psomopoulos et al. (2009) estimated that if the gasifier ash is landfilled, it would require 3% of area (volume) that otherwise would be needed for unprocessed MSW. Furthermore, gasifier ash does not contain organic dioxins, furans, and other toxins associated with incineration ash. It contains metals that can be recovered by magnets, and phosphorus extracted and used in agriculture as soil conditioner and fertilizer. However, proper methods of phosphorus and metals extraction must still be developed. Ash can be used as construction material. Hence, MSW gasification with recycling could lead to net-zero landfilling. However, toxic metal concentrations will occur in ash at much higher concentrations than in the feed stock. This will cause problems with the direct agricultural uses of ash as fertilizer. (See Chapter 9, "Nutrient Recovery.")

8.3   CONVERTING BIOGAS TO ELECTRICITY

In the first decade of this millennium, the technology used for electric energy production from the digester biogas produced in the integrated resource recovery facilities (IRRF) was thermal hydrolysis (TH), along with the combined heat and power (CHP) system (Figure 8.6). The excess heat can be used for heating both the digester reactor and buildings. In the TH process, the biosolids are heated under pressure using steam to about 165°C, held for about 20 to 30 minutes, and then transferred to anaerobic digestion. The key

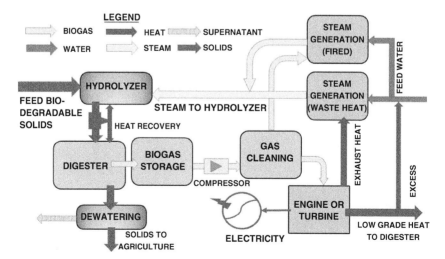

Figure 8.6. Schematic of a CHP power and heat generation power plant. Heat produces electric energy by combustion and warms biological and hydrolysis reactors. *Source:* Adapted from Knight et al. (2015).

benefits are enhancement of the digestion process, increased methane yield by about 50%, improved conversion of volatile solids in the digestive process, better COD removal, and better dewatering capability. Cooking of solids at a high temperature results in biosolids that can be used for land agricultural applications and a drier cake (Knight et al., 2015; USEPA-CHPP, 2011).

In the CHP process shown schematically in Figure 8.6, gas produced in digesters and other anaerobic units is converted to energy by combustion engines or turbines and converted to energy by a generator. The produced steam is used in the hydrolyzer. Table 8.6 presents the overall efficiencies of the current energy recovery systems. Because generator efficiency is more than 90%, most of the heat is emitted from the combustion engine or boiler/turbine cooling and steam condensation cycle.

The US EPA CHPP (2011) report listed the CHP installations in the US; 52% of installations used reciprocating engines with generators for electricity production, 32% used microturbines or combustion turbines, and 13% used fuel cells. Gas cleaning is provided to protect the engines and turbines and to satisfy the air pollution criteria and it is also very important for fuel cells. Gas cleanup includes removing hydrogen sulfide (H_2S), moisture, and siloxane (Knight et al., 2015). If steam production from waste heat is insufficient to meet the process needs of the TH plant, then supplementary natural gas–fired steam production is needed to make up biogas available to be burned in the engines or turbines to produce electricity, thus reducing the overall system efficiency for electricity production. However, this situation is unlikely with hydrogen fuel cells because the cell is a steam producer in excess of water used to reform methane and syngas. Table 8.6 includes the data for internal combustion engines (IC) that operate like heating systems of automobiles and turbines with and without heat recuperators.

The USEPA-CHPP report also estimated that on average CHP 0.22 kW-h/m^3 of influent flow to the IRRF in cold zones to 0.14 kW-h/m^3 in hot states of Texas to Georgia and Florida is needed to heat the digesters. However, it should be pointed that all treatment plants evaluated by the US EPA-CHPP were traditional biological aerobic treatment plants and had traditional anaerobic digesters. Therefore, it may be more convenient to convert the digester heating to W-h/liter of produced gas. A typical US WWTP processes 378 L (100 gallons) per day of wastewater for every person served, and produces approximately 0.0283 m^3 (1.0 ft^3) of anaerobic digester gas per person per day (Metcalf and Eddy, 2003), of which 70% is methane. The heating range is then 3.05 kW-h for cold states to 1.94 kW-h in warm states per m^3 of digester gas.

The analysis of traditional WWTP revealed that a substantial amount of surplus heat for space heating is available only in warm and hot climates, where demand for space heating

Table 8.6. Energy balance for heat and power CHP options.

Energy Flow	IC Engines	Mids Size Turbine with Recuperator	Small Turbine Without Recuperator	Small Turbine with Recuperator
Input energy	100%	100%	100%	100%
Electricity	38%	35%	23%	32%
High-grade heat[*]	27%	46%	63%	39%
Low-grade heat (hot water)	17%	0%	0%	0%
Energy loss	18%	19%	14%	29%

(Source Knight et al., 2015)
[*]Exhaust cooled to 100°C.

is minimal, except in cold winter months. In cooler climates, in many cases, there is no heat left for space heating. The cost to generate electricity using CHP at traditional wastewater treatment facilities ranged from 1.1 to 8.3 cents per kW-h, depending on the CHP prime mover (turbine or engine) and other factors. Current retail electric rates range from 3.9 to more than 21 cents per kW-h. The unit cost to generate electricity decreased with the increasing size of the primary user population.

Steam Methane Reforming (SMR) to Syngas and Then to Hydrogen

Steam reforming of methane produced by digesters and syngas carbon monoxide conversion to hydrogen followed by a hydrogen fuel cell would make the process of energy recovery more efficient than combustion and would produce more valuable hydrogen, which per one kg of weight has substantially more energy than gasoline. One kilogram of hydrogen has about the same amount of energy as 3.8 liters (one gallon) of gasoline that weighs about 2.8 kilograms. Furthermore, not only is hydrogen more potent fuel, but the conversion of hydrogen into electric energy by hydrogen fuel cell is about 60% efficient (increasing to 65% soon), as compared to about 35% efficiency of methane combustion and turbine and less than 20% efficiency of incineration. The US Department of Energy (2014b) estimated that the total baseline cost of producing hydrogen from biomass in 2011 was $2.5/kg of H_2, about half of the cost of producing hydrogen by electrolysis of water. According to the National Hydrogen Association (2010), hydrogen production cost was $1.61/kg. Thus, even then hydrogen fuel was competitive with gasoline.

Steam reforming using methane and other hydrocarbons (e.g. methanol) as feed is the most common industrial process to produce hydrogen, starting with the Haber-Bosch process. Table 4.2 in Chapter 4 presented the characteristics of the methane, carbon monoxide, and dioxide gases. As presented in Table 8.1, reforming methane to hydrogen in the SMR is a two-step process. The first process of reforming methane to carbon dioxide and hydrogen (syngas) in low oxygen atmosphere (Reaction 3 in Table 8.1) is mildly exothermic but yields only two molecules of hydrogen. Furthermore, the produced gas contains about 40% syngas ($CO + 2\ H_2$) and 60% of nitrogen from air used to provide oxygen This significantly reduces the calorific value of the produced gas. In the steam methane reforming (SMR) mildly endothermic reaction (Reaction 4 in Table 8.1), methane is mixed with high-temperature steam. In the presence of a metal-based catalyst, methane and steam in a high-temperature shift (650-800°C) produce syngas with three hydrogen molecules (moles) of hydrogen and one mole of carbon monoxide per each mole of methane, and no nitrogen is present in the produced syngas. Because of the higher hydrogen yield, the SMR process yields more energy than the low oxygen endothermic process (Process 3 in Table 8.1) and the produced gas does not have nitrogen and other gases from air.

In the second exothermic reforming step, by adding more steam, CO in a lower temperature ($\geq$ 230°C) is converted to CO_2 and more hydrogen. The overall SMR process shown in Table 8.1 is mildly endothermic but produces a lot of high-energy hydrogen and the needed energy can be partially or fully provided by solar energy (Bakos, 2005) or other excess green or waste (e.g., power plant cooling) heat sources. The steam reforming of methane can produce for each mole of CH_4 four molecules of hydrogen that theoretically will have significantly more energy than the original methane entering the process, especially when a portion of the heat needed for the reaction is provided by solar power.

The steam gas shift of carbon monoxide to hydrogen also emits one mole of carbon dioxide per one mole carbon monoxide reformed from one mole of methane. After separation of H_2, the syngas shift unit would be the only unit from which a concentrated CO_2 gas would be emitted if it is not passed to the hydrogen fuel cell or to a carbon dioxide-hydrogen separator. A special separator unit can be installed on the effluent from the gas shift unit that separates CO_2 from hydrogen (Yang et al., 2008) but it may not be needed if the produced syngas is conveyed to a hydrogen fuel cell. See the sections on carbon sequestering and on the IRRF power plant.

SMR needs very clean steam that in the proposed systems will be provided by the hydrogen fuel cell. Otherwise, in commercial hydrogen-producing operations water must be cleaned, deionized, and boiled, which would increase energy use and cost. The synergy of SMR and hydrogen fuel cells that emit high-temperature clean steam may avoid the need for high-efficiency treatment of water and using auxiliary fuel for boiling. Furthermore, if hydrogen fuel cells produce electricity from the produced hydrogen, the second gas shift of carbon monoxide in the syngas to hydrogen can occur internally in the anode compartment of the hydrogen fuel cell, which produces hot steam from the hydrogen supplied to the cell. This is discussed more thoroughly in the subsequent section describing hydrogen fuel cell.

8.4 MICROBIAL FUEL CELLS (MFCs) AND MICROBIAL ELECTROLYSIS CELLS (MECs)

In the section on fundamentals of anaerobic treatment (Figure 7.1) it was documented that in the digestion and co-digestion processes, the end products of anaerobic fermentation are acetates and hydrogen. This dark fermentation performed by obligate and facultative microorganisms is then followed by methanogenesis, in which methanogens convert acetic acids and reduce carbon dioxide to methane by using hydrogen as electron donor. The methanogenesis process is sensitive to pH, so to complete the process of methane production pH should be maintained in the range of 6.6 to 7.6 (Tchobanoglous, Burton, and Streusel, 2003). The bacteria in the process are also sensitive to low pH caused by sulfate conversion to sulfide and/or to the presence of toxic compounds such as toxic metals, and to temperature.

The theoretical hydrogen production potential by anaerobic decomposition of biodegradable compounds was shown to be 12 moles of H_2 from hexose ($C_6H_{12}O_6$). However, in a complete anaerobic digestion approximately three molecules of methane and three molecules of carbon dioxide in a form of bicarbonate alkalinity and carbonic acid are produced, which decreases pH. Then, using the energy values of methane and hydrogen listed in Tables 4.2 and 8.1 (789 KJ/mole for methane and 283.6 KJ/mole for hydrogen) the theoretical yield of energy from hydrogen would be $12 \times 283.6 = 3403.2$ kJ/mole of hexose and that from methane would be $3 \times 789 = 2367$ kJ/mole, respectively, Obviously, the real values in either case will be less but if the efficiency of conversion of the biogas to energy is considered, the advantages of hydrogen would become even more attractive. Currently, using the abiotic hydrogen fuel cells, the efficiency of hydrogen conversion to energy is ≥60% for electricity and 90% for both electricity and heat, while that of methane combustion is at most 38% for electricity and 80% total (Table 8.1). Clearly, hydrogen electric energy potential is better. However, to overcome the dead-end acetate "barrier" and produce more hydrogen, electron gains and losses of the molecules must be manipulated, and energy must be added to the process (Logan, 2008; Cheng and Logan, 2007, 2008; Verstraete et al., 2010), which can be accomplished by:

- Electrogenesis: Electricity production using microbial fuel cells (MFCs)
- Electrohydrogenesis: H_2 production from biomass using the microbial electrolysis cell (MEC), also reported as bioelectrochemically assisted microbial reactor (BEAMR) processes

Microbial Fuel Cells (MFCs)

A microbial fuel cell (MFC) is a bio-electrochemical process that produces electricity directly from the anaerobic oxidation of biodegradable organic substrates by electrogenesis. Exoelectrogenic microbes, most of which are bacteria, in the anode compartment produce electrons and protons from the oxidation of organic matter, with CO_2 and biomass as final products (Pham et al., 2006; Logan, 2008; Ahn and Logan, 2012; Arends and Verstraete, 2012; Mohan Venkata et al., 2008; Li, Yu and He, 2014; Rabaey and Verstraete, 2005; Rabaey et al., 2005; Capodaglio et al., 2013). In an MFC, microorganisms, as elsewhere in decomposition processes, oxidize organic matters to obtain energy. Under normal circumstances, the electrons are transferred to the terminal electron acceptors (TEA), such as oxygen, nitrate, and ferric ion, which may diffuse into the cells through the cell membrane and become reduced. However, bacteria used in MFCs can transfer electrons exogenously (outside the cell) to a TEA. Electrons, protons, and oxygen react in the cathode compartment, producing water (Rabaey et al., 2005; Cheng and Logan, 2007; Logan, 2008). The concept of this process has been known for more than 100 years since the publication of the article by Potter (1911) on the electrochemical effects of anaerobic decomposition. The phenomenon of microbial electricity production was then studied off and on until the first decade of the twenty-first century (Arends and Verstraete, 2012).

In their analysis, Li, Yu, and He (2014) highlighted the adaptability of MFC technology to a sustainable pattern of wastewater treatment as follows:

1. MFC enables direct recovery of electric energy as a value-added product.
2. It can achieve good effluent quality and low environmental footprint because of effective combination of biological and electrochemical processes.
3. It is inherently amenable to real-time monitoring and control, which provides good operating stability.

Microbial fuel cells are different from batteries and chemical fuel cells that use a typically unsustainable primary source of energy (e.g., by plugging the cells into fossil fuel–derived electricity outlets) to become an energy source. Unlike batteries, MFCs produce electricity directly, can use organic waste sources as a source of energy, and, in this sense, are attractive sources of "green electric energy" (non-fossil CO_2 is emitted). Li, Yu, and He also stated that unlike other energy products in used water treatment and resource recovery such as CH_4 or H_2 produced in anaerobic digestion (AD) processes, electricity from MFC is a cleaner and more widely utilizable form of energy. Moreover, MFCs can work well at ambient temperature and thus consume less energy for temperature maintenance than AD reactors. This gave an impetus to the great interest of researchers in this century in developing MFCs.

Figure 8.7 shows the principle of MFCs schematically. In the chamber on the left, bacteria use biodegradable organics as electron donors, oxidize them, and transfer the electrons to the "soup" in the reactor. In doing so they produce protons, which penetrate the membrane to the cathode. The electrons pass through a load (e.g., a resistor) and reach the cathode in the chamber on the right, where they reduce oxygen and form water. The two chambers are separated by a membrane allowing the migration of protons produced in the anode chamber. This process requires exoelectrogens. Microbiology researchers have tried to find the best microbiological population that would not only produce electricity but also efficiently treat the wastewater. The mechanisms of exoelectrogenic pathways are still not well known.

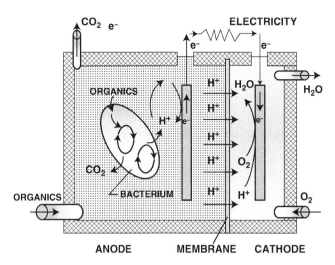

Figure 8.7. Microbiological fuel cell schematics. *Source:* Adapted and replotted with permission from Rabaey and Verstraete (2006), *Trends in Biotechnology* 23(6): 191–198; Copyright © Elsevier.

The benchmark of the research is to achieve better energy yield and performance than traditional anaerobic treatment and digestion producing methane as a source of energy. Arends and Verstraete (2012) estimated the typical net energy production via methane generation in an anaerobic digestion process being about 1 kW-h/kg of COD removed; therefore, to be reasonably competitive the MFC should roughly produce the same amount of electricity per 1 kg of COD removed in the cell and the daily rate of conversion should be about 1 kg of COD removed per m^3 of the anode compartment volume in which COD is converted to CO_2. The cathode compartment converts H^+ ion by oxidation to water.

However, in 2017 the state of the art of MFC producing electricity was still far from even entering the pilot demonstration project phase and most of the treatability studies had been performed in one-liter-size reactors. Christgen et al. (2015) focused on the economics of MFC and reported that the cost of membranes is very high and would have to be reduced by an order of magnitude to recover the capital cost by produced electricity. In analyzing MFC feasibility Li, Yu, and He (2014) and Arends and Verstraete (2012) used the generally accepted rule of thumb target value for energy conversion of 1 KW-h/kg per kg of COD or m^3 unit volume and pointed out that the maximum power output in the liter-scale MFCs are in the order of several W-h/m^3 or per one kg of COD removed, or even less (Mohan Venkata, et al., 2008). This is two to three orders of magnitude less than 1 kW-h/m^3, the target value. This level of energy production has economic ramifications; current prices for electricity, as well as projected future prices based on economic models, are too low to allow MFCs to recover the necessary capital costs for membranes and other appurtenances required for energy production.

Li, Yu, and He also compared capital and present values of operation costs of the traditional digestion and MFC, which are also greatly unfavorable by an order of magnitude, for MFC technology. They confirmed that the cost of membranes and electrodes was high, although new and cheaper materials are being discovered and tested. It is estimated that, even with relatively cheap carbon cloth electrodes and a nonwoven fabric separator that have been demonstrated in several studies, the overall capital costs of an air-cathode MFC for municipal wastewater treatment would still reach $3 /kg COD (or approximately $1.5/$m^3$ municipal wastewater). This capital cost is much higher than that of a conventional activated sludge system. The complexity of the used water and sludge without pretreatment, as well as low conductivity of the treated feed stock, is adding to the problems with MFC.

However, analyses focusing on only tangible benefit/cost analyses using current data were often misleading in the past and may not be valid in the next generation because of the worldwide thrust to reduce GHGs and search for clean green energy. Also, membranes and electrodes will become much cheaper with increased production, as happened in the case of solar photovoltaics that in 2016 became economically positive and affordable, and in one or two decades wind power may become the cheapest and dominant source of electric energy. Although many recent studies have resulted in a much deeper understanding of limitations and challenges in MFCs, more studies with real wastewater as substrate will be needed to develop low-cost, efficient components and reactor designs leading to commercial use of electricity producing MFCs for wastewater treatment in the future.

Modifications of MFCs to MECs for Hydrogen Production

Instead of generating small amounts of electricity, adding a small external DC power source potential (>0.25 V) will "assist" the bacteria at the anode to generate hydrogen abiotically at the cathode. This unit process uses *microbial electrolysis cell (MECs)*, which is a promising method of generating hydrogen energy from organic matter, including used water and other liquid wastes. MECs were discovered in the first decade of the twenty-first century by two independent research teams at Pennsylvania State University led by Professor Logan (Chang and Logan, 2008; Logan, 2008, Logan et al., 2008) and at Wageningen University in The Netherlands led by Professor Vestraete (Aelterman, et al., 2008). Figure 8.8 shows a laboratory setup of the MEC. Electrochemically stimulated active bacteria in the anode compartment of the MEC oxidize organic matter and generate CO_2, protons (H^+), and electrons (e^-). Because no air is used (the process is completely anaerobic), CO_2 is pure gas dissolved in the effluent supernatant. Recall the equation for the bicarbonate equilibrium:

$$CO_2 + H_2O \leftrightarrow H^+ + HCO_3$$

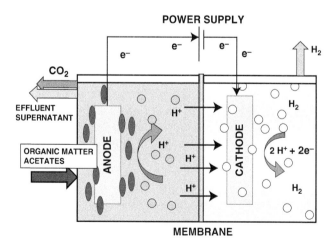

Figure 8.8. Concept of the two-chamber MEC. The Penn State Research Foundation holds several patents on MEC cells. *Source:* Adapted with permission from Logan et al. (2008), copyright © American Chemical Society (2008); and Kadier et al. (2016), open access review.

which represents that under a favorable pH, MEC will generate additional protons that subsequently could pass through the membrane to the cathode compartment to form more hydrogen.

The electrons are transferred by the bacteria to the anode, from which they travel externally to the cathode with a help of added current, and protons are released to the solution of the anode compartment. Because of the proton (pH) potential between the cathode and anode compartments separated by a membrane, protons permeate through the membrane to the cathode compartment, where they combine with the electrons to produce hydrogen. This process is not spontaneous; in order for the electrons to travel from anode to cathode they need a small amount of externally supplied DC current ($\geq 0.2 - 0.8$ V). If acetate is used as solution, the following equations describe the process (Kadier et al., 2016):

$$\text{Anode} \quad C_2H_4O_2 + 2\,H_2O \rightarrow 2\,CO_2 + 8\,e^- + 8\,H^+$$

$$\text{Cathode} \quad 8\,H^+ + 8\,e^- \rightarrow 4\,H_2$$

Of note is the theoretical hydrogen production. The approximate COD of acetate is about 1.07 grams per gram of acetate. Then one gram of acetate can produce in a MEC 66% of the theoretical H_2, which is $4 * (MW\ H_2)/(MW\ of\ acetate) = 4 * 2/60 = 0.133$ grams of H_2 per gram of acetic acid or $0.133/1.07 = 0.125$ g of H_2/g of COD or acetate. However, earlier in this chapter in the section on hydrogen production by acid fermentation, it was shown that one mole of hexose can produce by fermentation two moles of acetate and because the composition of C, H, and O of MSW ($C_6H_{10}O_4$) is not far from that of hexose, the same assumptions can be made for approximate production of H_2 from typical biodegradable MSW. It will subsequently be shown that the measured H_2 yields are close to the theoretical calculation. The molar concentration of carbon dioxide (alkalinity) produced in conversion to hydrogen is one-half of the produced moles of hydrogen.

If one assumes that one mole of hexose or mole representation of MSW can produce 2–4 moles, realistically 3 moles, of H_2 in the fermentation phase and 8 moles in MEC the total production of H_2 in the fermentation-MEC hybrid will be 11 moles of H_2. The COD of hexose is 1.07 kg O_2/kg, then *a hybrid fermentation/MEC process can realistically produce up to 11 moles of hydrogen per mole (90%) of a sacchariferous or biodegradable MSW substrate entering the first acid fermentation reactor*. This would represent about:

$$(3 + 8) * 2/[(6 * 12 + 12 + 6 * 16) * 1.07] = 0.114 \text{ g } H_2/\text{g hexose COD}$$

The theoretical maximum hydrogen yield, assuming 12 molecules of H_2 is produced, is 0.125 g H_2/g hexose COD, which is the value reported by Ditzig, Liu, and Logan (2007). The complete information on COD, TOC, accompanied by H and O concentrations in sludge from the anaerobic water reclamation plants in the literature has not been easy to find. Interestingly, it can be argued that no H_2 molecule is lost in the IRRF digestion/gasification process because the unrecovered H_2 in sludge will be sent to the gasifier and recovered as H_2 therein after syngas reforming. However, for the time being, the unaccounted H_2 is a safety factor.

The relatively high-energy outputs in MEC studies with pure substances were promising but several glitches appeared soon after this discovery. For example, Professor Logan's team at Pennsylvania State University (Ditzig, Liu, and Logan, 2007) tested at about the same time a graphite-packed BEAMR (bioelectrochemically assisted microbial reactor) with a plain carbon electrode treating domestic used (waste) water with relatively low COD concentrations, but the hydrogen yield was relatively small, reported as 0.0116 kg of H_2/kg of COD, which was about 9% of the potential H_2 yield. Later, Call, Wagner, and Logan (2009) reported that under optimal conditions that would prevent methane formation in the MEC, almost all (97%) of the substrate energy in MEC input can be recovered as hydrogen. To make more hydrogen from acetates and other organic acids, Professor Logan's team estimated that 0.4 V of the supplemental electric current is needed, of which 0.2 V can be supplied by the bacteria. This is an order of magnitude less than the amount of electricity needed to split water in electrolysis. As shown in Figure 8.8, because of pH in the anode being between 6 and 6.5, most of the MEC carbonic content will be roughly divided between carbonic acid and bicarbonate; therefore, it may not be released as carbon dioxide gas and may yield more protons to increase H_2 production. In any case, CO_2 will be completely separated from the hydrogen produced in the cathode compartment.

The MEC research results with pure feed stocks (hexose, acetic acids) were promising but not so with mixtures such as organic wastewater. Soon thereafter, it was realized that the MECs are efficient mostly for more refined products of dead-end fermentation (acetates, some organic acids), less for glucose and cellulose, and inefficient for raw used water or sludge solids. The efficiency increases with increased influent COD concentrations (Ditzig, Liu, and Logan, 2007). Unlike the dark fermentation that is inefficient in decomposing proteins, these compounds can be decomposed and converted to hydrogen by MECs (Lu et al., 2010).

The first MEC reactors, called the BEAMRs, had anode and cathode chambers separated by a membrane allowing permeation of protons. The arrangement was similar to that for an MFC, shown in Figure 8.8. Chang and Logan (2007) reported that by improving the materials and architecture of the modified MFC, hydrogen gas was produced with a yield of up to 3.95 moles of hydrogen per mole of acetic acid, which is equivalent to the theoretical yield and also equal to the theoretical H_2 yield of the dark fermentation of glucose or cellulose to acetic acid. The applied voltage was 0.6 V. The overall energy yield was almost three times the energy applied. Similar high hydrogen yields were demonstrated using glucose, several volatile acids, and cellulose.

The source of MEC external power can be (1) solar photovoltaic panel (Bakos, 2005), (2) a parallel or separate MFC (not efficient yet), (3) a follow-up hydrogen fuel cell (see the next section) that will provide a small portion of the current generated by hydrogen to MEC, or a (4) DC battery or transformed AC current. In contrast to MFCs that needs oxygen to be supplied by aeration or pure oxygen, MEC produces more than three times more energy than the electricity used to extract it. The MEC process may not need specific exoelectrogens; it may be accomplished by bacteria present in the conventional anaerobic digestion process. These bacteria operate best in a temperature range of 20–30°C.

Because of the promising results of the hydrogen-producing MECs, in 2010 there was worldwide research on these technologies, and researchers in many countries have been feverishly looking into more efficient designs, looking at possibilities for developing MEC without membranes, continuous operations, and finding the best microbial population (see the reviews by Logan et al., 2008; Kadier et al., 2016; Li et al., 2014; and others). MEC arrangements used for continuous operation reviewed by Li et al. include:

- Upflow, tubular-type MEC with inner graphite bed anode and outer cathode
- Upflow, tubular-type MEC with anode below and cathode above; the membrane is inclined
- Flat plate design where a channel is cut in the blocks so that liquid can flow in a serpentine pattern across the electrode

- Single-chamber system with an inner concentric air cathode surrounded by a chamber containing graphite rods as anode
- Stacked MEC (MFC), in which separate MECs are joined in one reactor block

There are new numerous designs of MECs and the applications are expanding into other fields. For example, the design from Cusick et al. (2014) uses a fluidized bed reactor as a cathode to remove COD and phosphate, producing struvite (see Chapter 9).

Hybrid Fermentation and the MEC System

The symbiosis of the dark fermentation producing organic acids and acetates and some hydrogen and relatively high hydrogen outputs from MEC where the substrates are the products of dark fermentation is directing researchers to focus on *hybrid fermentation–MEC systems,* in which the two processes complement each other. The advantages and drawbacks of fermentation and MEC used individually to treat mixed substrates such as concentrated used water and effluents from co-digestion were outlined above. The most logical arrangement is a two-step process (shown in Figure 8.9) consisting of:

1. *The dark fermentation reactor* is treating or pretreating the mixed sludge from used water, MSW biodegradable solids and high organic concentration liquid waste at pH ~6 with HRT for about 8–10 hrs (Hawkes et al., 2002), which is much shorter than HRT of traditional methane-generating digesters. Organic solids, carbohydrates, celluloses, and perhaps lignin will be broken down by anaerobic fermentation in this unit and the end product of this process is organic acetic acids, acetates, and hydrogen. Methanogenesis should be prevented by low pH. Limited mass of hydrogen will be produced in the fermentation reactor along with some intermediate volatile organic fatty acids, which should be minimized (Dahia and Venkata Mohan, 2018) and, if needed, separated from H_2. Hydrogen has very low solubility in water, three orders of magnitude less than carbon dioxide. The amount of hydrogen could be about one-quarter to one-third of the total hydrogen potential in the influent. If it becomes possible to separate hydrogen from other gases, H_2 should be sent directly

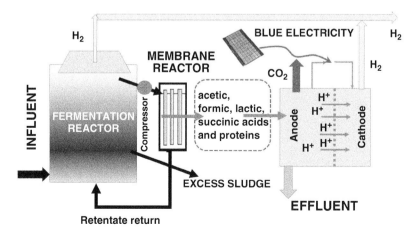

Figure 8.9. A potential schematics of the low pH UASBR dark fermentation–MEC hybrid system. *Source:* Inspired by the Penn State presentation to the National Renewable Energy Laboratory, June 19, 2014.

to the hydrogen fuel cell along with H_2 produced in the MEC. The effluent from the fermentation reactor containing acetates, organic acids, and proteins will be separated in the membrane reactor, collected, and sent for cleaning and further processing in the microbial electrolysis cell (MEC) for additional production of hydrogen. The excess solids yield should be small and should be removed and sent for dewatering and gasification and the rest returned. The type of the reactor could be either AnMBR or a completely mixed fermentation reactor of the traditional two-step digestion process. Blue or green electricity DC current stimulates bacteria in the anode compartment to free electrons from the biomass and produce protons.

2. *MEC for processing acetates and proteins.* Because the influent into the MEC is pre-processed by fermentation, removing the suspended solids and carbohydrate organics are converted into acetates, organic acids, and proteins in the fermentation reactor, the hydrogen yield will be much higher than that from the MEC treating the entire mixed influent without preprocessing. The recovered hydrogen could be close to the total H_2 potential in the influent into MEC. The output from the MEC reactor will be H_2 from anode and carbon dioxide (bicarbonate) from cathode compartments. Nutrients can be separated from the supernatant. The location of the membrane reactor is not known, either after the fermentation reactor or after the MEC.

Small masses of waste solids accumulation in the cathode compartment of MEC can be periodically or continuously removed for final treatment. As of 2015 these hybrid systems were intensively studied (e.g., at Penn State University in the US, at Wageningen University in The Netherlands, and at Chinese universities), but results have not appeared in the literature and the most optimum system design will still need to be developed. But this system is logical and should work.

In some instances, to improve fermentation and increase hydrogen or methane yields *a hydrolyzer unit* can precede the digestion reactor (Veolia, 2017; Wyman et al., 2005) or, by the same reasoning, the fermentation reactor. Hydrolysis has been used extensively in forming ethanol and other bioproducts from cellulose. In the hydrolyzer reactor the biomass is heated, pressure-cooked with water with or without acids or enzymes to temperatures of 160–260°C, which can be achieved by steam injection. This requires energy that can be almost fully recuperated and the hydrogen-based IRRF produces plenty of hot steam. The hydrolysis process breaks down cellulose, hemicellulose, and lignin to sugars described as:

$$(C_6H_{10}O_5)_n + n\ H_2O \rightarrow n\ C_6H_{12}O_6$$

No CO_2 is emitted. If hydrolysis reactors precede fermentation or digestion, the content must be subsequently cooled to mesophilic temperature or be followed after cooling by thermophilic digestion. Lignocelluloses materials include manure and swine waste, straws, leaves, papers, sorted refuse, and textiles. Research and pilot testing should decide whether this process is needed before fermentation or digestion, or unbroken cellulosic material can be sent as waste sludge to the indirect gasifier, where under much higher temperature (900°C) it can be almost fully converted to syngas. In the overall energy balance the impact of using hydrolysis would not be great but, could enhance the hydrogen yield.

8.5   HYDROGEN YIELD POTENTIAL BY INDIRECT GASIFICATION

Energy from waste has been at the forefront of interest among water and energy professionals and societies since the end of the last century. Incineration of urban solid waste and later the addition of dried sludge from water reclamation plants has been practiced for decades and gasification of coal has been known and implemented for light and heating in European cities for almost two centuries. However, in traditional wastewater treatment plants of the last century, after processing sludge by anaerobic digestion emitting methane, the biogas was wasted or flared or used for heating digesters. Only at the end of the last century methane biogas from used water sludge digestion and solid waste digestion processes in landfills was used for producing energy by combustion in some water reclamation facilities, and some combustible municipal solid waste was incinerated. 50% MSW incineration in Sweden was an exception at the time of writing this handbook. Incineration is not an efficient source of energy, but it could be a serious source of toxic air pollution.

Logan et al. (2008) and Van Lier et al. (2008) documented (see also the section on anaerobic fermentation) that anaerobic breakdown of one mole of hexose could theoretically yield 12 moles of H_2 or, stoichiometrically:

$$C_6H_{12}O_6 + 6\,H_2O \rightarrow 12\,H_2 + 6\,CO_2$$

It was also pointed out that hydrogen production by anaerobic fermentation is restricted by an acetate formation barrier, which limits the hydrogen yield by fermentation and in the fourth and final step of the digestion process, methanogenesis, methanogens scavenge hydrogen and carbon dioxide and decompose acetates. If fermentation is followed by a microbial electrolysis cell, the approximate realistic H_2 yield potential of the hybrid acid fermentation–MEC sequential units is 0.125 kg H_2 per 1 kg of COD. Also, the overall COD removal in the hybrid system can be as much as 90% or even more. In contrast, COD removal and methane production in a traditional anaerobic methane producing reactors hover around 50% and the rest of the COD is in the digestate (Figure 7.3) that must be returned to the influent.

The formula for the maximum anaerobic (or any other) hydrogen production from organic compounds presented by Logan (2008) and Van Lier et al. (2008) is for hexose (fructose). In the breakdown (decomposition) of the organic compound, the final product is carbon dioxide, whereby each atom of carbon needs two atoms of oxygen. Oxygen is available from the air in aerobic processes, but in an anaerobic environment this type of free oxygen (for example, gaseous atmospheric two atoms oxygen or three atoms ozone) is not available and oxygen atoms will originate from the inside of the molecule and/or from another oxygen-containing compound, which in most cases is water (steam) that adds more hydrogen. Hence, the maximum number of molecules of hydrogen recoverable from any biodegradable mass is related to the number of atoms of oxygen needed to convert organic carbon to carbon dioxide, which have external and internal origin, plus the internal number of hydrogen atoms. This is illustrated in the following examples (some already presented in Table 8.1 and elsewhere in the text and previously in this chapter) assessing hydrogen and energy potential of MSW:

Org C indirect gasification $C + H_2O \rightarrow CO + H_2$ $CO + H_2O \rightarrow CO_2 + H_2$

   Maximum yield 2 H_2 moles/mole of C = 4/12 = 0.3 kg of H_2/kg of C

Methane steam reforming $CH_4 + H_2O \rightarrow CO + 3H_2$ $CO + H_2O \rightarrow CO_2 + H_2$

   Maximum yield 4 H_2/mole of CH_4 = 8/16 = 0.5 kg H_2/kg of CH_4

Glycol fermentation + MEC $C_2H_6O_2 + 2\,H_2O \rightarrow 2CO_2 + 5H_2$

   Maximum yield 5 H_2/mole of glycol = 10/(2 * 12 + 6 + 2 * 16)

      = 0.15 kg of H_2/kg of $C_2H_6O_2$

However, most organic compounds from which hydrogen can be retrieved are not pure carbohydrates or saccharides. The majority of organic compounds derived from photosynthesis and from abiotic processes producing; for example, cyanides contain other elements that during the breakdown of the molecules would combine with hydrogen, such as organic nitrogen and sulfur. During anaerobic breakdown, N combined with hydrogen forms ammonia, and sulfur (sulfate) will metamorphose into sulfide. Hence, the general formula for mining hydrogen from biodegradable and combustible organic compounds by indirect gasification as presented previously in this chapter is:

$$C_mH_nO_pN_xS_y + z\,H_2O \rightarrow m\,CO_2 + x\,NH_3 + y\,H_2S + [z + 0.5 * n - (1.5x + y)]H_2$$

where $z = 2m - p$ is the molar water use in the gasification reaction.

Maximum hydrogen yield is then:

$$H_2 \text{ yield} = 2 * [z + 0.5 * n - (1.5x + y)]/[m * 12 + n + p * 16 + x * 14 + y * 32]$$
$$\{g\,H_2/g \text{ of VSS}\}$$

From the same chemical balance formula, the water use to convert the compound to hydrogen is:

$$\text{Water use} = 18 * (z)/[m * 12 + n + p * 16 + x * 14 + y * 32] \quad \{g\,H_2O/g \text{ of VSS}\}$$

The above formulae may also be applied to complex heterogenous organic mixtures such as sludge, biodegradable and combustible MSW, algae mixtures, vegetation residues, manure, and so on. In such instances the subscripts (atomic counts) m, n, p, x, and y may not express the number of atoms of an element in a molecule; they are molar weight fractions between the key elements, normally C, H, O, N, S, and P, in a unit of weight of the heterogenous mixture. Other elements may also be present but generally they might be negligible, irrelevant for the hydrogen yield and may account for less than 3% of the total mass of the combustible (biodegradable) compound. Noncombustible compounds such as glass, metals, or sand are not included. The effect of phosphorus on yield of hydrogen is not clear.

The subsequent calculations will be derived from percent composition of the elements that will be normalized to C_6 (hexa C) base. This normalization will not affect the calculations of hydrogen yield nor water consumption because normalizing C_x to Hexabase C_6 along with the same normalization (multiplying by $6/C_x$ ratio) of the other elements fractions appears in both nominator and denominator of the H_2 yield and water use formulae and the results are dimensionless. This normalization was suggested and included in the NYC MSW chemical composition estimates published by Themelis, Kim, and Brady (2002).

Sources of Energy Hydrogen

Estimating hydrogen potential is important when mixed feed is to be gasified in order to estimate the hydrogen yield and water (steam) supply for the gasifier.

Algae as Source of Energy Hydrogen. *Chemical composition of algae and cyanobacteria* was investigated by Vijayakumar et al. (2013). The authors studied algae *Chlorella* sp., and cyanobacteria *Anabaena* sp., *Synechocystis* sp., and the marine algae *Nannochloropsis.*

Nannochloropsis was studied because its potential for biofuel production. When this algae and similar marine species are grown in nitrogen deficient-conditions, they can yield approximately 50% of its biomass as oil (Gouveia and Oliveira. 2009). This topic is beyond the scope of this book because the objective of the IRRF is to remove and recuperate hydrogen energy. Cyanobacteria and *Chlorella* are suitable for nutrient removal as well as for biomass production, from which hydrogen can be retrieved. *Anabaena* sp. removes nitrogen and phosphorus down to very small nutrient residuals but can also fix atmospheric nitrogen.

The chemical compositions of algae and cyanobacteria reported by Vijayakumar et al. are presented below, along with the stochiometric formulae. The molecular weights of elements in the stoichiometric formula are estimated by dividing the percent of the compound in a unit of mass by the atomic weights, which are $C = 12$, $O = 16$, $H = 1$, $N = 14$, $P = 31$, and $S = 32$.

Anabaena $C = 44.6\%$ $H = 7\%$ $O = 25\%$ $N = 7.68\%$ $P = 0.3\%$ $VSS = 86\%$

Stochiometric formula: $C_{3.7}H_7O_{1.6}N_{0.6}$ Base C_6: $C_6H_{11.3}O_{2.6}N$

$$\text{Max } H_2 \text{ yield} = 2 * (2 * 3.7 + 0.5 * 7 - 1.6 - 1.5 * 0.6)/$$

$$((3.7 * 12 + 7 + 1.6 * 16 + 0.6 * 14) * 0.86\} = 0.23 \text{ kg } H_2/\text{kg VSS}$$

The biomass of anabaena consists of 25–30% of carbohydrates, 43–56% of proteins and 4.7% of lipids.

Synechocystic $C = 51.4\%$ $H = 6.1\%$ $O = 27.5\%$ $N = 11.3\%$ $VSS = 96\%$

Stochiometric formula: $C_{4.3}H_{6.1}O_{1.7}N_{0.8}$ Base C_6: $C_6H_{8.5}O_{2.4}N_{1.1}$

$$\text{Max } H_2 \text{ yield} = 2 * [2 * 4.3 + 0.5 * 6.1 - 1.7 - 0.75 * 0.8]/$$

$$[(4.3 * 12 + 6 + 1.7 * 16 + 0.8 * 14) * 0.96\} = 0.23 \text{ kg } H_2/\text{kg VSS}$$

Chlorella $C = 60\%$ $H = 9\%$ $O = 21\%$ $N = 6.5\%$ $P = 0.8\%$ $VSS = 97\%$

Stochiometric formula: $C_5H_9O_{1.3}N_{0.46}P_{0.02}$ Base C_6: $C_6H_{10.8}O_{1.6}N_{0.6}S_{0.2}$

$$\text{Max } H_2 \text{ yield} = 2 * [2 * 5 + 0.5 * 9 - 1.3 - 0.75 * 0.46]/$$

$$\{[5 * 12 + 9 + 1.3 * 16 + 0.46 * 14] * 0.97\} = 0.27 \text{ kg } H_2/\text{kg VSS}$$

The biomass of *Chlorella* is 12–17% carbohydrates, 51–58% proteins, and 14–22% lipids. *Chlorella* is the best source of H_2 of the three-microorganisms' species. Nitrogen has a small effect and phosphorus has no effect on hydrogen yield.

Used Water Sludge as a Source of Hydrogen   Werle and Dudziak (2014) analyzed sludge from two municipalities in Poland and provided data on sludge composition. The calculation of the H_2 yield is as follows:

Chemical composition (average of two samples) expressed per dry weight)

$C = 50\%$ $H = 4.36\%$ $O = 19.2\%$ $N = 4.23\%$ $S = 1.74\%$ $VSS = 60\%$

Stochiometric formula: $C_{4.2}H_{4.4}O_{1.6}N_{0.3}S_{0.05}$ Base C_6: $C_6H_{7.4}O_{2.85}N_{0.4}S_{0.07}$

$$\text{Max } H_2 \text{ yield} = 2 * [2 * 4.2 + 4.4/2 - 1.6 - (1.5 * 0.6 + 0.05)]/$$

$$\{[4.2 * 12 + 4.4 + 1.6 * 16 + 0.45 * 14 + 0.05 * 32] * 0.60\}$$

$$= 0.31 \text{ kg } H_2/\text{kg ds}$$

Food Waste as a Source of Hydrogen.   The Wales Centre of Excellence for Anaerobic Digestion (Esteves and Devlin, 2010) analyzed many food waste samples obtained in Wales, UK; 5 kg samples were collected from Welsh localities and dried for moisture content estimations. The samples were reported for summer and winter periods. The summary information here is for dry solids.

	Summer	Winter
Carbohydrates g/kg	93.3	156
Lipids	48.8	59.3
Proteins	77.2	44.3
VSS %	87	93
Carbon % (AW)	45.75 (3.8)	49.32 (4.1)
Hydrogen % (AW)	6.26 (6.3)	6.53 (6.5)
Oxygen % (AW)	35.06 (2.2)	37.14 (2.3)
Nitrogen % (AW)	3.36 (0.2)	3.17 (0.2)
Sulfur % (AW)	0.71 (0.02)	0.35 (0.01) (Negligible)

Source: The Wales Centre of Excellence for Anaerobic Digestion (Esteves and Devlin, 2010).

Stochiometric formula: Summer $C_{3.8}H_{6.3}O_{2.2}N_{0.2}$ Base C_6: $C_6H_{9.8}O_{3.4}N_{0.3}$

$$\text{Max } H_2 \text{ yield} = 2 * [2 * 3.8 + 6.3/2 - 2.2 - 0.75 * 0.2]/$$

$$\{\{[3.8 * 12 + 6.3 + 2.3 * 16 + 0.2 * 14] * 0.93\} = 0.20 \text{ kg } H_2/\text{kg VSS}$$

Stochiometric formula: Winter $C_{4.1}H_{6.5}O_{2.3}N_{0.2}$ Base C_6: $C_6H_{9.5}O_{3.4}N_{0.3}$

$$\text{Max } H_2 \text{ yield} = 2 * [2 * 4.1 + 6.5/2 - 2.3 - 0.75 * 0.2]/$$

$$\{\{[4.1 * 12 + 6.5 + 2.2 * 16 + 0.2 * 14] * 0.87\} = 0.21 \text{ kg } H_2/\text{kg VSS}$$

Themelis, Kim, and Brady (2002) reported general stochiometric formula for New York City mixed food wastes as being $C_6H_{9.6}O_{3.5}N_{0.28}S_{0.2}$. Similarity between Welsh and NYC food waste chemical compositions is remarkable. Consequently, the estimate of the H_2 yield of 0.21 kg H_2/kg VSS may best characterize the food waste limit.

Combustible MSW as a Source of Hydrogen.   Themelis, Kim, and Brady (2002) calculated the molecular formula corresponding to each of the combustible components of New York City MSW and also provided extensive information on sources, character, and composition. The composition of NYC solid waste is given in Table 8.5. The results in hexa C format were presented by the authors as:

Mixed paper, textiles, diapers: $C_6H_{9.6}O_{4.6}N_{0.036}S_{0.01}$	39.4%
Mixed plastics, rubber and leather: $C_6H_{8.6}O_{1.7}$	9.1%
Mixed food wastes: $C_6H_{9.6}O_{3.5}NO_{0.28}S_{0.2}$	12.7%
Yard wastes and wood: $C_6H_{9.2}O_{3.8}NO_{0.01}S_{0.04}$	6.3%
Miscellaneous combustible: $C_6H_{9.6}O_{3.8}$	7.8%
Total %	75.3

This MSW composition would yield the overall formula $C_6H_{9.4}O_{3.9}N_{0.06}$

$$\text{Max } H_2 \text{ yield} = 2 * [2 * 6 + 9.4/2 - 3.9 - 1.5 * 0.06]/\{\{[6 * 12 + 9.4 + 3.9 * 16 + 0.06 * 14]\} = 0.18 \text{ kg } H_2/\text{kg of dry combustible solids.}$$

The authors also provided an overall chemical composition for the general formula NYC solid waste as $C_6H_{10}O_4$. For this composition the hydrogen yield was calculated in the previous section of this chapter (gasification of combustible MSW) was also Max H_2 Yield = 0.18 kg H_2/kg of combustible MSW. It appears that $C_6H_{10}O_4$ Max H_2 yield of 0.18 can be used as a good representation for MSW from large communities.

Plastics contain little or no oxygen and therefore need proportionally more water to be gasified, which yields more syngas hydrogen. The general formula for plastics in MSW was presented in the preceding section. For example, polyethylene (PE), the most common plastic for making bottles, plastic bags, and the like, is a chain of carbons and hydrogens with a formula $(C_2H_4)_n$ and specific density around 970 kg/m³. Other widely used plastics are polypropylene $(C_3H_6)_n$, specific density 946 kg/m3, and polyethylene terephthalate $(C_{10}H_8O_4)_n$, specific density 1.38 kg/m³. Gasification of polyethylene was studied by Cipriani, De Fillipis, and Pochetti (1998). In their experiments they gasified by hot steam eucalyptus biomass with added PE and found that increasing percentage PE in the biomass/PE mixture increased the H_2 percentage in syngas and calorific value of the gas. For example, when the gasifier temperature was 920°C, the percentage of H_2 was 37% and that for 30% PE in the biomass was 44%, respectively. Hence, polyethylene plastics appears to be a great source of hydrogen, as illustrated by the following equation (assuming 3% ash):

$$H_2 \text{ yield} = 0.97 * 2 * [2m + n/2]/[m * 12 + n] \text{ kg } H_2/\text{kg of plastic}\}$$

$$\text{Polyethylene (PE) } C_2H_4 + 4 H_2O \rightarrow 2 CO_2 + 6 H_2$$

$$\text{Maximum } H_2 \text{ yield of PE} = 0.97 * 2 * [2 * 2 + 4/2]/[2 * 12 + 4] = 0.42 \text{ } H_2/\text{kg}$$

$$\text{Polypropylene (PP) } C_3H_6 + 6 H_2O \rightarrow 3 CO_2 + 9 H_2$$

$$\text{Maximum } H_2 \text{ yield of PP} = 0.97 * 2 * [2 * 3 + 6/2]/[3 * 12 + 6] = 0.42 \text{ } H_2/\text{kg}$$

$$\text{Polyethylene terephthalate (PETP) } C_{10}H_8O_4 + 16 H_2O \rightarrow 10 CO_2 + 20 H_2$$

$$\text{Maximum } H_2 \text{ yield of PETP} = 0.97 * 2 * [2 * 10 + 8/2 - 4]/[10 * 12 + 8 + 4 * 16]$$
$$= 0.20 \text{ } H_2/\text{kg}$$

In general, plastics make good to excellent high-energy feedstocks for gasification, thereby reducing the amount of those unrecyclable materials that would otherwise end up in a landfill (Lee, Chung, and Ingley, 2014), oceans, and lakes. Compared with biomass or coal, plastics have a higher hydrogen potential, higher heating value, and a lower char yield in pyrolysis, which ensures a higher hydrogen production potential in syngas from the gasification of plastics (Kannan et al., 2012). However, recycling plastics is a more sustainable alternative than gasification (see the section "Circular Economy" in Chapter 2).

Gasification of halogenated plastics and formation of hydrogen is a challenge. The mechanism of thermal degradation of these plastics is not yet completely clear. This will be illustrated on the most common halogenated plastic polymer, polyvinyl chloride (PVC), which has a formula $[CH_2 - CHC-]_n$ or C_2H_3Cl. During gasification of PVC dehydrochlorination occurs at a temperature as low as 250–320°C. Chloride in indirect gasification combines with water as hydrochloric acid. Progressing to high gasification temperatures (~900°C), hydrogen-enriched but HCl-contaminated syngas is produced as:

$$CH_2 - CHCl + 2\, H_2O \rightarrow 2\, CO_2 + 3H_2 + HCl$$

and the maximum H_2 yield after reforming CO to CO_2 and H_2 is $5H_2$ (specific density of PVC is about 1.4 kg/m^3) and the maximum yield (assuming ash 5%) is:

$$\text{Max } H_2 \text{ yield} = 0.95 * 2 * (2 * 2 + 0.5 * 3)/[2 * 12 + 3 + 35.45]$$
$$= 0.16 \text{ kg } H_2/\text{kg of PVC}$$

HCl must be removed because of corrosivity and its ecologically damaging persistence in the environment. HCl represents 55% of the plastic waste, which may be impossible to handle in the gasification process. Furthermore, per weight, PVC is not a great producer of hydrogen. Borgiani et al. (2002) investigated and suggested adding sodium carbonate (soda ash) to the blend containing PVC and refuse-derived fuel (RDF) while Cho et al. (2015) used in the pilot study Ni loaded activated carbon. Hence, research must continue to find the best method of dechlorination of the produced syngas and, meanwhile, large quantities of PVC should be taken out from the RDF and recycled or, simply, not be produced.

Rubber (tires) can also be gasified as documented in Lee, Chung, and Ingley (2014). They reported a formula for rubber as $CH_{1.05}O_{0.03}$ (hexa formula $C_6H_{6.3}O_{0.2}$), which indicates that rubber can also be a good source of hydrogen in indirect gasification. The average compacted density of tires ranges from 650 kg/m^3 to 840 kg/m^3 (average 745 kg/m^3). The gasification of rubber is approximately represented by:

$$CH_{1.05}O_{0.03} + 1.97\, H_2O \rightarrow CO_2 + 2.5\, H_2$$

Maximum H_2 yield $= 2 * (2 + 0.5 * 1.05 - 0.03)/(12 + 1.05 + 0.03 * 16) = 0.38$ kgH_2/kg of pure rubber

Gasification of tires is suspect because tires are not pure rubber. Tires contains 31% rubber (natural and synthetic), 28% carbon black, 14–15% steel, 16–17% fabrics and fillers, and so on. Apparently, ash percentage is more than 10%. However, Lee, Chung, and Ingley (2014) gasified tire shredding successfully with a high final H_2 yield in the gas.

Abiotic Hydrogen Production from Biomass and Organic Matter by Photocatalytic Process.

Wakerley et al. (2017) reported a light-driven photo reforming of cellulose, hemicellulose, and lignin to H_2 using semiconducting cadmium sulfide quantum dots in alkaline aqueous solution. The system that operates under visible light can reform to hydrogen unprocessed lignocellulose, such as wood and paper, under solar irradiation at room temperature. This process, which was also reported by Hallenbeck and Benemann (2002), could represent an alternative to gasification and/or could be a prerequisite to a hybrid acid fermentation/MEC process. Photo reforming requires a photocatalyst able to break lignocellulose and uses the resultant electrons to reduce aqueous protons to H_2. At the end of the second decade of the 2000s, this process of H_2 production was demonstrated only in laboratory conditions.

Maximizing Hydrogen Energy Yield by Selecting the Proper Technologies

This chapter has introduced feasible, established, and extensively tested methodologies by which hydrogen can be derived from biodegradable and combustible used water and solid waste compounds. This section further describes their attributes:

- COD removal in anaerobic fluidized bed membrane reactors and full digestion of organic solids from the primary and secondary sludge and co-digestion with food waste, glycols, and other biodegradable components of MSW in a conventional digester to yield methane gas, followed by the two steps of methane steam reforming to syngas and syngas to hydrogen.
- The hybrid fermentation–MEC system, which produces hydrogen in the first fermentation reactor along with acetates, bicarbonates, and dissolved carbon dioxide. In the second MEC step acetates and proteins are decomposed to hydrogen and bicarbonates with the help of external electron manipulation by DC current.
- Indirect gasification producing syngas that is subsequently reformed to hydrogen and carbon dioxide either in a separate steam reforming unit or internally in the gasifier and/or subsequently in the expanded hydrogen fuel cell.

8.6 HYDROGEN FUEL CELLS

The US Department of Energy in the promotion of the technology states that fuel cells are an energy user's dream: an efficient, combustionless, virtually pollution-free power source, capable of being sited in downtown urban areas or in remote regions that runs almost silently and has few moving parts. This energy-making process is becoming very popular and has made already major breakthroughs, especially in automobile industry (see Chapter 4). There are several types of the hydrogen fuel cells; however, at the end of the second decade of this century two high-temperature cells received the most attention from researchers, energy agencies, and providers: Molten carbonate fuel cell (MCFC) and solid oxide fuel cell (SOFC). The key reason for interest has been that at these high temperatures, besides using hydrogen, these fuel cells can internally reform and use as fuel hydrogen-rich hydrocarbon gases and liquids such as carbon monoxide (syngas), methane, alcohol, and others. Producing energy by these cells is a non-combustion process that produces electricity chemically from hydrogen reacting with an electrolyte in the MCFC or carbon monoxide and hydrogen with solid oxide in the SOFC. The reactants flow in the unit and the reaction products flow out while the electrolyte remains within. The "waste" is clean water and concentrated CO_2 that can be reused. MCFC and SOFC are high-temperature types that allow the use of low carbon gaseous hydrocarbon fuels in addition to hydrogen (Milewski et al., 2018; Zhang, 2018). The gas entering the cells must be cleaned to remove sulfur, chlorine, and other compounds that can poison the cells.

Lower-temperature phosphoric acid hydrogen fuel cells operating at temperature range of 150–220°C are also candidates to be used in future integrated water, energy and solids systems. Using hydrogen as fuel and producing clean electricity and heat may lead to a change of the paradigm of generating and providing electricity to urban and rural areas from large regional energy providers to distributed energy productions. Under the new 2050 sustainable energy paradigm the regional grid will not be one or two large power plants providing energy to a large region; it will be a cyber center managing a grid with a number of smaller (1 MW, mainly photovoltaics and wind turbines) and medium-size (2–20 MW, such as IRRFs) energy providers. Large nuclear and hydro energy power plants as well as large offshore wind energy farms (>100 MW) will also be a part of the grid of the future. The advantages of distributed energy production were brought to forefront in Puerto Rico in the aftermath of the category 5 Hurricane Maria in 2017 when the centralized power plant and the grid for the entire island was severely damaged and it took months to bring energy back to the island.

Molten Carbonate Fuel Cells (MCFCs)

Figure 8.10a and b present concepts, not designs, of the most popular high-temperature fuel cells. Figure 8.10a shows a conceptual design that uses molten carbonate as electrolyte. Large molten carbonate fuel cells have been manufactured in the US by Fuel Cell Energy (2013), located in Danbury, Connecticut, other companies outside of the US (Canada, Korea, Europe), plus dozens of companies making smaller fuel cells for automobiles. Many components and designs are covered by patents awarded to the manufacturers. The original configuration was adapted herein to process three different energy carriers generated by the IRRF – methane from digestion, hydrogen from MEC (microbial electrolysis cells), and syngas (hydrogen and carbon monoxide) from gasification (Cherepy et al., 2005; Ciccoli et al., 2010). Fuel Cell Energy is also developing MCFCs that can concentrate CO_2 from flue gases of fossil fuel–burning power plants and combustion engines (see Section 8.7 on carbon dioxide sequestering). Apparently, all hydrogen molten carbonate fuel cells are suitable.

MCFC contains carbonate salts of alkali metals as electrolyte. Breeze (2017) described the most common electrolyte as a mix of lithium and potassium carbonates, which is solid in normal temperatures but becomes liquid at the fuel cell operating temperatures of greater than 650°C. Due to the highly corrosive nature of the electrolyte, various countermeasures are being developed. MCFCs are appropriate for high-efficiency power generation systems using hydrocarbon fuels, such as natural gas and coal gas and obviously are applicable to processing biogas from digestion (Ciccoli et al., 2010). The nickel anode incorporated in the MCFC design is an excellent catalyst for the "shift reactions" converting carbon species (ultimately carbon monoxide) and water into hydrogen, which then releases the electrons that generate the electric current. Hence, the expanded cell concept presented in the figure combines the fuel cell with a steam methane reforming unit that converts methane to syngas, which then enters the anode compartment of the MCFC and reforming syngas to hydrogen and carbon dioxide is done internally in the anode compartment of the cell. As shown in Table 8.1, syngas reforming to hydrogen is an exothermic reaction that provides a portion of heat to keep high temperature within the cell. Carbon dioxide is a source of carbonate ion. Breeze (2017) also stated that the MCFCs are economical for power plants wit power capacity greater than 100 kW.

The advantage of MCFC catalysts as compared to platinum catalysts used in other designs is the cost. The MCFC consists of an ion-conducting electrolyte matrix (carbonate salt), two electron-conducting electrodes, and an electron conducting separator plates. The fuel cell is basically an electrolyte sandwiched between two negatively and

A. Molten Carbonate Fuel Cell

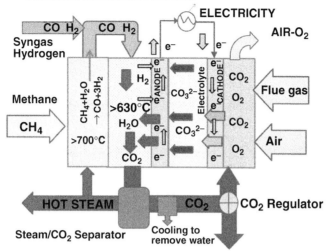

B. Solid Oxide Fuel Cell

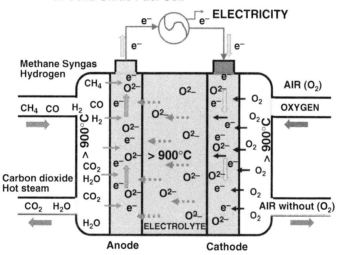

Figure 8.10. (a) Enhanced molten carbonate fuel cell schematic combined with external steam methane and internal syngas re-forming: the key components of the IRRF power plant. (b) Solid oxide fuel cell schematics. In addition to hydrogen this cell can accept and internally reform methane and syngas to produce electricity. *Source:* (a) Adapted from US Department of Energy (2004); Fuel Cell Energy (2013), and Wikipedia Commons. (b) US Department of Energy (2004).

positively charged electrodes that utilizes chemical reaction rather than combustion to produce electric energy. The key reactions in the cell are (US Department of Energy, 2004; Takizawa, 2013; McPhail et al., 2015):

$$\text{Cathode: } 2CO_2 + O_2 + 4e^- \rightarrow 2\ CO_3^{2-}$$

$$\text{Anode: } CO + H_2O \rightarrow CO_2 + H_2 + \text{Heat}$$

$$2H_2 + 2\ CO_3^{2-} \rightarrow 2\ H_2O + 2CO_2 + 4e^-$$

$$\text{Overall reaction: } 2H_2 + O_2 \rightarrow 2\ H_2O$$

The oxygen in the outside air flowing in the cell upon contact with a negatively charged electrode combines with extra electrons and carbon dioxide to form carbonate ion that diffuses through the electrolyte. The carbonate ions travel then through the positive charged electrode and react with hydrogen, losing the extra electrons to form water (steam). The electrons generate electricity (US DOE, 2004; Fuel Cell Energy, 2013; Takizawa, 2013; McPhail et al., 2015; Cassir et al., 2016; Breeze, 2017). The electric power production efficiency of fuel cells is expected to reach 65% in the near future and, when combined with heat recovery, the overall energy efficiency can be as high as 90% or even more (Fuel Cell Energy, 2013; McPhail et al., 2015). The limits are unknown.

The minimum temperature for low steam shift of CO in syngas reforming to hydrogen and CO_2 is less than the temperature in the MCFC (Table 8.1); hence, if CO is added to the anode side of the MCFC, it will be reformed by the steam produced in the cell to H_2 and CO_2. Furthermore, the carbon monoxide reforming with steam provided by the cell is exothermic and adds heat to the unit and a lot of heat is also provided by the 900°C hot syngas feed produced by gasification. Consequently, CO mixed with steam produced on the anode side can serve as fuel in MCFC. Takizawa (2013) stated that the minimum temperature requirement for reforming methane or natural gas to syngas should be reformed to carbon monoxide and hydrogen externally before entering the anode compartment of MCFC. However, research by Dicks (1996) as well as Fuel Cell Energy designs confirmed that internal reforming of methane (natural gas) is also possible because the temperature in the SMR compartment can be increased. While the hot steam generated by the MCFC will be more than sufficient to provide all steam needed for syngas and methane reforming to hydrogen, a separate SMR unit may need small amount of additional heat to keep the unit temperature greater than 700°C. This will be provided by the subsequent design of the power plant. The current (2018) designs and commercial applications of the MCFC power plans can use methane gas as feed (Breeze, 2017).

In steam reforming of methane (natural and digester gas, biogas) one molecule of CH_4 produces three molecules of H_2 and one molecule of carbon monoxide itself is subsequently reformed to another H_2 molecule (see Table 8.1); therefore, in the MCFC four molecules of H_2O steam are produced from one molecule of methane while only three molecules of steam are needed for methane/carbon monoxide reforming to hydrogen. The excess water may be on par with high-quality distilled/deionized water. The simple chemical reactions describing the MCFC show that the mass of water produced in the cell is five times the weight of hydrogen entering the cell and it should be larger than the mass of water needed in reforming syngas and methane to hydrogen. The hybrid fermentation followed by microbial electrolysis processes produce hydrogen (and CO_2); this gas can also directly enter the anode compartment of the MCFC and be converted to electricity and water. Carbon dioxide brought from the other process units will be more than the MCFC needs because MCFC recycles CO_2 between the cathode and anode sides of the MCFC. Therefore, a CO_2 flow regulator on the return CO_2 separates the carbon dioxide needed for MCFC from the hot CO_2 exhaust. Both steam and heat are needed to operate the IRRF; hence a steam/CO_2 separation may need to be installed on the exhaust line. The hot carbon dioxide left after steam separation is recycled back into the cathode compartment and excess goes to sequestering and reuse.

Input gas to the MCFC anode compartment must be clean. For example, sulfurous compounds are highly poisonous for the fuel cells and they need to be extracted before the electrochemical process inside the MCFC. When present in the biogas, sulfur has multiple negative effects on the performance and durability of an MCFC. These issues and remediation of sulfur contamination in the biogas from used water IRRF are discussed in Ciccoli et al. (2010).

According the US Energy Information Agency, at the end of 2016, the United States had 56 medium-size fuel-cell-generating units greater than 1 megawatt (MW), totaling 137 MW of net summer capacity. Most of this capacity (85%) has come online since 2013. The largest 59-MW hydrogen fuel cell power plant in 2018 was built in the Republic of Korea. Fuel cell power plants provide electricity to large department stores, military bases, many companies, treatment plants, and cities. Germany was planning to install thousands of stationary fuel cells by 2020 with a target cost of 1,700 euros (2,100 US $) per kW capacity (Zhang, 2018).

Stationary fuel cell systems also take up much less space in proportion to other clean energy technologies. For instance, the Fuel Cell & Hydrogen Energy Association claims that a 10-MW fuel cell installation can be sited in a about 0.4 ha of land. This is compared to about 3–4 ha of solar PV panels required per MW of solar power and about 20 ha of land per MW of wind farm (http://www.fchea.org/stationary/).

Solid Oxide Fuel Cells (SOFCs)

SOFC (Figure 8.10b) is similar to the MCFC (US DOE, 2004; Ormerod, 2002; Ferges et al., 2009; Milewski et al., 2018). SOFC is an electrochemical conversion device that produces electricity directly from oxidizing a fuel, which is hydrogen, syngas, methane, or even coal. It uses a solid ceramic oxide electrolyte rather than the liquid one used in MCFC. SOFC operating temperature is between 700 and 1000°C. The solid electrolyte selectively conducts negatively charged oxygen ions (O^{2-}) at high temperatures from the cathode to the anode. The high operating temperature allows internal reforming to hydrogen of both carbon monoxide in syngas and methane in small amounts from gasification or from digestion. The cells use oxygen from air as oxidant that enters the cell at the cathode. The electrode reactions are as follows (Ormerod, 2002; Mikulski et al., 2018):

Cathode	$O_2 + 4e^-$	$\rightarrow$	$2\,O^{2-}$
Anode	$H_2 + O^{2-}$	$\rightarrow$	$H_2O + 2e^-$
	$CO + O^{2-}$	$\rightarrow$	$CO_2 + 2e^-$
Overall	$H_2 + CO + O_2$	$\rightarrow$	$H_2O + CO_2 + \Delta E$

ΔE is a sum of electric and heat energies produced by the cell. Modeling and technology of SOFCs have been extensively covered in Mikulski et al. (2018).

In the second decade of the 2000s, state-of-the-art SOFC modules were nominally rated at ≤100 kW. Modules on order of 250 kW to 1 MW are generally considered to be required for cost-effective integration with central station-scale power generation, gas cleanup, and heat recovery technologies (US DOE, 2013). However, at the same time, the SOFC had a very high energy yield, similar to that expected from MCFC. The US Department of Energy 100-kW research SOFC unit had electricity efficiency greater than 60% and overall heat and electricity energy efficiency of 89% (Ghezel-Ayagh, 2018). In contrast, concurrently, MCFCs power plants were already planned and built with a power capacity exceeding 100 MW by fuel cell manufacturers.

The elevated operating temperatures of the MCFC and SOFC, in addition to the electrical power, produce high-temperature heat as a by-product. Steam from PAFC has a temperature just above the boiling point. This high-quality heat should not be wasted, but can be used in various ways, such as in combined heat and power systems to drive a turbine to generate more electricity. This significantly increases the overall efficiency of the MCFC and SOFC compared to lower-temperature variants. However, high temperatures inside MCFC and SOFC require more insulation. Another disadvantage of SOFC is that while the SOFC generates concentrated CO_2 from the fuel, it cannot be used for concentrating CO_2 from an auxiliary flue gas. Currently, only MCFC has this capability. Heat from these high-temperature fuel cells can be used directly to provide a part of the endothermic heat to maintain the temperature at 900°C inside the indirect gasifier and supply hot steam. Furthermore, because the high temperatures in MCFC and SOFC both steam methane reforming to syngas and reforming of syngas to hydrogen and carbon dioxide can be done internally in the anode compartment of the cells.

Phosphoric Acid Fuel Cell.   The phosphoric acid fuel cell (PAFC) is a lower-temperature fuel cell, manufactured by Doosan Fuel Cell America, located in South Windsor, Connecticut. Electrolyte is highly concentrated or pure liquid phosphoric acid (H_3PO_4) saturated in a silicon carbide matrix (SiC). Operating range is about 150–220°C (Breeze, 2017). Although the PAFC can tolerate carbon monoxide up to 1.5% concentrations in the fuel, it cannot internally convert it to energy, and higher CO concentrations poison the process. It cannot tolerate sulfur because this will damage the electrode, but this is a problem of all fuel cells. The operating reactions are:

$$\textbf{Anode} \quad 2H_2 \rightarrow 4H^+ + 4e^-$$
$$\textbf{Cathode} \quad O_2 + +4H^+ + 4e^- \rightarrow 2\,H_2O$$
$$\textbf{Overall} \quad 2H_2 + O_2 \rightarrow 2\,H_2O + \Delta E$$

Commercial units incorporate a reformer that will convert natural gas into hydrogen, using heat from the fuel cell to drive the reaction. The fuel cells can also be operated with biogas from waste to energy systems (Breeze, 2017). Methane and syngas reforming is included in the PAFC by Doosan but it is outside of the anode. Because CO is an intermediate of the methane reforming to hydrogen, these fuel cells can be adapted to accept syngas from the gasifiers. Furthermore, if indirect gasification is used the conversion of CO to hydrogen can be accomplished by the gasifier by increasing the steam input (Lee et al., 2014), but this may not be energy efficient because it would require heating by added steam from the PAFC (~200°C) to the internal temperature of the indirect gasifier (~900°C). Reforming syngas into hydrogen and carbon dioxide is accomplished by a low temperature shift of 250–400°C. PAFC tolerates CO_2 but currently cannot concentrate nor accept CO_2 from auxiliary flue gases.

Indirect gasification producing syngas followed by electricity production in the MCFC or SOFC or PAFC does not produce toxins. Also, these fuel cells, unlike traditional incinerators and gas turbines, produce no nitrous oxides (NOx), no volatile organic compounds (VOCs), no fly ash, nor particulates in fly ash, and no toxins. The hot exhaust gas from the hydrogen fuel cells connected to indirect gasifiers contains about 70% carbon dioxide and the rest is steam.

Producing Hydrogen and Oxygen by Electrolysis

Students is many high school chemistry and physics laboratories conduct a simple experiment of splitting water with DC current into hydrogen, forming as bubbles near the cathode, and oxygen, forming near the anode. This simple process uses significantly more energy than is produced in hydrogen. Electrolyzers, like fuel cells, consist of an anode

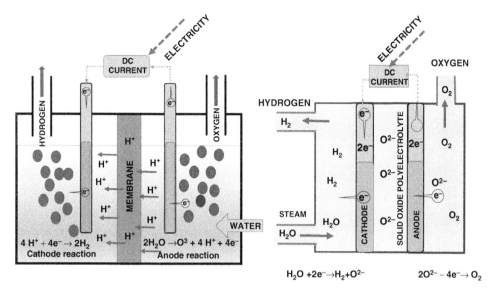

Figure 8.11. Concepts of electrolyzers. Left: Low-temperature membrane electrolyzer making hydrogen and oxygen from water. Right: High-temperature solid oxide cell as electrolyzer making H_2 and O_2 from hot ($>900°C$) steam. *Source:* Concepts based on US Department of Energy sources.

and a cathode separated by an electrolyte or membrane. The difference between various electrolyzers is in the electrolyte and membranes that separate the cathode and anode compartments. The physical concept of the simple low temperature electrolyzer that makes hydrogen and oxygen from water is shown in Figure 8.11 (left). Under the current paradigm of deriving energy from fossil fuels, use of the electrolyzers for producing hydrogen energy fuel would be counterproductive. However, under the new paradigm wherein the sources of energy are blue (wind, solar, hydro), which inherently sometimes produce excess energy, there is a synergy with renewable energy wherein excess hydrogen produced from renewable sources by splitting water or even carbon dioxide, can be stored and used later. Furthermore, produced oxygen has a commercial value and can be even used in the IRRF, for example, for producing ozone for disinfection.

The US Department of Energy (2018) listed the following types of the electrolyzers:

A *polymer electrolyte membrane (PEM) electrolyzer* (Figure 8.11, left) operates at lower temperatures at $70°–90°C$; hence, hot steam would have to be cooled down to water. The electrolyte is a solid specialty plastic material. In this electrolyzer:

- Water reacts at the anode to form oxygen and positively charged hydrogen ions (protons).
- The electrons flow through an external circuit and the hydrogen ions selectively move across the PEM to the cathode.
- At the cathode, hydrogen ions combine with electrons from the external circuit to form hydrogen gas.

Anode reaction: $2\,H_2O \rightarrow O_2 + 4H^+ + 4e^-$
Cathode reaction: $4H^+ + 4e^- \rightarrow 2H_2$

Efficiency of the low temperature water electrolyzer is less than 50%.

An *alkaline electrolyzer* operates via transport of hydroxide ions (OH^-) through the electrolyte from the cathode to the anode, with hydrogen being generated on the cathode side. These electrolyzers have been commercially available for many years. Alkaline electrolyzers operate at 100–150°C.

A *solid oxide electrolyzer* was previously presented as an alternative to the hydrogen fuel cell (Figure 8.11, right). Seemingly, this very efficient fuel cell can work in reverse direction (Wang et al., 2016) and produce syngas from a mixture of H_2O (g) + CO_2 or:

$$H_2O + CO_2 + \Delta E_e \rightarrow H_2 + CO + O_2.$$

ΔE_e is electric energy needed to split water and carbon dioxide, which is greater than the energy in the produced syngas (hydrogen). By adding electrons from the DC current, incoming CO_2 and H_2O are split to CO and H_2 gases and O^{2-} oxide ions. Oxide O^{2-} ions pass through the electrolyte to the anode where electrons are removed, which forms O_2 gas. No air is entered in the cell but CO, H_2, and O_2 gases are formed. Syngas (CO and hydrogen) and oxygen are produced in separated compartments of the cell and must be kept separated to prevent explosion. Solid oxide electrolyzers must operate at temperatures high enough for the solid oxide membranes to function properly (about 800–900°C). The heat could be provided by the hot steam released from the hydrogen fuel cell. The efficiency of this new application of solid oxide cell is greater than 50%.

Gas Separation

In the future IRRF, several gases will be produced, including CO_2, CO, H_2, H_2S, CO, plus the "waste" gases in the flue gas from combustion engines that include nitrogen and oxygen. Separating CO_2 and H_2 may be easier because of the large difference in the specific density of the two gases (i.e., 0.09 kg/m^3 for H_2 and 1.8 kg/m^3 for CO_2). These two gases do not interact so H_2 is always on the top and heavier-than-air CO_2 will stay below. There is no air in the anode compartment if indirect gasification and/or methane steam reforming produced syngas is feeding the cell.

Chemical separation may also be used by selecting compounds that adsorb the gases and then rerelease them later, but today and, in the future, it may be best achieved by special membranes that have very small pores sizes in Ås (Å = angstrom = 0.1 nanometers) and are selective for each gas. Membrane separations of O_2 and/or CO_2 from N_2 in flue gases, CO_2 from CH_4, H_2 from CO_2, and separation and collection of other greenhouse gases, have been extensively researched (Shindo and Nagai, 2014). Specifically, gas separation membranes are already widely used for hydrogen recovery in petroleum refineries and ammonia plants, and for CO_2 removal in natural gas processing. Membranes have competitive advantages against other gas separation technologies.

H_2/CO_2 and CO_2/H_2O (steam) separation will be important in the IRRF power plant for separating hydrogen from CO_2 gas after shifting methane to syngas followed by the CO in syngas shift to a mix of CO_2 and H_2 (and possibly other waste gases), and CO_2 separation from steam in the exhaust/recycle pipe of the expanded MCFC. It may be very premature to describe actual designs now because the speed of development of such separators is very rapid. Based on the 2010s state of the knowledge, the following membrane gas separators are being developed and some are already industrially working (National Energy Technology Laboratory, 2009):

- Silica membranes (De Vosa et al., 2015; Khatib et al., 2011; Gu et al., 2012; Khatib and Oyama, 2013)

- Carbon membranes (Paranjape et al., 1998; Saufi and Ismail, 2004)
- Zeolite membranes (Korelskiy et al., 2015)
- Polymer membranes have limited application (Lau et al., 2010)

Silica membranes have shown very high gas permeance for small molecules, such as H_2, CO_2, N_2, O_2, and CH_4. These are the gases produced and emitted by the IRRF processes and power plant. Khatib et al. (2011) and Khatib and Oyama (2013) identified silica-based membranes as promising materials due to their high permeation rates, high selectivity, hydrothermal stability, resistance to poisons, and mechanical strength and good performance at high temperatures. Hybrid organic–inorganic H_2 selective single or dual-layer membranes of silica incorporating organic aromatic group on a porous aluminum support have a good hydrothermal stability at high temperatures and exhibit high permeance for hydrogen while preventing the passage of other larger molecules such as CO_2 or CH_4. The pore size of these membranes is in the range of 5 to 8.5 Å (Gu et all., 2012).

Carbon membrane materials are also emerging for gas separation due to their higher selectivity, permeance, and stability in corrosive and high-temperature situations (Saufi and Ismail, 2004). The pore size is again in less than nanometers (6 to 7 Å) in the carbon matrix supported by a microporous aluminum support (Paranjame et al., 1998). These membranes can very efficiently separate CO_2–H_2 mixtures even at a low feed gas pressure of about 0.4 MPa (mega pascals).

Zeolite and polymeric membranes are also effective for CO_2 separation and have been industrially used for CO_2 separation from gas mixtures. Ceramic zeolite membrane separation systems are more compact than older polymeric membranes (Korelskiy et al., 2015).

There are two major factors that affect the overall gas transport in membranes: (1) sorption (solubility selectivity) and (2) diffusion. While sorption describes the interactions between gas molecules and the membrane surface, diffusion deals with the rate of the gas passage through the membrane (Khatib et al., 2011). Due to extremely small size of pores in membranes one would expect that extremely high pressure would have to be applied to push the gas through the membrane; however, the high pressure across the membrane is achieved by the hydrophilic nature of the membrane created by adding complex organic compounds during preparation of the membrane (De Vosa et al., 2015) and by strong capillary suction. The cost of gas separation membranes during the second decade of this century was somewhat higher but the research and technology development was rapid and accompanied by price decreases.

8.7 THE IRRF POWER PLANT

Figure 8.10a presented the molten carbonate fuel cell combined with steam methane reforming (SMR) that converts methane from the AnFBM and/or UASB reactors and a digester, if it is a part of the IRRF, to hydrogen-enriched syngas. The MCFC (or SOFC), SMR and the gasification unit producing syngas from MSW, sewage sludge and other nondigestible biomass, constitute the virtual "power plant" shown in Figure 8.12a and b. SOFC and in some applications MCFC also internally shifts methane to syngas and syngas to hydrogen by steam reforming; hence, methane collected from landfills could be added to the power plant input. Theoretically, no supplementary source of steam (e.g., boiler) is needed except for converting recuperated heat to additional electricity. All steam needed

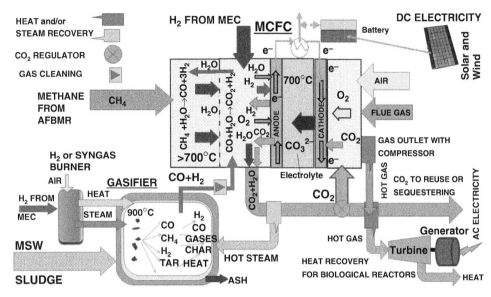

Figure 8.12a. Possible schematics of the IRRF Power Plant consisting of indirect gasification unit connected to MCFC or SOFC with internal SMR unit performing steam gas re-forming shift from methane to syngas followed by carbon monoxide to hydrogen shift, heat recovery, and electricity production. H$_2$ or syngas external or internal burner is needed to provide endothermic heat to indirect gasification. This is a concept and not actual design. Fuel Cell Energy in Danbury, Connecticut, has a patent for additional energy production from waste heat and several other patents for commercial molten carbonate fuel cells. *Source:* Inspired by Campanari et al. (2013), Fuel Cell Energy (2013), Richardson et al. (2015), and Ormerod (2003).

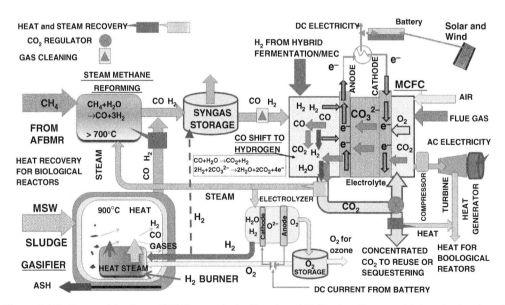

Figure 8.12b. Proposal for AnovaTNZ Power plant with external SMR of methane and internal reforming of syngas in MCFC that enables recovery of excess hydrogen and oxygen for commercial and IRRF internal uses. This power plant generates internal endothermic heat for gasifier by internal combustion of hydrogen and oxygen from electrolyzer that will not emit any GHGs. Syngas is stored, and excess blue energy is used to produce more hydrogen from clean steam from the hydrogen fuel cell by excess blue power and oxygen for hydrogen burner. Some MCFC design components are covered by patents awarded to Fuel Cell Energy.

for reforming methane and syngas and for converting biomass and combustible solids in the indirect gasifier can be recovered from MCFC. Because of the high temperature, the change of the power plant capacity is slow; therefore, possible solutions are installing on-site biogas and/or hydrogen storage tanks or storage in the digester or connection, as a last resort, to an auxiliary supply of natural gas in time of biogas shortage. Note that MCFC components and designs are covered by patents awarded to the manufacturers. Furthermore, the designs of the future power plants processing gases from IRRF can be only anticipated.

The power plants employing MCFC presented in Figure 8.12a and b solve the problem with the endothermic indirect gasifier design by combusting a portion of the produced hydrogen in an auxiliary or internal burner, which, along with the heat recovery unit, could increase the gasifier temperature to more than 900°C. Recall that both indirect gasification and SMR are endothermic (Table 8.1). Burning syngas or methane with air in a furnace emits CO_2 diluted with nitrogen from air and other gases, which would be difficult to separate and sequestrate and the efficiency of the heat recovery is between 40 and 60% (Knight et al., 2015); however, it would not be counted as GHG but without a CO_2 credit. The MCFC hydrogen fuel cell was selected because at the time of this writing it was the most advanced technology that had reached the MW capacities suitable for medium-size cities. The power plants were versatile and could internally reform the hydrocarbon fuels. Hydrogen fuel cell technology at that time was at the point that photovoltaic plants were at before 2010. Worldwide capacity is increasing rapidly.

The new IRRF power plant could combust hydrogen from the hydrogen outflow line of the MEC/fermentation unit (Figure 8.12a) or from the outflow from the syngas conversion unit, or hydrogen and oxygen can be made by an electrolyzer from hot steam released from the MCFC (Figure 8.12b). Locating hydrogen–pure oxygen burning inside the gasifier (the proper design had not been known during the time of writing this treatise) enables release of the heat energy by the heat content of the hot steam and combustion heat energy by radiation energy directly into the gasifier biomass. In contrast to CO_2 emitted from methane or syngas combustion, the hydrogen-oxygen burner produces only hot steam and radiation heat, which can be directly returned to the gasifier without contributing to CO_2 emissions. However, if air is used to provide oxygen, the steam will be mixed with nitrogen gas from the air (Figure 8.12a). This problem is resolved by making hydrogen and oxygen with a high-temperature electrolyzer. Smaller reverse SOFC can be used. (Figure 8.12b). Excess hydrogen made by the electrolyzer during the time of excess energy production can be returned to the syngas storage. Theoretically, the radiation heat and steam recovery of the alternative is shown in Figure 8.12b. Producing hydrogen and oxygen by electrolysis is more efficient than the H_2 or syngas burning alternative shown in Figure 8.12a, which is about 60% (Knight et al., 2015). Producing energy by electrolysis has energy recovery efficiency of about 80% and burning hydrogen with the oxygen produced by the cathode of the electrolyzer produces steam and heat that enters the gasifier. Oxygen produced at the anode of the electrolyzer also has an economic benefit and value for the IRRF because it can be converted on site with electricity from the IRRF power plant (DC current) to ozone for disinfection.

Syngas is also an important raw material for biofuel and organic chemical compounds, including plastics instead of oil. The heat recovery unit on the gasifier effluent line can also provide heat to the SMR unit to bring its temperature to more than 700°C. Gasifier releases some heat and in combination with heat released from the hydrogen fuel cell there will be enough heat to operate the entire IRRF power plant and units. Only a small fraction of sensible heat is needed for SMR because the heat released with the gas from the gasifier may be more than that needed for the SMR unit. Syngas produced by indirect gasification and stem reforming digester methane should not contain nitrogen.

The hot exhaust gas from the MCFC or SOFC contains concentrated CO_2 and steam and, to recover high-quality water from the 700–1000°C hot gas mixtures, it must be cooled down to less than 100°C and the sensible heat should be recuperated. The heat extraction could be like recovering heat by thermal solar panels from concentrated solar rays or recuperating heat from hot exhaust stacks. The heat can be also carried by silica sand, ceramic balls, or other mostly patented heat-carrying compounds. Methane to syngas reforming process is also endothermic and needs some heat and hot steam, which will be available in the anode compartment of MCFC if the reforming process is internal. The indirect CO to CO_2 and hydrogen shift is exothermic (Table 8.1) and will not require heating and may add some heat to the anode and heat management system of the IRRF. The feed gas from the SMR unit may contain some CO_2, which can be separated by a membrane CO_2 separator unit or another CO_2 separation method (Yang et al., 2008). The mixture of CO, CO_2 (if present in the feed), and hydrogen can enter the anode compartment of the MCFC. On the other hand, CO_2 added to the cathode compartment improves the MCFC performance (McPhail et al., 2015) so there are many issues to research and optimize. Separated CO_2 will be concentrated. If SOFCs are developed for larger power capacity, they can be used and provide the endothermic heat to the gasifier without a burner because of the higher temperature inside the SOFC.

The gas mixture from fermentation/MEC hybrid reactors may produce CO_2 but hydrogen and CO_2 are produced in different compartments of the MEC. Gaseous CO_2 from MEC, if any, can be diverted to the CO_2 recycle line in the power plant and in this way the hot concentrated CO_2 released from the hydrogen fuel cell will be the only place in the entire IRRF energy and resource recovery plant where concentrated CO_2 would be emitted and regulated. Finally, CO_2 can be recovered from the MCFC exhaust and can be sequestered or reused for neutralization of high pH flow from the struvite producing reactor (see Chapter 9), used for growing algae and other beneficial purposes that were described in Chapter 4. Recuperated heat is used for production of electricity and heating the mesophilic biologic reactors. Because the auxiliary energy production is by the recovered high-quality heat and not by burning, there are no CO_2 emissions from the secondary electricity production, as shown in Figure 8.11a. However, turbine and generator make AC current while MCFC makes DC current, hence, a transformer should be a part of the power plant, which may slightly reduce the efficiency to generate electricity. The more advanced power plant in Figure 8.11b generates only DC current.

Hydrogen-CO_2 Separator

At the time of this writing, H_2–CO_2 separation had been researched and was already being implemented in industrial applications and was being developed and patented by fuel cell manufacturers and by government, academic, and private research facilities. The reason the separation may be needed is the realization that directing a mix of CO, CO_2, and H_2 after steam shift from methane into the anode compartment of the MCFC would lead to loss of hydrogen into exhaust gas and subsequent problems with pure CO_2 recovery for reuse. The earlier fuel cell energy MCFC power plants installed in California used pressure swing adsorption (PSA) common in large-scale applications, which recovers high-purity H_2 (99.99%) and recovered approximately 70% of the available hydrogen. Alternative techniques for hydrogen separation, such as membrane separation and electrochemical separation (described in the preceding section) could achieve similar integration with carbon recovery at higher net electrical efficiencies. However, ongoing and future research should establish whether a CO_2–H_2 separation unit is needed.

Separation of Steam (Water) from CO_2 Exiting MCFC. The power plant needs smart heat and steam (water) management. In maximizing energy and hydrogen yield, water (steam) is needed in the steam methane and carbon monoxide units and in the indirect gasification. However, power production units requiring water (SMR, conversion of CO to CO_2 and H_2) and partially indirect gasification would have an adequate source of water and heat in the anode compartments of the MCFC where hydrogen gives up electron and two protons combine with oxygen from carbonate ion passed from cathode compartment through the carbonate electrolyte to form water. Water is also brought into the power plant by the moisture of MSW and sludge solids feeding gasifier. Because of high temperature in the MCFC, water is in the gaseous form of steam. By weight, stoichiometrically, the exhaust from the anode is 60–70% CO_2 and 30–40% H_2O (g). The produced water is more than needed for gasification and CH_4/CO reforming to hydrogen but because the feed into the gasifier has moisture content, the clean water production in the MCFC should result in a water production excess.

Lastly, the new power plants should be identified by an acronym. While that in Figure 8.12a uses an already known concept that can be altered, Figure 8.12b shows a power plant with a lot of originality, so let us call it ANovaTNZ. Nova is a star rapidly formed from an old dying star into a new, more efficient star with more complex composition and a constellation that also includes planets. Our Sun was a nova for billions of years. TNZ stands for triple net-zero, as defined previously, and means that the plant will save and generate its own water, will sequester carbon dioxide with no GHG emissions, and will reduce or eliminate the need for landfills.

Use of fuel cells in used water treatment plants is expanding and at the beginning of the second decade of the 2000s, a 2.8-MW hydrogen fuel cell power plant manufactured by Fuel Cell Energy, Inc., was installed in a San Bernardino County, California, water and resource reclamation plant serving 850,000 people. This power plant uses sludge digester biogas and not syngas from MSW. Figure 8.13 shows a 1.4-MW hydrogen PureCell® 400 system producing energy from a digester biogas at the Naugatuck, Connecticut, water treatment plant, installed by Doosan Fuel Cell America in 2017. This unit has energy (electricity and heat) recovery efficiency of 90% and electricity recovery above 40%, which, in combined heat and energy systems, could be increased. Most current traditional plants processing MSW to syngas by gasification and/or incineration use a combustion engine or a boiler producing steam, followed by a traditional turbine and generator to produce electricity with overall energy efficiency of 25% or less (Belgiorno et al., 2003, McCarty, 2011; Knight et al., 2015).

The use of electricity of an average US household in 2015 was about 900 kW-h/month; hence, to provide all electricity to 10,000 households, the power capacity of the IRRF power plant would have to be 12.5 MW. The good news is that both Fuel Cell Energy and Doosan hydrogen fuel cell manufacturers can produce power plants with this or greater capacity. The US first large MCFC 14.9-MW power plant has already been built in Bridgeport, Connecticut, by Fuel Cell Energy, and the largest 59-MW hydrogen fuel cell plant has been finished in Hwaseong, South Korea. These power plants run on natural gas, which is converted to electricity, heat, and water. At the end of the second decade of this century, larger 100-MW plants were planned to be installed in the US and South Korea.

Figure 8.13. 1.4-MW hydrogen PureCell® 400 system producing energy from a digester biogas at the Naugatuck, Connecticut, water treatment plant, installed by Doosan Fuel Cell America in 2017. *Source:* Doosan Fuel Cell America.

Including Batteries in the IRRF Systems.   Including batteries for storing excess electricity during off-peak energy production hours from waste and production of blue energy by wind and solar PV power would further improve their efficiencies toward becoming net-zero carbon operations. In the simplest applications, in 2017 several water utilities in California began to install large batteries alongside their pumping plants and water treatment facilities to store energy in the batteries overnight when energy is cheaper. Then during the daytime, when power is more expensive, the water agency can tap the battery power for its routine operations. The largest battery systems of 100-MW capacity were installed by municipalities in Australia by Tesla and 200 MW in China (see Chapter 4).

Finally, the power plant also needs a power convertor to convert the DC current it produces to AC current of the regional grid. The Doosan PAFC has an internal transformer that converts produced DC current to AC.

Carbon Dioxide Sequestering in an IRRF

Any used water biological treatment and reclamation process, aerobic or anaerobic, produces carbon dioxide. However, because of pH of the liquids in the process units being around the neutral pH ~7 and the presence of calcium, magnesium, and other cations in used water that provide buffering, carbon dioxide may be converted to dissolved CO_2 and bicarbonate alkalinity and may not be emitted. Indirect gasification does not emit CO_2 except if a portion of the produced syngas is burned for heating the gasifier because the process is mildly endothermic and/or the indirect gasifier overshoots the conversion of biomass into syngas by adding more water, increasing temperature inside the gasifier, and extending the residence time (Lee, Chung, and Ingley, 2014). The main product of gasification and SMR reforming is syngas. Using hydrogen produced by the IRRF for heating does not emit any GHG gases, only water and concentrated CO_2 (70%) if syngas is the feed of the MCFC. Reforming carbon monoxide will produce carbon dioxide and hydrogen and will emit CO_2, which in the proposed IRRF power plant can be done internally under controlled conditions and undiluted CO_2 can be sequestered or even reused.

The flue gas from the hydrogen-based MCFC in the IRRF power plant will not be the same as flue gases emitted from coal fired thermal power plants or MSW incinerators. Following the discussion in Chapter 4, unmitigated flue gas from fossil fuel power plants may contain sulfur oxides (SO_x), nitrogen oxides (NO_x), nitrogen gas (N_2), oxygen, carbon monoxide (CO), carbon dioxide (CO_2). CO_2 and NO_x are counted as GHGs per IPCC. The CO_2 content is relatively small, about 11% in the flue gas of coal combustion power plants, and about 3–5% from natural gas fired plants, respectively (Zeverenhoven and Kilpin, 2002). Flue gas from coal-fired power plants also contains fly ash, toxic metals (e.g., mercury, aluminum, arsenic, cadmium, vanadium, nickel, etc.), halide compounds (including hydrogen fluoride), unburned hydrocarbons, and other volatile organic compounds. Mercury contamination of fish tissues in the Everglades National Park is attributed to near and regional sources of atmospheric mercury by the large power plants burning coal in the southeastern US and waste incineration (Krabbenhoft, 1996). Fly ash together with bottom ash, known as coal ash removed from the bottom of the boiler, has been commonly dumped into landfills and liquids from scrubbers with fly ash into lagoons, which sometimes failed with serious adverse and lasting environmental consequences. In the US flue gases from power plants must be treated by scrubbing, which removes fly ash and sulfur oxides but does not remove carbon dioxide. Flue gases from power plants today burning natural gas are much cleaner but still contain CO_2 classified as GHG.

Although extensive research and monitoring data from the facilities resembling the proposed hydrogen-based IRRFs were not available at the time of this writing, it is possible to hypothesize that the flue gas emitted from the proposed hydrogen IRRF power plant will be hot, contain concentrated and relatively clean carbon dioxide and steam, and might contain only small amounts of residual pollutants such as unoxidized hydrocarbons but no toxins. Excess steam in most cases will be converted by cooling into clean, close to deionized water, and heat of cooling is recoverable. The opportunities for CO_2 recovery have grown considerably and pure captured CO_2 is a potential valuable feedstock for the synthesis of chemicals and synthetic fuels (Rinaldi et al., 2015). Capture and reuse of CO_2 would make the MCFCs net-zero carbon pollution sources. Because the small quantity of gasifier ash does not contain toxins (with the exception of some toxic metals that can be removed), it does not have to be landfilled. MCFC does not produce any ash.

Because the CO_2 side product of the H_2-based IRRF is concentrated carbon dioxide, which would not be classified as GHG, it would not require tall chimneys or flue gas stacks. In the latest design of hydrogen fuel cell power plants, the exhaust of CO_2 and used air is by a simple exhaust pipe and not by tall flue stacks and chimneys characterizing coal-fired power plants. If the heat emitted by the MCFC is not recovered for reuse, the units must be cooled by air. The produced water is super clean and could come as a hot steam and reused for heating or, after cooling, as hot water. Note that in Figure 8.13 the exhaust is difficult to locate. No tall flue stacks: just a simple, relatively short (in meters) pipe that in small district operations may not be much different from exhaust pipes of the modern natural gas-burning furnaces. This simple release is not classified as GHG and would be carbon neutral because IRRF is not processing fossil fuels.

Chapter 4 presented several reasons why the concentrated CO_2 gas would be a valuable sequestration commodity, such as:

- By entering the gas into the pipelines that provide pressurized carbon dioxide to enhanced oil recovery companies, CO_2 becomes a commodity and will generate revenues for the gas and a credit to buyers for replacing fossil CO_2 by a "green" gas.
- Carbon dioxide can be injected into supernatant return flow after struvite precipitation and/or high pH ammonia removal to neutralize high pH.
- By injecting into underground coastal aquifers under higher pressure, CO_2 will dissociate into bicarbonate and will provide a barrier against salt water intrusion. Similarly,

injecting CO_2 into basalt geological formations will form a solid carbonate mineral (see Chapter 4).

- The gas can be purified and used for making dry ice. Because of the very high temperature, the released gas has no pathogens, but it might contain some trace pollutants that would have to be removed to obtain food-quality ice.
- By providing CO_2 along with recovered nutrients to algal farms producing biofuel (see Chapter 9).

The power plant schematics exhibited in Figure 8.12a and b also recommend heat and steam recuperators. Heat can be recovered by heat exchangers or gas-to-gas heat pumps (Novotny et al., 2010a). Cooling down the anode exhaust recycle to less than 100°C will condensate steam to liquid water that can be separated, and a portion reinjected as steam into gas shift reactors. Water (steam) is needed for steam shift of methane and carbon monoxide to hydrogen. Condensation also releases significant amount of heat, the same as the heat of vaporization. This will result in the exhaust gas having close to 99% CO_2. Without heat recovery, MCFC would have to be cooled by air, as in H_2 cars.

Carbon Dioxide Capture and Concentration by the Molten Carbonate Fuel Cell

Development of the MCFC had one surprising side benefit (Bullis, 2013; Campanari et al., 2013) of concentrating CO_2. The cell processes CO_2, which is passed from cathode to anode along with the oxygen from the air. Plenty of CO_2 is produced by the IRRF to sustain the process. However, the cell can accept additional diluted CO_2 in flue gas in the cathode and pass it with oxygen through the cell (Figure 8.12a and b), which also increases the cell efficiency. The CO_2 gas emitted from the anode compartment is concentrated; typically, the exhaust gas has CO_2 content of 70% and the rest is steam. Apparently, only a molten carbonate fuel cell has this capability (Campanari, 2002). Rinaldi et al. (2015) found that the highest CO_2 concentration can be achieved by three MCFCs in series. Hence, MCFC fuel cells can be used as a concentrator of CO_2 from gas turbines, combustion engines, and for CO_2 capture. Research by the University of Milan has shown that such plant configuration can capture 70–85% of CO_2, with small efficiency penalties compared to the combined cycle and increasing by about 20% the overall power output (Campanari, Manzolini, and Chiesa, 2013). However, molten carbonate fuel cells (or any other fuel cells) do not produce extra energy from CO_2; the cells just make CO_2 in the exhaust from power plants 10 times more concentrated, so it could be better sequestered or commercialized and makes sequestration less costly and the energy production by the cell more efficient.

The US Department of Energy (2013) also counts on gasification and fuel cells to save the coal industry. Instead of BAU burning coal to produce (black) energy, the program of the US DoE Office of Fossil Fuel is researching and planning first to convert it into clean high-hydrogen synthesis gas (syngas) by coal gasification, that in turn can be converted with higher efficiency into electricity by the solid oxide fuel cell (or MCFC) and the produced concentrated CO_2 sequestered.

Thus it can be concluded that modern gasification in a synergistic combination with the high-temperature hydrogen fuel cells is the most efficient and environmentally friendly system of producing energy and recovering resources not only from waste but also from other sources such as harvested biomass, or even from natural gas (a GHG). Gasification has been around for more than a century, and has been tested and continuously improved. This combination has a potential to achieve net-zero or even negative carbon emissions systems.

NUTRIENT RECOVERY

9.1 THE NEED TO RECOVER, NOT JUST REMOVE NUTRIENTS

Fertilizing nutrients that include nitrogen, phosphorus, potassium, and trace element are responsible for producing almost all organic/living matter on the Earth. Over the past billion years or so, the planet's ecological system established a natural balance of recycling nitrogen and phosphorus between living and nonliving systems and between mineral and organic matter. This system persisted until the first half of the last century but thereafter the natural cycling has been distorted with the introduction of industrially made nitrogen fertilizers through the Haber–Bosh process that converts atmospheric nitrogen and industrially made hydrogen from fossil raw materials (methane) into ammonium. This happens at a great expense of energy from fossil fuel (natural gas) with emission of GHG CO_2 because to make hydrogen, methane is steam reformed into hydrogen and carbon dioxide. The produced ammonium was used to make fertilizers and also explosives during WWII. After the war, this resulted in dramatic increases of agricultural production relying on industrial fertilizers needed to satisfy food needs for the rapidly increasing population. The process of converting natural farming to high-yield agriculture relying on manufactured chemical fertilizers is known as the *green revolution* (Novotny, 2011b). Today, twice as much nitrogen is introduced into the earth, cycling from anthropogenic sources and the industrial production of nitrogen fertilizer though the Haber-Bosch process and enhanced nitrogen fixation enabled humans to feed the world.

In nature, the most common phosphorus-containing mineral (ore) is apatite. Phosphorus is a vital component of all living organisms as a building substance of DNA and RNA (in form of phosphate ion PO_4^{3-}) and plays essential roles in energy transfer through living cells as a component of adenosine triphosphate (ATP) and, as phospholipids, contributes to the formation of cell membranes. A considerable amount of phosphorus is contained in living beings' bodies, mostly as calcium hydroxyapatite $Ca_{10}(PO_4)_6(OH_2)$. Around 650 g of phosphorus can be found in the average human body; only calcium and the four organic elements of C, O, H, and N are higher than that (original references in Daneshgar et al., 2018). Unlike nitrogen fertilizer resources, phosphate rock (PR) is a finite, irreplaceable, nonrenewable resource and there is a danger that the human consumption of phosphorus will lead to the exhaustion of ore resources. Daneshgar et al. pointed out that there was a large anthropogenic per capita increase in phosphate use after World War II through the 1970s, attributed to the "green revolution" in agriculture (Novotny et al., 2010b). These high levels stabilized from 1975 to 1991 at an average of 30.1 kg PR capita^{-1}year^{-1}. After that time, use was reduced to about 22.8 kg P capita^{-1}year^{-1} ore use in 2006. These estimates do not consider phosphorus recovery from used water and solids.

Of note are the virtual energy use and GHG emissions from producing ammonium and phosphate industrial fertilizers. As quoted in McCarty et al. (2011), the energy requirement for production of nitrogen fertilizer by the Haber-Bosch process is 19.4 kW-h/kg N produced and that for phosphate is 2.11 kW-h/kg of P, respectively. Producing synthetic nitrogen fertilizer by the H-B process uses approximately 1.2% of the world's fossil fuel and contributes about 2% to greenhouse gas emissions in a form of residual methane and carbon dioxide releases from the production and agriculture uses of industrial fertilizers. The nitrogen cycling process is also distorted by combustion processes and traffic that emit nitric oxides (NO_x). In the US alone during the year 2002, 6.4 Tg (Tera-gram = 10^{12} grams or 10^6 tons) of reactive nitrogen (N_r) was added from natural sources, 15.3 Tg was added through fertilizer and non-fertilizer production, 7.7 Tg through enhanced agricultural nitrogen fixation, and 5.7 Tg as NO_x from stationary and transportation fossil fuel combustion (Burke, 2015). Because, both N and P are needed to grow crops, increased use of fertilizer made from atmospheric nitrogen also resulted in overmining of the mineral phosphate (apatite) to the point that economically extractable mineral P resources will become scarce in the next 100 years and exhausted in 170 to 300 years, depending on which population increase is considered and at what rate the phosphorus will be used (Childers et al., 2011; Steen, 1998; Smil, 2000; Vaccari, 2009, Daneshgar et al., 2018). Ore resources in the US and China will be exhausted in a shorter time and thereafter most of P ore will be in Morocco. It is therefore necessary, in the nutrient management paradigm shift, to recover and recycle nutrients from used water and wastes, instead of the current practice of removing and replacing.

Because the IRRF outlined in this treatise will also process biodegradable MSW by co-digestion (about 0.5 kg capita^{-1} day^{-1}; see Figure 3.2) and also the combustible nonbiodegradable fraction by gasification (about 1 kg capita^{-1} day^{-1}), the nutrient content of MSW should be also considered in the recovery process. The literature data document that the nutrient content per unit of MSW dry weight is less than that of residual solids from used water treatment and recovery, but in the unit per capita load is about the same. Mato et al. (1994) and Fuentes et al. (2006) reported nutrient values for the biodegradable (compostable) fraction of MSW in % of dry weight of solids as being 0.2–0.6% (1–3 g capita^{-1} day^{-1}) for phosphorus and 1.2 to 2% (6–10 g capita^{-1} day^{-1}) for nitrogen, respectively. Nutrient content in combustible nonbiodegradable MSW is much less; however, only phosphorus is recoverable from the gasification ash. Figure 9.1 shows the variability of nutrients and other compound contents in the MSW. Unit loadings of nutrients of used water and biodegradable (compostable) municipal solid waste are given in Table 9.1.

Without separation and water conservation, nutrient concentrations in typical municipal used water effluents are low for efficient and economic nutrient recovery, yet even at these low levels they are harmful to the environment. Ammonium in a typical municipal wastewater treatment effluent imposes oxygen demand by nitrification on the receiving water bodies. Both nitrogen and phosphorus stimulate eutrophication process, and in excessive concentrations (more than 0.2 mg TP/L) can cause damaging hypertrophic conditions and harmful algal blooms (Novotny et al., 2010b; Novotny, 2011b). While the O_2 demand to oxidize 1 g of biodegradable COD is 1 g (by COD definition), biological oxidation of 1 g of NH_4^+ to nitrate requires about 4.3 g of O_2.

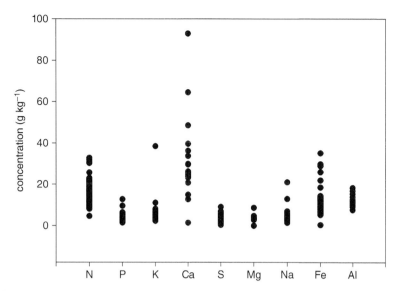

Figure 9.1. Content of nutrients in MSW, compiled by Hargreaves et al. (2008); data references are listed in the source article. *Source:* Copyright Elsevier, 2007; reprinted with permission.

Table 9.1. Per capita nutrient content in used water and MSW.

	P	N	Source
	g capita^{-1} day^{-1}		
Raw sewage	2.5	13	Imhoff and Imhoff (2007)
Black water	2	11	Henze and Comeau (2008)
After biological treatment	1.1	7.6	
Incorporated into sludge	0.7	1.7	
Biodegradable (compostable) MSW	1–3	5–10	Mato et al. (1994) Fuentes et al. (2006)

9.2 BIOLOGICAL NUTRIENT REMOVAL AND RECOVERY

Traditional Nutrient Removal Processes

Traditional wastewater treatment processes may remove but do not recover nutrients except for N and P in sludge, which in most cases used to end in a landfill or soils with limited agricultural production. Increasing the waste sludge phosphorus content is possible and is presented in the next section. Landfill and soil disposals of waste materials are most common, but suitable landfill sites are rapidly disappearing and landfilling violates the third goal of sustainable urban water/solids/energy management, which is no future landfilling of waste.

In the more advanced traditional processes of the last century, nitrogen was removed from used water flow by sequential nitrification-denitrification processes in which organic nitrogen was converted by ammonification to ammonium, which was then nitrified to nitrate in an aerobic unit and subsequently in an anoxic unit nitrate was converted to nitrogen gas. The second process requires organic carbon originally provided by methanol. The organic nitrogen fraction is recorded as Total Kjeldahl Nitrogen (TKN), which is a summation of ammonium and organic nitrogen.

In the Bardenpho nutrient-removing process the treatment units are reversed (Figure 7.1C). The first unit is anoxic in which organic N is converted to NH_4 (Figures 7.1c and 9.3A) and nitrogen is removed in the anoxic reactor by converting nitrates (NO_3^-) to nitrogen gas (N_2), which volatizes by aeration. Nitrogen is not recovered with the exception of nutrients incorporated in the excess sludge biomass. To convert incoming TKN to nitrate, bacteria in the anaerobic (if included in the EBPR) and anoxic reactors convert TKN to ammonium (NH_4^+), which is then oxidized in the aerobic reactor(s) first to nitrite and then to nitrate by two nitrifying microbial species:

$$2\,NH_4^+ + 3\,O_2 \rightarrow 2\,NO_2^- + 2\,H_2O + 4\,H^+ \qquad 2\,NO_2^- + O_2 \rightarrow 2\,NO_3^-$$

Nitrosomonas   *Nitrobacter*

28g of N + 96 g of O_2 28 g of N + 32 g of O_2

To convert one gram of ammonium nitrogen into nitrate would theoretically require 4.5 grams of oxygen but the amount is slightly less (4.3 g of O_2/g of NH_4^+-N) because of small amount of O_2 is available from the chemotrophic denitrification reaction. Conversion of nitrate to nitrogen gas occurs in anoxic conditions; hence, the nitrate-rich content of the aerobic reactor is then with a high recycle rate ($\sim$80%) pumped back to the anoxic reactor where NO_3^- is converted by facultative chemotrophic micro-organisms from nitrate back to nitrite, and nitrogen gas. A large group of heterotrophic chemotrophic facultative denitrifying anaerobic bacteria obtain their energy from decomposing reduced organic carbon, which instead of methanol in the original process (CH_3OH) in the Bardenpho is the incoming COD in the influent. A simple schematic of denitrification is illustrated on the methanol decomposition (oxidation), where denitrification provides the electron to anaerobic methanol oxidation:

$$6NO_3^- + 5CH_3OH \rightarrow 3N_2 + 5CO_2 + 7H_2O + 6OH^-$$

Optimum pH values for denitrification are between 7.0 and 8.5. Denitrification increases alkalinity and pH. The intermediate products are nitrous oxides NO and NO_2 which are considered GHGs:

$$NO_3^- \rightarrow NO_2^- \rightarrow NO + N_2O \rightarrow N_2 \text{ (g)}.$$

The process may emit GHG nitrogen oxides NO_x, which are intermediate byproducts of denitrification. Because of the high oxygen demand to nitrify ammonium, nitrification/denitrification processes require 40–60% more aeration energy than conventional activated sludge treatment plants.

Anammox

This acronym refers to the recently discovered process of *anaerobic ammonium oxidation* (Figure 9.2), which is a new method of ammonium removal and its conversion to nitrogen gas (Van de Graaf et al., 1995; Van Niftrik and Jetten, 2012). Anammox is a more efficient, less costly, and less energy-demanding alternative to the traditional nitrification-denitrification biological processes. Anammox bacteria oxidize ammonium directly to nitrogen gas using nitrate as an electron donor, which can be represented by the following simple chemical equation:

$$NH_4^+ + NO_2^- \rightarrow N_2 + H_2O$$

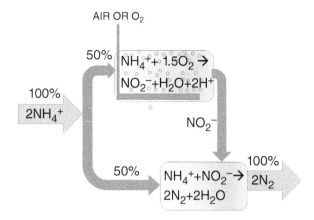

Figure 9.2. Nitrogen compounds and metamorphosis in the Anammox.

Although this process was mostly unknown or overlooked until it was fully analyzed and described in The Netherlands at the end of the last century (Van de Graaf, et al., 1995), it is apparently the main mechanism by which nitrogen fixed by cyanobacteria has been lost from the large water bodies with anoxic/anaerobic sediments. This process is a key component of the nitrogen cycle on Earth. The Anammox process does not require organic carbon for denitrification; it has a low biomass yield and requires substantially less aeration (Abma et al., 2007; Chamchoi, Nitisoravut, and Schmidt, 2008). The major obstacle to implementation of Anammox has been the slow growth rate of bacteria, which implies a long startup time of bacteria in the Anammox reactors (Ni and Zhang, 2013). As strictly anaerobic microorganisms, Anammox bacteria are inhibited even by low concentrations of dissolved oxygen, and nitrite (NO_2^-) concentration exceeding 100 mg-N/L are also inhibitory (Lackner et al., 2014). Nevertheless, research has been conducted at a high pace in The Netherlands and elsewhere and first full-scale Anammox plants appeared at the beginning of this century in Holland. Since 2005 the number of the full-scale installations has been increasing exponentially to 150 in 2015. Anammox process can operate at loadings up to 1.5 kg N m^{-3} day^{-1}. The most common technologies are sequencing batch reactors (SBR) (more than 50%), followed by granular systems and moving bed biofilm reactors (MBBRs) (Lackner et al., 2015).

The aeration needs for the Anammox anaerobic oxidation can be illustrated by an equal split of the influent flow as shown in Figure 9.2, where 50% of the inflow is oxidized by air (or pure oxygen) to produce nitrite that would require one and a half molecules of oxygen to oxidize one molecule of ammonium, in contrast to the full nitrification demand of two O_2 molecules in 100% of flow as presented on the preceding page. Then the nitrite stream is combined in a second reactor with the second 50% of the stream, whereby nitrite would oxidize ammonium. Thus, aeration requirements would be reduced by about 60% and sludge production by about 90% (Siegrist et al., 2008; Van Loosdrecht and Salem, 2006; Lackner et al., 2015). The best location of the Anammox and struvite-forming reactors is on the supernatant return line in biological nutrient removal treatment because Anammox has the best efficiency when N concentrations are high. The supernatant side-steam flow rate is very small but contains very high concentrations of nutrients (Table 9.2). It also contains some calcium and magnesium to form struvite and precipitate calcium phosphate, but not enough to remove all phosphate. For precipitating phosphorus as struvite, magnesium addition is needed in most cases. The fact that Anammox and Bardenpho processes do not reclaim nitrogen may not matter because they convert reactive nitrogen that

Table 9.2. Concentrations of COD and nutrients in the supernatant of the BNR and conventional activated sludge systems.

Parameter	Units	BNR Sludge	Excess Activated Sludge
Total COD	mg/L	15 320	13 640
Total solids	mg/L	12 620	11 080
Total nitrogen	mg/L	486	399
Total phosphorus	mg/L	335	143
Soluble phosphorus	mg/L	55	34
Total calcium	mg/L	686	247
Total magnesium	mg/L	108	39
pH	Units	7.2	7.4

Source: From Parsons et al. (2001); copyright © Taylor & Francis.
BNR: biological nutrient removal

could otherwise damage environment as a nutrient-stimulating eutrophication to nonreactive gaseous nitrogen gas. Anammox does it with less energy use.

Phosphorus Biological Removal and Limited Recovery

New developments of biological nutrient concentration processes focus on employing bacteria in the *enhanced biological phosphorus removal (EBPR)* process to accumulate nutrients as luxury uptake in the anaerobic step before anoxic step in the Bardenpho process (Figures 7.1C and 9.3A). *Phyto-accumulation* using adapted microalgae, polyphosphate-accumulating bacteria (PAOs), purple nonsulfur bacteria (PNSB), algae, and water hyacinths have added benefits by producing additional biomass for energy and biofuel production. Nutrients will then be transferred into ash or residual solids that can be used as fertilizer or soil conditioner. Algae and cyanobacteria are effective for treating a wide range of nutrient concentrations, including the diluted content of nutrients typically associated with centralized municipal water reclamation plants (Khunjar et al., 2015). Biological systems can remove/recover 70–90% of N and P and recover methane (red arrow in Figure 9.3A).

Theoretically, the biological phosphorus concentration process consists of the two sequential reactors, anaerobic and aerobic, with a large recycle, which is known as the enhanced biological phosphorus removal. The phosphate accumulating (micro) organisms (PAO) in the first anaerobic unit outcompete the anaerobic fermentation microorganisms for phosphorus, which they store as a *luxury uptake* energy reserve, releasing it later in the aerobic unit and enhancing the microorganism's growth. The microorganisms are recycled from the aerobic reactor to the anaerobic reactor to grow and reduce COD and accumulate phosphorus. Sludge solids separated from the EBPR process have twice as much phosphorus content (about 5%) than that in sludge from the traditional Bardenpho plant (about 2–3%) (Tchobanoglous et al., 2003; WEF, 2007). Therefore, they yield a better raw material for fertilizer production. This simple two-reactor EBPR process does not remove nitrogen beyond incorporating N into sludge solids.

The most popular EBPR system of the last century added an anaerobic influent receiving unit ahead of the anoxic unit in the traditional Bardenpho process (Figure 9.3A). This process performed primarily by heterotrophic bacteria reduces oxidized forms of nitrogen (nitrates and nitrites) DPB in response to the oxidation of an electron donor such as organic COD. In this system nitrogen is removed but not recovered. The added anaerobic reactor receives recycle from the anoxic zone, while the anoxic zone receives recycle of the

A) Bardenpho MBR system with anaerobic unit - EBPR

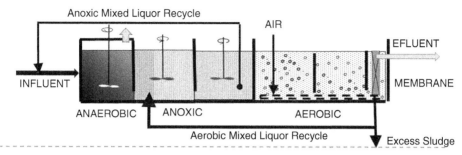

B) Sequential anoxic/anaerobic membrane bioreactor

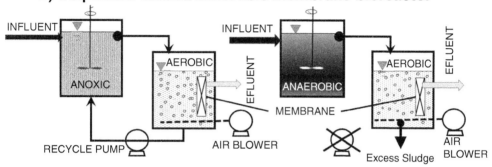

Figure 9.3. Two enhanced biological phosphorus removal (EBPR) systems. (A) Enhanced Bardenpho process, and (B) sequencing batch reactors. *Source:* (A) Water Environment Federation (2007). (B) Adapted and replotted from Ahn et al. (2003). Copyright Elsevier, 2003.

mixed liquor and sludge from the aerobic zone. The phosphorus luxury uptake by PAO microorganisms takes place in the anaerobic reactor and the accumulated phosphorus is then removed with the excess activated sludge. The aerobic basin also contains a membrane compartment to separate permeate from the completely mixed sludge in the reactor. The double sequential compartment anoxic and aerobic reactors in Figure 9.3A provide better removals of COD and nutrients than single completely mixed reactors.

In contrast to the continuous flow EBPR incorporating Bardenpho nutrient removal with membrane effluent separation that consists of three or more sequential completely mixed reactors (four or six if the membrane unit is separated), sequencing anoxic/anaerobic membrane (SAM) batch reactor system needs only two reactors (Figure 9.3B). Creation of anaerobic or anoxic conditions is controlled in the first reactor by the timing of the recycle of the oxidized content from the aerated second reactor. Hence, SAM has two modes of operation, anaerobic and anoxic, as shown in Figure 9.3B. This system uses a sequencing batch reactor (SBR) described by Kuba et al. (1993) as a process that uses nitrate as an electron donor instead of oxygen in biological phosphorus removal processes. SAM system differentiates from SBR by separating aerobic/membrane reactor. In the traditional SBR, four operations are carried out sequentially in a single reactor.

Kuba, van Loosdrecht, and Heijnen (1996) studied an anaerobic/anoxic (A1/A2) sequencing batch reactor followed by a nitrification (aerated) SBR. In this system, denitrifying phosphorus-removing bacteria (DPB) and nitrifiers were completely separated. Denitrifying and dephosphatizing enabled the removal of nitrogen and phosphorus by a luxury uptake with minimal use of COD, minimal oxygen consumption, and minimal surplus sludge production. It also required less aeration energy because aeration was only necessary for nitrification. The nitrified substrate was recirculated from the nitrification SBR to the A2 SBR, where nitrate was utilized by DPB as an electron acceptor for phosphorus removal. In this way, the oxygen requirements and sludge production can be decreased by about 30% and 50%, respectively

Ahn et al. (2003) in the double SAM system shown in Figure 9.3B treated semi-artificial wastewater (sugar and acetic acids were added to the feed) that had the following average composition: TKN = 7 mg/L, Total phosphorus = 3.60 mg/L, and COD = 344 mg/L, respectively. The concentrations of TKN and TP in this research were lower than those found in typical used water and much lower than those in black water. The sequencing anoxic/aerobic (A1/A2) conditions were achieved by intermittent recycle of the mixed liquor from the aerobic reactor into the A1/A2 reactor. In the time of recycle, the anoxic conditions in the A1/A2 reactor were induced by the recycle, while no recycle would quickly create anaerobic conditions therein. The anoxic conditions for denitrification was provided for 1 hour every 3 hours with the recycle that equaled six times the influent flow. The anaerobic conditions for phosphorus uptake was possible to be maintained for 2 hours of the 3-hour interval as long as there was no recycle from the aerobic reactor. Influent was continuously fed into the A1/A2 reactor. They achieved 93% phosphorus removal, corresponding to 0.26 mg of P/L in the effluent. The time interval of sequencing anoxic/anaerobic conditions in the first reactor was done in minute intervals by turning the recycle pump on and off from the aerobic reactor. This SA system also achieved about 67% nitrogen removal by nitrification/denitrification processes and 97% removal of COD. Efficient recovery of methane from the first reactor is doubtful.

In the traditional aerobic activated sludge process, total P (TP) removal into sludge can also be enhanced by adding coagulating chemicals such as iron or aluminum salts, which also enhances settling of solids. This process can reduce the TP content in the effluent to less than 1 mg/L but increases the volume and mass of sludge and adds cost for increased energy use, and transporting and purchasing the chemicals.

MEC Can Recover Struvite

Cusick et al. (2014) conducted experiments with the microbial electrolysis cell (MEC – Chapter 8) to simultaneously produce hydrogen and precipitate struvite. To promote bulk-phase struvite precipitation and minimize cathode scaling, a two-chamber MEC was designed with a fluidized bed to produce suspended particles and inhibit scale formation on the cathode surface at the cathode pH between 8.3 and 8.7 under continuous flow conditions. Soluble phosphorus removal using digester effluent ranged from 70 to 85%. This research enhances the versatility of MEC beyond hydrogen production.

A *hybrid phyto/MFC phosphorus concentration* (laboratory) system was described by Tse et al. (2016), who combined microbial fuel cells (MFC – see Section Microbial Fuel and Electrolysis Cells in Chapter 8) with membrane photobioreactor producing algae, and this combination achieved 92 to 97% soluble COD removal and almost 100% removal of ammonia. At the time of their article was published, these systems were in their infancy, but the idea is sound and based on the tested technologies. The laboratory system consisted of two identical MFCs followed by one membrane phytoreactor producing algae. The membrane phytoreactor was a glass tank containing two hollow-fiber membrane modules. The laboratory feedstock was made by mixing tap water with sodium acetate, ammonium salts, potassium phosphate, and some trace elements. The influent COD was 186 ± 15 mg/l and influent ammonium concentration was 37 ± 3 mg N/L.

9.3 UNIT PROCESSES RECOVERING NUTRIENTS

The new Anammox (anaerobic ammonium oxidation) process reduces the aeration energy because about half of the influent ammonium is oxidized only to nitrite (NO_2^-) and not to nitrate (NO_3^-) and the produced nitrite oxidizes the remaining ammonium. Neither Bardenpho and Anammox processes, however, reclaim nitrogen or phosphorus beyond incorporating them into excess sludge organic matter.

The modern nutrient recovery from used water is a three-step process (Latimer et al., 2012; Khunjar et al., 2015):

1. Removal of organics, nutrients, and other pollutants for further processing
2. Concentrating/accumulating nutrients to high concentrations in a small liquid flow
3. Extraction and recovery of nutrients as a chemical nutrient product

Multiple options and unit processes are available for each step.

Technologies to increase phosphorus fertilizer recovery from waste streams (sewage, food, domestic and industrial waste, wastewaters, and MSW), include:

- Anaerobic digestion and processing excess sludge as fertilizer
- Composting
- Thermal treatment of sludge (e.g. drying, followed by granulation, pelleting, or pulverizing
- Incineration, followed by acidic treatment or chemo-thermal treatment at high temperatures)
- Gasification with ash recovery
- Chemical processes (e.g., precipitation of struvite and ammonium sulfate)
- Concentration by membrane technologies
- Thermochemical processes (e.g. pyrolysis, hydrothermal carbonization)

Some specific technologies such as AshDec (pelletizing ash from the incineration of sewage sludge), PEARL (a specific struvite production process), and EcoPhos (soft digestion by hydrochloric or phosphoric acid) were cited in the literature and presented at specialty conferences.

Urine Separation

Separation of urine in the large house or district level scales can substantially reduce the N and P content in the effluent and recover more concentrated nutrients. Urine contains about 80% of the nitrogen, 70% of potassium, and about 50% of phosphorus per capita loads in less than 1% of the used water flow and has no pathogen content (Mitchel, Abeysuriya, and Fam, 2008). Nitrogen and phosphorus recovery is relatively simple and urine separation is possible and economical in public toilets or those in large buildings (for both males and females); household urine separation toilets are also available in Japan. Obviously, urine separation requires another (yellow) pipe and separate collection and processing system. Because of relatively small volume, separated urine can be collected, stored on site, and hauled away for processing, reuse, and nutrient recovery.

The first pilot study on collecting and agricultural use of urine sponsored by the Water Environment Research Foundation (WERF) in the US was conducted on 200 participating house and stand-alone urinals in Vermont and experimentally in Hampton Roads Sanitation District office in Virginia (Noe-Hays et al., 2016). Urine was sanitized and analyzed for chemical content, including nutrients (nitrogen, phosphorus, potassium) and pharmaceutical chemicals. The research confirmed that in comparison to phosphorus, the nitrogen and potassium contents were very high for the agricultural use and the proportion between N and P in urine were not conducive to growing crops and should be adjusted before use as a fertilizer. Either N content should be reduced (removed) or P should be supplemented by industrial chemicals.

Urine collection and reuse systems are nothing new. In ancient Rome and Pompeii, almost two thousand years ago, urine was considered a valuable resource, collected in public toilets, baths, and large houses of aristocracy, then processed and even taxed. There were many preindustrial-age uses of urine for laundry bleach, processing and cleaning leather, and, obviously, as a fertilizer.

Nutrient Separation

Nutrients can be concentrated during treatment into the liquid phase and into solids. Biological, physical, thermal, and chemical mechanisms can be used in each step to transform the nutrients present in domestic used water to a chemical nutrient product that has a secondary market value. As pointed out previously, the separation and concentration occur best in the distributed system whereby the used water flows are separated into black and gray water and, possibly, urine (yellow) flows. Most of the nutrients will be contained in the black and yellow water flow, which will also include particulate solids and colloids separated from gray water (Figure 5.5). A large portion of the recovered gray water on the cluster (neighborhood) or building level will be used in buildings for toilet flushing and, in this way, after receiving nutrient-rich urine and fecal matter, converted into black water. Nutrients in the black water flow are then transported to the regional integrated resource recovery facility IRRF. The nutrient content of gray water is small, about 20% of total N and 25% of total P, respectively, if nonphosphate detergents are used in the buildings (see Section 5.2 on gray water recovery in Chapter 5, and Tables 5.4 and 5.6).

If there is no separation on the house and/or cluster level, nutrient separation and concentration must be performed in the IRRF. Because of the dilution of nutrients in the total influent (including seepage and stormwater inflows), the separation process will be more tedious and not as efficient. The nutrient accumulation can be chemical (coagulation and separation) or biological. In the chemical separation/accumulation, the goal is to separate the nutrients into primary sludge before biological treatment. Almost all of N arriving into the IRRF is TKN (ammonium and organic N) and particulate. Even ammonium, which is a positively charged ion (NH_4^+), is mostly adsorbed on organic particulates. Phosphorus will come as particulate (or colloidal) organic P and as orthophosphate ion that can be dissolved and adsorbed on organic and clay particulates. Table 9.3 presents a summary of biological, physical, and chemical nutrient concentration processes.

Coagulating chemicals can be added to enhance settling of organic particulates containing both N and P and precipitate some phosphate. After primary separation, significant portions of remaining N and P will be incorporated in the biological step (preferably anaerobic) into waste sludge and processed by sludge handling and sent into effluent. Enough nutrients should be transported from the primary separation units into biological treatment to satisfy the nutrient needs of the bacterial growth. Excess sludge

Table 9.3. Summary of technologies for accumulating nutrients in centralized water reclamation plants.

| Technology Status | Medium | Operating Conditions | | Pre-treatment Required | Chemical Input | Nutrient Accumulated |
		Temp.°C	pH			
Embryonic	Microalgae	15–30	7.5–8.5			N and P
	Cyanobacteria	5–40	6.5–8		Inorganic carbon source	N and P
Innovative	Adsorption exchange	NA	<8	Solid-liquid separation	Adsorbent regeneration	
Established	EBPR	5–40	6.5–8		Inorganic carbon source	P only
	Chemical	25–40	6.5–10		Metal salts (Al or Fe)	P only

Source: Latimer et al. (2012).
Embryonic: technologies that are in the development stage (bench/pilot scale)
Innovative: developed technologies with limited full-scale applications
Established: commercially viable technologies with a proven history of success

growth is separated and diverted along with the primary sludge to a digester, or in the future into a bioelectrically enhanced digester unit or hybrid digester/microbial electrolysis unit (see the previous chapter).

In anaerobic digesters, biomass is sequentially broken by bacteria to fatty acids and then to acetates, carbon dioxide (bicarbonate alkalinity), and hydrogen, which are then converted by another group of bacteria to methane and carbon dioxide (see Chapter 7). In this process, organic nitrogen is converted to ammonium and phosphorus to phosphate. The digestion process produces a liquid supernatant that contains very high concentration of organic carbon and soluble nutrients.

Supernatant side stream flows are approximately 1% of the total influent hydraulic plant flow but could account for 15% to 40% of the total influent nitrogen load (Pugh and Stinson, 2012). Side stream flows N concentrations ranging from 900 to 1,500 mg/L as nitrogen (N) or more can increase the ammonia concentration in the plant effluent by 3 to 5 mg/L on an average day basis (Phillips, Kobylinski, Barnard, and Wallis-Lange, 2006). The concentration of phosphorus can be more than 100 mg/L. Supernatant cannot be discharged into the environment and must either be processed or, traditionally, returned into influent. The high concentrations of nutrients in the supernatant provide opportunity for nutrient recovery.

Phytoseparation of Nutrients

Algae and Phytoflora Farms. Microalgae and phytoflora provide the best prospects for algal farms processing IRRF effluents and nutrient recovery from effluents. They remove the nutrients even in lower effluent concentrations with nutrient recovery efficiencies exceeding 90% and can often achieve effluent dissolved N and P concentrations close to natural N and P concentrations (TP $\ll$ 0.1 mg/L and TN $\ll$ 1 mg/L). Furthermore, waste dissolved CO_2 from the IRRF power plant can be added (sequestrated) to the algal pond feed to stimulate photosynthesis that otherwise could be restricted by insufficient alkalinity in the feed water. However, the positive effect of adding CO_2 to increase growth of algae has been questioned because the growth of algae may be limited by nutrients and not by CO_2 or alkalinity.

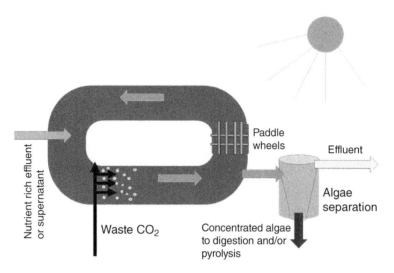

Figure 9.4. Standard circulating ditch used for algal production and carbon dioxide sequestering.

The most common reactors are recirculating ditches (Figure 9.4) used also in aerobic used water treatment plants to remove BOD. In the circulating ditch unit algae are dispersed. The second emerging and efficient algae growing technology is attached grow reactors.

Algal Turf Scrubber (ATS™) is a commercial technology developed by HydroMentia, Inc., that grows filamentous algal biomass "turf" on screen panels submerged in ponds and basins fed by CO_2 and nutrients (Adey and Loveland, 2007). This can significantly increase the biomass yields. The panels are periodically removed from the basin and algae mass is removed. This technology needs less energy than the circulating ditches. The overall biomass productivity from the ATS™ system ranges from 40–70 dry g m^{-2} day^{-1} in the summer depending on latitude, to 5–20 dry g m^{-2} day^{-1} during the winter months (Adey et al., 2011), and greatly outperforms typical yields of biofuel producing crops (Barnard 2007) at rates 5–10 times those of other types of land-based agriculture (Adey et al., 2011). Various algae species participate and grow in algal growing reactors or ponds, including diatoms, green algae, chlorella, and cyanobacteria.

Harvested algae can be used for several beneficial purposes, starting with supplementing animal food and providing additional biomass entering the IRRF digesters to be co-digested to increase production of methane and biofuel. Adding algae to sludge digesters also improves dewaterability of the digested sludge and the methane yield from co-digested algae biomass is about the same as that of used water sludge, i.e., adding algae is synergistic (Table 7.5) and enables further recovery of nutrients in a form of struvite, hydroxyapatite, and ammonium sulfate from the supernatant. After digestion, the solids can be gasified or, after drying and disinfecting, used in agriculture as a fertilizer. For producing biogas and gasification, algae with other solids (combustible MSW, sludge residues) can be fed directly into pyrolysis reactor or indirect gasifier. The latter reactor requires water; hence, added algal biomass does not have to be excessively dewatered.

Traditional methods of harvesting/separating algae include: (1) sedimentation (gravity settling), (2) membrane separation (micro/ultrafiltration), (3) flocculation, (4) flotation, and (5) centrifugation. However, harvesting algae by conventional methods may be inefficient and cost consuming because of their small size and small specific weight and, in higher concentrations, tendency to clog filters and membrane fouling. Centrifuges and especially hydrocyclones appear to be very efficient method of algae separation, followed by drying (Brotzman, 2015). Both require energy. These methods have also been used extensively in other industrial operations.

Cerff et al. (2012) and Vergini et al. (2015) researched and described magnetic separation as a powerful tool to capture algae by adsorption to submicron-size magnetic particles. In their study, they used silica-coated magnetic particles for the removal of fresh water and marine algae by high-gradient magnetic filtration. Hereby, separation efficiency depends on parameters such as particle concentration, pH, and medium composition, and separation efficiencies of >95% were obtained. Coagulation and flocculation using chemicals may not be recommended because the added chemicals could interfere with further processing of algae and degrade the final product. These methods are not suitable for IRRF. Auto-flocculation may occur when the CO_2 supply is interrupted, which causes algae to flocculate on its own; however, the efficacy may not be great and filtration and or centrifuges may be employed as the next step.

The algal biomasses grown on used water effluents and harvested from a range of locations, were characterized by Adey et al. (2013) and Hampel (2013) by measuring organic and inorganic carbon, nitrogen, phosphorus, and ash profiles. Algal biomass grown on wastewater effluent had the highest nutrient weight content (40 wt% org. C, 7.0 wt% N, >1.0 wt% P,) while that grown in brackish water had the lowest nutrient content (20-25 wt % org. C, 2.1 to 2.9 wt % N, 0.15 to 0.21 wt % P). Algae harvested from freshwater locations had less than 3.5 wt% biogenic silica while mixed fresh-salt water locations had biogenic silica content ranging from 10-27 wt%. Metals composed 0.045 wt% to 0.075% of the total dry algal biomass, with relative concentrations of As > Cu ≈ Cr > Co ≈ Mo > Cd. These values of nutrients and carbon content of algae grown in used water effluents can imply that algae not only produce a lot of biomass for energy but can also sequester significant amounts of carbon dioxide from the atmosphere in addition to that in the effluent. For example, if the phosphorus content of the effluent is 5 mg/L and P content of produced algae is 1% by dry weight and assuming 98% nutrient recovery into algal biomass, then the dry weight of the produced algae will be 5 ∗ 0.98 ∗ 99 = 485 mg/L and the carbon content of algae will be 40% or $0.4 \times 485 = 194$ mg dw/L. Since the atomic weight of carbon is 12 and the molecular weight of carbon dioxide = 12 + 2 ∗ 16 = 44 then the sequestered (incorporated into algal biomass) carbon dioxide will become:

$$CO_2 \text{ (sequestered)} = 194 * 44/12 = 711 \text{ mg/L}$$

or each kg of P taken up by algae and removed from the effluent will produce approximately on average 485/5 = 97 kg of biomass and sequester 97 kg ∗ (40%/100) ∗ 44/12 = 143 kg of CO_2. The carbon dioxide for sequestering is provided by the waste CO_2 emitted from the IRRF as alkalinity of used water and from the atmosphere. In the new IRRF power plants that include molten carbonate fuel cell, the CO_2 regulator on the CO_2 recycle line may be the only place where concentrated CO_2 would be released, as shown in Figures 8.12a and b. After dewatering the biomass can be added to the digester and co-digested with sludge and biodegradable MSW, to pyrolysis for producing biofuel and syngas and/or gasified and reformed to H_2 for production of electric energy. Growing algae and making energy with sequestering CO_2 could be characterized as a strongly negative CO_2 emission technology.

However, the energy balance of growing algae in ponds and circular ditches is not favorable. Murphy and Allen (2011) calculated the energy needed for growing algae, containing and moving and circulating water, replacing lost water by evapotranspiration within the cultivating systems (open ponds), algae separation in hydrocyclones, and compared it with the algae energy potential as biofuel. They calculated this balance for all 48 continental states and found that the energy of producing algae exceeds several times the yield of energy in the produced algal biomass. Therefore, the primary benefit of algal ponds and ditches may be nutrient removal and recovery and carbon sequestering first. The energy balance is expected to be better for algal turf systems.

Sequestering CO_2 from IRRF and Fossil Fuel Plants by Algae and Producing Biofuel.
ExxonMobil and several university research centers (e.g., Massachusetts Institute of Technology, MIT) have been studying CO_2 sequestering growing by algae in water and plants in greenhouses with increased loads of CO_2 to produce biomass and biofuel. The solubility of carbon dioxide in water is very high (Figure 7.6) and greenhouse atmosphere with CO_2 content that is 100 times greater than the ambient atmospheric CO_2 concentration is still safe for people to work therein. The problem is that under normal situation, CO_2 is not the limiting nutrient; nitrogen or phosphorus are, and adding CO_2 alone may not increase the algal and phytoflora yield. Hence, in many situations significant amounts of energy demanding industrial fertilizers would have to be added to sequester large masses of CO_2 by photosynthesis. Some cyanobacteria can fix atmospheric nitrogen but not phosphorus.

Wetlands.
Natural wetlands have limited efficiency to remove nutrients. They work best for removing nitrates (>85% efficiency) but their efficiency for removing TP is about ≤50%. Nutrients used by wetland vegetation for growth can be recovered by harvesting the vegetation and digesting it with the sludge and digestible MSW solids in the IRRF. Gasification of algae is also possible (Chapter 8). However, in the US natural wetlands are considered as protected natural water bodies and not treatment plants. Use of energy is minimal; however, if not designed properly wetlands can emit GHG methane produced by the anaerobic decomposition of organics in the sediments. Methane emissions cause elevated sediment oxygen demand and dissolved methane (Figure 7.4) can also impose biochemical oxygen demand. COD of dissolved methane, which is approximately 4 ∗ methane concentration emitted into oxygenated water above.

Wetland creation is biophilic design.
Constructed wetlands were introduced in Chapter 6 as a technology for treating urban runoff and wastewater and CSOs. The efficiency of wetland treatment is related to the hydraulic loading rate:

$$HLR \quad q = 100 \, Q/A$$

where Q is the averaged design flow in m^3/day, A is the surface area in m^2 and the unit for q is in centimeter/day. HLR q for free surface treatment wetlands was recommended as 2.5–5 cm/day and for submerged bed flow wetlands as 6 to 8 cm/day (Vymazal, 2010; Kadlec and Wallace, 2008). While the efficiency of constructed stormwater wetlands was high for nitrate, phosphorus removal efficiencies hovered generally around 50%.

After much research, South Florida Water Management District constructed large wetlands known as stormwater treatment areas (STAs) protecting the Everglades Protected Areas (EPA), which include Everglades National Park (ENP) (SFWMD, 2017; Kadlec and Wallace, 2008). The program has been carried out under the auspices of the Florida's Everglades Forever Act. It brought a new light on the constructed wetland performance parameters and design since the total phosphorus (TP) removal efficiencies of STAs were surprisingly high, more than 80%. STAs were constructed in 1998–2010 from abandoned agricultural lands between the Lake Okeechobee and the EPA and their role was to treat the medium to high TP concentrations in the stormwater runoff from the Everglades

Agricultural Areas (EAAs) and urban areas of South Florida to the levels that would protect oligitrophic to mesotrophic EPA and ENP areas. STAs are compartmentalized wetlands with planted emerged and submerged aquatic vegetation (EAV and SAV) (SFWMD, 2017). Hence, they must be sometimes hydrated by the Lake Okeechobee water to maintain submerged flow conditions to preserve the STA's flora and its TP removal efficiency. The inflow to the STAs is the runoff from the irrigated and non-irrigated EAAs where some best management practices have been implemented but still resulting in the TP concentrations in the STA inflows greatly exceeding 100 µg/L, which without STAs would be detrimental to the EPA water conservation areas and the ENP.

The performance of STAs in the period of May 2015–April 2016 is shown in Figure 9.5. The outflow flow weighted mean (FWM) TP concentrations for individual STAs in this period ranged from 12 µg/L (STA-3/4) to 36 µg/L (STA-1W) and the percent retained TP load ranged from 81% (STA-1E and STA-2) to 91% (STA-3/4). In this period STAs treated a combined 1.72×10^9 m^3 of water and safely retained 208 tons of total phosphorus (Chimney et al., 2017).

Hydraulic loading rates (HLR) in all STAs during water years 2016 (WY=May 2015 to April 2016) were q ≤ 2.2 cm/d. The corresponding phosphorus loading rates (PLRs) were ranging from 0.7 g/(m^2-yr) in STA-5/6 to 2.0 in STA-1W, respectively. The parameters of the best performing STA 3/4 shown in Figure 9.6 are further elaborated. This STA operational since 2004 is comprised of three parallel flow ways each with two treatment cells, first with rooted floating aquatic vegetation (rFAV) and the second with submerged aquatic vegetation (SAV). Total treatment area is 66.1 km^2. This area is very large and out of the scope for treatment and carbon sequestration of most IRRFs; however, the parameters presented

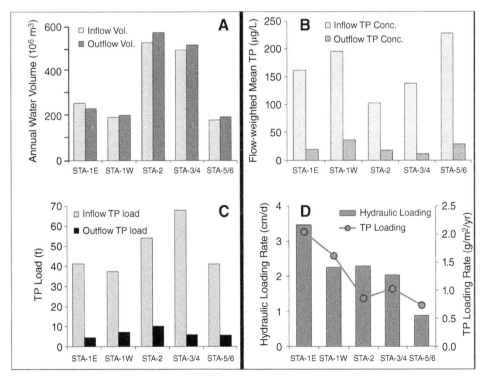

Figure 9.5. Performance of Stormwater Treatment Areas (STAs) constructed wetlands protecting Everglades National Park. *Source:* South Florida Water Management District; Chimney et al. (2017).

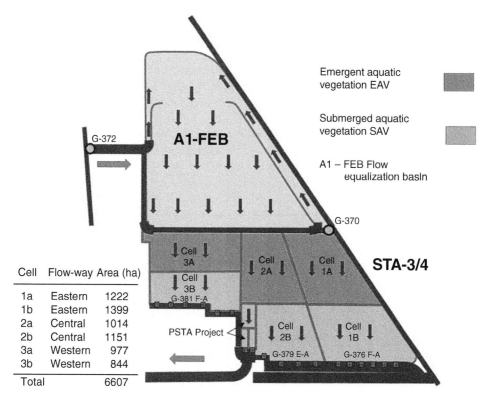

Figure 9.6. STA ¾-plan schematics. Cell areas are effective cell areas and exclude the areas of canals. *Source:* South Florida Water Management District.

Cell	Flow-way	Area (ha)
1a	Eastern	1222
1b	Eastern	1399
2a	Central	1014
2b	Central	1151
3a	Western	977
3b	Western	844
Total		6607

herein are per unit area such as the HLR in cm/day and PLR in g m^{-2} year^{-1}, respectively, which enables transposing these parameters to smaller installations. In WY 2016 the STA-3/4 had an inflow FWM TP concentration of 138 µg/L and produced an outflow geometric mean of the FWM TP concentration of 12 µg/L, which is the lowest annual outflow concentration recorded to date in any STA.

This facility retained in the year of observation 62 tons of TP, or 91% of the inflow TP load, which was the highest treatment efficiency observed in this STA to date and had a HLR and PLR of 1.25 cm/d and 1.06 g m^{-2} year^{-1}, respectively. Based on these metrics, STA-3/4 ranked as the best-performing STA in this water year. All cells in STA-3/4 were hydrated throughout the period of the record (Chimney et al., 2017). This high efficiency of STAs is impressive and better than that reported for many constructed wetlands in the literature. A big advantage, of course, is subtropical location without harsh winters and, consequently, absence of complete TP assimilation as it could happen in the temperate northern parts of the US (and Europe or China) and, above all, the availability of large tracks of land. The FWM outflow TP concentration from the STAs is smaller than that in the receiving waters, so it results in TP dilution downstream.

The highest HLR reported in Figure 9.5 of approximately 2 cm/day is similar to natural or free surface wetlands used for stormwater treatment. However, PLR is low when compared to design parameters for free surface constructed wetlands (US EPA 2000; Kadlec and Wallace, 2008). The US EPA wetland manual recommends TP loading in kg/(ha-day) that would be an order of magnitude greater after conversion to g m^{-2} year^{-1} than that presented on the Figure 9.5. There is also a difference between the parameters and wetland performance for wastewater treatment and cleaner surface runoff entering STAs. Nevertheless, typical constructed storm and waste treatment wetlands do not provide the degree of TP treatment that would be as good as TP removal in STAs. Fisher and Ackerman (2004) investigated the performances of many natural and constructed wetlands and reported that some but not all investigated natural wetlands had high TP removals up to 90% at PLR loads of less than 5 g m^{-2} year^{-1}. Hence, the high TP removals in STAs is apparently due to low PLR and good EAV and SAV vegetation. The study also noticed that the TP removal efficiencies are inversely proportional to PLR (a logical observation) and at PLR at 10 g m^{-2} year^{-1} or greater the average TP removals dropped to less than 50%, which would be detrimental to the EPA and ENP but it is typical for constructed wetlands treating stormwater runoff of providing post treatment for municipal used water effluents. Apparently, as the vegetation in the STAs is improving and stabilizing in time, the removal efficiencies are increasing.

In the subsequent water year 2018 report (James, Ivanoff, Piccone, and King, 2019) summarized the preliminary findings of the research on the effect of aquatic vegetation and design of the STA3/4:

1. SAV-dominated cells performed much better than rFAV (rooted floating aquatic vegetation) cells in removing total phosphorus.
2. STAs with cattail vegetation had low efficiency for removing phosphorus.
3. Bottom sediment capping by limestone increased phosphorus removal.
4. pH was higher in the SAV than in the rFAV (EAV) cells as a result of greater oxygenation activity, which precipitated more phosphorus and built denser flocculant layer in SAV cells.
5. Submerged vegetation in SAV cells released more photosynthetic oxygen into water, which kept the cells well oxygenated and prevented release of phosphorus from sediments. rFAV vegetation release photosynthetic oxygen in the air. Also CO_2 released by the SAV during nightly respiration periods is not emitted into the atmosphere but becomes alkalinity and ready to be picked up by the SAV during the daylight growth.

Consequently, proper selection of vegetation and its management is crucial for maintaining high TP removal efficiencies and both SAV and EAV should preferably include planted native wetland species. Fisher and Ackerman also noted TP removal efficiencies were reduced or were dropping for wetlands that were losing their substrate renewal capabilities attributed to the absence of a healthy emerging and submerged wetland vegetation. Dierberg et al. (2002) studied the efficiency of SAV and EAV in a pilot scale wetland emulating the Everglades' STAs and found that SAV communities exhibit phosphorus (TP) removal mechanisms not found in wetlands dominated by emergent macrophytes. This includes direct assimilation of water column P by the plants and pH-mediated P coprecipitation with calcium carbonate ($CaCO_3$).

Healthy and efficient wetland vegetation is composited mainly of emerging and submerged macrophyte vegetation (SAV and EAV) and periphyton. Algae should be limited and essentially not present to maintain high transparency needed to maintain healthy SAV, which is the most efficient TP removing vegetation. Loosing SAV and transparency, as it happened recently to Lake Okeechobee in Florida due to hurricane lifting sediment from the bottom and overloading the lake with nutrients, resulted in the loss of the TP removal efficiency, which in 2018 was close to zero. STAs maintained their very high TP removal efficiency, partially because they maintained high transparency and healthy SAV and EAV flora (South Florida Water Management District, 2019).

Widespread vegetation harvesting has often been suggested for STAs to manage vegetation in the STAs; however, harvesting was not under consideration in 2017, mainly because of the logistic of disposing the large masses of harvested vegetation. Harvested vegetation could be a valuable source of biomass for processing into biofuel. On the flip side of this issue, deposition of dead vegetation biomass and its conversion into the wetland substrate/soil may provide important adsorption sites for immobilizing the incoming phosphorus and other potential pollutants (e.g., toxic metals and herbicides). However, most of the TP removal in well-designed wetlands must result from biological processes, not from abiological processes.

Methane production may be a problem, considering that the GHG potential of methane is 25. Keeping water column oxygenated will reduce the CH_4 emissions from the substrate by forming an anoxic/oxic barrier in the top substrate layer that will oxidize the emitting CH_4, which will be exhibited as sediment oxygen demand (SOD), and in the water column rich with photosynthetic oxygen during daytime produced by the submerged vegetation. More oxygen to water and interstitial water/sediment layer is provided by submerged aquatic vegetation.

CO_2 Sequestration (Negative Emission) Benefits of STAs and Other Vegetated Wetlands.

In the preceding section, it was estimated that each kg of P taken up by algae and removed from the flow will produce approximately on average 100 kg of biomass and sequester 143 kg of CO_2. McJannet et al. (1995) analyzed 41 wetland plant species and found that the phosphorus concentrations can range from 0.2% to 0.4% of the dry weight of the wetland vegetation (which is less than that for algae). Boyd (1978) reported chemical composition of 28 species of wetland plants and found that organic C of the plants was 41+-0.7% (reported in Kadlec and Wallace, 2008). The TP loading rate for STA 3/4 in the Water Year 2016 (June 1, 2015–May 31, 2016) was 1.06 g m^{-2} year^{-1} and the STA retained 62 tons of TP, or 91% of the incoming TP. The STA was hydrated and aerobic, hence methane emissions were suppressed. Healthy submerged vegetation provides plenty of oxygen by photosynthesis to overlaying water that would restrict emissions of methane from the sediment and oxidize it biologically in the water column. This could moderately restrict the sequestering capacity of the wetland. Wetlands with submerged flow (dominated by EAV) have much smaller GHG sequestering capability.

To estimate the magnitude of CO_2 sequestered by the STA ¾ wetland, the following assumptions are made in this illustrative example:

1. 75% of removed TP is incorporated into tissue of submerged and emerged vegetation, 25% into sediment by precipitation and adsorption without assimilating CO_2.
2. After die-off, the vegetation organic carbon will become a part of the sediment substrate and subsequently, after decades and centuries of slow anaerobic processes, transformed into peat.

Total vegetation biomass growth (M_V) in a water year:

$$M_V = 0.75 * TP_{removed}[tons] \times 100/0.3[\%] = 0.75 * 62 \times 100/0.3 = 15\,500 \text{ [t]}$$

Organic carbon produced by photosynthesis:

$$OC \, [t] = 41 \, [\%] \, * \, M_V/100 = 41 \, * \, 15\,500/100 = 6\,355 \, [t]$$

Carbon dioxide sequestered by the STA ¾ in the water year 2016:

$$CO_{2\text{-seq}} = 6\,355 \, [t] \times (MW \, CO_2)/(MW \, C) = 6\,355 \times 44/12 = 23\,302 \, [t]$$

$$= 23\,302 \, [t]/66.75 \, [km^2] = 349 \, [t/km^2]$$

Chemical Removal and Recovery of Nutrients

Chemical coagulating metal salts have been used in treatment of municipal used water for more than forty years in the communities surrounding the Great Lakes. They can precipitate up to 85% of soluble phosphate. Years ago, steel washing waste liquids from nearby steel mills (iron sulfate "pickle liquor") were a cheap source of the coagulating chemicals added directly into aerated activate sludge basins in the Great Lakes communities (e.g., Milwaukee, Wisconsin) to reduce the effluent TP content below 1 mg/L. However, higher metal content of the sludge makes it less suitable as agricultural fertilizer. With the partial demise of steel production in the US (Ohio, Indiana), this cheap source of TP coagulating chemicals has become less available.

Struvite ($NH_4 . Mg . PO_4 \times 6\,H_2O$) Precipitation. Struvite is a known mineral that forms deposits and creates problems in sewers, treatment process pipes, pumps, and other equipment in water reclamation (wastewater treatment) facilities due to crystal formation on critical surfaces. The formation of struvite occurs under favorable conditions when concentrations of soluble magnesium (Mg^{+2}), ammonium (NH_4^+), and orthophosphate (PO_4^{-3}) exceed levels that promote the formation of crystals, referred as supersaturation. Uncontrolled struvite formation in pipes and pumps has been a problem for the utilities in the midwestern US, for example, in areas of the Great Lakes underlaid by geological formations of dolomite (calcium-magnesium carbonate, $CaMg\,(CO_3)_2$). In these regions, the pH of water is relatively high and there is plenty of magnesium and calcium ions in water combining with phosphate and ammonium in used water flows to form struvite crystals clogging the pipes and pumps. Insoluble calcium phosphate (hydroxyapatite, $Ca_5(PO_4)_3(OH)$) can also be formed. On the flip side, however, both struvite and hydroxyapatite are valuable fertilizers. As pH and temperature rise, the potential for struvite and hydroxyapatite precipitation increase. Therefore, digester supernatant or MEC side streams are amenable for struvite precipitation (Phillips et al., 2006). Struvite precipitation and recovery technology of simultaneously removing both N and P without energy from concentrated liquid used water and digester supernatant rich in nutrients is available (Barnard, 2007). Laboratory and pilot experiments documented that controlled precipitation of struvite from sludge supernatant could recover up to 90% of phosphate) and 50–60% of total P (Münch and Barr, 2001; Forrest et al., 2008).

However, Capdevielle et al. (2016) pointed out that organic matter in the supernatant can interfere with and slow down struvite precipitation due to interference of organic colloids. The influence of organic matter on struvite precipitation in acidified swine wastewater was negative on the reaction kinetics but positive on the size of the struvite crystals. Struvite precipitation was enhanced when the organic matter present in the solution was particulate. Attempts to precipitate struvite from sludge digestion supernatant of the public wastewater treatment plants in Czech Republic that had relatively low P concentrations yielded relatively low struvite precipitation, ranging from 6 to 39% in the precipitate after pH was increased to around 9 (Sýkorová et al., 2014), which might have been insufficiently low. Most of the phosphorus in the precipitate was in a form of hydroxyapatite (calcium phosphate). The low yield of struvite was attributed to the use of coagulating chemicals used to stabilize the sludge and relatively small P concentration of the liquid. Interestingly, pH increase in these experiments was achieved by blowing air that purged carbon dioxide from the CO_2 rich supernatant liquid, which sufficed to rise pH to more than 8.5 and not by adding base chemical like sodium and calcium hydroxides (NaOH or $Ca(OH)_2$). However, adding magnesium hydroxide $(Mg(OH)_2)$ and/or oxide (MgO), which also increase pH, are efficient and needed for supernatants that do not have enough magnesium to form struvite.

The struvite growth is impacted by nitrogen and was found to be surface controlled when the N:P molar ratio was greater than 1 and transport controlled by N:P molar ratio. The optimum molar ratio of N:P was found to be 3:1 (Capdevielle et al., 2013, 2014). At low concentrations of phosphate in the influent, the struvite formation is not efficient. The optimum pH is about 10.2 but effective P removals can be obtained if pH is greater than 9.2. There are several other parameters that influence the precipitation besides the pH: temperature, ionic strength of the solution, and magnesium ammonium and phosphate's molar ratios, The optimum loading of the reactor is supposed to be close to $Mg:NH_4^+:PO_4^{2-} = 1.7:1:1.1$ expressed as moles (Ezquerro, 2010; Sýkorová et al., 2014).

Figure 9.7 shows the upflow fluidized bed flocculation reactor that precipitates struvite floating crystals forming granules that are then continuously removed from the reactor. The study by the University of British Columbia reported 80% struvite recovery and P removal efficiencies reached 95% (Forrest et al., 2008). The application of a fluidized bed allowed for continuous reactor operation and self-classifying of the growing crystals to the point of harvest. Struvite crystals up to 5 mm in diameter have been produced for a supernatant-like concentrate. The most important parameters for precipitation and removals of struvite are pH and influent concentration of phosphate. The purity of the struvite was such that it easily qualifies as a commercial-grade fertilizer. In addition, the final product was low in

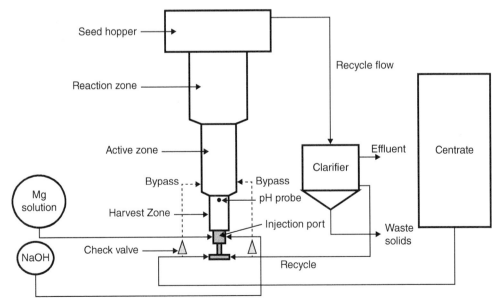

Figure 9.7. Struvite recovery by upflow fluidized bed reactor. *Source:* Reprinted with permission from Forrester et al. (2008); © ASCE.

impurities and was of a much higher quality than several forms of commercially available phosphorus-based products.

On a molar ratio basis, used water and side supernatant flows contain far more ammonium than phosphate; therefore, only about 10% of ammonium can be converted into struvite without adding phosphate. If calcium is present, phosphorus will also precipitate at a higher pH as calcium phosphate. Because struvite precipitates at a high pH, at pH = 9, about 50% of ammonia/ammonium mix is unionized gaseous NH_3 ammonia and at pH > 10 more than 90% is unionized ammonia gas, which can be removed also by volatilization. In the controlled recovery of struvite, magnesium is added to the struvite recovery process as magnesium hydroxide or magnesium chloride or sulfate.

Ammonium Sulfate Fertilizer.   Ammonium sulfate can be produced by spraying ammonia volatilized by high pH with sulfuric acid (Negrea et al., 2010, Jiang et al., 2010), used also in neutralization of the used water after struvite precipitation or by bubbling volatilized ammonia rich gas trough sulfuric acid. For this process to be effective, temperature should be raised to about 80°C. pH can also be adjusted back to neutral by carbon dioxide produced in the treatment and recovery processes. Liquid/liquid membrane reactors have been successfully applied for volatile ammonia removal from wastewater (Zarebska et al., 2012; Darestani et al., 2017). Chemical-physical methods of ammonia removal work best for high NH_3 concentrations, which also require high pH. Low concentrations of ammonium are commonly removed biologically by the EBPR modified Bardenpho process and partly recovered as organic N of sludge.

Phosphorus Flow in the Distributed Urban System

The purpose of this analysis is to assess the possibilities and approximate proportions between the output phosphorus forms and potential for reusable products. Figure 9.8 is a mass flowchart of phosphorus from urban sources through the district black and gray water separation and IRRF to the final points of reuse. Total phosphorus (TP) is a conservative substance that exists in used water in two forms: (1) P content of organic solids and (2) orthophosphate that can be either dissolved ion or adsorbed on organic and clay particulates. Phosphorus neither volatilizes nor decays.

On the house or district level, gray water is processed and reused. In this scheme, gray water contains 20% of P of used water which is treated in *fit-for-reuse* processing (filtration and disinfection). This process will retain most of P in the flow which then will be used for irrigation. Larger portion (80% of gray water) after filtration will be treated by nanofiltration or by reverse osmosis and more disinfection for in house reuse uses of toilet flushing, bath and laundry, as shown conceptually in Figure 5.5. In this treatment, most or all phosphorus is removed with reject water and conveyed to the IRRF with the black water flow. Reject water volume is a fraction of the total recovered gray water. Consequently, phosphorus concentration in the reject can be expected to be twice as high or higher than in the raw gray water. However, phosphorus recovery from reject water may be difficult on a district level.

As calculated in the preceding section of this chapter, the 1.8 g capita^{-1}day^{-1} of effluent TP assimilated by algae could result in 97 ∗ 1.8 = 175 g capita^{-1} day^{-1} of algal biomass that will have approximately (40%/100) ∗ 175 = 70 g capita^{-1} day^{-1} of organic C, which could be co-digested and became a significant source of blue energy as biogas. Table 7.5 shows that co-digestion of algal biomass with used water sludge is synergistic and improves dewaterability of the digested sludge. The digester may also receive 2 g TP capita^{-1} day^{-1} with

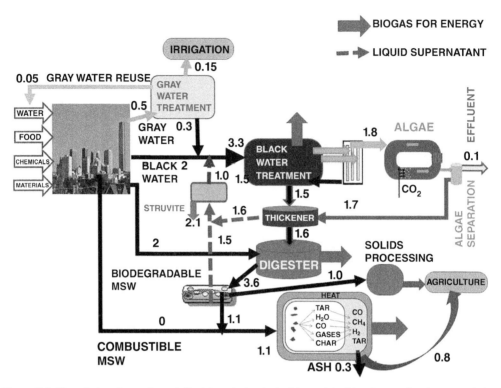

Figure 9.8. Flow of phosphorus through the integrated water/solids sustainable urban system in a generic integrated used water and MSW processing plant with nutrient separation by algae and biomass recovery. The numbers represent the phosphorus flow in g/(capita·day). Dark arrows from BW treatment, digester, and gasifier represent methane and syngas outlets to be diverted to the IRRF power plant.

organic biodegradable MSW and other highly bio-degradable organic compounds such as (suburban) manure. These additional sources of organic biodegradable carbon will require a larger volume of the digesters.

The digested sludge will be dewatered and the liquid supernatant along with the liquid from the predigester thickener will pass through the struvite forming unit where, after increasing pH and adding magnesium hydroxide, struvite will be formed, and after pH adjustment with CO_2 the supernatant will return to the inflow into IRRF. It is assumed that 65% of TP in the supernatant will be recovered as struvite fertilizer. Anammox nitrogen removal (not recovery) process can be added to remove excess ammonium not removed in the struvite and by volatilization. The residual solids provide biomass for producing quality soil conditioning solids that contain some organic carbon with low nutrient content that can be used in agriculture as soil conditioner and/or added to the combustible MSW and conveyed to the gasifier. In the overall mass balance, this integrated system can recover significant mass of phosphorus and remove nitrogen and, in doing so, can produce enough energy to satisfy most of the energy needs of the utility and the community, as it will be documented in the next chapter.

Nutrients in Gasifier Ash

This section deals with recovery of P from ash. Although all P in the solids entering the gasifier will go into ash, recovery of P from ash for use in agriculture as fertilizer was suspect in 2018 because of the metal content of ash and its slug consistency. Leaching experiments with lowering pH showed a possibility of recovering more than 70% of P in a form of usable phosphate. In Figure 9.7 it is assumed that 75% of ash P can be leached and became a medium concentration fertilizer.

Nutrients in char and ash from the gasification unit processing and from residual sludge solids, manure, vegetation, and food waste also contain the residual nutrients that were not separated and recovered in the IRRF or discharged with the effluent. The phosphorus in the feed solids (residual sludge and refuse derive fuel) cannot be incorporated into produced syngas. However, organic nitrogen oxides derived along with the inert gaseous N from air can end up in syngas and flue gas exhaust in direct gasification. Phosphorus stays in the char and ash but with incomplete air pollution control, small fractions can be lost with fly ash in direct gasification. Hence, with a good preprocessing, ash can be used as a slow release fertilizer and soil conditioner; without it, ash would have to be deposited in a landfill or ash lagoons (Belgiorno et al., 2003). Ash from incineration represents about 25% of the original mass of feed. The mass of ash from higher temperature gasification of biomass is less than 10% of MSW (Tum et al., 1998) and as low as 3% for high-temperature indirect gasification.

The solids in the pyrolysis to gasification undergo several metamorphoses. First at lower temperatures below 500°C char and tar are formed. Char and tar are rich with carbon and when the temperature in the gasifier increases above 700°C and more steam is added, carbon in char and tar is gasified to carbon monoxide and hydrogen and the residue is ash and/or slag. Both biodegradable MSW and residual sludge are anaerobically processed that removes a part of nutrients into supernatant as shown in Figure 9.7. Obviously, the flow of phosphorus exhibited on the figure is an expert guess and depends on how much TP will be removed and returned from the effluent and how much it will be converted to dry solids fertilizer.

When focusing on gasification ash in the IRRF that produces energy, all TP entering the gasifier will end up in the ash/slag output, which on the figure would be about 25% of the incoming TP. No phosphorus will be in leaving the gasifier in syngas. Considering that the mass of ash is less than 10% of the incoming MSW and sludge solids, the concentrations of TP can be high, about 5–14% (Table 9.4).

Table 9.4. Average concentrations of metals and phosphorus in ash from incineration and gasification of sewage sludge.

Element	Unit	German Incineration Ash Values	Gasification Ash Values Compiled from Literature
Al	g/kg	52	23–61
Fe	g/kg	99	88–123
As	mg/kg	17.5	N/A
Cd	mg/kg	3.3	<LOD
Cr	mg/kg	267	98–137
Cu	mg/kg	916	1,159–1 367
Ni	mg/kg	106	122–165
Pb	mg/kg	151	51–90
Zn	mg/kg	2 535	753–877
P	g/kg	90	51–149

Source: From Viader et al. (2017), using data from Krüger et al. (2015), Gil-Lagaluna et al. (2015), and Hernandez et al. (2014).
LOD =limit of detection

In indirect gasification, nitrogen can be converted to ammonia and sulfur to sulfide which is represented by the following equation (see also "Gasification of Digested Residual Used Water Solids with MSW" in Chapter 8):

$$C_mH_nO_pN_xS_y + (2m - p + 1,5x + y)H_2O \rightarrow m\,CO_2 + x\,NH_3 + y\,H_2S$$
$$+ [2m + n/2 - p - (1.5x + y)]\,H_2$$

This equation shows that the syngas produced from solids containing nitrogen and sulphur might in indirect gasification contain some ammonia and sulfur gases and the produced syngas should be cleaned before sending to the hydrogen fuel cell. Consequently, ashes from incineration and possibly gasification may not contain appreciable amounts of nitrogen, but phosphorus may be present in concentrations larger than those in sludge (Schiemenz and Eichler-Löbermann, 2010). These authors also claimed that ashes from combustion of biomass are the oldest mineral fertilizer in the world. In a subsequent book, Schiemenz et al. (2011) reported that P content of incineration ash depends on the source of the incinerated biomass and can range from 1 to 10%, which is less than that in gasification ash. There is a difference in P content in the ash from sewage solids (Table 9.4) and ash from incineration of a mixture of sewage solids, including those from food waste, ash from a mixture of sewage solids with combustible MSW which contain less N and P. Food and added sewage solids residuals are the major sources of P in solid waste.

Kalmykova and Karlfeldt Fedje (2013) found that the phosphorus content of the fly ash from municipal waste solids incineration is higher than in sewage sludge, but due to the trace metal content it is not acceptable for application to agricultural land in Sweden. They also claimed that phosphorus extracted from the MSW residues generated each year could meet 30% of the annual demand for mineral phosphorus fertilizer in Sweden. In their experiments, the P content in the ash from incineration of Swedish MSW was 3%. Because the volume of ash from gasification is smaller than that in the ash from incineration and metals and phosphorus remain in the ash, it can be hypothesized that concentrations of phosphorus in gasification ash will be greater than 3% and toxic metal content will be high.

However, ash from gasification does not have the same consistency as ash from incineration but not much was found in the scientific literature on P extraction from gasification ash besides few laboratory studies. This ash has a slag consistency, especially when higher than 1,000°C temperature gasification reactors are used (Materazzi et al., 2016), and the release of the P from it to soils is very slow at best. Hence, the research must focus on extraction of P from the residual ash and until an acceptable and economic method is found its application to agricultural soils is questionable. However, use of the ash/slag for construction purposes or as additive to concrete could be possible and should be investigated.

Nevertheless, ash/slag residuals have a potential to become a significant source of fertilizer phosphorus provided that fertilizer grade P can be retrieved, and the recovered fertilizer does not contain excessive concentrations of toxic metals. Most of the research focused on extraction of P by leaching from the incineration ash. Kalmykova and Karlfeldt Fedje (2013) conducted leaching experiments with MSW incineration ash on P extraction by acid and base leaching and precipitation procedures. They found that acidic leaching of the ash by hydrochloric acid (HCl) at very low pH (<1), followed by recovery of P from the leachate using sequential precipitation at pH = 3 and 4 resulted in a recovery of 70% of the ash P content. Atienza-Martínez et al. (2014) conducted research on recovering phosphorus from the ash obtained after combustion and after gasification at 820°C using a mixture of air and steam as fluidizing agent of char from sewage sludge fast pyrolysis carried out at 530°C. Almost all phosphorus present in gasification ash was leached after 2 h with both sulphuric and oxalic acid using an acid load of 14 kg per kg of P. Similarly, in their comparative research Viader et al. (2017) were extracting P from incineration and gasification sewage sludge ashes by electrodialytic (ED) methods. ED method accelerates pH reduction of the ash acid mixture by applying DC current between a cathode containing acid and ash mixture and an anode compartment containing distilled water separated from the cathode by a membrane. Adding DC current releases protons at the anode that penetrate the membrane from anode to cathode and decreases cathode pH. A product with a lower level of metallic impurities and comparable to wet process phosphoric acid was obtained from gasification ashes; however, this methodology was in the early stages of laboratory experimentation.

On the commercial side, Hermann (2018) presented a concept of P recovery from gasification ash which was heated in a separate high temperature (800–1,000°C) reactor with additive alkaline compounds. Under these conditions toxic metals are gasified and can be removed by the air pollution control system. The remnant is ash-borne phosphate solid concentrate containing up to 20–35% of P_2O_5 mass, similar to phosphate concentrates from ore.

Reducing metallic content of the gasification ash in a less costly way remains a problem, the solution of which should begin with identifying the sources.

10

BUILDING THE SUSTAINABLE INTEGRATED SYSTEM

10.1 ASSEMBLING THE SYSTEM

Concepts, Building Blocks, and Inputs

The previous chapters outlined the reasons for the new paradigm of urban water/energy/ resources management and described the building blocks of the systems. While the system concept may be new and revolutionary, most of the building blocks have been developed and tested for decades; some, like anaerobic treatment, have been used for more than a century; and urine separation and filtration and rainwater harvesting have been known for millennia. Other building blocks of previously less known or unknown (e.g., Anammox) technologies are now leaving the laboratories and are shovel ready to be implemented on a large scale. In some ways, the new system, while visionary, will represent new applications of much improved and efficient concepts of technologies that were running the economy of nations decades ago, such as gasification to produce fuel and gas, and decentralized water and wastewater management. Furthermore, hydrogen as the energy carrier is becoming more common and popular and it is expected that by 2030–2035, at least in some countries, it will provide most of the energy for automobiles and other vehicles and boats.

The conceptualization begins with a selection of the scope of the system that today relies mostly on regional centralized management with long underground pipelines or canals bringing water from large, sometimes subcontinental distances (from Northern to Southern California, from the Colorado River to Arizona and Southern California, desalinated sea water transported across the Arabic Peninsula, and South to North Canal in China) and sending used water to water reclamation plans often located great distances from the city. These systems are separated from the municipal solid waste collection and deposition into landfills or incineration. Such systems are not conducive to reuse, recycle, and resource recovery and have a high energy demand. To make the systems sustainable they must be broken down to smaller semiautonomous but interconnected cluster units and operated by smart cyberspace networks. The size of the system may also be determined by the optimal capacity of the power unit. At the end of the second decade of this century, the maximum size of the hydrogen fuel cell power plants was 100 MWatt but the optimal size may be smaller. Considering that IRRF power is primarily determined to replace a portion of the energy demand and the rest and sometimes more energy will be provided by blue solar, wind, and green energy, the size of the district served by one IRRF will be individually determined. Even today, larger cities have more than one water reclamation plant and future systems, as documented in Chapter 4, can be smaller and a part of neighborhoods rather than separated large water reclamation facilities and polluting incinerators.

Interconnectivity of water and energy in the water–energy nexus was covered in Chapters 2 and 5. The new systems will be energy producers but there might be times when they will be energy users. Sometimes they will produce more energy than the system could use. Although on a smaller house and district scale, rechargeable batteries are already available, and storage of excess hydrogen is feasible on a large scale, a connection to a smart regional energy grid network is necessary, and it is already attainable on a district or utility scale or even a regional scale (Cuthbertson, 2018). Large rechargeable 100-MW batteries are already on the horizon and as of 2017 have been installed in Australia and China (Chapter 4). Chapter 5 introduced decentralized hybrid systems of water/used water/energy. Regional energy storage can be also supplemented by storing excess syngas or hydrogen in the IRRF.

The natural landscape of the city with preserved or imitated and rediscovered streams as the backbone and lifeline of the city will be another feature of sustainability. The low impact development (LID) landscape (Chapter 6) in addition to aesthetics, imitation of nature, and proximity to natural hydrology, can also provide other necessary functions such as storage and treatment of captured rain, stormwater, and even treated gray water, cooling and heating in various forms such as sensible heat, vegetation shading, and carbon sequestration by horizontally and/or vertically planted vegetation. Some of the planted vegetation biomass – for example, in wetlands – can be harvested for biofuel and biogas production in the IRRF. The new Cities of the Future systems must also develop commercialization of the products such as fertilizers, compost, soil conditioners, biofuels, or even carbon dioxide for algal farms and, on a large scale, CO_2 for enhanced oil recovery or for creating aquifer barriers against salt water intrusion.

Following the urban metabolism concept introduced in Chapter 2, current cities are characterized by the linear metabolism processing five major inputs: (1) materials (raw materials, wood, building materials, and others); (2) food in various forms (cereals, animals for slaughter, fruits and vegetables); (3) energy (natural gas, electricity, oil); (4) water; and (5) industrial and household chemicals. Based on the old paradigm, the life and production processes within the city and in supply areas metabolize these inputs into (1) liquid waste impacting ground and surface waters and resources, (2) atmospheric emissions of harmful gases polluting air and contributing to adverse global climatic changes; and (3) solid waste taking more and more of land and polluting groundwater, oceans, and air. Generally, under the current paradigm the demand for resources exceeds their availability, resulting in diminishing availability and eventually exhaustion of resources (e.g., phosphorus). Energy production generates excessive amounts of greenhouse gas emissions. In the linear metabolism system, the liquid and solid wastes and emissions typically exceed the assimilative capacity of the environment, generate pollution, damage the environment, and contribute to global climate change.

Water/energy/solid waste management has been typically fragmented with limited or no interaction between the responsible municipal or regional governments and private utilities managing the system. The only connection between MSW management and water systems is a discharge of landfill leachate that sometimes finds its way into sewers and groundwater and terrible pollution of some water bodies and oceans by plastics and other garbage. Sometimes, water reclamation utilities dispose the sludge into landfills. The following section will present alternatives leading to sustainable water/energy/ resources management and recovery of water-centered communities.

The internal outputs from the metabolism of a sustainable hybrid/circular city to be further processed and recycled are:

Water
> *Used water* from homes, commercial and public buildings, and industries connected to urban drainage
>
> *Captured rainfall*, groundwater from basements and underground garages dewatering, and stormwater and snowmelt
>
> *Cooling water* and air conditioning condensate from large buildings and commercial establishments

Municipal solid wastes (Figure 3.2 in Chapter 3)
> *Biodegradable fluids and high-moisture solids*
>
> Food waste, breweries, slaughterhouses and stockyards, glycols from airport deicing, spilled and waste oils and grease, vegetation residues, biodegradable papers and materials, diapers, pet and other animal waste; also manure (liquid and solid) from suburban agriculture and animal operations like stockyard solid waste and liquids, wastewater from fish-processing plants; dairies
>
> *Combustible low-moisture compounds and materials*
>
> Lumber, woodchips, discarded furniture, cardboard, tires, plastic, recyclable organic trash (plastic cups and bottles), nonrecyclable textiles and leather, asphalt tar materials (not asphalt pavement)
>
> *Recyclable materials*
>
> Metals, glass, textiles, paper and cardboards, used batteries, bulbs
>
> Concrete and pavement materials (commonly recycled by pavement companies on site)

In a hybrid closed metabolism system and circular economy, the use of resources is sustainable, there is very little or no waste, and the outputs of urban metabolism become resources for reuse. This type of urban living and management was defined in Chapter 3 by the triple net-zero goals (no waste of water, no net GHG emissions, and no unsustainable landfilling of solid waste). Under this concept, used water and discarded solid wastes are resources that can produce valuable compounds and energy with minimum or no demand on landfill and without harmful excess discharges and emissions into the environment.

New urban developments that are now sprouting throughout the world should be built with sustainability and its triple net-zero goals incorporated in the design from the beginning. The challenge is how to convert, adapt, and rebuild the existing infrastructure in existing cities so that in a period of 20–30 years (or less) they reach the sustainability goals. The existing water infrastructure in many historic cities in developed countries is from 50 to more than 100 years old; it is crumbling and needs to be upgraded or replaced. In many developing countries, because of rapid growth, drainage and waste water conveyance infrastructure is either not functioning or works by uncontrolled gravity surface flows and contaminating aquifers.

The fundamental assumptions for conversion of traditional aerobic treatment water reclamation plants into sustainable integrated resource recovery facilities are:

1. Removal (treatment) of biodegradable organics (COD) is executed primarily anaerobically. Aerobic/anoxic units with much smaller energy use (for example, Anammox), or close to zero energy use (such as aerobic trickling filters), could be used for effluent

polishing (e.g., removal of dissolved methane or residual ammonium). Membrane filters and reverse osmosis are routine in modern water reclamation plants.

2. Stormwater collection, treatment, and conveyance should be decentralized and, wherever possible, on the surface, which will also include daylighting of buried streams (Chapter 6). Stormwater in the surface conveyance must be treated by mostly natural best management practices and could accept and dilute treated gray water. This will require implementation of LID concepts and must provide storage of the captured and treated stormwater on the surface and in the shallow aquifer for subsequent reuse, and daylighting buried streams. Such systems are possible even in densely built city sections (e.g., Portland, Oregon; Seoul, South Korea; Potsdamer Platz, Berlin; Ghent, Belgium).

3. Processing residual solids from water and solids reclamation facilities cannot be done by traditional incineration, which is inefficient and emits GHGs and toxins. The only energy that can be reclaimed from the incineration process is heat that must then be converted by boiling water to steam to run a turbine and generator. Overall efficiency of incineration to recover energy is low. Murphy and Keogh (2004) reported that the efficiency to produce electricity from MSW the traditional way by incineration is about 20% and the feed must be presorted and without plastics. Heat recovery from incineration is about 55%. In contrast, electric energy recovery by (indirect) gasification followed by hydrogen fuel cell power plant is twice as much as that from incineration (i.e., 40% to >60%) and is expected to grow, and plastics are therein considered as a good source of hydrogen energy (Chapter 8).

   Incineration of combustible solid waste and disposal of residual solids from water reclamation should be replaced by pyrolysis followed by gasification if both biofuel and biogas (syngas) are the desired products or just indirect gasification to maximize electricity production with minimum inert ash residues. In the future (the future begins now), syngas produced by gasification should be reformed to hydrogen, which can be done internally in the molten carbonate hydrogen or solid oxide fuel cell (MCFC and SOFC) or in the gasifier.

4. The water reclamation utilities should develop the system cooperatively with the municipal solid waste disposal and reuse companies and power companies. A typical situation in the US is a public regional utility being responsible for water supply and used water disposal and the community subcontracting MSW hauling, recycling, and disposal to a private company or doing it itself. Close collaboration of these utilities is essential, should decentralization into smaller districts be considered.

5. Collaboration with energy utilities is also essential. The future IRRF is to a great degree a power plant producing electricity, hydrogen, and heat that can be sold but can sometimes use some external energy during the times when the internal energy generation is insufficient. However, most of the produced energy by the IRRF (hydrogen fuel cells and photovoltaics) is the DC current that for some internal needs (e.g., machinery) and commercial distribution must be converted into AC current. Traditional generator produces AC current. DC to AC current conversions are common and should not pose a problem. Furthermore, DC power storage batteries for on-site storage are already available and economical. Excess electricity can be sold to the regional power grid of which the IRRF power plant should be a part. The utilities will also use local (rooftop and land) solar photovoltaic panels to increase DC current production for some process units; e.g., microbial electrolysis cells, electrolyzers, and other internal uses. This will require smart grid and cyber infrastructure, because

in the future a large number of low-voltage DC current–producing plants and power companies may locally return to the original Edison local DC systems connected to regional high-voltage AC grids, when most local power production will switch from rotating AC generators and transformers to static chemical DC power hydrogen fuel cells, hydrogen, wind, and solar power.

10.2 UPGRADING TRADITIONAL SYSTEMS TO CITIES OF THE FUTURE

At the beginning of this century several water reclamation facilities started implementing energy and resource recovery from used water and expanded processing by accepting other sources of biomass such as suburban manure (numerous utilities in Denmark, and in Madison, Wisconsin), food waste to be processed by the excess, and enlarged digester capacities. It was pointed out in Chapter 7 that used water alone does not have enough energy and would not even satisfy the energy needs of the utility. In order to become energy self-sufficient or an energy producer, the utilities must look for other sources of energy. Two examples will be presented herein.

Milwaukee (Wisconsin) Plan

Milwaukee Metropolitan Sewerage District (MMSD) is a large water reclamation utility serving a population of about 1 million. MMSD has two water reclamation facilities (WRF), a downtown plant (DWRF) that is one of the first aerobic treatment plants in the world, and a newer south shore plant. Until a complete reconstruction at the end of the last century, the older plant did not have primary treatment and for decades sludge has been processed by heat treatment into the popular lawn fertilizer Milorganite and the rest was disposed on land. The downtown plant also accepts combined and sanitary sewer overflows that are stored in a large underground tunnel, similar to that in Chicago (see Novotny et al., 2010a). The newer south shore WRF (SSWRF) has primary treatment followed by biological reactors that also accepts the primary sludge pumped from the downtown plant for biodegradation and gas production in the south shore plant digesters.

At the beginning of this century, it was realized that the digester capacity is underused; therefore the city was looking for other biodegradable waste sources and implemented collection of food waste and some other biodegradable solid waste. The city also had serious problems with the deicing waste liquid from the regional international airport that was discharged directly into Lake Michigan, which is also a source of drinking water. Research at Marquette University (Chapter 7, Section 7.4) has established that these waste liquids are synergetic or neutral and can be co-processed by the plant digesters and produce more biogas. Consequently, the utility has installed turbines and became an electric energy producer.

Later, the utility installed a gas cleaning facility at a nearby decommissioned landfill for captured landfill gas to produce more energy and continues to develop existing and new sources of renewable energy. The landfill gas capture and energy recovery have the following components:

> A landfill gas and energy recovery system is efficient in decommissioned landfill that are underlaid and covered by impermeable liners to prevent landfill leachate to contaminate groundwater and methane releasing to be into the atmosphere (Figure 10.1). The leachate collected by drainage system can be treated by the anaerobic bioreactors of the IRRF. Vertical wells are drilled into the landfill to collect the gas, which is then

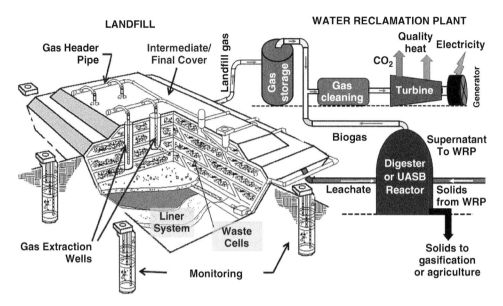

Figure 10.1. Landfill leachate and biogas capture for energy recovery. Leachate can be converted to biogas by co-digestion with the high-strength organic waste solid and liquids in the IRRF. *Source:* Wikipedia Creative Commons.

brought by the surface pipe to cleaning to remove moisture and pollutants that would pollute atmosphere. Clean landfill gas is brought by a pipeline to the water reclamation plant (DWRF), where it is mixed with the digestion gas and combusted by turbines to run AC generators. In the absence of energy recovery methane is flared; 60 to 80% of methane is collected and the rest finds its way into atmosphere.

Also, turbine waste quality heat represents more than 75% of all energy sources at the DWRF and will be captured. MMSD also attempts to use all the waste heat generated in the solids drying process and the DWRF boiler system. It is investigating whether to use all the waste heat generated by the new landfill gas-powered turbines. Similarly, MMSD continues to pursue new sources of high-energy strength waste as well as the mixing and storing of waste at the south shore WRF to help supplement the energy needs and is also seeking to eliminate the need to flare off excess gas at SSWRF by implementing gas storage and alternative uses (such as in fleet vehicles), and other renewable energy sources, including wind, solar (both PV and hot water, particularly at the headquarters building and an expansion of the drying and dewatering building at the DWRF). The utility is also implementing sewer-thermal heat recovery units. Through a combination of conservation and renewable energy, MMSD plans to become a net energy producer by the year 2035 (MMSD, 2012). A wastewater treatment plant in Strass, Austria, has already accomplished this.

MMSD is also involved in urban stream and wetland restoration (see Chapter 6), flood control of urban streams, dam removal, and LID drainage implementation with a goal of being a "water centric sustainable city" (Novotny et al., 2010a). MMSD is becoming an example of an Integrated Resource Recovery Facility.

Danish Billund BioRefinery

The IWA Compendium Report on Resource Recovery from Water (Holmgren et al., 2016) highlighted a visionary Danish project called Billund BioRefinery, which was realized by a collaboration of Veolia Water Technologies and BillundVand A/S. Other partners were the Danish Ministry of Environment, and the Foundation for Development of Technology in the Danish Water Sector (Veolia, 2017). Other Billund BioRefinery plants have already been commissioned and developed in Denmark and other countries.

This project integrates waste (used) water treatment with co-digestion of and biogas production from other organic wastes, specifically organic waste from industry, organic waste from households, and manure and organic waste from agriculture. The system has two key components, a *used (waste) water treatment* plant (WTP) accepting used water and producing treated water that would comply with the Danish criteria for discharges into surface waters, and sludge. The second component is *Energy Factory* (EF) that accepts the sludge from WTP and biomass from households, industries, and manure and organic waste from agriculture (see also Chapter 7). The first Billund bioenergy plant has a total capacity to process 4,200 tons of organic waste annually. The sludge from the WTP is delivered to the EF, where it is mixed with the other solids.

The Energy Factory is a combination of anaerobic digestion and thermal hydrolysis. The technology is called Exelys TM and Billund BioRefinery. The key processes were patented by Veolia. The first step in the EF is the waste pre-digestion and dewatering. The sludge is then pumped under pressure into a reactor tube. The sludge is heated in the reactor by steam injection and is exposed to pressure and high temperatures. The end product of the entire process is biogas (methane), which is converted to both electricity by combustion and heat, which is sold to the public district heating system and to local customers. This EF system combines co-digestion of sludge, biodegradable solids, and liquids.

Dockside Greens Integrated Resources Recovery (BC Ministry of Community and Rural Development, 2019; Novotny et al., 2010a). Dockside Green is an innovative residential and commercial development in central Victoria, British Columbia. The community is home to more than 2,500 people. The development features onsite wastewater treatment and a biomass gasification facility which provides heat to a district heating system. This infrastructure is helping Dockside Green to attain a LEED Platinum rating. Dockside Green provides an innovative example of Integrated Resource Recovery (IRR) in an urban setting. The development received LEED platinum rating for its sustainability characteristics. Its IRR systems include (used) water treatment systems, storm water collection and recovery system, and biomass gasification. The onsite wastewater treatment plant is integrated into the center of Dockside Green and is situated beneath some of the residential buildings (Figure 5.7). The reclaimed effluent from the treatment plant has quality better than potable water. Dockside Green reuses this water for flushing toilets, irrigation, and to supply a waterway which provides both aesthetic appeal and habitat for wildlife. Reuse of the treated water saves approximately 113 million liters of drinking water per year. Dockside Green also includes low-flow plumbing fixtures and appliances. In total, water consumption is 65% lower than in traditional developments. Biosolids are compacted and used as compost for landscaping and as feed in the gasification plant. Through the biomass gasification process, organic wastes (such waste lumber, biodegradable sludge and waste solids) are converted in to syngas and energy. In addition to Platinum LEED, Dockside Greens is a sustainable integrated TNZ community. By the same criteria and knowing the Swedish the result of Swedish Recycling Revolution, Hammarby Sjöstad is also TNZ community as there might be many other communities in several countries.

Figure 10.2 summarizes what can be done in the short period of one decade. The key is to enhance sludge digestion and accept biomass from other sources (Chapter 7, Section 7.4) and acquire turbines and MWatt electricity generators. Installing hydrogen fuel MWatt capacity cells in the water and energy reclamation plant has already become a reality in Connecticut and California (see Sections 8.6 and 8.7 in Chapter 8). It is important to maximize COD

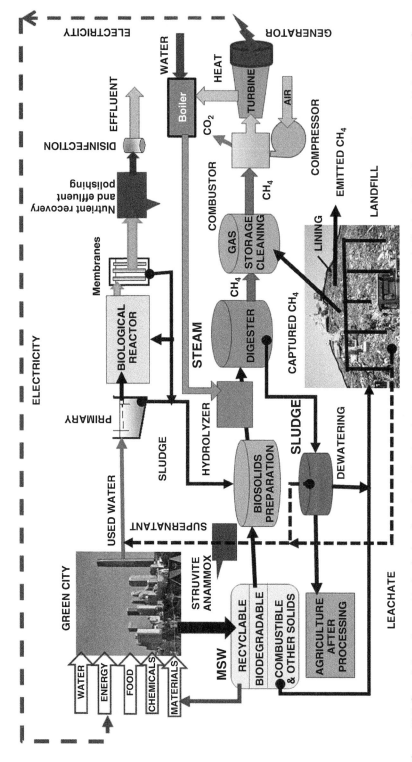

Figure 10.2. Compilation of present technologies of used water treatment, resource recovery, and energy production inspired by the Milwaukee Metropolitan Sewerage District and Danish Billund BioRefinery systems, both affiliated with Veolia designs and management.

removal in the primary treatment and send the sludge to the digesters to make methane instead of to aerobic reactors to be converted to CO_2.

Sludge hydrolysis (Chapter 8, Section 8.3) significantly increases COD removal by digestion and, consequently, biogas yield. Following the MMSD example, the turbines and generators can accept the landfill methane; however, the problem with combustibles and items that are difficult to biologically degrade (such as plastics) would not be resolved, but the mass of solids to landfill would be significantly reduced. Hence, the system in Figure 10.2 maybe be transitional but would make the utility energy self-sufficient and, by stopping the use of black energy, also a net-zero GHG carbon emitter.

Integrating MSW

The schematic in Figure 10.3 is presented for utilities that desire to adapt their existing water and solid waste management infrastructure, and for facilities to save and recover water for reuse in buildings and outdoor irrigation, and recover heat energy at the district level and some resources and more energy at the regional level. This alternative has the goals to modify, expand, replace, and adapt infrastructure and facilities in place at the end of the last century, using the existing components and relatively quickly replace the components that are inefficient and need replacement and overhaul. The goal should be to significantly reduce the very high water and energy use, limit landfilling, recover some resources such as fertilizing solids, and, in doing so, make the utility (but probably not the entire community) a net-zero carbon emissions facility. The "digester" unit in Figure 10.3 is a generic term defining the process. The key goal is also to implement with the solid waste management company or utility a step-by-step integration of water disposal and municipal solid waste (MSW) collection so that combustible and biodegradable fractions of MSW can be collected and processed symbiotically with IRRF used water and residual solids. The process has already started in Milwaukee, Dockside Greens, and possibly in several other cities and regional utilitiies Existing incineration plants should be decommissioned and replaced by gasification and additional gasification plant capacity should be built to replace landfills. The step-by-step process is outlined below:

1. *Implement water and energy conservation in buildings.* Chapter 5 covered numerous measures that are already being implemented to reduce water use. They are summarized in Table 5.2. Water-saving measures (low-flush toilets, low-flow showerheads) could bring the in-house water use in the US to less than 150 L/capita-day, which is already met in many cities in the developing world of Europe and Asia but not in the US or Canada. Outdoor water use can be reduced by changing landscape from irrigated grass lawns to xeriscapes. Installing solar panels on house roofs to dramatically reduce net energy use is already affordable and even profitable almost everywhere, even in developing countries, and has been attractive in the US because of subsidies and incentives. Water conservation has also a great beneficial effect on reduction of energy use and GHG emissions related to water processing and delivery.

2. *Water and energy recovery and reuse of used water on the district (cluster) scale.* Ideally, to recover water and energy from used water on the district scale, the sewer system should be separated into black and gray water collections (Figures 5.5 and 5.6). A cluster water reclamation filtration facility will produce water that will be reused in buildings for indoor non-potable uses and outdoor for irrigation, street washing and cooling and other uses. Treatment should be "fit for reuse." The treated gray used

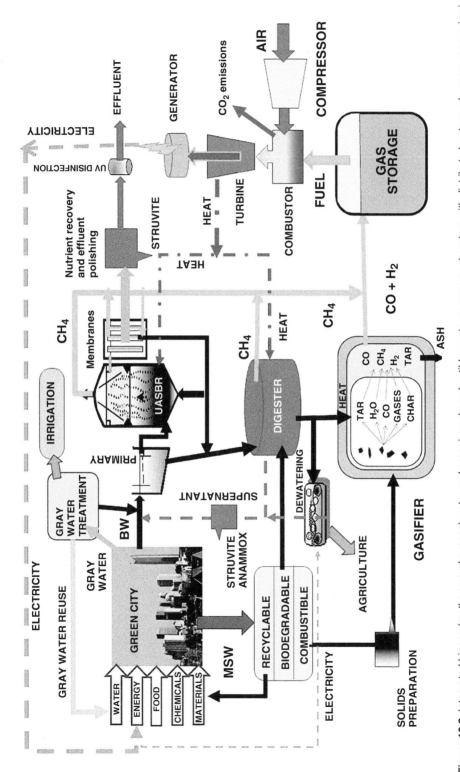

Figure 10.3. Integrated biogas (methane and syngas) water and organic and combustible waste management system with distributed water and energy recoveries in clusters and water/methane energy recovery facility in a regional IRRF using late twentieth-century technologies. UASBR = upflow anaerobic sludge blanket reactor.

water can be also mixed with other available water such as captured and stored rainwater, stormwater, and other sources of minimally contaminated water, or cooling water from power plants. Reject water from the gray water treatment is directed into a black water conduit and conveyed for resource recovery in a regional integrated resource recovery facility (IFRF).

Gray water is warm, around 40°C as compared to mixed sewage temperature in sewers of about 15°C, depending on the season and climatic conditions. Therefore, sensible heat can be recovered from the gray water and used in the cluster for heating or left in the recycled gray water flow to reduce heating demand for warming hot non-potable water. Sensible heat from sewer flow can be also done in districts. This recovery could be used in winter for warming pavements to keep them ice free instead of using salt. Hence, heat recovery can be done at house level or districtwide. Geothermal and solar heat harvesting in buildings also represents a significant heating (and geothermal cooling) source.

In the initial stages of the conversion of an existing traditional system into partially cyclic system, instead of a dual drainage, needed supplemental water can be obtained by withdrawing the necessary demand from an existing sewer and installing a packaged tertiary (physical) filtration treatment plant for providing recycled water (Figures 5.5 and 5.6). Such treatment facilities have been installed in the basements of buildings or in architecturally blending shelters within the neighborhood (for example, in the Dockside Green development in Victoria, British Columbia, shown in Figure 5.7, or the Potsdamer Platz complex in Berlin). Reclaimed water can be used for irrigation and toilet flushing but not for other in-house nonpotable uses (Asano et al., 2007).

The volumetric flow of black water with the reject water from the treatment of the district recycled flow will be significantly smaller than the flow existing in the sewers before being upgraded and sent to a regional IRRF. The existing pipes may stay but, in some places, may be upgraded by inserting smaller-diameter plastic pipes and the unused space can be rented to cable and phone companies.

3. *Implement low-impact development stormwater management practices in the districts.* Communities must abandon the notion that urban stormwater is waste that must be collected and disposed with some treatment as fast as possible. LID practices (Chapter 6) capture and store stormwater and reuse it by infiltration into aquifer and surface storage to provide water for irrigation and other uses (building cooling). In dry communities, captured rain and stormwater could also become after appropriate treatment supplement of potable water sources. Figures 5.5 and 5.10 in Chapter 5 show how urban stormwater and runoff management is incorporated into landscape to become a resource and even incorporated in the water energy nexus by providing cooling and heating energy.

4. *The modernized/upgraded/rebuilt Integrated Resource Recovery Facility* will treat black water separately but synergistically (with the solids in the reject water from the district gray water treatment), and liquid and semisolid biodegradable and solid combustible components of the municipal solid waste (BC Ministry of Community and Rural Development, 2019). Because water conservation and gray water recovery/reuse occur on the house and district levels, the black water flows will be smaller than the current used (waste) water flows. Hence, there will be no need to increase capacity of the treatment reactors that could be converted to anaerobic biologic membrane reactors, struvite recovery, and Anammox nitrogen removal

from the supernatant. The three line integrated water, energy, and resource recovery processes are

a. *Black water with solids and reject water* from gray water treatment, after pretreatment that removes settleable solids, is anaerobically treated in an anaerobic fluidized bed reactor followed by a separate membrane reactor. In a two-step process (UASBR followed by AnFBMR, Figure 7.12), granulated activated carbon can be added to enhance the COD and solids removal. Permeate from the membrane reactor could be post-treated aerobically to remove residual methane and nutrients (phosphate) recovery in a polishing aerobic unit (e.g., a trickling filter) that can be added. Excess solids will be removed from the anaerobic reactors and effluent polishing and after dewatering sent to a digester. pH of the supernatant from sludge dewatering could be raised by adding a caustic solution containing NaOH and $Mg(OH)_2$ to bring pH to more than 9.5 to remove struvite and calcium phosphate and the supernatant is then returned to the influent.

   Because the supernatant side flow is only a small fraction (1–2%) of the influent flow, a treatability study should determine whether after recovering struvite and removing nitrogen, the pH of the supernatant needs to be reduced because the anaerobic reactor maintains higher pH to complete methanogenesis. If there is a need to reduce pH, waste CO_2 can be used for this purpose. Moderately concentrated liquid waste from sources other than sewers could also be processed with black water in the anaerobic membrane bioreactors (Figures 7.10 and 7.12). Note that the COD concentration limit for the influent into the IRRF is about 20 times greater than the limits for the current aerobic activated sludge and Bardenpho processes (Lettinga et al., 1980; Lettinga and Hulshof–Pol, 1991). Stronger liquid wastewater/slurry (COD > 15 g/L) should be processed in the digester. Produced methane is diverted to a gas storage tank and, after mixing with other gas flows from the digester and gasifier, the biogas will run a turbine or a combustion engine to produce electricity in the generator.

b. *Digester capacity* may have to be increased to accommodate the liquid and high moisture/semi-liquid biodegradable wastes (resources) that currently have not been allowed to be discharged into sewers because they would overload the capacity of the current aerobic treatment units. Such contributions include glycol from airport deicing operations, industrial wastewater with high concentrations of biodegradable organics (brewery, animal slaughter operations, dairies, restaurant grease), and leachate from still functioning and even abandoned but still polluting landfills that have synergistic or neutral biodegradability (Table 7.5) when compared to the biodegradability of the residual solids from the municipal used water.

   Highly concentrated liquid and biodegradable semisolid fraction of the MSW conveyed directly to the digester(s) will be co-digested with the excess (removed) solids from the anaerobic treatment. Anaerobic treatment produces only a fraction of the waste solids mass when compared to traditional aerobic systems. The COD and mass of the biodegradable MSW is significantly larger than those of the BW solids, but the latter solids will provide necessary nutrients and liquid fraction for effective digestion. Additional biomass such as manure, algae, biodegradable paper, and vegetation residues can also be diverted to the digester for biogas production as well as for nutrient recovery from the digester supernatant.

   Supernatant from the digester and sludge dewatering will be returned to the inflow of the IRRF. Struvite recovery and Anammox reactors could be installed on the supernatant return flow to recover phosphorus and remove nitrogen, respectively.

The produced biogas (methane) should be diverted to the gas storage tank. A portion of the digested solids can be processed to make fertilizing and soil conditioning solids. However, such process will require heat to dry and disinfect the solids.

c. *Combustible MSW and residual digester solids* will be processed by a gasifier or a pyrolysis-gasification unit. Besides gasification there is no other sustainable way currently available for processing MSW except circular economy and maximum recycle and plasma technology. The indirect gasifier (or a combination of direct and indirect gasification) can process plastic to form syngas without forming toxic dioxins and furans. The gasifier will replace polluting and inefficient incinerators and will also receive digested solids from the digester. Unlike traditional incineration, which requires feed solids to be dried, because indirect gasification needs water and the mass of the combustible dry solids may be greater than the solid mass from the digester, expensive and energy-using full dewatering of the digester and MSW solids may not be necessary. A simple thickener may suffice for the residual digester solids from which the supernatant will be returned to the influent of the anaerobic membrane bioreactors and thickened solids could be pumped to the gasifier and mixed with drier combustible MSW. In Chapter 8 on gasification, the amount of water needed to yield maximum weight and volume of syngas from MSW was estimated as minimum 0.25 kg of water in a form of liquid or steam per 1 kg of dry combustible (New York type) MWS for syngas production to 1 kg of water 1 kg of dry combustible MWS for hydrogen production. Dry combustible MSW have a moisture content between 10% and 20%. Hence, in this alternative, most if not all needed water could be provided by the moisture in the feed of.

d. The product of the gasifier diverted to the gas storage tank is syngas, which is a mixture of carbon monoxide, hydrogen, and small amounts of other flammable gases (Chapter 8) and waste gases from the air if exothermic direct gasification is used. If endothermic indirect gasification is selected, a portion of the produced biogas will be used to generate the necessary heat for gasification. Heating can also be provided by produced methane from the anaerobic treatment and digestion processes. The burner heating the gasifier will emit CO_2. High-quality heat can also be recovered from the combustion turbine exhaust gases and used for heating digester and UASB and AnFBM reactors. Because air is used in turbine combustion, CO_2 in the exhaust gas will be diluted by nitrogen, small volumes of nitrogen oxides, and other gases and will be difficult to sequester. The NO_x may be counted as GHG. However, per ICCP classification, CO_2 emission is not classified as GHG but no credit is given; therefore, it will be classified as "green" energy.

Figure 10.4 shows the endothermic heat for the gasifier being provided by an external burner combusting a portion of the produced syngas. By combining direct gasification and indirect gasification compartments the endothermic heat can be generated internally (Grootjes et al., 2016). However, by introducing air into the reactor, the produced syngas will be diluted by nitrogen from air.

e. *Power plant.* In this traditional scheme, the syngas produced by the anaerobic treatment units, digester, and gasifier ran the turbine or combustion engine (depending on the scale) that powers the generator. A portion of the output heat (about 50% of the energy in the fuel gas) from the engines running the generator and the gasifier can be recovered as high-quality (hot steam) and low-quality (hot water) heat and used for heating the reactors and reducing moisture in the gasifier feed. It has already been persuasively established that the MSW incinerator with boilers, steam turbine, and generator is an inefficient and environmentally unacceptable process without expensive air pollution controls because of CO_2, fly ash, NO_x, SO_x, and toxins in the flue gas. Modern gasifiers do not emit toxins.

10.3 VISIONARY MID-TWENTY-FIRST CENTURY REGIONAL RESOURCE RECOVERY ALTERNATIVE

Figure 10.4 is the conceptual flow diagram of the anticipated new sustainable resource recovery system with IRRF that uses a conventional UASB (AnFBM) membrane reactor or UASB followed by AnFMM reactors in series, sludge hydrolysis followed by hybrid fermentation digester-MEC unit and an indirect gasifier as the focal units in the concept of the integrated IRRF hydrogen-based system. The IRRF is processing three resource streams from the community conveyed in the IRRF after district water reuse and energy recovery. This is a vision of a possible future system to be tested by a spreadsheet model, but it may not be the final design 30 years in the future. Figure 8.12b provided the basic concept for the power plant.

Biodegradable solids (manure, energy crops and vegetation residues, algae, and woodchips) can also be conveyed to this system fermentation/MEC line. This scheme will produce hydrogen, clean water, heat, and electricity that greatly exceed the IRRF energy needs and the excess can be commercialized. The system will recover resources in various forms, even in a form of hot concentrated reusable or sequesterable CO_2 gas.

Gray water recovery, treatment, and reuse was described in detail in Chapter 5. The goal of the overall gray and black water recovery is to reduce potable water use in the cluster (district, large building) to 100 L capita^{-1} day^{-1} or less, yet provide comfortable water availability within the recycle loop. This water conservation and reuse goal itself will significantly bring down the energy use with beneficial GHG reduction consequences if electricity for providing clean water is from fossil fuel power plants. The water management and reuse concept is presented in Figures 5.5, 5.7, and 5.10 in Chapter 5.

The district water management will also synergistically use rain and stormwater not only as nonpotable and even potable water supplements but also use the LID water stored in architecturally pleasing landscape storage and/or shallow underground aquifers (Figures 5.5, and 5.10) for cooling and heat energy. Reject water from district water recycle plants with almost all waste solids is diverted to the separated black water flow that is conveyed to the centralized IRRF for further processing and water, energy and resources recovery. A large portion, if not all, of home electricity will be provided by house or district solar panels and, when available, by wind, hydro, and geothermal power.

It has already been elucidated that the black water flow rate directed to the IRRF will be only a fraction of that in the current sanitary sewers and treated in the current water reclamation plants. The *IRRF system* will accept three streams of raw materials for processing: (1) black water flow from the districts; (2) biodegradable municipal waste solids and liquids, and excess organic waste solids from agricultural businesses (i.e., those waste solids that have not been reused directly by the farms as fertilizers or animal feed); algae grown by the IRRF to remove nutrients from the effluent or commercial algal farms, which also sequestrate CO_2, will be directed to this input; and (3) combustible items that are difficult to degrade (cardboard, wood) and nonbiodegradable (plastic, possibly tires) solids (BC Ministry of Community and Rural Development, 2019).

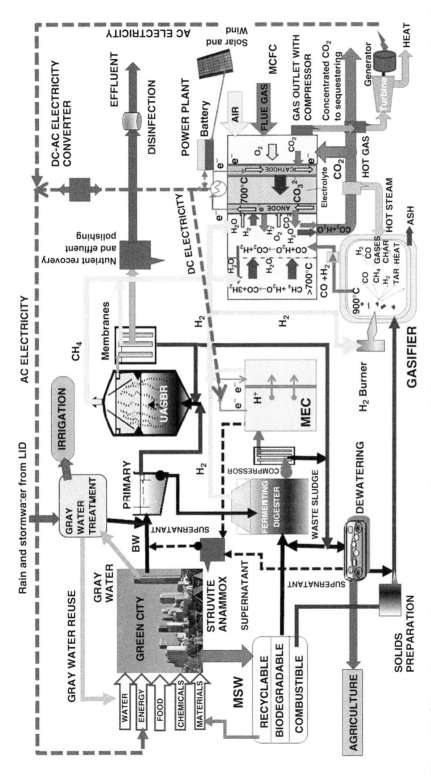

Figure 10.4. Schematic concept of the urban metabolism and integrated water/solids management and resource recovery based on methane and syngas reformed to hydrogen and hydrogen-fed molten carbonate fuel cell power plant. BW = black water flow, MSW = municipal solid waste, UASBR = upflow anaerobic sludge blanket. (Units not to scale.)

Used black water and other biodegradable liquids (COD < 15 g/L) from dischargers, such as brewery, dairies, cheese manufacturing, and fish processing that currently are required to have on-site supplementary anaerobic pretreatment to prevent overloads of the current water reclamation plants, will be directly conveyed to IRRF. The beginning steps are mandatory inorganic (higher-density) solids separation and optional primary settling that will separate organic solids and direct them to the fermentation digester. Instead of traditional circular or rectangular settlers, "lamella" settlers could be used (Figure 10.5). The inclined plates or beehive-shaped plastic modules inside the lamella settler induce better laminar settling conditions and the settler can be enclosed for odor control.

The core treatment unit for this line is the upflow anaerobic sludge blanket reactor (UASBR), heated when needed, with a separate membrane reactor, and/or an AnFBMR (Figures 7.10–7.12). This reactor system will ferment organics in the flow and produce methane. High concentrations of solids and large solids retention time (SRT), while decreasing the hydraulic retention time (HRT), will be provided by recycling the reject water from the membrane unit. To keep the membrane free of fouling, a portion of gas flow may be injected at the bottom of the membrane reactor and collected. Treatability studies must establish whether one UASB + membrane reactor will suffice to produce acceptable effluent or UASBR–AnFBMR reactor sequence will be necessary. Also, adding activated carbon granules may be considered to achieve high removal efficiencies of COD and other compounds and to remove dissolved methane. Anaerobic treatment is not efficient for nutrients removal (Total P and N) in excess of those incorporated in sludge. Note that the effluent flow will be much smaller (<100 L capita^{-1} day^{-1}). In contrast, typical effluent flows in the current US water reclamation plants is around 300 – 500 L capita^{-1}day^{-1}.

Excess sludge solids from the UASBR(s) will be directed to dewatering in a gravity thickener, possibly in a combination with a belt press, with other solids generated by the IRRF and, after separating the supernatant, the solids will be added to the feed of the gasifier. Sending the excess solids to a digester may not be efficient because most or all easily recoverable CH_4 has been already removed in the UASBR and AnMBR. However, this is now only a speculative hypothesis that should be proved or disproved by further pilot scale research.

A portion of dewatered solids after drying and disinfection could be sold as a low-grade fertilizer. The supernatant will be processed by struvite formation and harvesting unit to

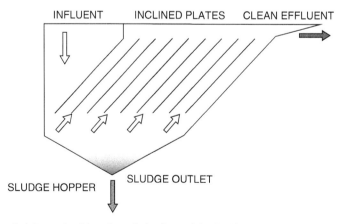

Figure 10.5. Inclined plates or beehive-shaped plastic modules inside lamella settlers provide better laminar flow for settling lighter organic solids than traditional circular or rectangular settlers; the settler uses less space and is typically enclosed.

recover phosphorus and nitrogen containing struvite. Forming struvite requires high pH, which is achieved by adding sodium and magnesium hydroxide. Neutralization of the high pH can be done by concentrated CO_2 from the MCFC power plant. Anammox removal reactors may be installed on the supernatant return flow to remove excess nitrogen. The Anammox units require some aeration energy for which energy can be provided by the IRRF power plant.

The produced methane collected from the UASBR and membrane reactors will be sent to the steam methane reforming (SMR) compartment of the MCFC or a separate SMR reactor where under a high temperature (>700°C) catalytic process CH_4 will chemically react with hot steam to form hydrogen-enriched syngas (see Chapter 8). Two molecules of water are needed to form one molecule of carbon dioxide and four molecules of hydrogen, which represents water use of 2.25 kg of water per 1 kg of CH_4; however, plenty of hot clean steam will be supplied by the molten carbonate fuel cell (MCFC) (hydrogen fuel cell) power plant. The produced methane could be also converted to biofuel and sold. The heat will be provided by produced heat energy in the hydrogen fuel cell and by recuperating heat from the exhaust of the gasifier (Figures 8.12a,b and 10.3). Gasifier temperature is greater or equal 900°C while the minimum SMR temperature is 700°C.

The permeate from the membrane reactor should be further processed to polish the quality of the effluent that may contain residual dissolved methane showing as COD and to remove phosphorus if it is discharged into eutrophication sensitive receiving waters. After disinfection, the effluent should be acceptable for discharge to maintain ecological flow in receiving waters or be reused for various beneficial purposes such as irrigation or production of algae or high-density aquatic vegetation. After an extended travel time and storage in the receiving water body or a shallow aquifer and further dilution with cleaner fresh water, the effluent could be reused with additional treatment for potable water supply.

Biodegradable municipal (and agricultural) semisolid waste and high organic content liquids and biomass plus the solids separated in the primary treatment of the liquid black water flow will be diverted to the dark fermentation–microbial electrolysis cell (MEC) hybrid reactors that produce hydrogen as the end-product. This novel approach shown in Figure 8.9 is promising but has not been fully researched on a pilot scale at the time of writing this manual. Organic solids, carbohydrates, celluloses, and lignin can be broken down at lower pH in the dark fermentation digester and the end product of this process will be liquid organic acids, hydrogen, and residual digested solids. It is estimated that about 50% of volatile suspended solids will be destroyed. By maintaining a pH in the fermentation digester between 5 and 6, methanogenesis should be prevented, which avoids the problem with the soluble methane showing as BOD and COD in the effluent.

The first step could be feed solids (sludge) hydrolysis reactor into which hot steam (900°C) could be brought from the hydrogen fuel cell that breaks the complex organic compounds and prepares the feed. Hydrolysis is followed by cooling (Veolia, 2017). Based on the current (2019) knowledge, the *fermentation digester* most likely could be a completely mixed tank reactor (CMTR) or UASBR followed by a membrane reactor (Figures 7.9a or c, 7.11, and 8.9); however, no scientific information was available regarding the type of the reactors and location of supernatant separation from the solids at time of writing this book. Up to one-quarter of the hydrogen potential can be recovered in the fermentation digester and sent to the hydrogen fuel cell (Logan, 2016). Because solids are returned from the membrane reactor and methanogenesis should not occur therein, the HRT and SRT will be much shorter than those in the traditional fermenting and methane producing digester. Waste solids from the fermentation-MEC reactors along with the excess solids from the membrane UASB reactor of the liquid flow will be diverted to dewatering and to the gasifier. A portion of the solids could be further dewatered, and heat treated to make a low-grade agricultural and landscape fertilizer. Supernatant from dewatering will be returned to the BW influent. Phosphorus and nitrogen can be removed from the supernatant to make struvite (ammonium magnesium phosphate) and more nitrogen can be removed but not recovered by the Anammox process (see Chapter 9, Section 9.3).

Permeate from the fermentation CMTR or UASBR is diverted to the *microbial electrolysis cell*, where the acetates, proteins, and other organic compounds will be broken down. A simple MEC presented herein for illustration of the concept has anode and cathode compartments divided by a membrane, as shown in Figure 8.8. Multiple-cell MECs have already been developed. In the MEC anode compartment, special bacteria decompose organic matter and generate CO_2, protons (H^+) and electrons (e^-). CO_2 could be dissolved and after pH adjustment become bicarbonate alkalinity, but if there is an excess gaseous CO_2 it could be sent to the common CO_2 exhaust of the molten carbonate fuel cell (MCFC). The electrons are transferred by the bacteria to the anode, from which they travel externally to the cathode compartment. However, for the electrons to travel from anode to cathode, they need a small amount of externally supplied DC current that was estimated at about 0.4V potential (see Chapter 8, Section 8.4). Because of the H^+ (pH difference) potential between the two compartments, protons will concurrently permeate through the membrane from the anode compartment to the cathode compartment, where protons are released to the solution and combine with the externally transferred electrons to form hydrogen gas, H_2. Hence, the formation process of carbon dioxide and hydrogen are completely separated and, theoretically, there should be no microorganisms in the cathode compartment participating in the process. Hydrogen is sent to the MCFC or SOFC as fuel. The DC current that excites electrons to travel from anode to cathode can be provided by a solar photovoltaic and DC current generated by the MCFC and stored in a battery.

Because the influent into the MEC might be preprocessed by removing the suspended solids and by converting carbohydrate organics into acetates and proteins in the fermentation reactor, the hydrogen yield will be much higher than that from the MEC treating the entire mixed influent without preprocessing. The hydrogen recovered in the cathode compartment after fermentation preprocessing could be about two-thirds of the total H_2 potential or more than 90% of the H_2 potential in the influent into MEC. No CO_2 should be produced in the cathode compartment.

The anode solution where acetates and proteins are converted to CO_2, electrons and H^+ protons are in the supernatant that contains nutrients and other poorly digestible pollutants; therefore, the effluent from the anode compartment should be sent with the supernatant from the thickener to struvite extraction and N removal by Anammox. It is likely that in a decade or so, other better methods extracting nutrient will leave the research laboratories. Most likely, the CO_2 will be dissolved or a part of bicarbonate alkalinity. At the time of writing it was not established whether the hybrid dark fermentation and MEC reactors would be completely mixed, plug flow, or SBR.

Combustible fraction of MSW is processed in the gasifier with the residual fermentation digester and UASBR solids at a high temperature (>900°C) to produce syngas. If biofuel and char production is desired, then a two-step process – pyrolysis and gasification – should be selected, as shown in Figure 8.2. Biofuel and char can be then separated in pyrolysis or char can be converted into syngas in the gasification process. Produced syngas does not need to be cooled because the steam reforming to hydrogen is carried out at a high temperature in the anode compartment of the MCFC. A part of the heat can be recuperated and used for heating the SMR unit.

Well-known, tested and continuously improving gasifiers (Figures 8.2, 8.3, and 8.12a, b) will replace inefficient and polluting incinerators and boiler furnaces to extract energy from combustible municipal solid waste combined with the residual organic solids from which energy was extracted in a form of methane or hydrogen. These solids will contain non-biodegradable organic carbon. MSW gasification is reappearing again in Europe, which has had a long tradition of gasifying low-quality coal and woodchips to syngas, especially during WWII and postwar periods when oil was not available in Central Europe. The energy product of gasification is syngas and the main residual is a small quantity of ash, far less than 20% of the incoming solids common to incineration. The gasifier ash may not have to be landfilled.

In Chapter 8, Section 8.2, example of gasification of combustible MSW, it was estimated that formation of syngas from combustible MSW needs 0.25 kg of water per kg of dry combustible MSW and an additional 0.75 kg of steam (provided by the hydrogen fuel cell) is needed to reform carbon monoxide in syngas to hydrogen, which represents about 1 kg H_2O/kg ds MSW.

MWS also contains hydrogen and oxygen in the molecules; therefore, the mass of hydrogen released by indirect gasification varies depending on how much hydrogen and oxygen is contained in the organic compounds. The hydrogen yields potential ranges from 0.166 kg H_2/kg of VSS for algae to 0.44 kg H_2/kg of plastic polyethylene and polypropylene (see Chapter 8, Section 8.6, "Hydrogen Yield Potential by Indirect Gasification"). Because water is used in the reaction, excessive dewatering and drying of the solids sent for gasification from the used water treating UASBR and fermentation digester or of MSW may not be needed. Additional steam is available in the anode of the hydrogen fuel cell. On the other hand, because water produced by the MCFC is very clean and in excess of the syngas and methane reforming demand it could be used for other purposes and water needed for gasification can be provided by the MSW and sludge moisture content. Energy-demanding full drying of MSW and sludge makes sense only if the gasifier is distant and the solids must be transported by trucks. SMR and syngas shift units should be a part of the IRRF power plant, as shown in Figures 8.12a and b.

The Power Plant

In a traditional and far less efficient configuration of process energy producing units, a power plant would include a gas combustion engine or turbine followed by a generator. In the new and more efficient clean power generation, the IRRF power plant of the year 2040+ could be an expanded molten carbonate (hydrogen) fuel cell (MCFC) or any better future hydrogen fuel cell that accepts syngas and converts it by a shift to hydrogen and CO_2 internally with an external or internal steam methane reforming (SMR) unit (Figures 8.12b and 10.3) or with a separated syngas shift reactor. If needed carbon dioxide can be separated from hydrogen in the H_2-CO_2 separator before hydrogen enters MCFC. Besides electricity this power plant will generate a concentrated hot stream of CO_2 that can heat the IRRF reactors, reused to grow algae, and the rest can be sequestered (Chapter 8) or commercialized.

The reason for the external and not internal steam methane reforming (SMR) unit is that SMR will be processing much smaller volume of the methane gas in higher temperature (greater than 700°C) than the MCFC processing syngas and hydrogen at a lower temperature of 630°C–700°C. Hot clean steam and heat will be provided from the MCFC, but additional heat will be needed to bring the SMR temperature to more than 700°C that could be provided by combusting some hydrogen produced by the power plant.

Theoretically, $1.12 \, kg$ of steam H_2O will have to be supplied from MCFC for each kilogram of CH_4 converted to H_2 enriched syngas ($CO + 3 \, H_2$). The gas may have to be cleaned because some waste compounds like sulphur (see Table 8.2 in Chapter 8) may damage the units.

MCFC as any other hydrogen fuel cell is not a biological unit. Electricity is produced by the electrochemical reaction and electron transfer. MCFC has three compartments: (1) the *anode*, which receives the feed hydrogen from fermentation reactor and microbial electrolysis cell (MEC) and syngas from SMR and gasifier units. CO_2 from steam shift of carbon monoxide in syngas may also be accepted by the cathode compartment of the MCFC and become a part of the CO_2 pool of the MCFC (McPhail et al., 2015); (2) the *middle compartment* between anode and cathode, which contains molten carbonate CO_3^{2-} salt electrolyte; and (3) the *cathode* compartment, which contains reacting oxygen (air) and carbon dioxide and receives the electrons from the anode compartment and the auxiliary power supply too from carbonate ion CO_3^{2}.

Because of the carbon monoxide reformed in the anode compartment of the MCFC to CO_2 and hydrogen and some carbon dioxide in the air and flue gases from turbine and combustion engines brought into the cathode compartment, MCFC will emit hot pressurized excess CO_2 and hot steam from the anode compartment. The exhaust gas should be clean. A part of the hot steam will be used in the anode compartment to reform carbon monoxide in syngas to carbon dioxide and additional hydrogen and a part will provide heat and steam for reforming methane in the SMR unit and potentially, if needed, some heat and steam to the indirect gasifier. Separated CO_2 is returned to the cathode compartment to form carbonate. If there is any residual gaseous CO_2 produced in the microbial electrolysis cell (MEC) line, it can be joined to the CO_2 return conduit of the MCFC and a CO_2 regulator placed on the return to cathode compartment. MCFC should produce an excess of steam that can be used for heating or after cooling, clean water can be used for other purposes. Hot stream of the excess carbon monoxide from the cathode compartment connection regulator can also provide heat and after cooling it can be sequestrated and used as a nutrient for production of algal biomass. CO_2 from the MCFC will be clean enough so that it can also be used for other beneficial purposes, even for making dry ice. Cooling the exhaust gas to less than 100°C will change the water phase from gas to liquid that can be easily separated and released heat, including the condensation heat can produce more electricity.

Apparently, the continuing research and development of hydrogen fuel cells is rapidly progressing, and large capacity and more efficient hydrogen fuel cell power plants fueled by syngas or natural gas are becoming available, as pointed out in Chapter 8.

Municipal solid waste landfilling is considered in Alternatives I and II. As presented in Chapter 3, Section 3.2 ("Zero Solid Waste to Landfill Goal and Footprint"), the GHG CO_{eq} emitted from landfill depends on (1) C sequestered in landfills, (2) C in CO_2 from waste decomposition, (3) C in CO_2 from collected and combusted CH_4, (4) C in CO_2 from CH_4 oxidation, and (5) C in CH_4 emitted into the atmosphere. Before 1975, landfills were not managed for minimization of GHG emissions nor for methane capture for energy. It was assumed that the Alternative II landfills were managed and in a portion of landfills methane was collected and converted to energy. However, in managed landfills the efficiency of methane capture has been between 60 and 90%. The captured biogas was either used for production of energy (650 landfills with a capacity of 2160 MW) or flared. Out of 232 million tons of MSW collected in 2016, 52.6% was deposited in managed landfills.

10.4 WATER–ENERGY NEXUS AND RESOURCE RECOVERY OF THREE ALTERNATIVE DESIGNS

Three alternative designs will be assessed and evaluated in the following section, which is an update of previous evaluations in Novotny (2012, 2013, 2015). However, significant progress has been made in the last five years; therefore, the evaluation must be periodically updated with the new information. The alternatives are as follows.

Three Alternatives

Alternative I. Typical community before the end of the last century (1975 level) representing average US households/communities with lawn sprinkler irrigation water demand, practising no water and energy conservation and discharging wastewater into a conventional sewer system connected to an activated sludge aerobic treatment plant with nitrification/denitrification that deposits residual sludge on land or in a landfill. Based on the AWWA-RF study (Table 5.2), water demand is 550 L capita^{-1} day^{-1}, of which 313 L capita^{-1} day^{-1} is for outdoor irrigation. Heated water was assumed to be 106 L capita^{-1} day^{-1} and there was no heat energy or resource recovery. In the second half of the last century only a small fraction of municipal wastewater treatment plants used methane from digesting sludge solids for internal energy uses. Co-digestion and coprocessing of biodegradable and combustible MSW was not practiced and was not included in the analysis of this alternative. MSW solids were landfilled. No landfills at that time recovered methane and only a fraction of landfills flared the methane emissions. Hence, the landfill methane emissions were considered in the emissions as GHG black emissions.

Alternative II. *Current to near future (2018 to 10 years ahead) environmentally conscious community* with households practicing indoor water conservation and outdoor xeriscape planting with minimal irrigation demand located in a cluster that has a capability to reclaim some used (tap-the-sewer) water and reuse it for toilet flushing and limited irrigation of public green spaces. Water conservation reduces the total indoor water demand to 140 L capita^{-1} day^{-1}, similar to the typical demand in Europe and Japan. Treatment and reuse of a portion of used water on the district level for toilet flushing could bring water demand further down to less than 120 L capita^{-1} day^{-1}. About 60 L capita^{-1} day^{-1} is reclaimed from the sanitary sewer and biologically treated (without nutrient removal) in satellite district plants for toilet flushing (20 L capita^{-1} day^{-1}) by a membrane bioreactor and rest, followed by micro-filtration and ozonation, for irrigation. Adding 20 L capita^{-1} day^{-1} of recycled water will reduce the grid water demand to 120 L capita^{-1} day^{-1} and sanitary sewer flow from the house to 100 L capita^{-1} day^{-1}. Providing treated effluent for irrigation and toilet flushing would require another (purple) water supply pipe for recycled non-potable water. The arrangement of the system is like that featured in Figure 5.8, except that in 2015 anaerobic treatment was generally not utilized and most MBR small (package) used water recovery plants were aerobic. This alternative needs a district-level separate piping for recycled toilet flushing water, storage, and a pump for the portion of flow treated in the district with a pressure tank for delivering reclaimed water to the toilets but does not separate black and gray used water flows.

The reject water from the district treatment with removed solids is diverted to the sanitary sewer and delivered to a regional aerobic activated sludge "water reclamation (treatment) plant" with nitrification and denitrification and phosphorus removal. This plant produces methane by anaerobically digesting sludge from the treatment and some additional municipal biodegradable municipal solid and liquid waste solids (about 0.25 kg capita^{-1} day^{-1} = 50% of daily biodegradable waste) and uses the biogas to run turbine for producing electricity and heat for heating of the digester and buildings.

Some combustible MSW is regionally processed by incineration that could also accept residual solids from the used water treatment facility or these solids can be deposited on land. However, only existing incinerators could be considered in the US with the outlook that they may be soon decommissioned. Because at the end of the second decade of this century solid waste collection and disposal was typically communal and not regional, an assumption was made that only 13% of MSW were incinerated; hence, there was a possibility to recover some energy in the form of hot steam for heating and electricity. The rest of MSW was diverted to landfills with flaring the landfill gas. While incineration can, with a lower efficiency, produce electricity that would replace energy from fossil fuel power plants, it emits CO_2 diluted by nitrogen and other gases, which would be very difficult to sequester. Incineration cannot process plastics and tires because that would result in emissions of toxins.

Landfills were actively receiving solid waste in quantities reduced by recycling and incineration. Methane collected in active and most likely in decommissioned landfills was flared although at the end of the second decade of the twenty-first century, there has been a push to produce energy from the collected methane and some landfills were already practicing waste to energy. However, realistically only about 60–80% of methane emitted by a landfill can be collected.

Alternative III. *A "virtual visionary 2040+sustainable community" with a hybrid distribution and used water collection systems* that on the cluster (ecoblock) level separates used water into a gray water (GW) recycle and black water (BW) flow, as shown in Figures 5.5 and 5.6. However, the system in Figure 5.5 tried to minimize the water demand from the grid to the absolute minimum that would be appropriate in areas of severe water shortage and drought emergencies such as the one that occurred in Cape Town, South Africa, in 2016–2018. This water recovery loop did not provide water for irrigation and ecological flow and assumed that a portion of bath and laundry water intake would be derived from the gray water recycle. The gray water loop with reuse (Figure 10.6) is an optimum system where the fresh water intake loop uses 85 L capita^{-1} day^{-1} from the grid but also provides more than 40 L capita^{-1} day^{-1} of treated gray water for irrigation and ecological flows.

The less than 100 L capita^{-1} day^{-1} water intake can be considered as desirable optimal water use in the Cities of the Future in water-stressed areas. It would provide 122 L capita^{-1} day^{-1} of fresh water inside the recycle loop, more if the leaks are controlled. Black water flow of approximately 50 L capita^{-1} day^{-1} with high COD and nutrient concentrations will be diverted to a regional IRRF for integrated coprocessing and energy and resource recovery, with the biodegradable and combustible MSW.

Volumetrically, this recycle system is similar to the "tap the sewer" system presented in Figure 5.8 and considered in the preceding Alternative II. However, the "tap-the sewer" system recycles a portion of the total used water flow that contains both black and gray water flow, while the above system recycles only highly treated gray water. In both systems, kitchen and bathroom potable water supply is fully provided from the potable water source; only laundry and shower will receive a mix of potable and warm recycled highly treated

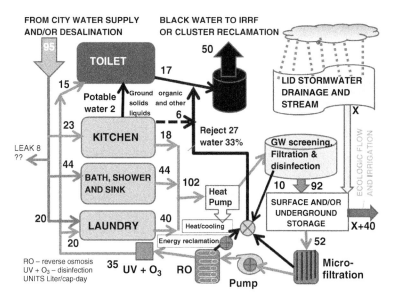

Figure 10.6. Mid-twenty-first century house/district water use and gray recycle loop for water stressed areas.

gray water. This warm potable quality recycle can be used for washing while the rinse cycle would use fresh water. Conceivably, low-capacity house RO filtration units easily installable under the laundry tub already available and affordable on the market could provide the final treatment for the laundry recycle. Which system will prevail in 2040 is not known. Figure 5.8 ("tap the sewer" system) could generate some energy in the district. The system in Figure 10.6 uses a small amount of energy for an RO unit requiring pressure of around 2 bar.

The BW flow includes water flow from toilet flushing and ground solids from the kitchen sink. It also receives reject water from the treatment of gray water and supplementary storm water. Stormwater should be added to the recovered gray water cycle as a makeup water to replace water lost in backwash and reject water of the filtration (including RO) units of the GWC, and for providing irrigation and ecological outdoor flow during dry weather growing period.

Admittedly, this is a visionary virtual concept still lacking prototype testing and parameter derivations, but if global warming reaches the dangerous levels forecasted by scientists, and available landfill sites disappear, the public will demand the triple net-zero urban living and water/solids and energy management.

The IRRF Alternative III presented in Figure 10.4 has three paths of processing the flow of liquids and solids (BC Ministry of Community Development 2009). The first path anaerobically treats the liquid used water and produces methane and less excess solids than a commensurable aerobic plant. It also accepts supernatant from the second path that recovers hydrogen by acid fermentation followed by microbial electrolysis fuel cell (MEC). The second path co-processes 0.56 kg capita^{-1} day^{-1} of high moisture biodegradable MSW solids with waste sludge from the anaerobic used water treatment. The IRRF also processes symbiotically by pyrolysis/gasification 1kg capita^{-1} day^{-1} of items that are difficult to biodegrade (waste lumber, woodchips) or nonbiodegradable (e.g., plastics, possibly tires) dry combustible MSW solids. Applying circular economy concepts, materials like textiles, paper, and carboard shall be recovered and recycled. Cardboard can be recycled several times. Tires can also be recycled and converted, for example, to construction materials or new tires. Plastics can be converted to many usable items. Currently, only 25% of plastic waste is recycled in Europe, less in the US. Recycling plastic uses 10–15% of the energy of producing new plastic made from the black energy primary sources natural gas, coal and oil that are also raw materials for plastic production.

This triple-path IRRF system produces methane from the liquid processing path, hydrogen from the biodegradable solids fermentation/MEC path, and syngas from gasification of nonbiodegradable combustible solids in the gasifier providing fuel to the IRRF power plant. The key central component of the power plant is the hydrogen fuel cell. Methane should first be steam reformed to syngas in a separate SMR unit at a temperature greater than 700°C; however, internal reformation in the anode compartment of the MCFC has also been considered and is possible. Syngas from the SMR and from the gasifier should be treated to remove sulphur and other impurities and processed by gas shift to more hydrogen and carbon dioxide. The mix of hydrogen and residual carbon dioxide from the acid fermentation and MEC cathode compartment may pass through the H_2–CO_2 separator if one is available. Thereafter hydrogen becomes the feed to the MCFC anode compartment. MCFC produces a lot of hot steam in the anode compartment, which will be used for methane and syngas reforming. It could also be used as a source of water and heat for gasification; however, enough water for indirect gasification may be provided by moisture in feed MSW and sludge solids. Water produced by MCFC should be very clean.

Phosphorus is recovered from the supernatant stream, which may also include Anammox reactors for nitrogen removal. Nutrients and phosphorus can also be recovered/removed from the effluent in polishing algae production facilities or submerged wetlands. Both produce additional biomass; however, this contribution is not quantitatively assessed in this analysis but was addressed in Chapter 9. Effluent polishing and nutrient removal by algae could be another opportunity to increase energy yield.

Unlike MSW incineration that cannot accept plastics and some other toxins producing combustible solids, these materials can be processed by indirect gasification to make syngas. Tires are questionable but future tire manufactures will be asked to develop recyclable and/or green power generating tires. It is expected that by 2035–2040 most if not all US MSW incineration facilities may be decommissioned and replaced by gasification, which is more energy efficient and does not emit flue gases with toxins, only syngas. The endothermal heat necessary for steam gasification will be provided by hydrogen and hot steam from MCFC. Produced syngas will be cleaned and steam reformed to hydrogen and carbon dioxide. It is expected that the alternative of redirecting the MSW from incineration to landfills will not be available. Therefore, in this zero–new landfill alternative at least 35% of MSW should be recycled (11% increase from the current recycling) and MSW load reduced by 10% or more by lightening materials or source reduction. This is realistic because some communities in the northeastern US (for example, Gloucester, Massachusetts; see Chapter 11) already recycle almost 40% of MSW and incinerate the rest because there are no landfills available nearby. The European Union requires that by 2035 only 10% of MSW will go to landfill and no landfill deposition by 2050 is expected.

Alternative III is visionary, at least partly, because at the time of this writing, the prototype hybrid acid fermentation/MEC unit processing biodegradable solids had not been built and tested in a full-scale format. But the concepts have been extensively tested in laboratories and are fundamentally sound and promising. Also, the entire integrated water/stormwater/MSW synergistic co-processing has not been implemented; therefore, Alternative III is a "2040–2050 vision" and not yet reality. It has not been fully tested and hence the author is not liable for any malfunctions if implemented without extensive research and testing. Other integrated systems may emerge, and other scientists and visionaries will propose and evaluate their ideas. Some proposed systems (e.g., plasma processing of solid waste) are still in the laboratory stage of preliminary pilot testing and making electricity directly in the anaerobic microbial fuel cells is still far from being economical.

This alternative would be a full triple or better than (= negative CO_2 emissions) net-zero system that could produce excess energy in the form of biogas, biofuel, and, above all, electricity and could recover resources such as nutrients, soil conditioners, and even hot concentrated carbon dioxide and, potentially, rare metals.

Inputs to the Analyses

The inputs and assumptions for the thee alternative analyses are shown in Table 10.1 and the outcomes of the spreadsheet assessment model analysis are graphically and tabularly presented at the end of this chapter. The spreadsheet calculation focused on evaluation of the three alternatives of water/energy/solid water management scenarios, considering the three net-zero assessment alternatives, which are: (1) no waste of water, (2) net-zero energy/carbon emissions, and (3) no new landfills. Regarding the water goal, the alternatives assessed the water demand and water conservation.

In Alternatives II and III, the calculations focused not only on water and electric energy; the balance was also carried out for heat, starting with home water heating but not space heating and cooling. Recuperation of heat from gray water and from black water was considered only in Alternative III calculations. Heat energy drives the TNZ Alternative III and the system must be heat energy positive (i.e., no net external energy sources should be required); therefore, the auxiliary energy sources must be less than energy produced by the system. This differentiates the Alternatives III and perhaps II from the current systems requiring auxiliary fossil fuel energy sources of electricity and natural gas for heating digesters and running blowers and pumps.

Regarding net-zero (equivalent) carbon emissions, the alternatives assessed carbon and methane energy uses and *waste-to-energy* options related not only to water but, in an integrated concept, also to disposal and processing the municipal solid waste (MSW). This integrated approach is important because MSW, when put in landfills, produces methane and smaller volumes of carbon dioxide. Landfills also discharge highly polluted liquid leachate. For graphic presentation, energy use and gains and carbon emissions were assigned colors defined in Table 4.1.

Blue energy includes produced energy that emits only small amounts of CO_2 when compared to energy produced or recuperated or does not emit any CO_2 nor any other GHG. This category includes sensible heat energy recuperated from warm gray water or from hydrogen energy produced in the MCFC under an assumption that heat energy from the produced hot concentrated CO_2 can be recuperated and a great portion of CO_2 can be sequestered or reused. GHG free energy to run the IRRF units provided by solar and wind power harnessed on the premises would also be classified as blue energy.

Green energy includes energy produced from methane by digestion of sludge or gasification or incineration of MSW or from collected methane in landfills that is not converted to hydrogen and emits CO_2, which is not counted as GHG and energy could be partially recuperated from heat generated in the production of electric energy. Emitted CO_2 is included in the overall balance as non-GHG.

Gray energy is energy from fossil natural gas (not oil or coal) used mainly for heating, which emits CO_2, but a portion of the heating energy can be recovered.

Table 10.1. Inputs into analysis of alternatives of urban water and energy management.

Parameter	Value				Source and Comment
COD load g/(capita-day)	130				Table 5.3; Henze and Comeau, (2008)
COD in suspended solids removed in primary clarifier g/(capita-day)	52				$40 * 130/100 = 52$ Figure 7.3
Energy for water services kW-h/m^3	1.87				Chapter 3, "Water Footprint"
Energy use of a traditional water reclamation plant without and with nitrification and aerobic microbial membrane reactor (MR) kW-hour/m^3	Population >50 000 25 000 <10 000	Without Nitrification 0.28 0.38 0.55	With Nitrification 0.42 0.5 0.7	MR 0.63 0.72 0.83	Metcalf and Eddy (2007); Novotny et al., (2010a)
Energy use kW-hour/m^3 microfilter (MF) reverse osmosis (RO)	MF 0.6	RO 2.0			Metcalf and Eddy (2007); see also Novotny et al. (2010a)
Energy use for ozonation of micro-filtration effluent kW-hour/ m^3	0.18				Ozone produced from air; Metcalf and Eddy (2007)
Emissions of kg CO_2 per kg COD removed in aerobic treatment plants	0.14–0.2				Metcalf and Eddy (2007)
Emissions of kg CO_2 per kg COD removed in aerobic treatment plants	extended aeration 1.5				Keller and Hartley (2003)
Digester methane production L GH_4/g COD removed	0.3				Table 7.1
Methane energy kW-h/m^3 of CH_4	9.8				Table 4.2
Heat pump efficiency coefficient of performance COP	4				COP = Energy acquired/energy applied to recuperate heat
COD/organic solids ratio kg COD/kg VSS	1.41				Metcalf and Eddy (2007)

(Continued)

Table 10.1. (Continued)

Parameter	Value	Source and Comment
COD of municipal solid waste kg COD/kg ds MSW	1.55	Chapter 8, "Gasification of Digested Residual Used Water Solids with MSW"
Incinerator efficiency to produce electricity and heat from MSW, %	20% electricity 12% with pollution controls 55% heat	Murphy and McKeogh (2004), without considering 2015 levels of air pollution control, with controls per USEPA (2016)
Electricity and heat from methane burning in a small turbine without a recuperator	23% electricity 63% heat	Table 8.3
Specific density of methane	$0.72 \, kg/m^3$ at $0°C$ $0.62 \, kg/m^3$ at $20°C$	Table 4.2 www .engineeringtoolbox.com
Theoretical incineration reaction heat yield kJoule/mole of OC	394	Table 8.1
Energy to run BW anaerobic treatment plant	$0.06 \, kW\text{-}h/m^3$	Table 7.2, per McCarty et al. (2011)
Methane production from COD by anaerobic digestion	0.2–0.4 L of CH_4/g of COD removed	Metcalf and Eddy (2007)
Energy for heating anaerobic digester of biogas depending on climatic region	1.94 to 3.05 $kW\text{-}h/m^3$	USEPA-CHPP (2011)
Theoretical H_2 recovery in fermentation/MEC process	0.19 kg H_2/kg VSS = 0.125 kg H_2/kg COD	90% of food waste and sludge potential yield
Molar formula for MSW and sludge	MSW: $C_6H_{10}O_4$ Sludge: $C_6H_{7.4}O_{2.8}$	Chapter 8, "Hydrogen Yield Potential by Indirect Gasification"
Heating energy to dry and convert to steam moisture of MSW with 50% water content	0.36 kW-h/kg MSW	Chapter 8, "Incineration"
Energy value of syngas from gasification	7 kW-h/kg	per Valkenberg et al. (2008)
Endothermic heat of gasification to be subtracted from the energy yield of combustible MSW	1.45 kW-h/kg	Gas. heat = 40 (AW_{OC}/kg) ∗ 131 (kJ/AW_{OC}) = 1.45 (KW-h/kg)

Table 10.1. (*Continued*)

Parameter	Value	Source and Comment
Endothermic heat loss in steam methane reforming	206 kJ/mole CH_4	Table 8.1
Exothermic heat gain from steam shift of CO to CO_2 hydrogen	41 kJ/mole CO	Table 8.1
Energy value of H_2	39.4 kW-h/kg H_2	Table 4.2
H_2 yield from gasification of combustible MSW to syngas	0.103 kg H_2/ kg	Chapter 8, "Hydrogen Yield Potential by Indirect Gasification"
Overall H_2 yield from combustible MSW	0.2 kg H_2/ kg	Chapter 8, "Hydrogen Yield Potential by Indirect Gasification"
Water requirement to process combustible MSW to syngas and hydrogen	0.35 L/kg MSW for syngas and 1.1 L /kg MSW for H_2	Chapter 8, "Hydrogen Yield Potential by Indirect Gasification"
Electric energy from MSW incineration per US EPA, considering also use of electricity for air pollution controls	$2720 * 365 * 24/28000000 =$ 0.851 MW-h/ton (kW-h/kg)	US EPA (2016); 2,720 MW-h power capacity per 28,000,000 tons MSW incinerated in a year, which is 12% of the energy value of MSW
Water use by SMR to convert CH_4 to syngas	$36/16 = 2.25$ g H_2O/g CH_4	2 moles H_2O/mole CH_4 Table 8.1
Water use to steam reform CO to H_2 & CO_2	$18/28 = 0.64$ g H_2O/g CO	1 mole H_2O/mole CO Table 8.1
Water produced in MCFC from H_2	$18/2 = 9$ g H_2O/g H_2	1 mole H_2O/ mole H_2 entering MCFC
Specific density of dry NYC MSW	1600 kg/m^3	Table 8.2
Specific heat of carbon dioxide gas, C	1.16 kJ/(kg–˚C) at 700°C	www.engineeringyoolbox.com
Landfill annual GHG emissions	1.1 kg GHGCO$_{2eq}$/Kg ds MSW	Chapter 3, "Zero Solid Waste to Landfill Goal and Footprint"

Black energy is the energy provided by fossil fuel power plants for running the blowers, pumps, lights, and so on, which is not recoverable. Energy provided by fossil fuel power plants that do not sequester GHG (almost all in 2015) is categorized as GHG energy and represents black energy use. Methane emitted from landfills is also listed as black emissions but not CO_2 emitted when landfill methane is collected and flared.

Unlike most previous analyses of water and energy uses related to water, this analysis also included estimates of water heating energy for households and in Alternatives II and

III also energy for heating of digestion reactors. MSW is considered a source of energy and a resource. Recuperation of heating energy was included in Alternative III. Inputs and variables for calculations are presented in Table 10.1.

All calculations were performed or related to per capita-day denominators. This enables quick estimates at any urban scale, although there are differences between large (regional) and smaller urban areas; in other words, the effect of the "economy of scale" cannot be fully overlooked. The used energy and carbon emissions were listed in the summary tables as positives and energy reduction and carbon credits as negatives.

CO_2/Kw-h Ratio for the Alternatives

The analyses and calculations were performed for an "average" US region. In calculating the effects of the alternative on carbon emissions it is necessary to find proper past, present, and future values of the ratio of carbon emissions in kg of $CO_{2\text{-equiv.}}$/kW-h of energy, either produced from the various fuels or emissions avoided and credited for green and blue energy produced by the systems. These estimates are different for the three alternatives.

Alternative I, around 1975. All energy provided by power plants was black energy. In the US, hydropower was less than 7% and was mostly limited to the Pacific West and the Tennessee Valley Authority operated several hydropower plants along with fossil fuel and one nuclear power plant. Nuclear power use was just beginning. Most of electric power in the US was provided by coal and the average CO_2 emissions therein were 0.95 kg CO_2/kW-hour. Water heating was mostly electric in rural areas and natural gas in the cities. CO_2 emission from natural gas heaters is 0.20 kg CO_2/kW-hour in fuel; hence the 50/50 split is 0.57 kg CO_2/kW-hour.

In 1975 few landfills were engineeringly managed with the goal of reducing landfill GHG emissions or producing energy. Emitted landfill biogas was not flared. $GHGCO_{2eq}$ emissions entered in the analysis were 1.1 kg $GHGCO_{2eq}$/kg ds MSW deposited in the landfill; 8% of the MSW was recycled. GHG emissions were increased by 20% to account for MSW collection and transportation to the landfill, tipping, and leachate pumping.

Alternative II, 2015. In Chapter 1 the GHG emission for power generation were presented (US DOE, 2016c) and proportions are shown in Figure 10.7.

	Kg of CO_2/kW-hour
Coal fired	0.96
Oil fired	0.89
Natural gas	0.60

About 30% of the energy was produced by power plants that do not emit CO_2. This resulted in the weighted average of the CO_2 emissions from power and credits for green-blue energy production for Alternative II as 0.61 kg of CO_2/kW-hour. For comparative purposes, in the same period reliance on black energy in France, Italy, and Austria was less than 10% and even less in Iceland. In the middle of the second decade of this century, US DOE (2016c) revised the proportions of energy and CO_2 emissions from power industry as follows:

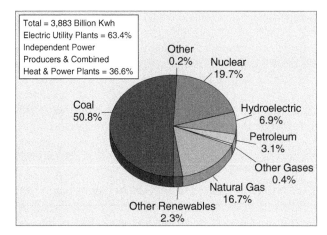

Figure 10.7. US electric power generation sources in the first decade of the 2000s. *Source:* U.S. Department of Energy, Energy Information Administration (2000).

Energy Source	Energy Production %	CO$_2$ Emissions %
Coal	31.6	70.9
Natural gas	31.6	27.5
Nuclear	20.3	–
Hydro	6.8	–
Other renewable	6.8	–
Other, including oil	0.8	1.6

Heating CO$_2$ emission split between electric and natural gas heaters for domestic water heaters was assumed to be 50/50 as in Alternative I; therefore, it was adjusted for the lower emissions by the electric power plants. Hence, the GHG emissions of domestic water heaters was 0.41 kg CO$_2$/kW-h. Heating of digester reactors was by the methane produced by the utilities and resulting in emissions of 0.2 kg CO$_2$/kW-h is not counted as GHG.

Alternative III, 2040+. Ascertaining the GHG emissions from the integrated urban water/used water/solid waste (MSW) system 25–40 years into the future is a difficult challenge. Water and solid waste management does not have a tradition of fast changes; changing complex heavy infrastructure takes time. However, most machinery and infrastructure have a lifetime of around 25 years and they should be repaid in that period and replaced thereafter. Therefore, Alternative III is a vision of what could and should be done to continue the progress and make these systems triple net-zero with no claims that all proposed changes will occur. But the pressures to change will be strong because the economies of green and blue power production and energy are rapidly changing and forecasted climatic changes cannot be ignored and are already occurring and BAU development would be catastrophic after 2050. The cost of power from photovoltaics and wind is rapidly dropping and is expected worldwide to be cheaper than fossil fuel energy by 2025, especially under the Paris 21 Agreement. On the flip side there is a chance that the US government attitude toward renewable GHG-free green power may be uncertain as far as until 2025. However, by 2025 France will be carbon neutral in the power industry (Carre, 2014) and many other countries will follow. Note that in 2019 Germany reported their 2018 green energy production totaled that produced from fossil fuels. Saudi Arabia and the Emirates are heavily investing in solar power, anticipating that the fossil oil energy demand will be decreasing and solar will be more profitable. Nevertheless, Saudi Arabia, the US, Kuwait, and Russia did not endorse the IPCC (2018) report.

The International Energy Agency (IEA) each year attempts to make forecast the worldwide energy situation. IEA (2016) projected that that under the optimistic scenario 60% of the power generated in 2040 will come from renewables, almost half of this from wind and solar PVs and many remaining black energy power plants may be sequestering a significant portion of their flue CO_2 gas. Hence, the power sector will be largely decarbonized in this scenario and the average GHG emissions from electricity generation per IEA could drop to 0.08 kg of CO_2/kW-h in 2040 worldwide, compared to 0.62 kg CO_2/kW-h in the US in 2010. To estimate the CO_2 emissions in 2040 and credits for the 2040+ integrated water/used water/MSW management systems, the following conservative but realistic assumptions were made for CO_2 emissions from power sources in Alternative III:

	% of Power	Kg of CO_2/kW-h	
Coal and oil	15	0.6[+]	(GHG)
Natural gas	25	0.4[+] (0.2)[++]	(GHG)
Hydro and nuclear	20	0	
Photovoltaics and wind	30	0	
Green (waste to energy, H_2)	10	0.2	(Not GHG)
Weighted average		0.21	

[+]Assumed 30% CO_2 sequestration.
[++]CO_2 from heating.

Heating by electric and natural gas heaters will result in 0.2 kg CO_2/kW-h emissions. Because of the marginal cost of electricity from the above sources in the 2040 economy, the CO_2 credit for blue and green energy excess will be applied to coal and natural gas emissions, which will be 0.5 kg CO_2/kW-h. The resulting balances of electricity and emissions are shown in Figures 10.8 and 10.9.

Note that in 2017, power generation in Massachusetts and Rhode Island were coal and oil free but not yet carbon neutral, followed thereafter by Washington State, Oregon, California, Iowa, and New York. Many European countries will also be soon coal free by that time and China's reliance on coal, oil, and natural gas will be significantly reduced.

Discussion and Results

Alternative I.   Alternative I reflects water and waste management in the middle of the second half of the previous century (1975). During that period, anaerobic sludge digestion and use of produced methane were not practiced. Wastewater was collected by separated storm and sanitary sewers and some historic sections of cities had combined sewers. In the US, urban stormwater was just classified as a point source of pollution that had to be conveyed from the area and with some treatment conveyed to the receiving waters. Some cities were planning to capture and store a portion of the combined sewer overflows in deep tunnels (Chicago and Milwaukee) and treat them in the regional water reclamation (wastewater treatment) plants, but large tunnels were not completed until after 1990. Sludge was dried and was either deposited in landfills, plowed into nonagricultural soils, or used as a fertilizer.

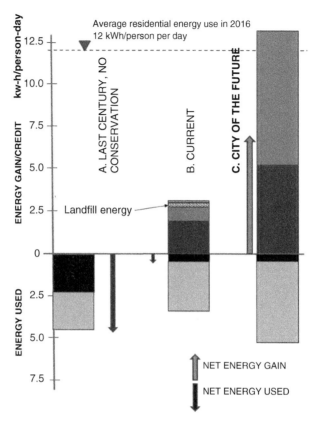

Figure 10.8. Energy balance between black and gray energy uses and green and blue energy gains in kW-hours capita^{-1} day^{-1} for three assessed alternatives.

Landfills were used for deposition of MSW, without collecting methane, which was vented but not flared or used for energy production.

As expected, Alternative I was greatly unsustainable because there were no mechanisms and means in place by which water and energy uses could be reduced, and resources recuperated and reused. The results described in calculation summaries and in Figures 10.8 and 10.9 are presented for black and gray energy use, and equivalent carbon dioxide emissions. Landfill emissions are also considered in the analysis but no "waste-to-energy" programs were implemented. Electric energy demand and use was divided between black energy provided by coal-fired power plants and gray energy used for heating. Some green (nuclear) and blue (hydro) energy was generated.

In the analysis, the per capita house water demand of 550 L capita^{-1} day^{-1} was divided between in-house water use of 237 L capita^{-1} day^{-1} and outdoor irrigation of 313 L capita^{-1} day^{-1}. Outdoor energy for irrigation included pumping irrigation water and an allowance for lawn mowers. The black energy use for delivering water was 1.87 kW-h/m^3, which resulted in 1.05 kW-h capita^{-1} day^{-1} and CO_2 emissions of 1.08 kg CO_2 capita^{-1} day^{-1}. The largest indoor water/energy nexus use was for water heating. Total heated water use was 120 L capita^{-1} day^{-1} that was heated by a water heater by 25°C to be used in laundry, bathing and showering, and for the dishwasher, which required 3.9 kW-h capita^{-1} day^{-1} 50% black and 50% gray energy use with corresponding black emission 1.07 kg CO_2 capita^{-1}

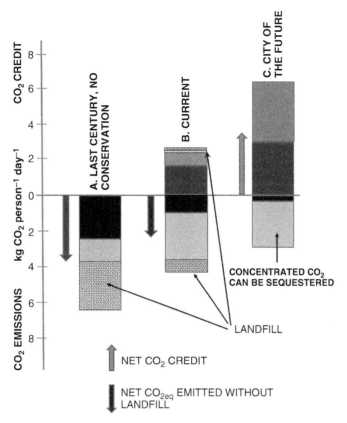

Figure 10.9. Balance of black and green carbon emissions and green and blue energy credits in kg of CO_{2-equv} capita^{-1} day^{-1}. Carbon dioxide equivalents from landfills are prorated averages of the US landfills in 1975 and in the first decade of this century.

day^{-1} and the same gray emissions. Water heating was by natural gas or electricity; however, the heat was not recovered. Nevertheless, heating could be counted as borderline black energy use emitting 0.6 kg CO_2/kW-h energy, 50% for methane and 50% for fossil coal energy. It should also be pointed out that water heater efficiency in the late twentieth century was not as good as it is now.

The used water from houses was conveyed by sanitary and/or in some cities by combined sewers to the regional water reclamation plant. It was assumed that infiltration-inflow (I-I) increased the influent flow 25% to 296 L capita^{-1} day^{-1}. Pumping in sewer used 0.03 kW-h capita^{-1} day^{-1} and resulted in black emissions 0.03 kg CO_2 capita^{-1} day^{-1}. The wastewater treatment plant (very little water was reclaimed during that period) consisted of traditional activated sludge treatment units with nitrification but no denitrification. The COD load to the plant (in all calculations) was 130 g of COD capita^{-1} day^{-1} and the resulting treatment energy use was 0.124 kW-h capita^{-1} day^{-1} and black CO_2 emissions were 0.117 kg CO_2 capita^{-1} day^{-1}. The total energy use related to water was then 5.12 kW-h capita^{-1} day^{-1} and corresponding CO_2 emissions were 3.41 kg CO_2 capita^{-1} day^{-1}. No water or energy were recuperated. Digester heating was not included in the analysis.

In the integrated analysis concept, the impact of landfilling cannot be overlooked. In this early baseline period of the late twentieth century, methane from landfills was mostly vented, not collected and converted to energy or flared. Figure 3.3 shows that building MSW incinerators in the US started around 1980 but no large new incinerators were built after 1995. In 1975 the mass of MSW deposited in landfills was 121 Mt, of which 8% was recycled. Population according to US Census Bureau was 216 million, which gives the daily MSW mass deposited into landfills as 1.41kg MSW capita^{-1} day^{-1} at that time. Total annual per capita GHGCO$_{2eq}$ emissions entered in the balance was 1.1 kg GHGCO$_{2eq}$ per kg of dry MSW, which was increased by 20% to account for trainspotting, tipping, and disposing MSW and leachate pumping. This resulted in black GHGCO$_{2eq}$ emission of 2.5 kg GHGCO$_{2eq}$ capita^{-1} day^{-1}. No biogas flaring or energy collection was practiced at that time. Table 10.2 presents the summary results.

Alternative II. Alternative II is an above-average 2015+ system that on district level withdraws and treat "fit for reuse" sewage for nonpotable in-house use (toilet flushing, possibly part of laundry) and outdoor irrigation. Current advanced used water reclamation (treatment) plants implemented some energy-from-waste recovery by solids incineration and flaring the rest of the methane from the landfills. Few landfills (12.7% in 2015) collected methane and converted it to energy by combustion. The use of incinerators for energy-from-waste systems in the US is limited by environmental regulations and very few installations were implemented after 1995. As reported in this manual, Sweden today incinerates 50% of the country combustible solid waste and even accepts, for a fee, combustible MSW from other countries. Similarly, until 2017, China was the largest importer of MSW from the US and other countries, used as resource by recycling and production of energy by incineration. However, most incinerators therein were emitting large amounts of GHGs, aerosols, and other air pollution. Consequently, after 2018, China drastically reduced the MSW import.

By reducing the water use from the water supply grid or source to 120 L capita^{-1} day^{-1}, the energy for providing water and corresponding black CO$_2$ emissions were reduced by almost 80% in comparison to the unsustainable amount of the previous century Alternative I. Implementing conservation on hot water use appurtenances, heating energy and using marginally green (natural gas) energy for heating, gray CO$_2$ emissions were reduced by 38%. Use of black energy for water supply was reduced to 0.24 kW-hour capita^{-1} day^{-1} with corresponding 0.14 of black emissions. Water heating is a large domestic energy. 75 L capita^{-1} day^{-1} was heated to elevate water temperature by 25°C by a water heater that used 2.43 kW-hour capita^{-1} day^{-1} of gray energy and corresponding CO$_2$ emissions of 1.38 kg CO$_2$ capita^{-1} day^{-1}. No house heat recovery from water was assumed but some heat was recovered in the regional IRRF facility as green energy.

Table 10.2. Summary results for Alternative I.

Energy	Energy Used or Credit kW-h capita^{-1}day^{-1}	Carbon Emissions Kg CO$_2$ capita^{-1}day^{-1}
Black (electricity from power grid)	2.28	2.34
Gray (heating by natural gas)	1.96	1.08
Landfill emissions of GHGCO$_{2eq}$		2.5
Total without landfill	4.24	3.42
Total with landfill GHGCO$_{2eq}$		5.46

60 L capita^{-1} day^{-1} of used water flow were assumed in this example to be taken from the sanitary sewer and treated fit-for-reuse in the district by a small manufactured aerobic biologic membrane treatment process unit with nitrification (not denitrification), similar to extended aeration, and by disinfection. The reason is that the recuperated water was for irrigation and toilet flushing and removing nutrients would not make sense. The treatment required 0.073 kw-h cap^{-1} day^{-1} of likely black energy and was responsible for 0.04 kg capita^{-1} day^{-1} of black CO_2 emissions. No energy recovery was performed at the district simply because in 2015 it was not practiced in the US. In this advanced system, 20 L capita^{-1} day^{-1} received additional treatment by microfiltration and ozonation (odor control and more disinfection) and recycled by pumping back to the houses in the district for toilet flushing, which increased the indoor water used to 140 L capita^{-1} day^{-1}.

80 L capita^{-1} day^{-1} were conveyed and pumped (lifted) in the sanitary sewer to the regional IRRF. The flow was increased by 25% to account for primary sludge and reject filtrates from the district water reclamation plant and for infiltration into sewer. The IRRF inflow was 100 L capita^{-1} day^{-1}. Pumping of used water in the sewers used an estimated 0.01 kW-h capita^{-1} day^{-1} of black energy. The assumed unit COD load was 130 g COD capita^{-1} day^{-1}, which was reduced by 95% by load treating and removing the COD load in the district water reclamation facility by $130 * 0.95 * 60/140 = 53$ g COD capita^{-1} day^{-1}. Of the 53 g of COD capita^{-1} day^{-1}, 65% (Figure 7.3) was incorporated into primary and secondary sludge that was sent to the reginal IRRF facility. Hence, the total IRRF COD load was $77 + 0.65 * 53 = 111.5$ g COD capita^{-1} day^{-1}, which resulted in IRRF influent COD concentration of 1115 mg/L.

The IRRF influent COD concentration is high and normally it would call for anaerobic rather than aerobic treatment or, at minimum, anaerobic pretreatment. However, in 2015 in the US and even in the most progressive water/energy nexus, the Republic of Singapore, there were no large-scale municipal anaerobic water reclamation plants, and aerobic bioreactors with membrane treatment were at the top of the practices used for regional treatment of used water. Hence, it was assumed that 55% of the incoming COD was removed by enhanced primary treatment and remaining COD was treated by Bardenpho anoxic/oxic completely mixed MBRs, which produce less sludge than the conventional activated sludge units. This process reduces incoming COD concentration in the anoxic reactor (without emitting methane) and the COD concentration in the influent to the oxic reactor could be reduced below 400 mg/L. The black energy use in the liquid treating Bardenpho treatment plant was estimated as 0.1 kW-h capita^{-1} day^{-1} and black CO_2 emissions of 0.06 CO_2kg capita^{-1} day^{-1}.

Because all removed sludge in the district water reclamation plant was sent to the IRRF for processing, the overall sludge COD yield was conservatively estimated as 65% of the total COD load which is $0.65 * 130 = 84.5$ g COD capita^{-1} day^{-1}. The sludge is processed by (a generic) mesophilic anaerobic digestion requiring heating. Figure 7.3b indicated that about 80% of the incoming COD in an anaerobic generic process is converted to methane., which corresponds to 0.3 L of CH_4/g of COD (Van Lier et al., 2008). The calculated gas yield from used water sludge was 20.2 L capita^{-1} day^{-1} and the energy yield and CO_2 emissions credit of 0.2 kW-h capita^{-1} day^{-1}.

However, in 2015 co-digestion of high-COD-concentration liquids and organic matter, mainly food waste and airport glycols, was already implemented by several regional utilities in the US and many more in Europe. After 2010, Metropolitan Milwaukee (WI) Sewerage District began to co-digest its primary and secondary sludge with food and airport deicing liquids to double its methane production and then added landfill methane to produce more energy. By converting landfill gas into energy, the district can produce most of its energy needed during dry weather. Therefore, to asses this contribution it was assumed in this analysis that 50% of biodegradable MSW or 0.25 kg capita^{-1} day^{-1} was co-processed in the anaerobic digester(s). This yielded additional 62.5 L capita^{-1} day^{-1} of biogas and 0.61 kW-h capita^{-1} day^{-1} of energy potential. The total biogas volume was 82.8 L capita^{-1} day^{-1} and energy potential was $0.20 + 0.61 = 0.81$ kW-h capita^{-1} day^{-1}. The data in Table 8.3 for small turbine energy production show that 25% can be converted to electric energy, which resulted in 0.25 kW-h capita^{-1} day^{-1} and the CO_2 credit for substituting black energy use with green energy production from COD was 0.15 kg CO_2 capita^{-1} day^{-1}. 63% of the energy potential could be recovered as heat but the heat for keeping the digester mesophilic must be subtracted, which resulted in net heat yield of 0.32 kW-h capita^{-1} day^{-1} and CO_2 credit of 0.13 kg capita^{-1} day^{-1}. 0.24 kg CO_2 capita^{-1} day^{-1} emitted from the turbine burning the biogas reduces the CO_2 credit to 0.15 kg CO_2 capita^{-1} day^{-1}.

Energy production by incineration is included in the analysis but with restricting assumptions. First, unlike Sweden, which incinerates and recycles 99% of its municipal solid waste, in 2015 the US incinerated only 12.7% of the total MSW collection, 17% when recycled MSW are subtracted. In 2015, 71 MSW incineration sites claimed to be energy-to-waste facilities with a total generating capacity of 2.3 gigawatts. Hence, since this analysis should represent an average US region, it was assumed that 13% of the total or about 0.27 kg of MSW capita^{-1} day^{-1} was incinerated in the "representative" region served by the IRRF. Digested solids generated by the digester from used water suspended solids and those from digested biodegradable MSW solids were then added to that amount. It was assumed that 50% of the solids entering the digestion process were removed in the digestion process. Heat energy to dry the solids was estimated, along with the organic (combustible) carbon. The solids are dried before incinerating, which required heating energy of 2.5 Watt-h/liter of biogas or 0.18 kW-h capita^{-1} day^{-1} of heat. Because the energy potential by incineration in Table 8.1 is given as 494 kJoule/mole of OC, the organic carbon content of solids to be incinerated was estimated as 52% of the dw of the solids. The calculated energy potential of the burned solids was then 2.53 kW-h capita^{-1} day^{-1}.

Both electricity (blue energy) and heat (green energy) credits were then estimated. The electricity production by energy from incineration was assumed to be 17% of the energy potential or 0.43 kW-h capita^{-1} day^{-1} and recuperated heat was assumed to be 55% of the energy potential minus energy for drying of solids, which was 1.57 kW-h capita^{-1} day^{-1}. Incineration also produces flue gas containing CO_2 (unrecoverable at that time), estimated as 0.81 kg CO_2 capita^{-1} day^{-1}. Overall, incineration added 2 kW-h capita^{-1} day^{-1} of blue and green energy and provided an energy CO_{2eq} credit of 1.07 kg CO_2 capita^{-1} day^{-1}. It is again interesting to compare the electric production from MSW incineration with that reported by the US EPA for the year 2015, which was 2163 MW for the year. If that energy value is converted to kW-h per day and divided by population (315 million), the result would be 0.451 kW-h capita^{-1} day^{-1} compared to 0.43 kW-h capita^{-1} day^{-1} calculated in this assessment.

In the US at the end of the first decade of this century, the energy potential and GHG emissions related to landfilling were lagging most of the advanced European countries, Japan, and Singapore. In the middle of the second decade of the 2000s 238 Mt (2 kg capita^{-1} day^{-1}) of MSW were produced in the US (Chapter 3, "Zero Solid Waste to Landfill Goal and Footprint") and 52% of that mass (Figures 3.2 and 3.3) or

124 MT $\approx$ 1.06 kg MSW capita^{-1} day^{-1} was deposited in landfills; 42% of landfills collected and combusted landfill biogas for electricity and 11% collected and flared the LFG. The remining 58% of US MSW landfills let the LFG seep from the landfills into the atmosphere. The efficiency of capturing the LFG is between 60 to 90%; 75% was assumed in the assessment calculation.

The biogas per capita production capture potential and GHGs were calculated for the total mass of deposited MSW. $CH_4 \approx 0.04 * 1.06 = 0.042$ kg CH_4 capita^{-1} day^{-1}, $CO_2 \approx 0.1 * 1.06 = 0/106$ kg CO_2 capita^{-1} day^{-1}, and GHG $CO_{eq} \approx 1.162$ kg capita^{-1} day^{-1}. Because 25% of landfill gas is not captured in all landfills, the emission of escaping gases was 0.011 kg CH_4 kg capita^{-1} day^{-1}, 0.026 kg CO_2 capita^{-1} day^{-1}, and 0.29 kg capita^{-1} day^{-1} of GHG CO_{eq}.

Energy production potential was calculated for 31% of electricity-producing landfills by multiplying energy potential of methane minus escaping methane by the energy value of methane which is 13.7 kW-h/kg resulting in 0.13 kW-h/kg. Efficiency of energy production was assumed to be 23%, which yielded 0.03 kW-h capita^{-1} day^{-1} and 0.02. 55% of adjusted energy potential goes as green energy to heat, 0.07 kW-h capita^{-1} day^{-1} and 0.04 kg capita^{-1} day^{-1} green GHG CO_{eq} credit. 0.01 kg CO_2 capita^{-1} day^{-1} was emitted in flue gas from combustion capita^{-1} day^{-1}, 0.02 kg CO_2 capita^{-1} day^{-1} by flaring the captured gas, and 0.45 kg GHG CO_{2eq} capita^{-1} day^{-1} of black (methane-containing) emissions came from 58% of landfill that allowed CHG gases to escape.

Consequently, as documented in Table 10.3, Alternative II, representing the situation in the US in the second decade of this millennium, would not meet the triple net-zero goal in two categories:

1. The water goal could be considered as satisfied. In the presented Alternative II scenario, the region implements water conservation and reuses a part of the used water flow, and by doing so significantly reduces the water demand to 120 L capita^{-1} day^{-1}, which in most communities would be considered sustainable use. The reclaimed water after treatment is not used for potable or direct human contact uses, only for indoor toilet flushing or outdoor irrigation. This elimination of water waste also affected black energy use and associated carbon dioxide emissions. However, Alternative II represents the best of what could have been done, not the actual situation in 2015 in

Table 10.3. Summary results for Alternative II[+].

Energy	Energy Used or Credit kW-h capita^{-1}day^{-1}	Carbon Emissions Kg CO_{2eq} capita^{-1}day^{-1}
Black (electricity from power grid)	0.41	0.99
Gray (heating by natural gas)	2.81	2.62
Landfill emissions (black)		0.71 (GHG)
Green credit (heat gain by IRRF)	−1.95	−1.70
(Landfill biogas combustion green energy)	(−0.07)	(−0.04)
Blue credit (electricity gain by IRRF)	−0.72	−0.46
(Landfill biogas combustion)	(−0.03)	(−0.03)
Total balance	0.56	2.16

[+]Positive values in the table denote energy use and carbon emissions, negative values are energy gains and carbon credits.

most US communities. While Europe, Singapore, and few other developed countries approached the water use of 120 L capita^{-1} day^{-1}, average use in the US was still very high and unsustainable.

2. The community and its generic utility would not meet their GHG CO_{2eq} goal. Alternative II uses 0.54 kW-h capita^{-1} day^{-1} more energy than it saves and emits 2.17 kg GHGCO$_{2eq}$ capita^{-1} day^{-1} more emissions than it can claim in credits for electricity and heat produced from digester methane, MSW incineration, and landfill gas capture. However, the main energy use is for heating water in homes and other buildings estimated as 2.4 kW-h capita^{-1} day^{-1} and carbon emissions of 1.3 kg CO_2 capita^{-1} day^{-1}, which is often beyond the utility control. If this use is subtracted, the utility still would not meet its carbon neutrality goal. Most of the energy is derived from limited MSW incineration heat and that from landfill gas combustion, which is questionable because of air pollution potential. Furthermore, it is not clear how the recuperated incineration heat would have been recycled to homes for heating. As demonstrated in Table 7.1, there is not enough energy in used water to cover the energy needs of the utility to process and reclaim water, most of the recuperated heat would come from processing MSW. Ideally, the MSW and water processing should have been integrated.

3. The region was still using extensive landfilling of MSW and produced large volumes of difficult to reuse ash from MSW incineration, which would fail the no-landfill impact goal. Although some energy was produced from the LFG, uncontrolled releases and seepage of methane were still unacceptable. While CO_2 emitted from LFG energy production and even flaring would not count as GHG per IPCC, 0.71 kg CO_{2eq} capita^{-1} day^{-1} of uncontrolled methane emission is large and counts as GHG.

New modern, less-polluting incinerators are expensive to build (large, modern facilities in Europe cost $150 million to $230 million) and to make a profit and repay investors, private incinerator operators need a guaranteed stream of waste. Furthermore, incineration is an oxidation process; therefore, despite plastics' high hydrogen energy value, incineration plants have problems with burning plastics because they cause dioxin pollution. Even in small concentrations, dioxins emissions are extremely toxic. If MSW are to be incinerated, plastics must be recycled.

Collecting landfill gas for energy moderately improves the carbon emission balance. However, extending incineration to all combustible MSW was not a realistic option in the US in 2015–2020 and would not eliminate landfilling because up to 25% of MSW mass would become ash that should go to landfill (Petrlik and Ryden, 2005). Bottom ash and fly ash are also toxic and may have to be deposited in special landfills. In the countries that incinerate a larger percentage of waste to energy, like Sweden, incineration handles about 50% of MSW and the rest is recycled, and for the foreseeable future the ash may be handled by existing landfills.

These conclusions do have caveats. For one, the effect of the solar and wind blue power generation and other energy savings were not considered because in 2000–2015 the number of houses and utilities in the US with installed blue power (solar, wind) was still relatively small but increasing exponentially after 2010. The analysis of Alternative II was strictly focused on water and MSW management and related energy use and carbon emissions, considering 2015 conditions. If the landfill problems had not been accounted for, the utility could have become a net-zero carbon emission if part of the energy (for example, heating of reactors) and electricity was provided by solar and wind power, but this aspect was not included in the above analysis.

In a future community wherein most buildings have solar panels and community wind turbines and practice water conservation, the overall carbon emissions will be favorably low and could be compensated for with an excess of generated green and blue power. However, having an incinerator as the main energy producer, sequestration of the emitted carbon dioxide would be expensive and/or inefficient because of CO_2 dilution by nitrogen and other gases in the flue gas.

Alternative III.   Alternative II has assessed what could have been done with 2015 tested technologies. It met one TNZ goal in the US but it could come close to passing triple net-zero goal in some advanced European countries that incinerate all combustible MSW, along with the solids from used water treatment; co-digest biodegradable MSW; and extensively recycle, with some help, producing solar and wind energy on the utility grounds. It has become clear that to achieve the triple neutrality goal and make the system economical, ideally making money, the water/stormwater/solid waste management must be integrated. However, incinerators' ash, 25% of the MSW mass, would still be problematic and would require landfilling if recycling is not greatly increased, which is a goal of several advanced countries.

Alternative III was set to find the maximum potential of sustainable water and energy recovery using technologies that are known and tested today but some may not yet be widely used. The goal is not just to become neutral in the triple net-zero goals but, overall, to do better and become a negative $GHGCO_{2eq}$ emitter, bringing more sustainability and economic benefits. The concepts and unit processes of this sustainable triple net-zero adverse impact alternative are presented in Figure 10.4.

District water and energy reclamation.   The assessment begins with reducing water demand by fully separating black and recycling gray water, as shown in Figure 10.6. This water reuse concept is at the top of what can be done and differs from Alternative II recovery by providing potable-quality water for all contact uses and nonpotable recycled water augmented by captured stormwater water for a portion of hot water for laundry and toilet flushing. Using recycled water for drinking, cooking, and dental hygiene may still not be acceptable to some. This reduces the demand of potable water from the grid or desalination to less than 100 L capita^{-1} day^{-1}, keeping the in-house water availability for kitchen, toilet, bath and shower, and laundry uses at a comfortable 120 L capita^{-1} day^{-1} (Table 5.2), more if estimated leaks and losses are reduced. The energy estimates for delivering water were 0.18 kW-h capita^{-1} day^{-1} of black energy and future CO_2 emission of 0.04 kg capita^{-1} day^{-1}, respectively, based on the estimates of future CO_2 emission per kW-h, which is more than an order of magnitude reduction from Alternative I. During dry weather, the district water supply also provides 40 L capita^{-1} day^{-1} for irrigation and stores and treats captured rain and stormwater.

The heating of water for kitchen, laundry, and bathing uses 2.78 kW-h capita^{-1} day^{-1} of gray (natural gas equivalent) energy with CO_2 emissions of 0.82 kg L capita^{-1} day^{-1}, like Alternative II. The gray energy classification is warranted because Alternative III assumes heat recovery, either by a heat pump or a heat exchanger transfer to the heated water. More than 50% of heat energy can be recovered and equivalent CO_2 emissions reduced. Treatment of recovered gray water, shown in Figures 5.5 and 10.5, may include sand or wetland filtration, followed by disinfection and surface and/or subsurface storage, where it could be mixed with captured and LID treated stormwater. An idea for the gray water–stormwater storage and energy loop was presented in Figure 5.10. From there the stored gray and rain/stormwater can be withdrawn for irrigation and overflow is discharged into receiving waters. In the western part of the US, reuse of stormwater may have legal hurdles because this water might already have been legally allocated to downstream users.

The water withdrawn from the storage for recycling in the district must be treated by microfiltration, possibly by reverse osmosis and disinfected by ozonation because some of this water will be used as hot water supplement for laundry. The unit energy uses in kW-h/m^3 are provided in Table 5.5. The energy use for treatment of gray water was estimated to be 0.23 kW-h capita^{-1} day^{-1} (including pumping) and corresponding (black) CO_2 emissions 0.05 kg cap^{-1}day^{-1}.

The estimated COD load is the same as for the previous alternatives (i.e., 130 g capita^{-1} day^{-1}). Because of the vigorous filtration and RO treatment in the district water and resources reclamation units, almost all COD is removed from gray water and sent with the black water (BW) to the IRRF. BW sent to the IRRF includes kitchen drain, toilet flush, and reject from gray water treatment. The estimate of BW flow (Figure 10.6) was 50 L capita^{-1} day^{-1}. This BW flow is about 15% of the flow estimated for the unsustainable last-century Alternative I. For these relatively small BW flows from the district it is envisioned that the smaller sewer pipes will be made of materials that do not leak nor accept infiltration and could be threaded into existing sewers. The consequence of the diminishing flow is the high IRRF influent COD concentration of 2,600 mg/L; similarly, other compounds of interest such as nutrients will be present in high concentrations, which is good for the recovery and removal because some methods, such as Anammox nitrogen removal and struvite production for phosphorus recovery, may be inefficient at low concentrations.

The proposed IRRF anaerobic treatment shown in Figure 10.4 has three paths. The third track brings combustible solids to the gasifier. The first track treating liquid black water includes primary settler (PS), followed by one-step UASB reactor with a separate membrane reactor, or a two-step UASB followed by AnFBM reactor. The latter scheme is preferred because it can achieve high COD removal. Besides the IRRF influent, the primary settler may also accept the supernatant flow from the acid fermentation/MEC reactor or the supernatant can be directed to the UASB reactor. It is estimated that 40% of COD (or more) and 10% of flow (5 L cap^{-1}day^{-1}) will be removed by the primary settler and diverted to the anaerobic fermentation and microbial electrolysis cell reactor (AF-MEC) processing of biodegradable solids.

After accepting the supernatant and subtracting the PS underflow, the UASB-AnFBM reactors inflow increases from 50 L capita^{-1} day^{-1} to 51 L capita^{-1} day^{-1}. The supernatant COD load originates from the AF-MEC process units that also accept biodegradable MSW (see the subsequent description of the AF-MEC hydrogen producing process). Hence, the supernatant return flow can be larger by 1–2 L capita^{-1} day^{-1} if the biodegradable MSW has a high moisture content. The COD load from the liquid BW to the UASB is 78 g cap^{-1} day^{-1}, which after accepting the supernatant COD is increased to 139 g cap^{-1} day^{-1}. It is expected that 95% of the COD will be removed in the UASB-AnFBM, which will result in the effluent COD 7.0 g capita^{-1}day^{-1}. The methane production is 0.4 L/g of COD removed or 53 liters of CH_4 or 2.38 moles capita^{-1} day^{-1}. The produced methane is conveyed to the steam methane reforming (SMR) unit. The sludge load is 5% of the COD removed or 4.7 g capita^{-1} day^{-1}. Because these solids underwent the digestion process they are sent directly to the gasifier. It was assumed that the reactor may need to be heated for half of the days by green heat energy produced by the IRRF. Approximate heating energy use is between 1.5 to 3.5 kw-h/m^3 of produced methane, or 0.07 kW-h capita^{-1} day^{-1}. Less heating, if any, is needed to operate UASB-AnFBM reactors than for traditional mesophilic digester and energy to run the UASB-AnFBM reactors is very small, negligible. Upflow fluidized bed reactors do not require mixing.

The *acid fermentation-microbial electrolysis cell (AF-MEC)* reactor is the second of the hydrogen-producing tracks in the IRRF. Hydrogen is also produced by steam reforming of methane from UASB-AnFBM reactors and by reforming carbon monoxide in the syngas from MSW (and sludge) gasification third track. In the first step, which could be an AnMBR-type reactor operating with pH between 5 to 6, the biodegradable MSW organic solids and liquids plus solids removed in the primary settler are anaerobically digested to produce acetates and some other easier-to-decompose derivatives, to produce about one-quarter of the hydrogen potential and some mostly dissolved CO_2. and dissociated bicarbonate ion. The solids grown in the AF reactor (less than 5% of the COD removed) could be either separated in the AnFBM reactor by membranes and the permeate directed to the MEC as shown in Figure 8.9 or the entire content of the AF reactor may be directed into the MEC reactor and the small excess solids can be withdrawn therein and send to dewatering and gasifier as shown in Figure 10.4. Research and treatability studies must determine the best arrangement and sizes of the reactors. In the MEC reactor, the acetates and remaining dissolved carbonic compounds (proteins) are converted to more hydrogen with a help of auxiliary DC current providing electrons.

This assessment assumes that, overall, about 90% of the hydrogen potential can be recovered as hydrogen gas in the hybrid AF/MEC reactors. The hydrogen potential for food and sludge was estimated to be between 0.2 to 0.25 kg H_2 per kg of dry solids, which is approximately 0.125 kg H_2/kg of COD removed (see Chapter 8, "Hydrogen Yield Potential by Indirect Gasification"). The incoming COD load was estimated as 70% of the biodegradable MSW load of 560 g cap^{-1} day^{-1}, which with added sludge solids form PS becomes COD load = $0.7 * 560 * 1.43 + 52 = 612$ g cap^{-1} day^{-1}. If 90% of the COD load is converted to H_2, the calculated H_2 yield is 0.062 kg H_2 $capita^{-1}$ day^{-1}. Multiplying the H_2 yield by the energy value of hydrogen in kW-h, the energy yield of the AF-MEC reactors was estimated as 2.44 kW-h $capita^{-1}$ day^{-1}.

MEC requires auxiliary DC energy 0.61 kW-h $capita^{-1}$ day^{-1}, which was conservatively estimated as 25% of the hydrogen energy output. This energy can be totally provided by photovoltaic panels with batteries for storing energy and/or by the electricity produced by the MCFC. Since this is internal electricity, no CO_2 credit was considered. Note that the electricity produced and used by the IRRF is all DC unidirectional flow of electric charge, unless some AC electricity is recovered from the waste heat produced by the fuel cell. Batteries with tens or even hundreds of MW capacity became available in 2017. Using outside AC current would require a transformer/convertor, which would result in some energy loss.

Steam reforming of methane to syngas and hydrogen. Hydrogen is the ultimate product of the IRRF. While syngas can be an input to molten carbonate fuel cell (MCFC) or solid oxide fuel cell (SOFC) and be reformed there to hydrogen, methane is reformed at higher temperatures than that recommended for the MCFC. In the two-step reforming process, one molecule of hot steam (water) converts methane to one molecule of carbon monoxide and three molecules of hydrogen (H_2 enriched syngas). This gas reforming is endothermic. In the second step, additional steam H_2O molecule converts CO to carbon dioxide and another hydrogen. The second step is exothermic and can occur at lower temperatures than the first; therefore, it can occur in the anode compartment of MCFC (see Chapter 8). 38 g CH_4 $capita^{-1}$ day^{-1} or 2.38 moles of methane per capita is produced in the UASB-AnFBM reactor. Based on the reforming reactions, the same molecular mass of carbon monoxide and three times greater molecular mass of hydrogen is produced in the first syngas-forming step and again the same molecular mass of carbon dioxide and additional molecular mass of hydrogen in the second step (8.1). The outcome is 9.52 of H_2 moles capita^{-1} day^{-1}, which represents 19 grams of H_2 capita^{-1} day^{-1} and 26.2 g CO_2 capita^{-1} day^{-1} emitted into gas flow from SMR units. Significant portions or even all CO_2 could be sequestered. The energy value of hydrogen produced from methane was calculated as 0.75 kW-h capita^{-1} day^{-1}. The methane steam reforming to syngas is an endothermic process and the reaction heat energy of 0.14 kW-h capita^{-1} day^{-1}, which also emits additional 42 g CO_2 capita^{-1} day^{-1}, must be included in the balance. The SMS heat can be provided by sensible heat from gasifier or by hot steam from MCFC (SMR needs steam), as shown in Figure 8.12b.

Gasifier for producing syngas from combustible MSW and IRRF waste solids. Indirect gasification is an airless process producing syngas by exposing organic carbon-containing compounds to high temperatures (>900°C) and to steam. No CO_2 is emitted from gasification and there is no flue exhaust gas. This process yields significantly more energy than incineration, producing only heat to boil water to obtain steam for running a turbine and emits CO_2, fly ash, nitrogen, and other gases, some toxic. The mass of solids to be processed by the gasification includes 1 kg ds capita^{-1} day^{-1} of combustible MSW (including plastic and other combustible materials that cannot be incinerated), small mass of solids from the acid fermentation/MEC reactors, plus waste solids from the UASB-AnFBM reactors. The formula for the MSW used in this assessment is $C_6H_{10}O_4$ and that for sludges is $C_6H_{7.4}O_{2.85}$. The total mass of solids entering the gasifier is 1.22 kg ds capita^{-1} day^{-1}. By mass balance between combustible solids and sludges the formula for the mix composition is adjusted to $C_6H_{9.5}O_{3.8}$. The production of excess solids by the anaerobic used water treatment and sludge fermentation is very small compared to the combustible MSW solids mass. Based on the formula, the molar weight of solids entering the gasifiers is 142 moles. Produced hydrogen is $2 * (6 + 0.5 * 9.5 - 3.8)/142 \approx 0.1$ and $CO = 28 * 6/142 \approx 1.18$ kg capita^{-1} day^{-1}. Carbon monoxide is converted by the second exothermic gas shift step either in the gasifier and/or in the anode of the hydrogen fuel cell to an additional 0.08 kg of H_2 capita^{-1} day^{-1} for the total hydrogen yield 0.18 kg of H_2 capita^{-1} day^{-1}, which has blue energy potential of 7.19 kW-h capita^{-1}day^{-1} and provides GHGCO$_{2eq}$ energy credit of 2.16 kg CO_2 capita^{-1}day^{-1}.

Gasification also produces 1.1 kg of CO_2 capita^{-1} day^{-1}, which is transferred with gas to the hydrogen fuel cell. Endothermic (gray) heat to produce syngas and hydrogen plus heat to dry the moisture to 900°C hot steam is Heat = 2.12 kW-h capita^{-1} day^{-1} and CO_2 emissions of 1.18 kg CO_2 capita^{-1} day^{-1}. Hydrogen and oxygen for heating the gasifier can also be produced by electrolyzer from steam from the hydrogen fuel cell, otherwise the fuel hydrogen or syngas can be provided from syngas produced by SMR and/or by the gasifier. If the produced hydrogen is used for heating the gasifier, there are no CO_2 emissions caused by heating the gasifier (see Figure 8.12b) because steam from H_2 burning can be added to the steam demand of the indirect gasification.

The total hydrogen mass produced in the system from UASB methane, AF/MEC, and syngas gasification is 0.26 kg H_2 capita^{-1} day^{-1}, which has energy potential of 10.38 kW-h capita^{-1}day^{-1} and 3.115 kg GHG CO_2 capita^{-1} day^{-1} credit, counted as blue energy.

A *molten carbonate fuel cell* is a hydrogen fuel cell that converts hydrogen and potentially other organic gases into blue electricity and water. The cell at the cathode compartment takes two molecules of carbon dioxide and oxidizes them to carbonate, which releases four electrons that are transferred externally to the anode compartment. Carbonate ions travel through the membrane and electrolyte to the anode compartment, where they combine with hydrogen to produce water (hot steam) and CO_2. Carbon dioxide is recycled externally from anode to cathode. These processes produce electricity and heat. Because of the high internal temperature inside the cell exceeding 750°C, heat is released with steam and excess CO_2 gas from which it can be recovered, and some converted into additional electricity. Otherwise, MCFC would have to be externally cooled by air, as in hydrogen-powered vehicles, or by water to produce hot steam.

The total hydrogen mass produced by SMR, MEC, and gasifier is 0.26 kg capita^{-1} day^{-1}, which has an energy potential of 10.4 kW-h capita^{-1} day^{-1}. The fuel cell industry, supported by research in government and university institutes, expects the near future energy efficiency of hydrogen or solid oxide fuel cells to be 65% for electricity and about 30% for heat, or 95% overall. Hence the electricity output from MCFC is 6.75 kW-h capita^{-1} day^{-1} and that of heat 3.1 kW-h capita^{-1} day^{-1}, respectively. The hot exhaust flue gas from anode is mostly concentrated CO_2 with more than 700°C hot steam without nitrogen and other gases normally present in the flue gas of incineration. This exhaust hot gas flow could be tapped for production of additional energy either the traditional way (i.e., heating a boiler producing steam, turbine, and generator), or by some other more efficient way now investigated in laboratories. In contrast to microbial fuel cells, the MCFC electricity is produced chemically. Additional 28% electricity or 0.87 kW-h capita^{-1} day^{-1} could be recuperated from the heat produced by the MCFC and hot CO_2 exhaust flue, making the total electricity output from MCFC 7.0 kW-h capita^{-1} day^{-1}. 3.1 kw-h capita^{-1} day^{-1} of green heat energy can be then used to provide the endothermic heat to the gasifier, heating reactors, and the rest for buildings.

Water balance, normally not done for traditional systems, should also be considered. The system uses water for producing hydrogen-rich syngas and for steam reforming methane and syngas to maximize hydrogen production. The MCFC system is a net producer of a small volume of ultra-clean water. All calculations regarding MSW assumed dry solids weight. But, in reality, there is plenty of water within the IRRF and the amount of water being transferred within the energy system (power plant) of the IRRF system in a form of steam is small, liters capita^{-1} day^{-1}, as shown below:

Water in the moisture of MSW (assumed 50% of ds)	+1.0 L capita^{-1} day^{-1}
Water needed to steam reform CH_4 to syngas	−0.09 L capita^{-1} day^{-1}
Water to steam reform CO to H_2 and CO_2	−1.18 L capita^{-1} day^{-1}
Water produced by MCFC	+2.37 L capita^{-1} day^{-1}
Water excess (+) or deficit (−)	+2.1 L capita^{-1} day^{-1}

The overall balance of black, gray, green, and blue energies along with the corresponding carbon dioxide emissions and credits are presented in Table 10.4 and in Figures 10.8 and 10.9. Because the CO_2 will be emitted only from one place (Figure 10.4) as hot concentrated gas, it can be sequestered and some reused, as outlined in Chapters 2 and 8.

The landfill effects were not estimated because employing gasification along with co-digestion and recycling of MSW will not generate biodegradable residues that could emit methane if deposited in a landfill. The only "waste" is inorganic phosphorus containing ash (less than 10% of processed MSW) that can be reused. Conceivably, old landfills may still pose some but diminishing and manageable GHGCO$_{2eq}$ problems.

The system is highly energy productive, as shown in Figures 10.7 and 10.8, and is also better than neutral (negative GHGCO$_{2eq}$ emissions) with respect to carbon emissions. Logically, it emits some black and gray CO_2 but, overall, it provides more CO$_{2\text{-eq}}$ credits than emissions, including items such as water heating and delivery. It could provide a large

Table 10.4. Summary results for Alternative III[+].

Energy	Energy Used or Credit kW-h capita^{-1}day^{-1}	Carbon Emissions Kg CO$_2$ capita^{-1}day^{-1}
Black (electricity from power grid)	0.89	0.33
Gray (heating by natural gas and hydrogen)	4.92[++]	2.62
Green credit (heat gain)	−5.92	−2.84
Blue credit (electricity gain)	−7.62	−3.50
Total energy and CO$_2$ balance	−6.82	−3.40
Overall heat energy balance	−0.1[++]	

[+]Positive values in the table denote energy use and carbon emissions; negative values are energy gains or excess and carbon credits.
[++]Subtraction of endothermic heats for SLR reactor and gasifier and reactor heating is included.

portion of residential energy (not traffic or commercial/industrial). Carbon emissions can be reduced, and energy balance significantly improved by producing more blue energy by photovoltaic panels and wind on the IRRF premises.

The overall heat balance is important. Heat is used for warming water in houses/apartments, heating of reactors on the BW treatment line, fermentation digester in the AF/MEC line, endothermic heat for methane SMR reactor, and endothermic heat for the indirect gasifier provided by hydrogen or syngas burning. Green heat is gained from the gray water by heat pump or heat exchanger, from exothermic conversion of syngas to hydrogen and, mainly from the MCFC. The overall heat balance excess is 0.1 Kw-h capita^{-1} day^{-1}, i.e., the system can provide all the heat it needs, including heating the solids. It will require a sophisticated heat transfer and management system. Auxiliary heat may not be needed because in emergencies the system can supplement the heat provided by burning some of its produced hydrogen or syngas.

Overall assessment. Only Alternative III has met (exceeded) the triple net-zero (TNZ) adverse impacts goals regarding water and landfilling and provides energy excess and better than net-zero impact (negative) carbon emissions (carbon credits are larger than emissions). The problem herein is to overcome unsustainable traditions and, unfortunately, the resistance of some citizens and fuel industry lobbying against the progress and the necessary changes to be made. This problem was recognized in the Shell Global (2018) Sky 2070 Scenario. The change to better than triple net water/energy/solid waste management scenario is revolutionary but realistic. The triple net-zero systems do not waste water, should derive more energy from green and blue renewable sources than from fossil black and gray sources, and eliminate deposition of harmful methane-producing municipal solid waste (MSW) into landfills or burning them in air pollution–creating incinerators. This book not only shows the way but it proves that it can be done.

The presented solutions should be assessed in the overall picture of the change toward stopping and hopefully reversing the harmful process of global warming. At first look, the effect of urban systems providing water may look small when compared to the other possible sources of greenhouse gases. IPCC estimated that providing water and disposing of used water represents only about 3% of the problem, about double in some parts of the US (California). This assessment included the key components of the integrated water/stormwater/MSW/energy nexus and found that integrating water and MSW systems and recovering energy symbiotically will have significant positive effects on achieving the global warming emission reduction goals that will far exceed the IPCC estimates. It will also provide resources such as biofuel, hydrogen, fertilizer, or char. There is a lot of energy in MSW and similar other high-energy wastes (glycols, manure, food industry, plastics, etc.).

One result that at first looks surprising is the amount of blue energy produced by Alternative III. Skeptics will be asking where the blue energy is coming from. The blue energy comes from water (steam) entering the indirect gasifier that generates blue hydrogen. Compare the two MSW handling gasifiers, both producing syngas and hydrogen, but the ratio between the carbon monoxide (green energy) and hydrogen (blue energy) masses in the two processes is different.

Alternative III can be considered by some as a costly luxury solution, some might even say unrealistic. But not all communities must treat all gray water to the quality of potable water. The first net-zero impact criterion does not specify this. It calls for no waste of water, which implies that water treated to potable water quality should not be used for grass irrigation, car washing, street washing, or toilet flushing, and aquifer and surface water resources should not be depleted. There are differences in water supply management between communities located on the Great Lakes in Canada and the US and water-hungry and deprived desert communities in the southwestern United States like Tuscan and Phoenix in Arizona and Orange County in Southern California, and in the Middle East, Australia, or northeast China. In the water-rich regions, the separation into black and gray water may not be needed and the demand for irrigation, toilet flushing, and some other nonpotable uses can by satisfied on a district level by satellite water recovery facilities that treat fit-for-reuse a portion of the used water from the sewer. This is important because it increases the COD concentration of used water, which improves efficiency of the anaerobic processes in the IRRF. If most of the water demand must be provided by energy-demanding desalination of sea water (Middle East countries, Southern California, even Florida), full separation between black and gray water is appropriate and makes sense.

Implementation of indirect gasification and switching from landfilling and incineration is a necessary step that must be implemented if the goals of restricting global atmospheric warming to less than 1.5°C by 2040–2050 and zero warming thereafter is to be achieved. Neither landfilling nor incineration can achieve these goals. MSW disposal and energy and resource recovery from municipal solid waste by pyrolysis/gasification appears to be the only solution for replacing landfilling.

CLOSING THE QUEST TOWARD TRIPLE NET-ZERO URBAN SYSTEMS

11.1 COMMUNITY SELF-RELIANCE ON TMZ SYSTEM FOR POWER AND RECOVERING RESOURCES

To assess the effect of triple net-zero management on energy, let us compare it to the average residential energy use. The question each community may be asking is what return on energy savings the TNZ solution will provide, considering that the capital cost may be (hopefully only initially) higher than the cost of the traditional (business as usual) embedded fourth paradigm approaches. Figure 11.1 may provide an answer. According to the US Energy Information Administration (EIA) estimates in 2017, about 1,410 billion kilowatt hours (kW-h) of electricity were used by the residential sector in the United States in 2016. Dividing by the US population of 323 million in 2016, the per capita daily residential energy use is about 12 kW-h capita^{-1} day^{-1}. While the overall energy gain by the proposed Alternative III TNZ system of 12.62 kW-h capita^{-1} day^{-1} exceeds the average energy consumption, the net gain is about 7 kW-h capita^{-1} day^{-1}. It means that the energy gained by implementing the TNZ system could cover about 60% of the total residential demand. According to Figure 11.1, air conditioning, space and water heating, and refrigeration – as well as providing and treating water – would be covered by green and blue energy provided by the integrated water/used water/MSW and energy system. Space and water heating is provided currently by electricity and natural gas from the grid that by 2050–2070 could be provided primarily by green and blue electricity and hydrogen.

Photovoltaics, wind, and hydro renewable sources were also introduced in this book but were not extensively considered in the energy balance of the Alternative III TNZ system, with exception of a small allowance for providing the DC electric current to the Microbial Electrolysis Cell reactor for producing hydrogen and providing electricity to the electrolyzer. However, Alternative III assessment revealed that the ANovaTNZ 2040 system would cover a large portion of the residential demand and the remaining demand could be provided by solar and wind powers. The photovoltaic area needed to provide all necessary residential power was estimated in Chapter 4 ("Solar and Wind Blue Power") as 34 m^2/person in the northeastern US, 25% less in California. Considering the TNZ contribution, this area could be cut to 13.5 m^2/person or about 45 m^2 per house in Massachusetts,

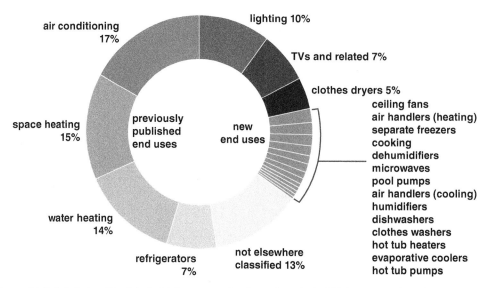

Figure 11.1. Detailed residential electricity consumption by end use in the US, in percent of total residential use, 2015. *Source:* Energy Information Administration.

which could be now installed (for free in Massachusetts in 2017) on many residential and commercial houses and the excess electricity could be used to charge electric cars. Several other US states provide grants and incentives to install photovoltaic panels on buildings. In addition, wind turbines can be installed or wind energy from regional land and offshore wind farms and photovoltaic power plants can be applied.

Example of TNZ management. The population of Gloucester, Massachusetts, in 2018 was almost 30,000 people. Two wind turbines already provide 4.3 MW of power to the city, which represents 3.45 kW-h capita^{-1} day^{-1} for its population (Figure 4.7). The blue energy estimate from Figure 10.7 of power production of a future TNZ city shows that about 7.5 kW-h capita^{-1} day^1 can be provided from the TNZ integrated used water/municipal solid waste (MSW) system with a gasifier for MSW and other solids and ANovaTNZ power plant. The green energy on Figure 10.7 is recovered heat energy that is used on site and compensates for the black and gray energy used by the IRRF and in the buildings. Gloucester is a compact, medium-density city. The power plant MWatt capacity would then be:

$$\text{Power} = 7.5 \ [\text{kW-h cap}^{-1} \ \text{day}^1] * 30\,000 \ [\text{population}]/24 \ [\text{hours}] = 9.4 \ \text{MW}.$$

In 2019, 10 MW hydrogen fuel cells were already commercially available, which means that by installing TNZ management with ANova power plant the community would not only treat all of their waste and eliminate their need for landfill, it could cover close to 100% of their residential and commercial energy needs by the TNZ system and already installed wind power. Another 2.5-MW wind turbine provides industrial power. Currently Gloucester recycles 33% of its solid waste and incinerates the combustible (wet and dry) MSW, including some plastic that cannot be recycled. Recycling is being reduced and is now limited by China's restrictions on importing recyclable solids. There are no landfills nearby. The wastewater (it is a waste) receives only primary treatment and is discharged by a long pipeline into the ocean.

The TNZ systems would solve the city's problem and could also save Gloucester's fish-processing industry, which today is severely restricted by a ban on discharging organic waste into the ocean and severely restricted solid waste disposal. Some organic waste and vegetation residues are composted locally. The energy contribution from the fishing industry waste connected to IRRF would be significant. Gloucester fish industry annual fish catch is about 45,000 tons, of which a significant portion turns into biodegradable fish-processing waste that can produce methane. The fish waste can be assumed as 30% of the catch weight, or 1.4 kg cap^{-1} day^{1} related to the inhabitants of the city. Fishing season in Gloucester is year-round. Fish waste also contains nutrients, and there is a private company in the city that converts fish waste to fertilizer but still produces difficult-to-dispose high-COD-concentration waste and uses energy rather than produce it. This biodegradable fish processing waste has similar energy yield potential as the city's MSW and could be symbiotically codigested with biodegradable MSW and sludge and residual solids could be gasified with combustible MSW or, because of the high nutrient content, processed to more nutrient rich fertilizer. Coprocessing could increase production of energy from waste and wind by more than 60% and reach total energy production of 16 kW-h capita^{-1} day^{-1}. Hence, the future city's IRRF and the existing associated wind power would make the city a 20-MW power source and this estimate does not include the photovoltaics that are already being installed on roofs of houses throughout the communities. The excess energy can be sold to surrounding communities or, better, the Cape Ann island communities of Gloucester and Rockport with a total population of 38,000 would join in one IRRF.

Economic benefits that would pay for the IRRF include electric power, dumping fees for solid waste, recycled water supply, and fertilizers. Electric power alone could bring annual income of $33 million and several millions dollars for accepting and processing MSW. Hydrogen fuel cell power plants with this power capacity and greater were already commercially available in 2018. A 14.9-MW power plant serving Bridgeport, Connecticut, was installed by Fuel Cell Energy in 2013. The FCE supplied five Direct FuelCell® stationary fuel cell power plants and turbines that convert waste heat from the fuel cells into additional electricity. This power plant uses natural gas as fuel but this type of power plant could also accept syngas because reforming syngas is easier and more efficient that reforming methane.

Other sources of bioenergy such as algae were also presented in this treatise but not included in the overall energy and GHGCO$_{2eq}$ balances. Wind turbines are already ubiquitous in Europe and even in some US states from Texas, Kansas, and Iowa to Massachusetts, and there are even medium-size cities in Texas that derive all their energy from solar and wind power. Recovering nutrient-rich fertilizer by struvite and fish waste industrial processing could also become a significant source of income by a private/public consortium managing energy and resource recovery.

By 2030 or shortly thereafter suitable landfill sites will disappear in most countries and environmentally unsafe incineration will be expensive, provide little benefit, and could be abandoned. Even today, landfills are already limited, and new landfill sites are not available. After a discourse, a decision must be made whether last-century methane-based energy and resource recovery using well-known and tested but less efficient digestion and inefficient combustion technologies are the solution, or the future is in clean and more efficient indirect gasification, MEC, and hydrogen fuel cells. Very high temperature plasma syngas/hydrogen producing gasifiers are still being developed in laboratories but may become available on pilot scale soon and have been used in refineries.

People may resist gasification near their residences because they might believe that gasifiers are the same air polluters as toxin-emitting incinerators. Indirect gasifiers should have no exhaust flue stacks emitting GHG and toxic gases if hydrogen and superheated steam from the hydrogen fuel cell are used to provide the gasification endothermic heat. Internal "burning" (i.e., oxidizing of hydrogen) produces heat and steam, which could be used by the gasifier to enrich syngas with hydrogen, but it would also bring nitrogen from the air into the system. Hence, a small outside burner should be used that would emit steam and residual unused air. Using pure oxygen made by electrolyzers for hydrogen burners is a possibility that, however, must be investigated regarding its economy. The traditional method (more than 120 years old) for producing oxygen and other gases liquifies air, which require a lot of energy. Pure oxygen produced by electrolyzers needs DC energy for production that is about 25% more than the energy in produced hydrogen. Excess green and blue energy is sometimes produced when it can be used to make H_2 and pure O_2 gases that can be stored. Hydrogen-powered indirect gasifiers also eliminate landfill deposition because the small amount of ash, 5–10% of the processed MSW, contains all phosphorus and could become a soil conditioner rather than being deposited into landfills, or phosphorus can be extracted from ash.

These new systems also provide the opportunity for resource recovery, additional biomass production for producing biofuel, commercial hydrogen, fertilizers, and eventually recovery of rare metals from the ash. This means that the new systems will be public or private (or both) commercial enterprises that can recover costs not only by collecting fees (taxes) for accepting used water and other biodegradable and combustible waste compounds such as excess manure, fish processing, stockyards, dairy, and MSW, but also by profiting from income for production of excess electricity, hydrogen, high-density compressed carbon dioxide, biofuel, fertilizers, and biosolids, as well as recyclables recovered from MSW.

However, these measures to minimize GHG emissions from the residential/commercial sectors should be viewed in the context of the overall GHG emissions from the terrestrial inhabited land. Figure 11.2 shows that the residential, commercial, and industrial power generation, wastewater disposal, and waste burning encompass about 50% of the total anthropogenic emissions. It was already alluded herein that the water/used water sector represents about 3% of total GHG emissions (per IPCC), but it receives, justifiably, a lot of public attention. Switching from fossil fuel–powered electricity-generating plants is being rapidly implemented in many countries, including most US industrial states (including Massachusetts, Washington, Rhode Island, California, and New York). The fossil power demand reductions in the sectors covered in this book are fully or partially reducing GHG emissions, but in the second decade of this century, this mostly involved replacing coal by natural gas. Mining fossil oil and methane by fracking from shell geological layers is now becoming one of the main sources of combined (Methane + CO_2) GHGCO$_{2eq}$ emissions due to the large CH_4 losses and flaring excess methane at the mining sites and during transportation by long pipelines. Agriculture is the largest source of methane and nitrous oxide, which accounts for about 12% of the combined GHGCO$_{2eq}$ emissions. While gradually switching from fossil fuels to renewables can be almost complete by 2050, the outlook for reducing agricultural emissions is less optimistic. These is no reason, except economy, why the newer methane power plants couldn't be adapted to burn hydrogen for energy and emit steam instead of carbon dioxide. However, the actions in the urban sector outlined in this book and in transportation and industrial sectors described by others could meet the Paris Agreement goal to keep the increase in global average temperature to way below 2°C above preindustrial levels. The IPCC panel meeting in Korea (IPCC, 2018) limited the manageable increase to 1.5°C and claimed that this goal is achievable.

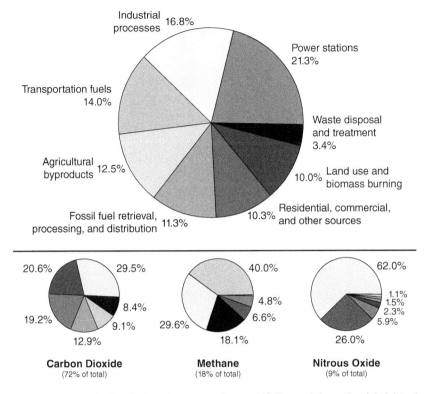

Figure 11.2. Annual GHG emissions by sector. *Source:* US Energy Information Administration.

11.2 ECONOMIC BENEFITS AND APPROXIMATE COSTS OF THE 2040+ INTEGRATED WATER/ENERGY/MSW MANAGEMENT

The systems presented in this treatise are based on the state-of-the-art technologies of the second decade of this century and some visionary assumptions for the future. However, the progress in the field of developing green and blue energy and their utilization in the water system and water energy nexus is so rapid that it is in no way possible to predict 100% which systems will be implemented in 2050. Will it be a system that derives the energy from biomass by traditional biological digestion or a more advanced hybrid system of fermentation/MEC or by a new abiotic production of hydrogen from biomass by photocatalytic decomposition?

Alternative III appears complex and sophisticated but when considering the cyber capabilities expected 20–30 years from now, it will be possible. The present cyber technologies are already so advanced that this is not an issue. Thirty years ago, there were no ultra-smart mobile phones, the internet was primitive, and the majority of cars were fossil fuel gas guzzlers. All technologies envisioned for the future triple net-zero water/energy system are known, tested, and in most cases already used today or ready to leave laboratories. They likely will be very economical when considering the tangible benefits of savings on landfill dumping fees, electricity and heat energy from auxiliary sources, selling excess energy and hydrogen, and making fertilizer, and the intangible benefits of reducing GHGs, and having a generally cleaner environment. Switching to hydrogen as an energy carrier makes sense and provides links to other economies such as automobile manufacturing, heating and cooling homes, and energy use.

Other reasons for optimism are rapidly changing costs of green and blue energy, and the increasing cost of tipping fees for deposition of MSW into landfills (today more than $50/ton of MSW, and in 25–30 years no new landfills will be available). The cost and even the type of fuel cells is still uncertain, and the differences in energy demand and production between the currently developed and developing countries (including China, India, and Africa) with much larger populations than they currently have is uncertain. It is like asking planners in 1985 – when the majority of phones were wired stationary dial landlines – to estimate the cost and use of telephones in 2015. The proposed system is revolutionary and currently none have been realized. However, even though the cost is still very uncertain, it is possible to outline the tangible and intangible triple bottom line benefits (Table 11.1). Tangible benefits are those that can be directly ascertained monetarily. Intangible benefits have social and environmental values. The majority of intangible benefits and costs can be monetized by assessing the willingness to pay for them ascertained by scientific surveys of affected people (see Chapter 5).

Cost of Green and Blue Energies Is Decreasing

The International Energy Agency (IEA, 2016) has estimated that, under an optimistic scenario, by 2050 or even before, green and blue energy will be so plentiful and less expensive that it will displace most fossil fuel energy. Meanwhile, the black energy GHG emissions that still exist must be sequestered. Similarly, a Stanford University study (Jacobson et al., 2017) and IEA predicted that by 2040–2050, most transportation by cars, trucks, boats, and trains will be powered either by electricity produced by green and blue energy or by blue hydrogen. Cars and boats powered by blue and green energy are already on the market; electric trains have been operating in Europe for a century, and hydrogen-powered trains and light rail have been leaving testing tracks in 2018 in China, France, and the Czech Republic. In 2018 the first hydrogen-powered trains manufactured in France were put into operation in Germany, followed soon thereafter in Great Britain, The Netherlands, Denmark, Norway, Canada, and France. Hydrogen-powered flotillas of fishing and other boats are operating in Iceland. Like hydrogen-powered cars, these trains and boats emit only water. Finally, hybrid smaller and medium-sized commercial planes took off in 2018. Hence, all means and tools for the world becoming green and blue by 2040–2050 are ready and the goal is achievable.

The article by the Stanford University team (Jacobson et al., 2017) was the most optimistic study of the blue energy outlook. The study analyzed and predicted a complete electrification (with hydrogen use) of 139 countries by blue water (hydropower), wind, and solar power by 2050 and compared it with the "business as usual" (BAU) future scenario that would obviously be highly detrimental to humans and ecology. The WWS (wind, water, solar) scenario eliminated gas and coal even with carbon capture, nuclear power, liquid and solid biofuel, and natural gas. Oil was not mentioned but could also be intrinsically eliminated. The outcome of this analysis was that the 100% clean and renewable WWS alternative is feasible and realistic in all 139 analyzed countries and would also provide less costly energy than the BAU alternative. Energy from waste was not included in the Stanford analysis, and it will be documented throughout this treatise that that it should have been.

The Stanford WWS study emphasized the benefits of meeting the Paris agreement and stated that fully implementing the Paris 2015 roadmap would by 2050 meet the 1.5°C global warming limit; eliminate millions of deaths from air pollution annually; create 24.3 million net new long-term, full-time jobs; and reduce energy costs to society. WWS would not

Table 11.1. Tangible and intangible benefits of the future integrated water/used water, energy/MSW nexus.

Action	Tangible Benefits	Intangible Benefits
Reduction of water demand by 75% by reuse and including treated stormwater	Reduction of cost of providing water, avoidance of desalination and bringing water from distant sources, energy savings	Improving ecological flow in surface waters, prevention of overdrafts and groundwater mining, reduction of GHG emissions.
Gray water – black water district separation and reuse	Reduction of water demand, heat recovery energy savings.	Reducing used water flow in sanitary sewers and eliminating sanitary sewer overflows (SSO), providing flow to urban streams and irrigation, providing nonpotable water in homes.
Tap the sewer district used water reuse	Possibility of district energy production, saving water and energy.	Reducing flow in sanitary sewers and eliminating SSO, providing nutrient-rich irrigation water and possible biomass from treatment wetlands providing flow to urban streams
Implementing LID district stormwater management	Providing supplemental water source to all district uses, greatly reducing flood damages, providing cooling and saving on heating.	Aesthetic, providing surface flow and groundwater recharge reduces GHG emissions, reduces urban heat island effects.
Anaerobic treatment of black water in IRRF	Dramatic reduction of cost of treatment, methane production and conversion to hydrogen, commercial production of struvite.	Significant reduction, almost eliminating GHG (power) emissions, reducing demand on nonrenewable phosphorus resources.
Coprocessing IRRF biodegradable MSW and suburban manure	Significant hydrogen production by the IRRF, reduction of MSW deposition in landfills, contributing phosphorus to struvite production, adding organic solids to gasification to produce syngas.	Significant reduction of GHG by IRRF power use and eliminating methane emissions from landfill, reducing demand on nonrenewable phosphorus resources. MSW is a "free" source of energy and resources.
Indirect gasification of combustible MSW with residual sludge	The largest source of syngas energy, dramatic reduction or elimination of waste solids to landfill, source of soil conditioning ash	No air pollution nor CO_2 emissions if hydrogen is used for adding heat to gasifier.

(Continued)

Table 11.1. (*Continued*)

Action	Tangible Benefits	Intangible Benefits
Methane and syngas reforming to hydrogen	Increase in efficiency electricity production, besides being a source of hydrogen syngas, has commercial value for producing biofuels and other products.	No GHG and CO_2 emissions, clean production.
Hydrogen fuel cell	Very efficient production of electricity and heat, concentrated CO_2 has a commercial value and is easier to sequester; produces clean water as byproduct. Accepting diluted CO_2 in flue gas from other operation and concentrating it for sequestering.	Clean production of energy, no air pollution.
Entire system	Produces significantly more energy than it uses, system provides concentrated CO_2 emission and credits that can be monetized, saves water, and eliminates the need for landfilling. Generates fertilizers, clean water, and excess hydrogen that can be commercialized. Millions of new jobs are created implementing and managing the new TNZ systems and developing new green and blue energy sources and distribution. Eliminates cost of air pollution related to water use and waste disposal.	The entire system is urban friendly and its components can be in neighborhoods or even in basements of large buildings. It provides more resiliency and significantly reduces the impacts of global warming. Industries producing biodegradable waste will benefit from integrating their expanding circular economy provided by the integrated system and reducing dramatically their waste that may also be a resource processed by the IRRF. Avoiding the dangerous global atmosphere temperature increase and meeting the Paris Agreement goal and health effects of air pollution.

only replace business-as-usual power demand but would also reduce it 42.5% because the work:energy ratio of WWS electricity would exceed that of combustion (23.0%). Obviously WWS energy requires no mining, transporting, or processing of fuels (12.6% less) and WWS end-use efficiency is assumed to exceed that of BAU (6.9%) and provide less costly energy (10% less). Most of the electric technologies proposed for replacing fossil fuel technologies are already commercial on a large scale today (e.g., electric heat pumps for air and water heating and cooling, induction cooktops, electric passenger vehicles and trains, electric induction furnaces, electric arc furnaces, dielectric heaters), but a few are still being designed for wide commercial use (e.g., electric and hybrid hydrogen fuel cell–electric aircraft).

The Stanford study was criticized and deemed unrealistic. The critiques focused on the proposal of greatly increasing hydropower and problems with storage of blue power. Furthermore, the solutions could not be universally applied in 139 nations situated in vastly different climatic, hydrologic, economic, and geographic regions. A follow-up article (Jacobson et al., 2018) divided the world into 20 regions and tried to solve the problems with storage, which made the study more realistic. In a web discussion forum, exclusions of biofuel and methanol from corn and sugar cane were critiqued. Most of this discussion is beyond the topics addressed in this book. Nevertheless, making biofuel from waste is a "green" option (see Chapter 8).

The Stanford study (Jacobson et al., 2017) is discussed further in this book. The WWS alternatives from the Stanford team focus strictly on 100% electrification by WWS blue power and do not consider other blue and renewable green sources with exception of geothermal power. Yet despite the critiques and adaptive corrections and improvements, the WWS Stanford study and those by the Carnegie Institute and Massachusetts Institute of Technology and the discussions among scientists are seminal because they prove that the pathways toward 100% blue energy are feasible by 2050 in most countries (specifically, the 139 countries assessed in the Stanford study), not just developed ones, and the energy switch and production are economical. However, it is herein argued that focusing just on producing electricity may not work. The problem must be attacked from many sides and the energy savings derived by water conservation and production from used water and waste solids outlined in this treatise are substantial but have not been included in the above WWS analyses. The results of Chapter 10 are seminal because they show that a significant portion of blue electricity can be derived from what is now waste. Instead of spending a lot of funds on damming rivers, dramatically changing their hydrologic regime, and building many storage reservoirs, the same amount of energy could be derived from MSW and used water.

If the outcome of the Stanford study is taken literally, it would mean that BAU alternatives in the field of used water and solid waste disposal would continue to emit carbon dioxide, methane, and toxins. Deadly air pollution in some megalopolises of Pakistan, India, and China, and even in Kabul, Afghanistan, is, in addition to traffic, partly caused by emissions from garbage incinerators, coal and animal fat burning during winter, and in developing countries by landfill fires, would also continue. Leaving landfills in place and using them for final disposal of solid waste would continue emitting volumes of methane and carbon dioxide. Fishing industry waste would remain a problem in fishing cities because of severe restriction on disposition of waste. The business-as-usual legacy of unsustainable water and waste solids use would also be felt in the increasingly high GHG concentrations in the atmosphere because of uncontrolled or inefficiently controlled emissions of methane. WWS energy production will not reduce legacy and continuing air pollution and $GHGCO_{2eq}$ emissions, but agriculture, reforestation, and CO_2 sequestering (negative $GHGCO_{2eq}$ emissions) will. Cheaper blue energy is the goal, but without circular economy and without energy from used water and solid waste, this energy would persist in producing plastics and chemicals that would continue to go into water bays and oceans. This book has documented that implementing TNZ concepts, technologies, and systems described herein would reduce the demand for black energy and provide 50% or more of the per capita residential energy demand (excluding industrial demands and transportation). The TNZ concepts also include circular economy of reducing virtual industrial and commercial energy use by increasing recycling and deriving resources (e.g., fertilizers) and energy from used water and municipal and suburban waste solids and obtaining raw feed for producing biodegradable and recyclable plastics, biofuels, and other materials and goods.

This book is in 100% agreement with the Stanford, Carnegie Institute, MIT, and other scientific studies with respect to switching entirely to green and blue energy production However, it was clearly documented that energy from used water and waste solids and liquids could be a significant part of the overall quest for 100% blue and renewable energy. Used water and waste solids and liquids must be treated and safely converted to energy and resources without unnecessary GHG emissions. There is no other feasible and sustainable option. At the beginning, hydrogen fuel cell energy will likely be costlier than WWS energy, but it is necessary to convert waste to sustainable energy and resources, otherwise plastics, garbage, and sludge would continue to accumulate. Including triple net-zero (water, carbon, and landfill) TNZ impact solutions proposed herein will significantly contribute to decarbonization of energy emissions, significantly reduce methane and toxic gas emissions, and will be a solution to the problem of plastics. The switch from methane-driven water/used water/energy management and energy recovery to hydrogen is essential and employing hydrogen fuel cells for producing energy is a key component. Consequently, another "S" should be added to the renewable energy sources water to make it "wind, water, solar, and solid waste" (WWSS), where "water" would include all water and used water energy sources, not just hydropower or tide, and "S" would include solid waste, as covered in this book.

Detz, Reek, and van der Zwaan (2018) identified and analyzed the cost of solar-driven processes with H_2O and CO_2, which are the basic feedstocks for producing "solar fuels" that could substitute their fossil-based counterparts. Chapter 4 of this book described hydrogen and production of syngas as the solar energy carrier, producer, and source derived with H_2O. H_2O and CO_2 together are used for producing methanol or diesel fuel by solar power and syngas and CO_2 from indirect gasification of MSW with plastics in the circular economy concept will be building components for the producing new "green" biodegradable plastic. In an optimistic scenario they projected that competitiveness with fossil fuels could be reached between 2025 and 2048 for all renewable energy production pathways that they investigated. Two techniques will reach break-even competitive costs before 2050 even in a conservative base-case scenario: H_2 production through electrolysis and diesel production by Fischer–Tropsch synthesis. Both processes use solid oxide electrolysis, which profits from rapid cost reductions and high efficiency. It can be deduced, but not yet proven, that electricity production from used water and waste with secondary incomes to the utilities for processing and selling the products will also be competitive.

The US National Academies Committee on Developing a Research Agenda for Carbon Dioxide Removal and Reliable Sequestration (NAC, 2018) has made a preliminary cost of achieving and implementing negative emissions technologies (i.e., technologies that would reduce GHG CO_2 emissions on a large geoengineering scale). These costs were related to the cost of one ton of CO_2 either removed from the atmosphere or prevented from being emitted. They assigned a ranking of low for the $1 to $20/ton of CO_2, medium for costs from

Table 11.2. Approximate costs of CO_2 removal from the atmosphere or preventing emissions.

Negative Emission Technology	Cost	Limiting Factor
Terrestrial carbon removal and sequestration: forest and urban plants management	L	• Available land given coastal development and land use • Understanding of future rates with sea level rise and coastal management
Terrestrial carbon removal and sequestration: practices to enhance soil carbon storage	L–M	• Limited per-hectare rates of carbon uptake by existing agricultural practices • Inability to fully implement soil conservation practices
Biomass energy production with carbon capture and sequestration (BECCS)	M	• Cost • Availability of biomass, given needs for food and fiber production and for biodiversity • Inability to fully capture waste biomass • Fundamental understanding
Direct CO_2 capture from air	H	• Cost greater than economic demand • Practical barriers to pace of scale up

Source: National Academies Committee (2018).

$20 to $100/ton of CO_2, and high if the cost is greater than $100/ton of CO_2. "Economical" means that the deployment would cost less than $100/t CO_2. A great part of the NETs would cost less than $20/t CO_2, as shown in Table 11.2.

Figure 11.3 is taken from the US Fourth National Climate Assessment report (USGCRP, 2018a). It presents annual economic impact estimates for labor and air quality. The bar graph on the left shows national annual damages in 2090 (in billions of 2015 dollars) for a higher scenario (RCP 8.5°F = 4.4°C in 2090) and lower scenario (RCP 4.5°F = 1.8°C in 2090). The difference between the height of the RCP 8.5 and RCP 4.5 bars for a given category represents an estimate of the economic benefit to the United States from global mitigation action. For these two categories, damage estimates do not consider costs or benefits of new adaptation actions to reduce impacts, and they do not include Alaska, Hawaii and US-affiliated Pacific Islands, or the US Caribbean. The maps on the right show regional variation in annual impacts projected under the higher scenario (RCP 8.5) in 2090. The top map shows the percent change in hours worked in high-risk industries as compared to the period 2003–2007. The hours lost result in economic damages: for example, $28 billion per year in the Southern Great Plains. The bottom map is the change in summer-average maximum daily 8-hour ozone concentrations (ppb) at ground level as compared to the period 1995–2005. These changes in ozone concentrations result in premature deaths: for example, an additional 910 premature deaths each year in the Midwest.

A scientific panel from the US, Canada, and Mexico (USGCRP, 2018b) estimated the cost of overall North America global warming prevention and control measures. Although activities in North America cannot alone reduce emissions enough to limit the total global temperature rise to 2°C, the estimated cumulative cost from 2015 to 2050 for the United States to reduce emissions by 80% relative to 2005 levels (an amount considered to be in line with the 2°C goal), by using a variety of technological options, is in the range of $1 trillion to $4 trillion (US$, 2005). The total annual cost in 2050 alone for climate

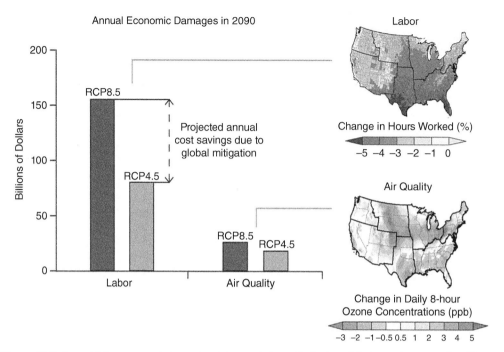

Figure 11.3. Annual future economic impact on labor and air quality for two global warming scenarios. The bar graph on the left shows national annual damages in 2090 (in billions of 2015 dollars) for a higher scenario (RCP 8.5 in 2090) and lower scenario (RCP 4.5 in 2090). The maps on the right show regional variation in annual impacts projected under the higher scenario (RCP 8.5) in 2090. *Source:* 4th National Climate Assessment report.

change damages across health, infrastructure, electricity, water resource, agriculture, and ecosystems in the United States is conservatively estimated to range from $170 billion to $206 billion. However, the goal is to limit the temperature rise to 1.5°C, so the above considerations must be upgraded.

The TNZ concept covered in this book will not benefit only water and MSW systems. The other side is energy. Many existing power plants using fossil fuels make electric energy using nineteenth-century steam power, where coal is burned to heat water in the boiler to produce steam that runs the turbine and the steam is then condensed in a heat exchanger back to liquid, which releases waste heat into the environment in a heat exchanger or a cooling pond or river. The waste heat is rarely reused. This is an archaic and inefficient process. The overall efficiency of standard fossil fuel power plants is about 40% of the energy in fuel; the rest is waste heat that causes thermal pollution if released into water resources, and, if released into air by cooling towers, it contributes more heat to the atmosphere and the heat is wasted either way. Furthermore, the emissions of diluted CO_2 with other toxic gases into the atmosphere by flue stacks is a serious GHG air pollution problem that also deposits toxins (e.g., mercury) into receiving waters and even water bodies far away, such as the Florida Everglades. Coal, natural gas, and oil are geologic organic minerals and compounds that were formed millions or billions of years ago when the Earth was much warmer and had far more CO_2 in the atmosphere. Continuing to mine these mineral solids and liquids for energy without sequestering the CO_2 would be very dangerous, even catastrophic, as alluded to by the IPCC (2018) and US National Climate Assessment (USGCRP 2018a) reports.

11.3   CAN IT BE DONE IN TIME TO SAVE THE EARTH FROM IRREVERSIBLE DAMAGE?

The reality of the danger of global warming has been known to climate scientists for more than 40 years, but not much was done for the last 30 years of the twentieth century. The latest warning has come from the recent IPCC and US National Climate Assessment reports and more warnings will come when global warming effects (hurricanes, heat records, droughts, Arctic and Antarctic and Greenland glacier melting, losses of coral reefs, and rising sea levels) will worsen and become more frequent. The last IPCC report (2018) predicted lasting cataclysmic effects of climate change by 2040 if the BAU GHG emissions continue. The Earth could go to another, more adverse ecological state (see Chapter 4) from which return would be impossible in this century and would require extremely expensive geoengineering actions.

Political-Economical Tools

The fight to reduce the effects of global warming and save society and the environment must be carried out in the policy domain. Global warming and pollution are affected by the "external diseconomy" dilemma, which states that polluters or emitters of harmful substances do not have an economic/market incentive to implement pollution controls and elimination measures on their own because their goal is to maximize profits (Solow, 1971). Overcoming the externality problem and incorporating the cost of repairing damage caused by unacceptable emission of harmful substances in the economic thinking of emitters and consumers using pollution-causing products (fossil fuels, plastics) should be the major objectives of abatement policies. Most efficient policy alternatives require a combination of regulation and enforcement with subsides and taxation.

Rogers and Rosenthal (1988) defined several policy imperatives for control of pollution and emissions of harmful substances, which in a modified adapted form are:

Equity. No group of individuals or population groups should bear a disproportionate cost of meeting environmental quality requirement or damages. The levels of environmental quality chosen should be such that no additional benefits can be derived by one group (energy producers, plastic makers) if making another group (people living in areas impacted by global warming) worse.

Irreversible impact. No action may be permitted that would irreversibly harm the people, environment, and natural resources. The concept of irreversible environmental degradation and consumption of nonrenewable resources has to be guarded against by society at large (intergenerational impact). This requirement is now known and implemented as *sustainable development*.

Regulation and standards. Due to the failure of the general (world and national) market to control the quality of the environment, and protect the population from harmful and irreversible effects of pollution and overproduction of harmful substances, there must be legislation and regulations.

Acceptance. There must be concurrence from people and groups that they will obey the regulations.

The above imperatives were proposed for controls of nonpoint (diffuse) pollution and applied to some degree to agricultural sources and emissions (acid rain caused by some coal power plants). Carbon CO_{2eq} emissions and even plastics pollution of the environment all have characteristics of diffuse pollution (Novotny, 2003a).

The following legislative tools have been proposed by the nations subscribing to the Kyoto and Paris (European Union) accords, and some have already been implemented in certain states (Massachusetts, California, Washington).

Legislative Goals. These goals have been proposed in the US and other countries to control carbon emissions (Harder, 2019; IPCC, 2018):

Limiting global warming to 1.5°C over the preindustrial period by developing, implementing and enforcing emissions standards

By 2050 or before, decarbonizing the economy and energy production

Implementing energy conservation in buildings

Reducing methane emissions from agriculture and fossil fuel mining operations

Pollution (Carbon) Taxes and Trading. This legislative tool imposes taxes on carbon emissions and rewards the emitters who perform better than the goal. Taxing economic activities that release carbon (and by this reasoning any harmful pollution) makes sense because these damage the environment, so the funds can be used to remedy the damages or provide incentives for switching from the products and eliminate the activities that produce the harmful emissions. Known as Pigouvian tax, this tool has been used to control pollution in rivers as well as emissions from power plants. In this context, the pollution control authority imposes uniform target limits related to the quantity of the harmful emission (for example, BOD/COD discharge from papermaking companies on a river or harmful emissions from power plants in the entire region).

In another tax scenario, the authority imposes a regional limit on the emissions that is linked to the assimilate capacity of the environment to absorb the pollutant without damage and issue a permit that specifies the limit to individual emitters. The emitter (discharger) will receive a credit if the emission is less than the limit and a penalty (tax) if it is exceeded. Then the emitters may negotiate among themselves on attaining the overall goal. In this process the dischargers can sell their credit to those who exceed their permits and could then avoid the tax. This concept is called *transferable discharge limits, or cap and trade*. Cap and trade is a market-based system where the government caps the amount of emissions (in this case greenhouse gases) and creates a trading system (Harder, 2019).

Imposing taxes on citizens for fuel use is questionable. First, the universal fuel tax is regressive and falls heavily on people who must travel by automobile to work. Second, if the tax is low, it may raise money for the government, but it will not discourage people from reducing fuel use. And if the tax is excessive, the general public may rebel, as it happened in France in 2018 where the cost of fuel was already high and people were mostly driving smaller, high-mileage vehicles; therefore, the population resented the increased cost of fuel, which led to violence but may not have led to significant lowering of gasoline use. Because the ratio of carbon emissions per kW-h of electric energy produced in France is already very low, a subsidy for switching from fossil fuels to electric automobiles paid from taxes on fossil fuel industries might have been more efficient.

Subsidies. A state or national government creates a system by which it rewards installation of pollution control or green/blue energy generation for homeowners and communities or for switching from high fossil fuel use to more efficient appliances, automobiles, or installing photovoltaics on houses. The system is funded from general taxes or from taxes imposed on emitters. This system has been successfully implemented in some US states and in other countries. In Massachusetts and several other states, the subsidy is essentially a low-cost loan that covers the cost of installing roof photovoltaic panels, which is then repaid, typically in five years, by the excess electricity produced by the homeowner. The payments are lower than the cost of electricity before installation. Buyers can also obtain rebates for buying more fuel-efficient hybrid and electric cars, water heaters, furnaces, and so on. Funding for subsidies can be derived from a general progressive tax and/or from the Pigouvian tax on emitters.

The Process to Achieve the Goals

The entire process toward the new paradigm may begin with converting garbage and sludge-processing incineration facilities, old coal-fired power plants, and landfilling of MSW into indirect gasification, followed by green and blue energy production and resource recovery and recycling. The current worldwide situation with garbage and plastics is catastrophic and BAU GHG emissions are dire, dangerous, and unacceptable. Wouldn't it be far better and environmentally friendlier if power companies returned to another older and tested but greatly modernized indirect gasification technology and use produced syngas to generate energy by fuel cells and produce hydrogen and other valuable commodities? Perhaps more modern energy production processes will become available soon. With sequestering of the concentrated and clean CO_2 byproduct in a permanent geological gas storage (see Chapter 4), the problem of GHG emissions by fossil fuels would be solved.

The story of the fight for action to control emissions of greenhouse gases and how the efforts of scientists and visionary politician in the US and worldwide almost succeeded in implementing regulations and actions 30 years ago and how and why these efforts collapsed in the US was extensively covered in a special issue of the *New York Times Magazine* (Rich, 2018). The major reason for the failure at the end of the last century was a paucity of hard evidence of the catastrophes caused by global warming; the public could not actually see and feel the impact of global warming and the media could not write about it to show extensive suffering (with exception of hurricanes). Between 1970 and 1990 there was only one very hot year (1988); there were hurricanes and droughts but nothing extraordinary that could be persuasively linked to global warming. The evidence about the possible catastrophic effects of global warming was derived from forecasting by computer models but, while most models indicated the effects, their predictions were not uniform. At the end of the last century, the very small measured temperature increases (a fraction of °C) were characterized as "natural" and measured sea level rises were very small (in centimeters). The cost of abandoning fossil fuels was deemed high and was opposed by coal and oil companies. A large portion of the public and the politicians in the US were not persuaded about the potential future catastrophes caused by the increases of carbon dioxide in the atmosphere and even the term *greenhouse* was misunderstood by some. "Greenhouses" have lush vegetation, healthy oxygen-rich air, and they are warm and humid but not hot. Hence, in 1989 the US government abandoned efforts leading toward a moderate (20%) reduction of CO_2 emissions and the US originally was not part of the "Kyoto protocol" agreement in 1992 among most United Nation member countries. It was the first international treaty among industrialized nations setting limits on greenhouse gas emissions.

The twenty-first century is a period of awakening because global warming is now occurring, and its adverse effects are increasing year by year. The hot temperatures of

1988 are now reached and exceeded almost annually. The temperatures in Europe in the summer of 2018 – even as far north as Scandinavia and above the Arctic Circle – exceeded 32°C (90°F) by great margins and mountain glaciers in Sweden and the Alps were melting rapidly along with the Arctic ice. In 2018–2019, warm African air from Sahara desert reached the Arctic and displaced the super-cold North Pole air vortex to the midwestern US and Canada, where temperatures in January 2019 reached –45°C (–49°F) and daily cold high temperatures stayed below –15°C (5°F) for weeks. This was not global "cooling," but was an effect of global warming of the North Pole. Summer temperatures in the southwestern US, China, India, the Middle East, North Africa are now unbearable. Forest wildfires are now widespread, extensive, and deadly in the western US, northern Europe, and the Mediterranean. Droughts are more frequent and severe. It is predicted that by the end of this century, many coastal densely populated areas (Miami, Florida; Venice, Italy; The Netherlands; parts of Boston and New York) and the Pacific Islands will be under water during high tides unless expensive engineered barriers are built to protect them, as in The Netherlands, where a large part of the country is already under the current sea level. Parts of Venice are now flooded several times each year. And the "monster" hurricanes like Maria (2017), Irma (2017), or the one-two punches of Florence (2018) in the Atlantic and the most monstrous of them, Matthew (2018) in the Gulf of Mexico and Dorian (2019) in Bahamas, may be expected to occur more frequently. Super Typhoon Yutu crossed the US territory of the Northern Mariana Islands in 2018, making it the strongest hurricane to hit any part of the US. Maximum sustained winds of 180 mph were recorded around the eye of the Yutu storm as it hit the island territory and in the Florida landing of Matthew and Dorian in Bahamas. These catastrophic hurricanes are now also annual meteorological events. Year after year the US will witness massive destruction not only from hurricanes but also from droughts and wildfires that are also linked to global warming.

Doing nothing (BAU) or doing little and continuing to argue is not acceptable and would lead toward a "Hothouse Earth" pathway that could prevent stabilization of the climate and places on earth would become uninhabitable even as human emissions are thereafter reduced. "Hothouse Earth" will lead to the end of life as we know it in parts of the Earth if strong geoengineering and local measures of carbon reducing climate interventions (some outlined in this book) are not implemented (Steffen et al., 2018). This prediction was confirmed by the IPCC (2018) and the US Fourth National Climate Assessment reports. The Steffen et al. study and the IPCC state that, currently, civilization and ecology are at the point where maximizing the chances of avoiding a "Hothouse Earth" requires not only reduction of carbon dioxide and other greenhouse gas emissions but also enhancement and/or creation of new biological carbon stores – for example, through improved forest, agricultural, and soil management; biodiversity conservation; and technologies that remove carbon dioxide from the atmosphere and store it underground, as outlined in the National Research Council report of the US Academies (NRC, 2015), National Academies Committee (NAC, 2018), US Global Change Research Program (2018a, b), Watts et al. (2013), and other publications. Apparently, society is now at the onset of the fifth revolution related to fuel energy (see Chapter 1) that will also lead to the fifth paradigm of urbanization.

So, according to the IPCC (2018) and the US Fourth National Climate Assessment, now is the last chance to act. The situation is now more favorable and the reasons for a modest optimism that the changes will come are:

1. Global warming is occurring as predicted by the models and the effects are already devastating in many parts of the world and are widely covered by the media.

2. Almost all countries in the world are united in the effort to do something about it. Even in the US, most populated and industrial states are involved and committed to reducing GHG emission and to avoiding "Hothouse Earth."

3. For economic reasons, power companies – including those in the industrial states of the US – will continue decommissioning old coal-fired power plants and replacing them with more efficient and less costly, less polluting, and less CO_2-emitting facilities powered initially by natural gas and by expanding distributed installation of green and blue power, so that the goal of 25% to 35% GHG emission reduction will be met therein by 2030. These plants could rapidly switch to or accept syngas and hydrogen.

4. The implementations of technologies producing green and blue energy are spreading and becoming cheaper than conventional fossil fuel. Installing PV panels on the roofs of buildings and homes makes money. Such installations are not only possible, but are affordable and already occurring even in developing countries. China is installing the largest PV solar power plants in the world, far larger than any such plants in the US or Europe.

5. The situation with the generation of solid wastes and plastics has become dire, and something must be done in a relatively short time. Indirect waste and residual solids gasification as a part of the integrated energy/water nexus solution outlined in this book, if implemented, would have a very beneficial effect and would bring dramatic reduction of GHG emissions from plastic and MSW. The technologies are known and tested and should be implemented immediately. Plastic waste is a great source of energy and recycled materials.

6. Occurrences of harmful algal blooms partially caused by global warming have become catastrophic in Florida (red tide and cyanobacteria blooms in many lakes, including the largest in Florida, Lake Okeechobee) and other US states (e.g., Clear Lake in California, Lake Erie), China, and some European countries, and will become worse under the BAU scenario.

7. Within a generation, rooftop photovoltaics will be common, people will be driving electric and hydrogen cars, riding hydrogen-powered public transportation (including airlines) and living in communities and regions where energy will be derived from green and blue sources (wind, water, solar, biomass, or waste).

8. The pressure on politicians magnified by the media reporting of continuing and increasing inconveniences and catastrophes caused by global warming will be strong.

This book addressed the problems and provided solutions for municipal water, energy, and solids management. It has focused on the urban part of the entire problem and provided engineering solutions leading to the triple net-zero adverse impact of urban water, energy, and solid waste management. These systems and solutions would significantly reduce GHG emissions from urban residential and commercial areas but would not solve the entire problem. Global engineering solutions leading to the avoidance of "Hothouse Earth" are broader and there is an increasing interest in geoengineering science and solutions (NRC 2015; NAC, 2018; Watts, 2013), which focuses on prevention and remediation of acidification of oceans, resilient and net CO_2 consuming agriculture, transportation without GHG emissions, forest management and reforestation (Shell Global, 2018), coastal hazard management, sustainability of water resources and changes in availability and distribution of water, expanding production and use of nuclear energy (somewhat questionable), and adaptation to global climate changes that are unavoidable. However, the recent reports by the National Research Council of the US National Academies and by the IPCC pointed out that geoengineering alone cannot solve the global climate warming problem without "slashing fossil fuel emissions," switching from fossil-fueled economies to green and blue power, and CO_2 sequestering.

The problems are large; however, they are not unsurmountable. Currently, a power-generating company pays for fossil fuel mining and transporting from larger distances and earns income from selling the energy. The "profits" on energy sales are generally regulated and kept low. Similarly, water reclamation (used water treatment) utilities collect fees for discharges of used water into sewers and treatment and, at best, save some money by using in-plant-produced methane as a source of energy. Landfilling of garbage today has high tipping fees: a minimum of $50/ton in the US in 2018, with added costs of collecting, processing, and transporting These costs will be rising. Implementing the TNZ designs outlined in this book will have win-win benefits for the utilities and for users. Gradually or even in a short time, the current power plants could be redesigned to accept local MSW and other carbonaceous waste energy sources – for example, green fuel solids and algae – and switch to green energy sources, partially at first and fully soon thereafter. Because the ANovaTNZ and similar integrated power plants with indirect gasification produce energy gas (syngas) and the MCFC power generation can accept methane and natural gas (mostly methane) along with syngas, energy production could continue using methane for some time. Later, stored hydrogen or syngas can be used for overcoming fuel shortages. MCFC power plants can also accept diluted CO_2 in flue gas from other facilities (for a fee) and concentrate it for sequestering or beneficial reuse (FuelCell Energy, Connecticut). The concentrated CO_2 has other circular economy uses such as production of polymers for plastics or dry ice or even reversion to syngas by SOFC electrolysis if excess blue energy is available.

Would it be possible for mining companies that know how to handle, transport, and sort coal and other minerals to accept local MSW and other green fuels, sort them, and reprocess them as fuel instead of sending MSW to China and India? Mining is a very dangerous profession due to black lung disease and methane explosions in the deep mines, which kill hundreds of miners every year. Former miners could also manage CO_2 sequestering in deep abandoned mines and other geological formations such as basalt. This would be a win-win job-creating situation that would bring a double income by collecting the waste solids (dumping fees) and selling them after processing as fuel to power companies, plus receiving income from recovering other recyclable materials. To some degree this is already happening in Sweden and Denmark, which accept MSW from other countries for green energy production and make a profit. It is very likely that these job opportunities will be more plentiful than the current job opportunities in subsidized coal mining. The alternative of garbage export is diminishing rapidly.

The new ANovaTNZ "power plants" may not pay for the MSW and sludge residue "fuel"; instead they could collect tipping fees on the entry side for accepting the MSW, including plastics and perhaps tires, for processing into fuel and other recycling raw materials and sell the produced electricity and recovered resources. Or presorting and preparing the MSW fuel may be another industry evolving from current recycling companies. The produced energy can be increased by installed PV panels and wind turbines on their premises and/or operate them in the community. They can also sell "credits" for green and blue energy to the enterprises that are not carbon neutral if there is a carbon tax. These "power recovery and recycle plants" can be large and region-wide, as in China, or smaller, serving a city or a county (district) like smaller power plants today. Besides electricity, the large plants processing combustible MSW and other combustible raw materials from agriculture and vegetation will produce hydrogen that, like natural gas, could be distributed to hydrogen recharging stations at airports and former gas stations, regional and local hydrogen fuel cell power stations, and even buildings for heating and cooking. Regional hydrogen fuel power facilities may also produce a little ultra-clean water.

These regional energy- and resource-producing facilities could be connected to or include a network of interconnected distributed water and resource reclamation plants that will process liquid black water and organic waste liquids and will produce methane, hydrogen, and residual organic solids that can be converted into fertilizers and soil conditioner and, above all, clean water. These facilities may also accept algal biomass from nutrient removal and CO_2 sequestering algal farms. Hence, reaching triple net-zero impact management of urban areas can be achieved multiple ways. This will bring radical changes to the ways people live; produce food, energy, and materials; and recycle and dispose of waste and will become the generational fifth energy revolution.

REFERENCES

Abma, W.R., C.E. Schultz, J.W. Mulder et al. (2007) Full-scale granular sludge Anammox process," *Water Science and Technology*, 55(8–9): 27–33.

Adey, W., and K. Loveland (2007) *Dynamic Aquaria*, 3rd ed., Academic Press, Elsevier, https://www.elsevier.com/books/dynamic-aquaria/adey/978-0-12-370641-6.

Adey, W.H., P.C. Kangas, and W. Mulbry (2011) Algal Turf Scrubbing: Cleaning Surface Waters with Solar Energy while Producing a Biofuel, *BioScience*, 61(6):434–441, https://doi.org/10.1525/bio.2011.61.6.5.

Adu, R.O., and R. Lohmueller (2012) The use of organic waste as an eco-efficient energy source in Ghana, *J. Environmental Protection*, 3:553–562 http://dx.doi.org/10.4236/jep.2012.37066.

Aelterman, P., M. Versichele, M. Marzorati, N. Boon, and W. Verstraete (2008) *Bioresource Technology* 99:8895–8902, DOI:10.1016/j.biortech.2008.04.061.

Ahern, J. (2007) Green infrastructure for cities: The spatial dimension, a chapter in *Cities of the Future: Towards integrated sustainable water and landscape management* (V. Novotny, and P. Brown, eds.), IWA Publishing, London.

Ahern, J. (2010) Planning and design for sustainable and resilient cities: Theories, strategies, and best practices for Green infrastructure, Chapter 3 in *Water Centric Sustainable Communities* (V. Novotny et al.), John Wiley & Sons, Hoboken, NJ.

Ahmed, I.I., and A.K. Gupta (2010) Pyrolysis and gasification of food waste: Syngas characteristics and char gasification kinetics, *Applied Energy*, 87:101–108.

Ahn, K.H., et al. (2003) Enhanced biological phosphorus and nitrogen removal using a sequencing anoxic/anaerobic membrane bioreactor (SAM) process, *Desalination*, 157:345–352, https://doi.org/10.1016/S0011-9164(03)00415-6.

Ahn, Y., and Logan, B.E. (2012) A multi-electrode continuous flow microbial fuel cell with separator electrode assembly design, *Appl Microbiol Biotechnol* 93(5):2241–2248, DOI: 10.1007/s00253-012-3916-4.

Ajjawi, I., et al. (2017) Lipid production in *Nannochloropsis gaditana* is doubled by decreasing expression of a single transcriptional regulator, *Nature Biotechnology* 35:647–652, https://www.nature.com/articles/nbt.3865?foxtrotcallback=true.

Akanyeti, I., H. Temmink, M. Trmy, and A. Zwijnenburg (2010) Feasibility of bioflocculation in a high-loaded membrane bioreactor for improved energy recovery from sewage, *Water Sci. Technol.* 61(6):1433–1439, https://doi.org/10.2166/wst.2010.032.

Allan, J.A. (2005) Water in the Environment/Socio-Economic Development Discourse: Sustainability, Changing Management Paradigms and Policy Responses in a Global System, *Government and Opposition*, 40(2):181–199, https://doi.org/10.1111/j.1477-7053.2005.00149.x.

Allen, L., J. Christian-Smith, and M. Palaniappan (2010) *Overview of Greywater Reuse: The Potential of Greywater Systems to Aid Sustainable Water Management*, Pacific Institute, Oakland, CA 94612 http://www.pacinst.org/wp-content/uploads/sites/21/2013/02/greywater_overview3.pdf, accessed July 2015.

Angelidaki, I., and L. Ellegaard (2003) Codigestion of Manure and Organic Wastes in Centralized Biogas Plants: Status and Future Trend, *Appl Biochem Biotechnol* 109(1–3):95–105. https://doi.org/10.1385/ABAB:109:1-3:95.

Antonopoulou, G., H.N. Gavala, I.V. Skiadas, K. Angelopoulos, and G. Lyberatos (2008) Biofuels generation from sweet sorghum: Fermentative hydrogen production and anaerobic digestion of the remaining biomass, *Bioresour. Technol.* 99:110–119, DOI:10.1016/j.biortech.2006.11.048.

Ara, E., M. Sartaj, and K. Kennedy (2015) Enhanced biogas production by anaerobic co-digestion from a trinary mix substrate over binary mix, substrate, *Waste Management & Research* 33(6): 578–587.

Aranda, G., A. van der Drift, B.J. Vreugdenhil, H.J.M. Visser, C.F. Vilela, and C.M. van der Meijden (2014) Comparing direct and indirect fluidized bed gasification: Effect of redox cycle on olivine activity, *Environmental Progress & Sustainable Energy*, 33(3)711–720, https://onlinelibrary.wiley.com/doi/abs/10.1002/ep.12016.

Archer, D., M. Eby, V. Brovkin, A. Ridgwell, L. Cao, U. Mikolajewicz, K. Caldeira, K. Matsumoto, G. Munhoven, A. Montenegro, and K. Tokos (2009) Atmospheric lifetime of fossil fuel carbon dioxide, *Annual Review of Earth and Planetary Sciences* 37:117–134.

Arena, U. (2012) Process and technological aspects of municipal solid waste gasification. A review, *Waste Management*, 32:625–639, DOI:10.1016/j.wasman.2011.09.025, accessed 4/2018 https://pdfs.semanticscholar.org/2256/fd8b42004147443b6f7a9ff69db6cfa5ba99.pdf.

Arends, J.B.A., and W. Vestraete (2012) 100 years of microbial electricity production: three concepts for the future, *Microbial Biotechnology*, 5(3):333–346, https://doi.org/10.1111/j.1751-7915.2011.00302.x.

Arlington, A. (2017) Looming Chinese import ban creates U.S. Recycling Bottleneck, *Bloomberg Environment and Energy Report*, https://news.bloombergenvironment.com/environment-and-energy/looming-chinese-import-ban-creates-us-recycling-bottleneck, accessed January 2018.

Asahi, S. et al. (2017) Two-step photon up-conversion solar cells, *Nature Communications*, Article number: 14962. DOI: 10.1038/ncomms14962.

Asano, T, F.L. Burton, H.L. Leverenz, R. Tsuchihashi, and G. Tchobanoglous (2007) *Water Reuse: Issues, Technologies, and Applications, Metcalf & Eddy/WECOM*, McGraw Hill, New York.

Atienza-Martinez, M., G. Gea, J. Arauzo, S., R.A. Kersten, and A.M.J. Kootstra (2014) Phosphorus recovery from sewage sludge char ash. *Biomass & Bioenergy*, 65:42–50, https://doi.org/10.1016/j.biombioe.2014.03.058.

Atrea, S.K. (2009) The mystery of methane in Mars and Titan, *Scientific American*. January 2009, https://www.scientificamerican.com/article/methane-on-mars-titan/.

AWWA-RF (1999) *Residential End Uses of Water, American Water Works Association*, Research Foundation, Denver, CO.

Bakos, J. (2005) Renewable solar hydrogen production, *Environmental Science and Engineering*, May 2005; http://www.esemag.com/archive/0505/solar.html, accessed October 2009 .

Barnard, J. (2007) *Elimination of Eutrophication through Resources Recovery*, 2007 Clarke Lecture, National Water Research Institute, Fountain Valley, California.

Bartáček, J., and V. Novotny (2015) Recovery of energy from municipal used water, Proc. *Water and Energy 2015:Opportunities for Energy and Resource Recovery in the Changing World*, WEF-IWA Symposium June 7–10, 2015, Washington, DC.

BC Ministry of Community Development (2009) *Resource from Waste: A Guide to Integrated Resources Recovery*, ISBN 978-0-7726-6116-6, Victoria, BC, Canada, https://www2.gov.bc.ca/assets/gov/british-columbians-our-governments/local-governments/planning-land-use/resources_from_waste_irr_guide.pdf.

BC Ministry of Community Development (2019) *Integrated Resource Recovery Case Study: Dockside Green Mixed Use Development*. Victoria, BC, Canada https://www.waterbucket.ca/gi/sites/wbcgi/documents/media/271.pdf.

Beatley, T. (2011) *Biophilic Cities: Integrating Nature into Urban Design and Planning*, Island Press, Washington, DC.

Beatley, T. (2017) *Handbook of Biophilic City Planning and Design*, Washington, DC: Island Press.

Beecher, N., and Y. Qi (2012) *DRAFT Biogas Study Report – Use of Biogas at Wastewater Treatment Plants in the United States*. Water Environment Research Foundation (WERF), Alexandria, VA.

Belgiorno, V., G. De Feo, C. Della Rocca, and R.M.A. Napoli (2003) Energy from gasification of solid wastes, *Waste Management* 23:1–15, DOI:10.1016/S0956-053X(02)00149-6.

Biello, D. (2014) Can Carbon Capture Technology Be Part of the Climate Solution? *Yale Environment 360*, Yale University School of Forestry and Environmental Studies, New Haven, CT, http://e360 .yale.edu/features/can_carbon_capture_technology_be_part_of_the_climate_solution.

Benemann J.R, (1996) Hydrogen biotechnology: progress and prospects. *Nature Biotechnol.* 14: 1101–1103.

Biggs, R. et al. (2012) Toward principles for enhancing the resilience of ecosystem services, *Annual Review of Environment and Resources*, 37:421-448, https://www.annualreviews.org/doi/abs/ 10.1146/annurev-environ-051211-123836.

Bingemer, H.G., and P.J. Crutzen (1987) The production of methane from solid wastes, *Journal of Geophysical Research* 92(D2):2181–2187, http://onlinelibrary.wiley.com/doi/10.1029/JD092iD02p0 2181/pdf.

Blunt, F., J, Fayers, M. Franklin, and J. Orr (1993) Carbon dioxide in enhanced oil recovery, *Energy Conversion and Management* 34(9–11):1197–1204.

Bogner, J., and E. Matthews (2003) Global methane emissions from landfills: New methodology and annual estimates 1980–1996, *Global Biogeochemical Cycles*, 17(2): 34-1 to 34-18, http:// onlinelibrary.wiley.com/wol1/doi/10.1029/2002GB001913/full, accessed June 2017.

Borgiani, C., P. De Fillipis, F. Pochetti and M. Paulucci (2002) Gasification process of wastes containing PVC, *Fuel* 81(14):1827–1863, https://doi.org/10.1016/S0016-2361(02)00097-2.

Bouška, V. (1981) *Geochemistry of Coal*, Elsevier Science & Technol., Oxford, UK.

Boyd, C.E. (1978) Chemical composition of wetland plants. In: *Fresh Water Wetlands: Ecological Processes and Management Potential*, (R.E. Good, D.F. Whigham, and R.L Simpson, Editors), Academic Press, New York, pp. 99–114.

Breeze, P. (2017) The molten carbonate fuel cell, *Fuel Cells,* Academic *Press*. https://doi.org/10.1016/ B978-0-08-101039-6.00006-6.

Bridgwater, A.V. (1994) Catalysis in thermal biomass conversion, *Applied Catalysis A: General* 116:5–47.

Brotzman, R. (2015) Hydrocyclone Separation of Targeted Algal Intermediates and Products, Argone National Laboratory presentation, https://www.energy.gov/sites/prod/files/2015/04/f21/ algae_laible_133100.pdf.

Brundtland, G.H. (1987) *Our Common Future – Report of the World Commission on Environment and Development,* United Nations, http://www.un-documents.net/our-common-future.pdf.

Buchanan, R, and T.R. Carr (2011) Geologic sequestration of carbon dioxide in Kansas, *Kansas Geological Survey*, Publ. Circular 27, http://www.kgs.ku.edu/Publications/PIC/PIC27_2011.pdf.

Bullis, K. (2013) Fuel cells could offer cheap carbon dioxide storage, *MIT Technology Reviews*, https:// www.technologyreview.com/s/515026/fuel-cells-could-offer-cheap-carbon-dioxide-storage/, Cambridge, MA, accessed November 2017.

Burke, D.A. (2015) A profitable process to recover ammonia nitrogen and phosphate from anaerobic digestate, *Proceedings Residuals and Biosolids Conference 2015: The Next Generation of Science, Technology, and Management*, Session 9, WEF, IWA, Washington, DC, June 7–10.

Call, D.F., R. Wagner, and B.E. Logan (2009) Hydrogen production by Geobacter species and a mixed consortium in a microbial electrolysis cell, *Appl. Environ. Microbiol.* 75(24):7579–7587, DOI:10.1128/AEM.01760-09, https://aem.asm.org/content/aem/75/24/7579.full.pdf.

Callegari, A., P. Hlavinek, and A.G. Capodaglio (2018) Production of energy (biodiesel) and recovery of materials (biochar) from pyrolysis of urban waste sludge, *Ambiente & Aqua: Journal of Applied Science*, 13(2), DOI:10.4136/ambi-aqua.2128.

Campanari, S. (2002) Carbon dioxide separation from high temperature fuel cell power plants, *Journal of Power Sources* 112(1):273–289, https://doi.org/10.1016/S0378-7753(02)00395-6.

Campanari, S., G. Manzolini, and P. Chiesa (2013) Using MCFC for high efficiency CO_2 capture from natural gas combined cycles: Comparison of internal and external reforming, *Applied Energy*, 112: 772–783, https://doi.org/10.1016/j.apenergy.2013.01.045.

Capdevielle, A., E. Sýkorová, B. Biscans, F. Béline, and M-L. Daumer (2013) Optimization of struvite precipitation in synthetic biologically treated swine wastewater – determination of the optimal process parameters. *J Hazard Mater.* 244–245:357–369.

Capdevielle, A., E. Sýkorová, F. Bélinne, and M.L. Daumer (2014) Kinetics of struvite precipitation in synthetic biologically treated swine wastewaters, *Environmental Technology*, 35(10):1250–1262, DOI: 10.1080/09593330.2013.865790.

Capdevielle, A., E. Sýkorová, F. Bélinne, and M.L. Daumer (2016) Effects of organic matter on crystal-lization of struvite in biologically treated swine wastewater, *Environ Technol.* 37(7):880–92. DOI: 10.1080/09593330.2015.1088580.

Capodaglio, A.G., D. Molognoni, E. Dallago, A. Liberale, R. Cella, P. Longoni, and L. Pantaleoni (2013) Microbial fuel cells for direct electrical energy recovery from urban wastewaters, *Scientific World Journal*, 2013: 634738, DOI: `10.1155/2013/634738`, https://www.ncbi.nlm.nih.gov/pmc/articles/PMC3881690/.

Caprotti, F. (2014) Eco-urbanism and the Eco-city, or, Denying the Right to the City? *Antipode*, 46(5): 1285–1303, accessed July 2017. http://dx.doi.org/10.1111/anti.12087.

Carpenter, S, R., et al. (2012) General Resilience to Cope with Extreme Events, *Sustainability* 4:3248–3259, DOI:10.3390/su4123248.

Carre, F. (2014) France's perspective on sustainable energy and integrating distribution, presentation at the 2014 University of Wisconsin Energy Summit, Madison, WI, October 29, 2014, accessed January 2015, https://energy.wisc.edu/sites/default/files/Carre.pdf.

Carlson, R.E. (1977) A trophic state index for lakes, *Limnol. Oceanogr.* 22:361–369.

Carson, Rachel (2002) *Silent Spring*, Houghton Mifflin Harcourt (first published 1962).

Cassir, M., A Meléndez-Caballeros, A. Riguedé, and V. Lair (2016) Molten carbonate fuel cells Chapter 3 in Compendium of Hydrogen Energy, Volume 3 of Hydrogen Energy Conversion Series, Elsevier, https://doi.org/10.1016/B978-1-78242-363-8.00003-7.

Cerff M., N. Morweiser, R. Dillschneider et al. (2012) Harvesting fresh water and marine algae by magnetic separation: screening of separation parameters and high gradient magnetic filtration. *Bioresource Technology* 118:289–295, https://www.ncbi.nlm.nih.gov/pubmed/22705536?dopt=Abstract.

Chamchoi N., S. Nitisoravut, and J.E. Schmidt JE (2008) Inactivation of ANAMMOX communities under concurrent operation of anaerobic ammonium oxidation (ANAMMOX) and denitrification, *Bioresource Technology*, 99 (9):3331–3336, https://www.ncbi.nlm.nih.gov/pubmed/17911013.

Chang, S, (2014) Anaerobic membrane bioreactor (AnMBR) for wastewater treatment, *Advances in Chemical Engineering and Science*, 4:56–61, DOI:10.4236/aces.2014.41008.

Chen, H.B. et al. (2010) Investigation od domestic wastewater separately discharging and treating in China, in *Water Infrastructure for Sustainable Communities: China and the World* (X. Hao, V. Novotny, and V. Nelson, eds.) IWA publishing, London, UK.

Cheng, S., and B.E. Logan (2007) Sustainable and efficient biohydrogen production via electrohydro-genesis. *PNAS*, 104(47):18871–18873, https://www.engr.psu.edu/ce/enve/logan/publications/2007-Cheng&Logan-PNAS.pdf.

Cheng, S., and B.E. Logan (2007) Sustainable and efficient biohydrogen production via electrohydro-genesis. *PNAS*, 104(47): 18871–18873.

Cherepy, N.J., R. Krueger, K.J. Fiet, A.F. Jankowski and J.F. Cooper (2005) Direct Conversion of Carbon Fuels in a Molten Carbonate Fuel Cell, *Electrochem. Soc.*, 152 (1):A80–A87, DOI: 10.1149/1.1836129J.

Childers, D.I., J. Corman, M. Edwards, and J. Elser (2011) Sustainability challenges of phosphorus and food: Solutions from closing the human phosphorus cycle, *BioScience* 61 (2): 117–124.

Chimney, M. et al. (2017) Performance and operation of the Everglades Stormwater Treatment Areas (Chapter 5B), *2017South Florida Environmental Report*, vol. 1, South Florida Water Management District, West Palm Beach, FL, www.sfwmd.gov/sfer.

Cho, M.H., Y.K. Choi, and J.S. Kim (2015) Air gasification of PVC (polyvinyl chloride)-containing plastic waste in a two-stage gasifier using Ca-based additives and Ni-loaded activated carbon for the production of clean and hydrogen rich producer gas, *Energy* 87(1):586–593, https://doi.org/10.1016/j.energy.2015.05.026.

Christgen, B., K. Scott, J. Dolfing, I.M. Head, and T.P. Curtis (2015) An evaluation of the performance and economics of membranes and separators in single chamber microbial fuel cells treating domestic wastewater, *PLOS ONE* 10(8): e0136108. https://doi.org/10.1371/journal.pone.0136108.

Ciccoli, R., et al. (2010) Molten carbonate fuel cells fed with biogas: Combating H_2S, *Waste Management*, 30(6):1018–1024, https://doi.org/10.1016/j.wasman.2010.02.022.

Cipriani, P., P. De Fillippis, and F. Pochetti (1998) Solid waste gasification: energy recovery from polyethylene biomass mixtures, *J. Solid Wastes Technol. & Management*, 25(2):77–81, https://www.researchgate.net/publication/289131689_Solid_waste_gasification_Energy_recovery_from_polyethylene_biomass_mixtures.

Circle Economy (2015) http://www.circle-economy.com/circular-economy/?gclid=CNjYprLDzMkCFQsjHwodj2sPkA.

City of Cape Town (2018) Commercial Water Restrictions Explained, accessed June 2018. http://www.capetown.gov.za/work%20and%20business/commercial-utility-services/commercial-water-and-sanitation-services/Commercial-water-restrictions-explained.

City of Portland, Oregon (2014) *Kenton Living Building,* Bureau of Planning and Sustainability, accessed January 2014, https://www.portlandoregon.gov/bps/article/437414.

Clark, D.E., V. Novotny, R. Griffin, D. Booth, A Bartošová, M.C. Daun and M. Hutchinson (2001) Willingness to pay for flood and ecological risk reduction in an urban watershed. Wat. *Sci. & Technol.*, 45(9):235–242, 2001.

Committee to Assess the Scientific Basis of the TMDL Approach to Water Pollution (2001) *Assessing TMDL Approach to Water Quality Management*, National Academy Press, Washington, DC.

Committee on Health Effects of Waste Incineration (2000) Chapter 3 Incineration Processes and Environmental Releases, in *Waste Incineration & Public Health*, National Research Council, National Academies Press, Washington (DC): https://doi.org/10.17226/5803.

Cotterill, S., E. Heindrich, and T. Curis (2016) Microbial Electrolysis Cells for Hydrogen Production, Chapter 9 in *Microbial Electrochemical and Fuel Cells: Fundamentals and Applications* (K. Scott and E.H. Yu, eds.), pp. 287–319. Woodhead Publishing Series in Energy. Elsevier, DOI:10.1016/B978-1-78242-375-1.00009-5.

Cusick, R.D., M.L. Ullery, B.A. Dempsey, and B.E. Logan (2014) Electrochemical struvite precipitation from digestate with a fluidized bed cathode microbial electrolysis cell, *Water Research* 54: 297–306, http://dx.doi.org/10.1016/j.watres.2014.01.051.

Cuthbertson, A. (2018) Tesla is turning 50,000 homes in south Australia into a giant battery, *Newsweek,* February 5, http://www.newsweek.com/tesla-turning-50000-homes-south-australia-giant-battery-799273.

Dahia, S., and S. Venkata Mohan (2018) Selectivity control of volatile fatty acids production from food waste by regulating biosystem buffering: A comprehensive study, *Chemical Engineering Journal*, https://doi.org/10.1016/j.cej.2018.08.138.

Daigger, G. (2009) Evolving urban water and residuals management paradigms: Water reclamation and reuse, decentralization, resource recovery, *Water Environment Research* 81(8):809–823.

Damkjaer S., and R. Taylor (2017) The measurement of water scarcity: Defining a meaningful Indicator, *Ambio*, 46:513–531, DOI 10.1007/s13280-017-0912-z.

Daneshagar, S., A. Callegary, A.C. Capodaglio, and D. Vaccari (2018) The potential phosphorus crisis: Resources conservation and possible escape technologies: A review, *Resources* 7(37), DOI:10.3390/resources7020037.

Darestani, M. et al. (2017) Hollow fibre membrane contactors for ammonia recovery: Current status and future developments, *Journal of Environmental Chemical Engineering* 5(2):1349–1359, https://doi.org/10.1016/j.jece.2017.02.016.

de Jong, M., S. Joss, D. Schraven, C. Zhan, C., and M. Weijnen, M. (2015) Sustainable smart–resilient–low carbon–eco–knowledge cities; making sense of a multitude of concepts promoting sustainable urbanization. *Journal of Cleaner Production*, 109: 25–38, December 2015, accessed July 2017, http://www.sciencedirect.com/science/article/pii/S0959652615001080?via%3.

de Vosa, R.M., W.F. Maierb and H. Verweija (2015) Hydrophobic silica membranes for gas separation. *Journal of Membrane Sciences* 158(1–2):277–288.

Detz, R.J., J.N.H. Reek, and B.C. van der Zwaan (2018) The future of solar fuels: when could they become competitive? *Energy Environ. Sci.*, Advance Article accessed July 2018, http://dx.doi.org/10.1039/C8EE00111A.

Diamantis, V., Eftaxias, A., Bundervoet, B., Verstraete, W., 2014. Performance of the biosorptive activated sludge (BAS) as pre-treatment to UF for decentralized wastewater reuse. Bioresour. Technol. 156, 314–321, DOI: 10.1016/j.biortech.2014.01.061.

Dicks, A. (996) Hydrogen generation from natural gas for the fuel cell systems of tomorrow, Jpurnal of Power Sources 61(1–2):113–124, https://www.sciencedirect.com/science/article/pii/S0378775396023476?via%3Dihub.

Dierberg, F.E., T.A. DeBusk, S.D. Jackson, M.J. Chimney, and K. Pietro (2002) Submerged aquatic vegetation-based treatment wetlands for removing phosphorus from agricultural runoff: response to hydraulic and nutrient loading, *Water Research* 36(6):1409–1422, http://dx.doi.org/10.1016/S0043-1354(01)00354-2.

Dietz, M.E. (2007) Low Impact Development Practices: A Review of Current Research and Recommendation for Future Directions. *Water Soil and Air Pollution* 186:351–363.

Ditzig, J., H. Liu, and B.E. Logan (2007) Production of hydrogen from domestic wastewater using a bioelectrochemically assisted microbial reactor (BEAMR), *Internat. J. Hydrogen Energy*. 32(13), 2296–2304.

Dodds, F., and J. Bartram (2016) *The Water, Food, Energy and Climate Nexus: Challenges and an Agenda for Action*. Taylor and Francis, DOI: 10.4324/9781315640716.

Dowaki, K. (2011) Energy Paths due to Blue Tower Process, Biofuel's Engineering Process Technology, Marco Aurelio Dos Santos Bernardes (Ed.), InTech, http://www.intechopen.com/books/biofuel-s-engineering-process-technology/energy-paths-due-to-blue-tower-process.

Dreiseitl, H., and D. Grau (2009) *Recent Waterscapes: Planning, Building and Designing with Water*, Birkhäuser Basel, Boston, Berlin.

Dvořák, L., M. Gómez, J. Dolina, and A. Černín (2016) Anaerobic membrane bioreactors: A mini review with emphasis on industrial wastewater treatment: Applications, limitations and perspectives, *Desalination and Water Treatment* 57:19062–19076, DOI: 10.1080/19443994.2015.1100879.

Dyke, P., and C. Foan (1997) A review of dioxin releases to land and water in the U.K. *Organohalogen Compounds* 32:411–416.

Ecocity Builders (2014) Ecocity definition. Available at: http://www.ecocitybuilders.org/why-ecocities/ecocity-definition/, accessed July 2017.

Eide, J. (2013) Rethinking CCS – Strategies for Technology Development in Times of Uncertainty, Master of Science in Technology and Policy, Massachusetts Institute of Technology, Cambridge, MA.

Eisenberg, B., K. Collins Lindow, and David R. Smith (2015) *Permeable Pavements, ASCE manual*, American Society of Civil Engineers, https://www.asce.org/templates/publications-book-detail.aspx?id=15418.

Ellen MacArthur Foundation (2017) Concept: What is a circular economy? A framework for an economy that is restorative and regenerative by design, https://www.ellenmacarthurfoundation.org/circular-economy/concept.

Elkington, J. (1997) *Cannibals with Forks: The Triple Bottom Line of 21st Century Business*, Capstone Publishing Ltd, Oxford, UK.

Engle, D. (2007) Green from top to bottom, *Water Efficiency*, 2(2):10–15.

Eriksson, E., K.Auffarth, M. Henze, and A. Ledin (2002) Characteristics of grey wastewater, *Urban Water* 4(1): 85–104.

Esteves, S., and D. Devlin (2010) *Food Wastes Chemical Analysis*, Final Report, Project code: COE-P029-09/COE-P036-10, Wales Centre for Excellence, WRAP, Wales, UK, http://www.wrapcymru.org.uk/sites/files/wrap/Technical_report_food_waste_characterisation_Wales_2009x2.9086.pdf.

Eurelectric (2003) *Efficiency in Electricity Generation*, Union of the Electricity Industry: EURELECTRIC, Brussels, Belgium.

Exxon Mobil (2018) Advanced biofuels and algae research, https://corporate.exxonmobil.com/en/research-and-technology/advanced-biofuels/advanced-biofuels-and-algae-research#/section/1-algae-for-biofuels-production, accessed September 2018.

Ezquerro, A. (2010) Struvite Precipitation and Biological Dissolution, TRITA-LWR Degree Project 10:22, 36p, http://www.diva-portal.org/smash/get/diva2:503781/FULLTEXT01.pdf.

Farr, D. (2008) *Sustainable Urbanism: Urban Design with Nature*, John Wiley & Sons, Hoboken, NJ.

Farzad, S., M. Ali Mandegari, J.F. Görgens (2016) A critical review on biomass gasification, co-gasification, and their environmental assessments, *Biofuel Research Journal* 12:483–495, https://www.biofueljournal.com/article_32132_c1804a20c9aec0867882b6412908bc05.pdf.

Fergers, J.W., R. Hui, X. Li, D. Wilkinson, and J. Zhang (2009) *Solid Oxide Fuel Cells, Materials, Properties and Performance*, CRC Press, Boca Raton, FL.

Field R. (1986) Urban stormwater runoff quality management: Low-structurally intensive measures and treatment. In *Urban Runoff Pollution*; (H.C. Torno, J. Marsalek, and M. Desbordes, eds.), Springer Verlag, Heidelberg, Germany and New York, pp. 677–699.

Fisher, J., and M.C. Acreman (2004) Wetland nutrient removal: a review of the evidence, *Hydrology and Earth System Science* 8:673–682, https://www.hydrol-earth-syst-sci.net/8/673/2004/hess-8-673-2004.pdf.

Folke, C., S. Carpenter, T. Elmqvist, L. Gunderson, C.S. Holling and B. Walker (2002) Resilience and Sustainable Development: Building Adaptive Capacity in a World of Transformations, *Ambio* 31 (5):437–440.

Forrest A.L., K.P. Fattah, D.S. Mavinic, and F.A. Koch (2008) Optimizing Struvite Production for Phosphate Recovery, *Journal of Environmental Engineering* 134(5):395–402© ASCE. DOI:10.1061/(ASCE)0733-9372 (2008)134:5(395).

Fraunhofer Institute (2014) New world record for solar cell efficiency at 46% French-German cooperation confirms competitive advantage of European photovoltaic industry, accessed July 2017, http://web.archive.org/web/20150823133519/http://www.ise.fraunhofer.de/en/press-and-media/press-releases/press-releases-2014/new-world-record-for-solar-cell-efficiency-at-46-percent.

Fredén, J. (2018) *The Swedish Recycling Revolution*, Swedish Institute, Sweden, Sverige, accessed September 2018, https://sweden.se/nature/the-swedish-recycling-revolution/.

Friedler, E. (2004) Quality of Individual Domestic Greywater Streams and its Implication for On-Site Treatment and Reuse Possibilities, *Environmental Technology* 4(9):997–1008.

Friedler, E., R. Kovalio and N.I. Galil. (2005) On-site greywater treatment and reuse in multistory buildings. *Water Science & Technology*, 51(10): 187–194.

Friedman, M. (2002) *Capitalism and Freedom: Fortieth Anniversary Edition*, University of Chicago Press, ISBN-13: 978-0226264219.

Fuel Cell Energy (2013) Equipment Profile: profile: Biogas-powered fuel cell, *Biomass magazine. Com*, July 2013, http://biomassmagazine.com/articles/9154/equipment-profile-biogas-powered-fuel-cells, accessed 2-2015.

Fuentes, B., N. Bolan, R. Naidu, and M. de la Luz Mora (2006) Phosphorus in organic soil-waste systems, *J. Soil Sc. Plant. Nutr.* 6(2):64 – 83, http://mingaonline.uach.cl/pdf/rcsuelo/v6n2/art06.pdf.

Gano, F. (2016) *Composition of the Atmosphere with Special Reference to its Oxygen Content*, Wentworth Press.

Ganguli, P., D. Kumar, and A.R. Ganguly (2017) Future droughts will severely impact power production, *Scientific Reports*, 7:11893, www.nature.com/scientificreports, accessed September 2017, https://scienceblog.com/496641/future-droughts-will-severely-impact-power-production/.

Gasification Technology Council (2014) *Gasification: The Waste-to-Energy Solutions*, http://www.gasification-syngas.org/uploads/downloads/GTC_Waste_to_Energy.pdf.

Gerdes, J (2012) *UC Merced's Triple Zero Commitment: Zero Net Energy, Zero Landfill Waste, and Climate Neutrality By 2020*, http://www.forbes.com/sites/justingerdes/2012/06/19/uc-merceds-triple-zero-commitment-zero-net-energy-zero-landfill-waste-and-climate-neutrality-by-2020/.

Gershman, Brickner & Bratton, Inc. (2013) *Gasification of Non-Recycled Plastics from Municipal Solid Waste in the United States*, Report Prepared for American Chemistry Council, Fairfax, VA, accessed June 2017, https://plastics.americanchemistry.com/Sustainability-Recycling/Energy-Recovery/Gasification-of-Non-Recycled-Plastics-from-Municipal-Solid-Waste-in-the-United-States.pdf.

Ghezel-Ayagh, H. (2018) SOFC Development Update at FuelCell Energy, 19th Annual Solid Oxide Fuel Cell (SOFC) Project Review Meeting Washington, DC, UD Department of Energy, accessed November 2018, https://www.netl.doe.gov/File%20Library/Events/2018/sofc/FE1-19th-Annual-SOFC-Workshop-FCE-Team.pdf.

Gianico, A., G. Bertanza, C.M. Braguglia, M. Canato, A. Gallipoli, G. Laera, C. Levantesi, G. Mininni (2015) Enhanced anaerobic processes on waste activated sludge: methane yields, hygienization potential and techno-economic feasibility, *Proceedings Residuals and Biosolids Conference 2015: The Next Generation of Science, Technology, and Management*, Session 9, WEF, IWA, Washington, DC, June 7–10.

Gil-Lalaguna, N., J.L. Sanchez, M.B. Murillo, and G. Gea (2015) Use of sewage sludge combustion ash and gasification ash for high-temperature desulphurization of different gas streams, *Fuel*, 141: 99–108.

Goldsmith, W., B. Barnhart, M. Same, and A.H. Same (2011) *Implementing a DOD Net-Zero Strategy, The Military Engineering March–April 2011* No. 670, pp. 73–74.

Gómez-Barea, A., and B. Leckner (2010), Modeling of biomass gasification in fluidized bed, *Prog. Energy Combust. Sci.* 36:444–509.

Goodside, M.E., and S. Sirku (eds.) (2017) *Green Defense Technology: Triple Net Zero Energy, Water and Waste Models and Applications*, NATO Science for Peace and Security Series C: Environmental Security, Springer International Publishing.

Gouveia, L., and A.C. Oliveira (2009) Microalgae as raw material for biofuel production, *J. Industrial Microbiol. & Biotechnology* 36(2):269–74, https://www.ncbi.nlm.nih.gov/pubmed/18982369.

Grootjes, A.J., B.J. Vreugdenhil and R.W.R. Zwart (2016) Indirect gasification of waste to create a more valuable gas, *Proc. 6 th International Conference on Engineering for Waste and Biomass Valorisation* – May 23–26, 2016, Albi, France, https://www.ecn.nl/publications/PdfFetch.aspx?nr=ECN-M--16-051.

Grydehøj, A., and Kelman, I. (2016) Island smart eco-cities: Innovation, secessionary enclaves, and the selling of sustainability. *Urban Island Studies*, 2:1–24, http://www.urbanislandstudies.org/UIS-2-Grydehoj-Kelman-Island-Smart-Eco-Cities.pdf, accessed July 2017.

Guest, J.S., et al. (2009) A New Planning and Design Paradigm to Achieve Sustainable Resource Recovery from Wastewater, *Environmental Science & Technology* 43:6126–6130.

Gupta, D., and S.K. Singh (2012) Greenhouse gas emissions from wastewater treatment plants: A case study of Neida, *Journal of Water Sustainability* 2(2):131–139.

Gouveia, J., F. Plaza, G. Garralon, F. Fdz-Polanco, and M. Peña (2015) Long-term operation of a pilot scale anaerobic membrane bioreactor (AnMBR) for the treatment of municipal wastewater under psychrophilic conditions, *Bioresour. Technol.* 185:225–233.

Guardian (2016) Soak it up: China's ambitious plan to solve urban flooding with "sponge cities," https://www.theguardian.com/public-leaders-network/2016/oct/03/china-government-solve-urban-planning-flooding-sponge-cities, accessed September 2017.

Gu at al. (2012) Hybrid H_2: Selective silica membranes prepared by chemical vapor deposition, *Separatopm Science and Technology*, 47(12):1698–1708.

Gűngőr-Demirci, G., and G.N. Demirer (2004) Effect of intial COD concentration, nutrient addition, temperature and microbial acclimation on anaerobic treatability of broiler and cattle manure, *Bioresource Technology*, 93:109–117, http://users.metu.edu.tr/goksel/environmental-biotechnology/pdf/42.pdf.

Hallenbeck P.C., and J.R. Benemann (2002) Biological Hydrogen Production: Fundamentals and limiting processes. *Intl. J. Hydrogen Energy* 27:1185–1193.

Halmann, M., and M. Steinberig (1998) *Greenhouse Gas Carbon Dioxide Mitigation: Science and Technology*, CRC Press, Boca Raton, FL.

Halvadakis, C.P., A.P. Robertson, and J.O. Leckie (1983) *Landfill methanogenesis: Literature review and critique*, Stanford Univ. Dep. of Civ. Eng. Tech. Rep. 271, Stanford Univ., Stanford, CA.

Hamadek, C.M.W., N.G. Guilford, and E.A. Edwards (2015) Chemical Oxygen Demand analysis of anaerobic digester content, *STEM Fellowship Journal, Chemical Engineering* 2(1):2–5, http://journal.stemfellowship.org/doi/pdf/10.17975/sfj-2015-008.

Hamilton, D.W. (2012) Anaerobic digestion of animal manures: Methane production potential of waste materials, BAE-1762, Oklahoma Cooperative Extension Service, http://pods.dasnr.okstate.edu/docushare/dsweb/Get/Document-8544/BAE-1762web.pdf.

Hampel, K. (2013) The Characterization of Algae Grown on Nutrient Removal Systems and Evaluation of Potential Uses for the Resulting Biomass, Dissertation 183, Western Michigan University, http://scholarworks.wmich.edu/dissertations/183.

Harder, A. (2019) The anatomy of climate-change policy: A primer, *Axios January* 22, 2019, https://www.axios.com/anatomy-of-climate-change-policy-primer-54b73843-8221-4bf8-9119-eee1f6f82ef1.html.

Hargreaves J.C, M.S. Adl, and P.R. Warman (2008) A review of the use of composted municipal solid waste in agriculture, *Agriculture, Ecosystems and Environment* 123:1–14, DOI:10.1016/j.agee.2007.07.004, http://docshare01.docshare.tips/files/9254/92546447.pdf.

Harwood, V.J.; Levine, A.D.; Scott, T.M.; Chivukula, V.; Lukasik, J.; Farrah, S.R.; Rose, J.B. (2005) Validity of the indicator organism paradigm for pathogen reduction in reclaimed water and public health protection. *Appl. Environ. Microbiol.*, 71 (6): 3163–3170.

Hawkes, F.R., H. Dinadale, D.I. Hawkes, and I. Hussy (2002) Sustainable fermentative hydrogen production: challenges for process optimization, *Interntl. J. Hydrogen Energy*, 27(11–12): 1339–1347.

He, M., et al. (2009) Hydrogen-rich gas from catalytic steam gasification of municipal solid waste (MSW): Influence of catalyst and temperature on yield and product composition, *Interntl. J. of Hydrogen Energy*, 34:195–203.

Heaney, J.P, L. Wright, and D. Sample (2000) Sustainable urban water management, Chapter 3 in *Innovative Urban Wet-Weather Flow Management Systems* (R. Field, J. P. Heaney, and R. Pitt, eds.) TECHNOMIC Publ. Comp., Lancaster, PA.

Heaney, J.P. (2007) Centralized and decentralized urban water, wastewater, & storm water systems, Ch 15, in *Cities of the Future: Towards Integrated Sustainable Water and Landscape Management* (V. Novotny and P. Brown, eds.), pp. 236–250, IWA Publishing, London.

Henze, M., and Y. Comeau (2008) Wastewater concentrations, in *Biological Wastewater Treatment: Principles, Modeling, Design* (M. Henze, M. van Looschrecht, G.A. Ekama, and and B. Brdjanovic, eds), IWA Publishing, London, pp. 33–52.

Hermann, L. (2018) OUTOTEC Modular Energy and Phosphorus Recovery Process, accessed June 2018 https://phosphorusplatform.eu/images/Conference/ESPC2-materials/Hermann%20poster %20ESPC2.pdf.

Hernandez, P., and P. Kenny (2010) From energy to net zero energy buildings: Defining life cycle zero energy buildings (LC-ZEB), *Energy and Buildings* 42(6):815–821.

Hernandez, A.B., J.H. Ferrasse, P. Chaurand, H. Saveyn, D. Borschneck, and N. Roche, N. (2011) Mineralogy and leachability of gasified sewage sludge solid residues, *Journal of Hazardous Materials*, 191(1–3), 219–227.

Herzog H. (2011) Scaling up carbon dioxide capture and storage: From megatons to gigatons. *Energy Economics*, 33:597–604.

Herzog, H., and J. Eide (2013) Rethinking CCS: Moving forward in times of uncertainty, *Mining Report*, 149(5):318–323.

Hill, K. (2007) Urban ecological design and urban ecology: an assessment of the state of current knowledge and a suggested research agenda, in *Cities of the Future: Towards integrated sustainable water and landscape management* (V. Novotny and P. Brown, eds), IWA Publishing, London.

Hoekstra, A.Y., and A.K. Chapagain (2008) *Globalization of Water: Sharing the Planet's Freshwater Resources.* Blackwell Publishing, Oxford, UK.

Holling, C.S. (1973) Resilience and stability of ecological systems, *Annual Review of Ecology and Systematics* 4:1–23.

Holmgren, K.E., H. Li, W. Verstraete, and P. Cornel (2016) *State of the Art Compendium Report on Resource Recovery from Water*, IWA Resource Recovery Cluster, International Water Association, London, http://www.iwa-network.org/publications/state-of-the-art-compendium-report-on-resource-recovery-from-water/.

Horton, R.P. (2017) The rise of urban vertical forest, *Urban Gardens,* accessed October 2017, http://www.urbangardensweb.com/2017/07/15/urban-vertical-forest/.

Howarth, R.B. (2007) Toward an operational sustainability criterion, *Ecological Economics*, 63:656–663.

ICF Consulting (2005) *Determination of the Impact of Waste Management Activities on Greenhouse Gas Emissions: 2005 Update Final Report*, submitted to Environment and Natural Resources of Canada, http://www.rcbc.ca/files/u3/ICF-final-report.pdf.

ICF International (2015) *Documentation for Greenhouse Gas Emission and Energy Factors Used in the Waste Reduction Model (WARM)*, https://www3.epa.gov/warm/pdfs/WARM_Documentation .pdf; Report prepared for the U.S. Environmental Protection Agency Office of Resource Conservation and Recovery, Washington, DC.

Imhoff, K. (1931) Possibilities and Limits of the water-sewage-water cycle, *Engineering News Report*, May 28.

Imhoff, K., and K.R. Imhoff (2007) *Taschenbuch der Stadtentwässerung (Pocket Book of Urban Sewage)* 28th ed., Oldenburg Verlag, Munich, Germany.

International Energy Agency (2016) *World Energy Outlook 2016*, Organization for Economic Cooperation and Development (OECD), Paris, France, accessed August 2017, https://www.iea.org/ newsroom/news/2016/november/world-energy-outlook-2016.html.

International Energy Agency (2017) 2016 Snapshot of Global Photovoltaic Market, Report IEA PVPS T1-31:2017, Organization for Economic Cooperation and Development (OECD),

Paris, France, http://www.iea-pvps.org/fileadmin/dam/public/report/statistics/IEA-PVPS_-_A_ Snapshot_of_Global_PV_-_1992-2016__1_.pdf, accessed July 2017.

IPCC (2007) *Summary for Policy Makers, Climate Change 2007: The Physical Scientific Basis*, Fourth Assessment Report, Intergovernmental Panel on Climatic Change, Working Group (WG 1), Geneva.

IPCC (2013) *Climatic change 2013, The Physical Science Basis,* WMO, UNEP, accessed January 2015, http://www.ipcc.ch/report/ar5/wg1/.

IPCC (2018) *Global Warming of 1.5°C*, Intergovernmental Panel on Climatic Change http://www.ipcc .ch/report/sr15/.

Island Press and the Kresge Foundation (2015) *Bounce Forward: Urban Resilience in the Era of Climatic Change*, Island Press, http://kresge.org/sites/default/files/Bounce-Forward-Urban-Resilience-in-Era-of-Climate-Change-2015.pdf.

Jacobson, M.Z, and M.A, Delucchi (2009) A path to sustainable energy by 2030, *Scientific American* 301(5):58–65, DOI:10.1038/scientificamerican1109–58.

Jacobson, M.Z. et al. (2017) 100% Clean and Renewable Wind, Water, and Sunlight All-Sector Energy Roadmaps for 139 Countries of the World, *Joule* 1:108–121, September 6, 2017 Elsevier, https://www.sciencedirect.com/science/article/pii/S2542435117300120?via%3.

Jacobson, M.Z., M.A. Deluchi, A. Cameron, and B.V. Mathiensen (2018) Matching demand with supply at low cost in 139 countries among 20 world regions with 100% intermittent wind, water, and sunlight (WWS) for all purposes, *Renewable Energy*, 123:236–248, https://doi.org/10.1016/j.renene.2018.02.009.

James, B.D., G.N. Baum, J. Perez, and K.N. Baum (2009) *Technoeconomic Boundary Analysis of Biological Pathways to Hydrogen Production*, Final Report NRE NREL/SR-560-46674 L, *National Renewable Energy Laboratory*, Boulder, CO, accessed January 2015, http://www.nrel.gov/docs/fy09osti/46674.pdf.

James, T., D. Ivanoff, T. Piccone, and J. King (2019) *Chapter 5C: 1 Restoration Strategies 2 Science Plan Implementation*, 2019 South Florida Environmental Report, vol. 1, South Florida Water Management District, West Palm Beach, FL, www.sfwmd.gov/sfer.

Jeníček, P., J. Bartáček, J. Kutil, J. Zábranská, and M. Dohanyos (2012) Potentials and limits of anaerobic digestion of sewage sludge: Energy self-sufficient municipal wastewater treatment plant? *Water Sci. Technol.* 66(6):1277-1281. DOI:10.2166/wst.2012.317.

Jepsen, S.E. (2005) Co-digestion of animal manure and organic household waste: The Danish experience, Danish EPA, Ministry of Environment and Energy, accessed February 2015, http://ec.europa .eu/environment/waste/compost/presentations/jepsen.pdf.

Jiang, A., T. Zhang, Q. Zhao, C. Frear, and S. Chen (2010) Ammonia recovery technologies in conjunction with dairy anaerobic digestion, *CSNAR Research Report 2010-01, Climate Friendly Farming,* Ch. 8, pp. 1–19, http://csanr.wsu.edu/wp-content/uploads/2013/02/CSANR2010-001.Ch08.pdf, accessed January 2015.

Johnson Foundation at Wingspread. (2014) *Charting New Waters: The Structure and Scale of Urban Water Management: Integrating Distributed Systems*, http://www.johnsonfdn.org/sites/default/files/ reports_publications/CNW-DistributedSystems.pdf.

Joss, S., R. Cowley, and D. Tomozeiu (2013) Towards the "ubiquitous eco-city": An analysis of the internationalisation of eco-city policy and practice. *Urban Research & Practice*, 6(1): 54–74. http:// dx.doi.org/10.1080/17535069.2012.762216.

Judd, S., and C. Judd (eds.) (2011) The MBR Book: Principles and Applications of Membrane Bioreactors for Water and Wastewater Treatment. 2nd Edition, Elsevier, Amsterdam.

Julian, P., B. Gu, G. Redfield and K. Weaver (2016) Chapter 3B: Mercury and Sulfur Environmental 1284 Assessment for the Everglades. In: *2016 South Florida Environmental Report*, vol. 1, West Palm Beach, FL, www.sfwmd.gov/sfer.

Kadam, S.R., V.R. Mate, R.P. Panmand, and L.K. Nikam, M.V. Kulkarni, R.S. Sonawane and B.B. Kale (2014) A green process for efficient lignin (biomass) degradation and hydrogen production via water splitting using nanostructured C, N, S-doped ZnO under solar light, | *RSC Adv.* 4:60626–60635, DOI: 10.1039/c4ra10760h.

Kadier, A., Y. Simayi, P. Abdeshahian, N. Farhana Azman, K. Chandrasekha, and M.S. Kalil (2016) A comprehensive review of microbial electrolysis cells (MEC) reactor designs and configurations for sustainable hydrogen gas production, *Alexandria Engineering Journal* 55:427–443, https://doi.org/10.1016/j.rser.2016.04.017.

Kadlec, R.H., and S. Wallace (2008) *Treatment Wetlands*, 2nd ed., CRC Press, Boca Raton, FL.

Kalmykova, Y., and K. Karlfeldt Fedje (2013) Phosphorus recovery from municipal solid waste incineration fly ash, *Waste Management*, 33(6):1403-1410, https://www.sciencedirect.com/science/article/pii/S0956053X13000743?via%3Dihub.

Keller, J., and K. Hartley (2003) Greenhouse gas production in wastewater treatment: Process selection is the major factor, *Water Sci. Technol.* 47(12):43-48.

Kennedy, C., J. Cuddihy, and J. Engel-Yan (2007) The changing metabolism of cities, *Journal of Industrial Ecology* 11(2):43-59, https://doi.org/10.1162/jie.2007.1107.

Kenny, S. (2016) Singapore's "Gardens of the Bay" and Milan's "Vertical Forests" Stole the Show in Planet Earth 2, MPORA, accessed November 2017, https://mpora.com/environment/singapores-gardens-bay-milans-vertical-forests-stole-show-planet-earth-2.

Khatib, S.J., S.T. Oyama, K. de Souza, and F.B. Naronha, (2011) Review of silica membranes for hydrogen separation prepared by chemical vapor deposition, Chapter 2 in *Membrane Science and Technology* 14:25–60.

Khatib, S.J., and S.T. Oyama (2013) Silica membranes for hydrogen separation prepared by chemical vapor deposition, (CVD), *Separation and Purification Technology* 111:20–42.

Khunjar, W., R. Latimer, and S, Jeyanayagam (2015) Nutrient recovery as a sustainable nutrient management tool, *Proceedings Residuals and Biosolids Conference 2015: The Next Generation of Science, Technology, and Management*, WEF and IWA, Washington, DC.

Kim J., Kim K, Ye H, Lee E, Shin C, McCarty PL, Bae J. (2011) Anaerobic fluidized bed membrane bioreactor for wastewater treatment, *Environ Sci Technol.* 45(2):576–81.

Kim, H.W., S.K. Han and H.S. Shin (2003) The optimisation of food waste addition as a co-substrate in anaerobic digestion of sewage sludge, *Waste Manag. Res.* 21(6):515–26.

Knight, G., C. Polol, P. Scanlon, G. Shimp, D. Long, and J. Takerstall (2015) Combined heat and power, for thermal hydrolysis: Optimization, *Proceedings Residuals and Biosolids Conference 2015: The Next Generation of Science, Technology, and Management*, Session 10B, WEF, IWA, Washington, DC, June 7–10.

Korelsliy, C., S. Fouladvand, S. Karimi, E. Sjöberg, and J. Hedlung (2015) Efficient ceramic zeolite membranes for CO_2/h_2 separation, *Journal of Material Chemistry A*, 3(25)12500-12506 DOI 10.1039/C5TA02152A.

Krabbenhoft, D.P. (1996) Mercury studies in Florida Everglades, U.S. Geological Survey, accessed 12/2018, https://pubs.usgs.gov/fs/0166-96/report.pdf.

Krüger, O., A. Grabner, and C. Adam (2014) Complete Survey of German Sewage Sludge Ash. *Environmental Science & Technology* 48(20):11811-11818.

Kuba, T., G. Smolders, M.C.M. van Loosdrecht, J.J. Heijnen (1993) Biological Phosphorus Removal from Wastewater by Anaerobic-Anoxic Sequencing Batch Reactor, *Water Sci. & Technol.*, 27 (5-6) 241-252, https://doi.org/10.2166/wst.1993.0504.

Kuba, T., M.C.M. van Loosdrecht, and J.J. Heijnen (1996) Phosphorus and nitrogen removal with minimal COD requirement by integration of denitrifying dephosphatation and nitrification in a two-sludge system. *Water Research*, 30(7):1702–1710, https://doi.org/10.1016/0043-1354(96)00050-4.

Kwon, E., K.J. Westby, and M.J. Castaldi (2010) Transforming municipal solid waste (MSW) into fuel via the gasification/pyrolysis process, *Proc. 19th Annual North Amer. Waste-to-Energy Conference*, NAWTEC18, May 11–13, 2010, Orlando Fl. accessed June 2015 http://www.seas.columbia.edu/earth/wtert/sofos/nawtec/nawtec18/nawtec18-3559.pdf.

Lackner, S., E.M. Gilbert, S.E. Vlaeminck, A. Joss, H/ Horn, and M.C.M. van Loosdrecht (2014) Full-scale partial nitritation/Anammox experiences: An application survey, *Water Research* 55:292–303, https://www.sciencedirect.com/science/article/pii/S0043135414001481?via%3Dihub.

Lantz, E., B. Sigrin, M. Gleason, R. Preus, and I. Baring-Gould (2016) *Assessing the Future of Distributed Wind: Opportunities for Behind-the-Meter Projects*, Technical Report NREL/TP-6A20-67337, National Renewable Energy Laboratory, US Department of Energy, Washington, DC, accessed July 2017, https://www.nrel.gov/docs/fy17osti/67337.pdf.

Lanyon, R. (2007) Developments towards urban water sustainability in the Chicago metropolitan area, In *Cities of the Future: Towards integrated sustainable water and landscape management* (V. Novotny and P. Brown, eds), IWA Publishing, London, pp. 8–17.

Latimer, R., Khunjar, W.O., Jeyanayagam, S., Mehta, C., Batstone, D., Alexander, R., (2012) *Towards a Renewable Future: Assessing Resource Recovery as a Viable Treatment Alternative* (NTRY1R12). Water Environment Research Foundation. Alexandria, VA.

Lau, C.H., B.T. Low, L. Shao, T.-S. Chung (2010) A vapor-phase surface modification method to enhance different types of hollow fiber membranes for industrial scale hydrogen separation, *Int. J. Hydrogen Energy* 35: 8970–8982.

Lee, U., J.N. Chung, and H.A. Ingley (2014) High temperature steam gasification of municipal solid waste, rubber, plastic and wood, *Energy & Fuel* 28:4573–4587, *Am. Chem. Soc,* dx.doi.org/10102/ef500713.

Lee, U., J. Han, and M. Wang (2017) Evaluation of landfill gas emissions from municipal solid waste landfills for the life-cycle analysis of waste-to-energy pathways, *Journal of Cleaner Production* 166:335–342, https://www.sciencedirect.com/science/article/pii/S0959652617317316?via%3Dihub.

Leichenko, R. (2011) Climate change and urban resilience, *Current Opinions in Environmental Sustainability* 3:164–168, Elsevier.

Leopold, A. (2001) *Sand County Almanac*, Oxford University Press, Oxford, UK.

Leopold. L.B., M G. Wolman, and J.P. Miller (1992) *Fluvial Processes in Geomorphology*, Dover Publications, New York.

Lettinga, G., A.F.M. Van Velsen, S.W. Hobma, and W. de Zeeuw and A. Klapwijk (1980) Use of upflow sludge blanket (USB) reactor for biological wastewater treatment, *Biotechnol. and Bioeng.* 22:699–734.

Lettinga, G., and L.W. Hulshoff–Pol (1991) UASB-process design for various types of wastewater, *Water Sci. & Technology* 24(8):201A–208A.

Levine, A.D., and T. Asano (2004) Recovering sustainable water from wastewater, *Environmental Science and Technology, June* 1, 2004, 201A–208A.

Li, D.H., W.L. Yang, and J.C. Lam (2013) Zero energy and sustainable development implications: A review, *Energy* 54(1):1–10.

Li, W.W., H.Q. Yu, and Z. He (2014) Towards sustainable wastewater treatment by using microbial fuel cells-centered technologies, *Energy & Environmental Science* 7:911–924, https://pubs.rsc.org/en/content/articlepdf/2014/ee/c3ee43106a.

Liao, B.Q., J.T. Kraemer, and D.M. Bagley (2006) Anaerobic membrane bioreactors: Applications and research directions, *Critical Reviews in Environmental Science and Technology* 36(6): 489–530, DOI: 10.1080/10643380600678146.

Logan, B.E., D. Cal, S. Cheng, H.V.M. Hamelers, T.H.J.A. Sleutels, A.W. Jeremiasse, and R. A. Rozendal (2008) Microbial electrolysis cells for high yield hydrogen gas production from organic matter, *Environ. Sci. Technol.* 42(23):8630–8640, https://doi.org/10.1021/es801553z.

Logan, B.E. (2008) *Microbial Fuel Cells*, Wiley-Interscience, Hoboken, NJ.

Lorrain, D., C. Halpern, and C. Chevauché (2018) *Villes sobres: Nouveaux modèles de gestion des resources* (Sober Cities: New Models of Resource Management) France. Presses de Sciences Po.

Lovins, A.B. (2005) *Twenty Hydrogen Myths*, White paper published by Rocky Mountain Institute, Snowmass, CO, www.rmi.org/sitepages/art7516.php.

Lucey, P., and C. Barraclough (2007) Acceleration adoption of integrated planning & design: A water-centric green value and restoration economy, Power point presentation at Northeastern University, Boston (MA), Aqua-Tex, Victoria, Canada.

Lu, L., N. Ren, D. Xing, and B.E. Logan, B.E. (2009) Hydrogen production with effluent from an ethanol–H2-coproducing fermentation reactor using a single-chamber microbial electrolysis cell, *Biosens. Bioelectron.* 24:3055–3060.

Lu, L., N. Ren, T. Xie, D. Xing, and B.E. Logan (2010) Hydrogen production from proteins via electrohydrogenesis in microbial electrolysis cells, *Biosen. Bioelectron.* 25(12):2690–269.

Luo, S., and C. Yi (2012) Syngas production by catalytic steam gasification of municipal solid waste in fixed-bed reactor, *Energy* 44(1:391–395.

Mahmoud, N., G. Zeeman, H. Gijzen, H., and G. Lettinga (2004) Anaerobic sewage treatment in a one-stage UASB reactor and a combined UASB-Digester system, *Water Res.* 38, 2347–2357.

Malkow, T. (2004) Novel and innovative pyrolysis and gasification technologies for energy efficient and environmentally sound MSW disposal, *Waste Management* 24(1):53–79, https://www.sciencedirect.com/science/article/pii/S0956053X03000382.

Maimon, A., A. Tal, E. Friedler, and A. Gross. (2010) Safe on-site reuse of greywater for irrigation – A critical review of current guidelines. *Environmental Science and Technology* 44:3213–3220.

Maršálek, J., R Ashley, B Chocat, M R Matos, W Rauch, W Schilling, and B. Urbonas (2007) Urban drainage at cross-roads: Four future scenarios ranging from business-as-usual to sustainability. In *Cities of the Future: Towards Integrated Sustainable Water and Landscape Management* (V. Novotny and P. Brown, Eds), Iwa Publishing, London, pp. 339–356.

Marszal, A.J. et al. (2011) Zero energy building: A review of definitions and calculations methodologies, *Energy and Buildings* 43(4):971–979.

Massachusetts Institute of Technology (MIT) (2007) *The future of coal: Options for a carbon constrained world*. MIT interdisciplinary report. Cambridge, MA.

Materazzi, M., P. Lettieri, R. Taylor, and C. Chapman (2016) Performance analysis of RDF gasification in a two stage fluidized bed–plasma process, *Waste Management* 47(Part B):256–266 https://www.sciencedirect.com/science/article/pii/S0956053X15004304.

Mato, S., Orto, D., Garcia, M. (1994) Composting of < 100mm fraction of solid waste, *Waste Management & Research* 12(4):315–325, https://doi.org/10.1177/0734242X9401200404.

McCarty, P.L., J. Bao, and J. Kim (2011) Domestic wastewater treatment as a net energy producer – can this be achieved? *Environ. Sci. & Technol.* 45:7100–7106.

McGrail, B.P. et al. (2017) Field validation of supercritical CO_2 reactivity with basalts, *Environ. Sci. Technol. Lett.*, 4(1): 6–10, DOI: 10.1021/acs.estlett.6b00387.

McJannet, C.L., P.A. Keddy, F.R. Pick. (1995) Nitrogen and phosphorus tissue concentrations in 41 wetland plants: a comparison across habitats and functional groups, *Functional Ecol.* 9:231–238, http://drpaulkeddy.com/pdffiles/McJannet%20et%20al.%201995%20wetland%20plant%20tissue%20nutrients.pdf.

McKay, G. (2002) Dioxin characterisation, formation and minimisation during municipal solid waste (MSW) incineration: review, *Chemical Eng. J.*, 86:343–368, http://nswaienvis.nic.in/Waste_Portal/Research_papers/pdf/rp_may15/Dioxin%20characterisation,%20formation%20and%20minimisation%20during%20MSW%20incineration.pdf.

McMahon, K.D., P.G. Stroot, R.I. Mackie, and L. Raskin (2001) Anaerobic codigestion of municipal solid waste and biosolids under various mixing conditions, II. Microbial population dynamics, *Water Research* 35(7):1817–1827.

McPhail, S.J., L. Leto, M. Della Pietra, V. Cigolotti, and A. Moreno (2015) *International Status of Molten Carbonate Fuel Cells Technology, Annex 23 – MCFC*, ENEA National Agency for New Technologies, Energy and Sustainable Economic Development, Rome, Italy, http://www.enea.it/it/pubblicazioni/pdf-dossier/2015_MCFCinternationalstatus.pdf.

Meda, A., J.J. Henkel, and P. Cornel (2012) Comparison of processes for greywater treatment for urban water reuse: Energy comparison and footprint, Chapter 14 in *Water, Energy Interactions in Water Reuse* (V. Lazarova, K.H. Choo, and P. Cornel, eds.), p. 203–211, IWA Publishing, London, https://www.iwapublishing.com/books/9781843395416/water-energy-interactions-water-reuse.

Mentens, J., D. Raes, and M. Herny (2003) Greenroofs as a part of urban water management. In *Water Resources Management II* (C.A. Brebbia, ed,), VIT Press, Southampton, UK, pp. 35–44.

Mekonnen, M.M., and A.Y. Hoekstra (2011) National Water Footprint Accounts: The Green, Blue And Grey Water Footprint of Production and Consumption Volume 1: Main Report, Research Report Series No. 50, UNESCO-IHE Institute for Water Education, Delft, The Netherlands, http://waterfootprint.org/media/downloads/Report50-NationalWaterFootprints-Vol1.pdf.

Metcalf and Eddy (2003) *Wastewater Engineering: Treatment and Reuse*, 4th ed., McGraw-Hill, NY.

Milewski, J., K. Świrski, M, Santarelli, and P, Leone (2018) Advanced Methods of Solid Oxide Fuel Cell Modeling, Springer, London, https://doi.org/10.1007/978-0-85729-262-9.

Mitchel, D. (2017) Google headquarters to headline Toronto's plan for a high-tech waterfront community, Global News, accessed November 2017, https://globalnews.ca/news/3808401/google-headquarters-to-headline-torontos-plan-for-a-high-tech-waterfront-community/.

Mitchell, X., K. Abeysuriya, and D. Fam (2008) *Development of Quantitative Decentralized System Concept: For the 2009 Metropolitan Sewerage Agency*, Institute for Sustainable Cities, University of Technology, Sydney, Australia.

Mitsch, W.J., and J.G. Gosselink (2015) *Wetlands*, 5th ed., J. Wiley & Sons, Hoboken, NJ.

Milwaukee Metropolitan Sewerage District (2012) *SeWeR: Sustainable Water Reclamation*, https://www.mmsd.com/application/files/9314/8416/1452/Sustainability_Plan.pdf.

Minnesota Pollution Control Agency (2019) Minnesota Stormwater Manual, PCA, St. Paul, MN, https://stormwater.pca.state.mn.us/index.php?title=Main_Page.

Moddemeyer, S. (2010) Yesler Terrace Sustainable District Study, Collins and Woerman & Gibson, Seattle, WA, https://www.seattlehousing.org/sites/default/files/YT_Sustainable_District_Study.pdf.

Moddemeyer, S. (2016) Urban Resilience, Lecture, Lake Como School for Advanced Studies. Como, Italy, May 2016.

Mohan Venkata, S., R. Saravanan, S. Veer Rghavulu, G. Mahahakrishna, and P.N. Sarma (2008) Bioelectricity production from wastewater treatment in dual chambered microbial fuel cell (MFC) using selectively enriched mixed microflora: Effect of electrolyte, *Bioresource Technology* 99:596–603.

Monteith, H.D., H.R. Sahely, H.L. MacLeon and D.M. Bagley (2005) A rational procedure for estimation of greenhouse-gas emissions from municipal wastewater treatment plants, *Water Environment Research* 77(4):390–403.

Morello-Frosch, R., M. Pastor, J. Sadd, and S. Shonko (2009) *The Climate Gap Inequalities in How Climate Change Hurts Americans and How to Close the Gap*, Program for Environmental and Regional Equity (PERE), Los Angeles, https://dornsife.usc.edu/assets/sites/242/docs/The_Climate_Gap_Full_Report_FINAL.pdf.

Montusiewicz, A. (2014) Co-digestion of sewage sludge and mature landfill leachate in pre-augmented system, *Journal of Ecological Engineering* 15(4)98–194.

Mosbergen, D. (2018) China no longer wants your trash. Here is why that's potentially disastrous, Huffington Post, January 24, https://www.huffingtonpost.com/entry/china-recycling-waste-ban_us_5a684285e4b0dc592a0dd7b9.

Munier, N. (2006) *Handbook on Urban Sustainability*, Springer Netherlands, Technology & Engineering.

Muradian, R. (2001) Ecological Thresholds: A Survey, *Ecological Economics* 38(1):7–24, DOI: 10.1016/S0921-8009(01)00146-X.

Murphy, C., F., and D.T. Allen (2011) Energy-water nexus for mass cultivation of algae, *Environ. Sci. & Technol.*45(13):5861–8, DOI:10.1021/es200109z.

Murphy, J.D., and E. McKeough (2004) Technical, economic and environmental analysis of energy production from municipal solid waste, *Renewable Energy* 29(7):1043–1067, http://www.sciencedirect.com/science/article/pii/S0960148103003951.

Münch, E.V., and K. Barr (2001) Controlled struvite crystallization for removing phosphorus from anaerobic digester sidestream, *Water Research* 35(1):151–159.

Nandi, R, and S, Sengupta (1998) Microbial production of hydrogen: an overview, *Critical reviews in microbiology.* DOI:10.1080/10408419891294181.

Nasr, P., and H. Sewilam (2015) Forward osmosis: an alternative sustainable technology and potential applications in water industry, *Clean Techn Environ Policy*, Springer-Verlag Berlin Heidelberg, 17-7, DOI 10.1007/s10098-015-0927-8.

National Academies Committee on Developing a Research Agenda for Carbon Dioxide Removal (2018) *Negative Emissions Technologies and Reliable Sequestration: A Research Agenda*, National Academies Press, Washington, DC, DOI: https://org/10.17226/25259.

National Energy Technology Laboratory (2009) Integration of H_2 Separation Membranes with CO_2 Capture and Compression, DOE/NETL 401/113009, US Department of Energy, accessed December 2017, https://www.netl.doe.gov/energy-analyses/pubs/H2_Mmbrn_Assmnt.pdf.

National Hydrogen Association (2010) Hydrogen and Fuel Cells: Then US Market Report, Washington, DC, http://www.ttcorp.com/pdf/marketReport.pdf, accessed February 2015.

National Research Council (2001) *Assessing the TMDL Approach to Water Quality Management*, Committee to Assess the Scientific Basis of the Total Maximum Daily Load Approach to Water Pollution Reduction, National Academy Press, Washington, DC, https://www.nap.edu/catalog/10146/assessing-the-tmdl-approach-to-water-quality-management.

National Research Council (2015) *Climate Intervention: Carbon Dioxide Removal and Reliable Sequestration*. Washington, DC: National Academies Press. https://doi.org/10.17226/18805.

National Science and Technology Council (2008) Federal Research and Development Agenda for net-zero Energy, High Performance Green Buildings, Office of the President of US, Washington, DC, https://www.wbdg.org/files/pdfs/fedird_netzero_energy_hp_green_buildings.pdf.

Navaneethan, N., P. Topczewski, S. Royer, and D. Zitomer (2010) Blending anaerobic co-digestates: Synergism and Economics, Proc. 12th IWA Specialty Congress on Anaerobic Digestion, Guadalajara, Mexico, October 2010.

Navigant (2016a) *Navigant Research Leaderboard Report: Global Wind Turbine Vendors* https://www.navigantresearch.com/reports/navigant-research-leaderboard-report-global-wind-turbine-vendors.

Negrea, A., L. Lupa, P. Negrea, M. Ciopes, and C. Muntean (2010) Simultaneous removal of ammonium and phosphate ions from wastewater and characterization of the resulting products, *Chemical Bulletin of "Politehnica,"* University of Timisoara, 55(69):136–142, https://www.researchgate.net/profile/Lavinia_Lupa/publication/265312325_Simultaneous_Removal_of_Ammonium_and_Phosphate_Ions_from_Wastewaters_and_Characterization_of_the_resulting_Product/links/54b768b50cf24eb34f6ea02a/Simultaneous-Removal-of-Ammonium-and-Phosphate-Ions-from-Wastewaters-and-Characterization-of-the-resulting-Product.pdf.

Nelson, B. (2015) Can the Bullitt Center prove that it pays for buildings to go "deep green"? *Guardian* 23, April 2015, http://www.theguardian.com/sustainable-business/2015/apr/23/deep-green-buildings-bullitt-center-living-building-challenge.

Ni, S.Q., and J. Zhang (2013) Anaerobic ammonium oxidation: from laboratory to full-scale application *BioMed Research International*, Volume 2013, Article ID 469360, https://www.hindawi.com/journals/bmri/2013/469360/.

NOAA (2015) *The Paleo Perspective on Global Warming -Instrumental Record of Past Global Temperatures*, National Oceanic and Atmospheric Administration, accessed January 2016, http://www.ncdc.noaa.gov/paleo/globalwarming/instrumental.html.

Noe-Hays, A. et al. (2016) Resource recovery done simpler and better, *Water Environment & Technology* 28(1):35–39.

Norstrom, R.J. (2002) Understanding bioaccumulation of POPs in food webs: Chemical, biological, ecological and environmental considerations, *Environmental Science and Pollution Research* 9(5):300–305.

Novotny, V., K.R. Imhoff, M. Olthof, and P.S. Krenkel (1989) *Karl Imhoff's Handbook of Urban Drainage and Wastewater Disposal*, John Wiley & Sons, Hoboken, NJ.

Novotny V., et al., (1998) Cyanide and metal pollution of urban snowmelt: Impact of deicing compounds, *Water Sci & Technol* 38(10):223–230.

Novotny, V., and H. Olem (1994) *Water Quality: Prevention, Identification and management of Diffuse Pollution*, Van Nostrand-Reinhold, New York (distributed by John Wiley & Sons, Hoboken, NJ).

Novotny, V. (2003a) *Water Quality: Diffuse Pollution and Watershed Management*, John Wiley & Sons, Hoboken, NJ.

Novotny, V. (2003b) Incorporating diffuse pollution abatement into watershed management: Watershed vulnerability, *Proc. 7th Intern. IWA Conf. on Diffuse Pollution and Basin Management*, Dublin, Ireland August 17–22, 2003.

Novotny, V. (2007a) Effluent dominated water bodies – their reclamation and reuse to achieve sustainability, in *Cities of the Future: Towards Integrated Sustainable Water and Landscape Management* (V. Novotny and P. Brown, eds.), pp. 191–214, IWA Publ. Co., London.

Novotny, V. (2007b) The new paradigm of integrated urban drainage and diffuse pollution abatement in the Cities of the Future, *Proc. XIth IWA International Conference on Diffuse Pollution*, Belo Horizonte, Brazil, August 26–31, 2007; also to be published in *Water Science and Technology*.

Novotny, V. (2008) Sustainable urban water management, In *Water and Urban Development Paradigms* (J. Feyen, K. Shannon, and M. Neville, eds.), pp. 19–31, CRC Press, Bocca Raton, FL, DOI 10.1201/9780203884102.pt1, https://www.researchgate.net/publication/271586464_Sustainable_Urban_Water_Management.

Novotny, V., and E.V. Novotny (2009) Water centric ecocities. Towards macroscale assessment of sustainability, *Water Practice and Technology* 4, no. 4.

Novotny, V., J.F. Ahern, and P.R. Brown (2010a) *Water Centric Sustainable Communities: Planning, Retrofitting and Constructing the Next Urban Environments*, John Wiley & Sons, Hoboken, NJ.

Novotny, V., X. Wang, A.J. Englande, D. Bedoya, L. Promakasikorn, and R. Tirado (2010b) Comparative Assessment of Pollution by the Use of Industrial Agricultural Fertilizers in Four Rapidly Developing Asian Countries, *Environment, Development, and Sustainability* 12:491–50.

Novotny, V. (2011a) Water and energy link in the Cities of the Future – Achieving net zero carbon and pollution emission footprint, Proc. World Congress, International Water Association, Montreal (CA), September 19–24, 2010, *Water Science & Technology* 63(1):184–190.

Novotny, V. (2011b) Danger of hyper-eutrophic status of water supply impoundments resulting from excessive nutrient loads from agricultural and other sources, *Journal of Water Sustainability* 1(1):1–22.

Novotny, V. (2012) Water and energy link in the cities of the future: Achieving net zero carbon and pollution emission footprint, Chapter 3 in *Water-Energy Interactions of Water Reuse* (K.H. Choo, P. Cornel, and V. Lazarova, eds.), IWA Publishing, London.

Novotny, V. (2013) Water energy nexus: retrofitting urban areas to achieve zero pollution, *Buildings Research & Information (UK)* 41(5):589–609.

Novotny, V. (2015) Integrated Urban Water/Used Water/Solids Management Recovers Energy and Resources and Achieves Triple Net Zero Adverse Impacts Goals in Future Cities, *Proc. Water and Energy 2015: Opportunities for Energy and Resource Recovery in the Changing World,* WEF and IWA Specialty Conference, Washington June 8–10.

NSTC (2008) *Federal Research and Development Agenda for Net-zero Energy High Performance Buildings*, National Science and Technology Council, Committee on Technology, Office of the President, Washington, DC.

O'Connor R.P., E.J. Klein, and L.D. Schmidt (2000) High yields of synthesis gas by millisecond partial oxidation of higher hydrocarbons, *Catalysis Letters*, 70(3–4): 99–107.

Olsson, J. (2018) Co-Digestion of Microalgae and Sewage Sludge: A Feasibility Study for Municipal Wastewater Treatment Plant, PhD Thesis, Mälardalen University Press Dissertations No. 26, Sweden, https://mdh.diva-portal.org/smash/get/diva2:1204273/FULLTEXT02.pdf.

Olsson, G. (2012) *Water & Energy*, IWA Publishing, London.

Organisation for Economic Cooperation and Development, (1982) Eutrophication of Waters, Monitoring, Assessment and Control. Paris, OECD.

Ormerod, R.M. (2003) Solid Oxide Fuel Cells, *Chem. Soc. Rev.* 32:17–28, DOI:10.109/b105764m www.researchgate.net/publication/10890897_Solid_Oxide_Fuel_Cells.

Owen, W.F., D.C Stuckey, J.B. Healy, L.Y. Young, P.L. McCarty. (1979) Bioassay for Monitoring Biochemical Methane Potential and Anaerobic Toxicity. *Water Res*, 13:485-492.

Ozgun, H., J.B. Gimenez, M.E. Ersahin, Y. Tao, H. Spanjers, J.B. van Lier. (2015) Impact of membrane addition for effluent extraction on the performance and sludge characteristics of upflow anaerobic sludge blanket reactors treating municipal wastewater. *J. Membr. Sci.* 479, 95–104.

Pacific Environmental Services (1996) Background Report AP-42 Section 6.18 Ammonium Sulfate, Prepared for US Environmental Protection Agency, accessed February 2015, http://www.epa.gov/ttnchie1/ap42/ch08/bgdocs/b08s04.pdf.

Panigrahi, S., A.K. Dlai, and N.N. Bakhshi (2002) Production of syngas/high BTU gaseous fuel from the pyrolysis of biomass derived oil, *Fuel Division Chemistry Preprints 2002* 47(1):118–122.

Paranjape, M., P.F. Clarke, B.B. Pruden, D.J. Parrillo, C. Thaeron, and S. Sircar (1998) Separation of bulk carbon dioxide-hydrogen mixtures by selective surface flow membranes, *Adsorption* 4:355–360, Kluver Ac. Publ., The Netherlands.

Paris COF 21 (2015) *Framework Convention on Climatic Change: Paris Agreement*, United Nations, http://www.cop21.gouv.fr/wp-content/uploads/2015/12/l09r01.pdf, accessed January 2016.

Parris, K.M. et al. (2018) The seven lamps of planning for diversity in the city, *Cities*, https://doi.org//10.1016/j.cities.2018.06/007.

Parsons, S.A., F. Wall, J. Doyle, K. Oldring, and J. Churchley (2001) Assessing the potential for struvite recovery at sewage treatment works, *Environmental Technology* 22:1279–1286, http://citeseerx.ist.psu.edu/viewdoc/download?doi=10.1.1.508.6039&rep=rep1&type=pdf.

Paudel, S., Y. Kang, Y. Yoo, and G.T. Seo (2015) Hydrogen Production in the anaerobic treatment of domestic-grade synthetic wastewater. *Sustainability* 7(12):16260–16272, http://www.mdpi.com/2071-1050/7/12/15814/htm.

Perez, G.K. (2013) Anaerobic codigestion of cattle manure and sewage sludge: Influence of compositions and temperature, *International Journal of Environmental Protection* 3(6):8–15.

Petrlik, J., and R.A. Ryden (2005) After Incineration: The Toxic Ash Problem, IPEN, Manchester, Prague, http://www.ipen.org/documents/after-incineration-toxic-ash-problem.

Pham, T.H., K. Rabaey, P. Aelterman, P. Clauwaert, L. De Schamphelaire, N. Boon, and W. Verstraete (2006) Microbial Fuel Cells in Relation to Conventional Anaerobic Digestion Technology, *Engineering in Life Sciences*, Special Issue, 6(3):285–292, https://onlinelibrary.wiley.com/doi/pdf/10.1002/elsc.200620121.

Phillips, H.M., E. Kobylinski, J. Barnard, and C. Wallis-Lage (2006) Nitrogen and phosphorus – rich side streams: managing the nutrient merry-go-round, Prc; WEFTEC 2006, Water Environment Foundation, Alexandria, VA, pp. 5282–5304.

Phillips, K. (2018) A dead sperm whale was found with 64 pounds of trash in its digestive system, *Washington Post*, April 11, https://www.washingtonpost.com/news/speaking-of-science/wp/2018/04/11/a-dead-sperm-whale-was-found-with-64-pounds-of-trash-in-its-digestive-system/?utm_term=.3e580b70eb19&wpisrc=nl_az_most&wpmk=1.

Possner, A., and K. Caldeira (2017) Geophysical potential for wind energy over the open oceans, *PNAS,* October 9, https://doi.org/10.1073/pnas.1705710114, accessed July 2018.

Potter, M.C. (1911) Electrical effects accompanying the decomposition of organic compounds. *Proceedings of the Royal Society of London* 84:260–276.

Prince George County (1999) *Low-Impact Development Design Strategies: An Integrated Design Approach*, http://www.epa.gov/owow/nps/lid/lidnatl.pdf, accessed December 2009.

Prince George County (2014) *Stormwater Management Design Manual* accessed August 2015, http://www.princegeorgescountymd.gov/sites/DPIE/Resources/Publications/design/Documents/DPIE.StormwaterDesignManual_%2010.9.14.pdf.

Psomopoulos, C.S., Bourka, A., Themelis, N.J. (2009) Waste-to-energy: a review of the status and benefits in USA. *Waste Management* 29:1718–1724, DOI:10.1016/j.wasman.2008.11.020.

Publicover, B. (2017) Panda Green Energy finishes 50 MW in China, PV Magazine, accessed July 2017, https://www.pv-magazine.com/2017/06/30/panda-green-energy-finishes-50-mw-in-china/.

Pugh, L., and B. Stinson (2012) Sustainable Approaches to Sidestream Nutrient Removal and Recovery, Presentation 2012 MWEA Wastewater Administrators Conference Frankenmuth, Michigan, https://www.mi-wea.org/docs/Lucy%20Pugh%20PP%201-19-12.pdf.

Puget Sound Energy (2014) *Customer Handbook for Climate Change*, accessed December 2014, https://pse.com/aboutpse/Environment/Documents/4405_Climate_Change_Handbook.pdf.

Quina, M.J., J.C.M. Bordado and R.M. Quinta-Ferreira (2011) Air Pollution Control in Municipal Solid Waste Incinerators. Chapter 16 in *The Impact of Air Pollution on Health, Economy, Environment and Agricultural Sources (M.K. Khalaff, ed.)*, pp. 332–358, INTECH Publishers, Rjeka, Chroatia, DOI: 10.5772/17650, https://www.intechopen.com/books/the-impact-of-air-pollution-on-health-economy-environment-and-agricultural-sources/air-pollution-control-in-municipal-solid-waste-incinerators.

Rabaey, K., G, Lissens, and W, Verstraete (2005) Microbial Fuel Cells: Performances and Perspectives. in *Biofuels for Fuel Cells: Renewable Energy from Biomass Fermentation* Pp. 377–399, (Piet Lens, P. Westermann, M. Haberbauer, and A. Moreno Eds), Iwa Publishing, London.

Rabaey K, Verstraete W. (2005) Microbial fuel cells: novel biotechnology for energy generation. *Trends in Biotechnology.* 23(6):291–298, https://www.ncbi.nlm.nih.gov/pubmed/15922081.

Ramesh, A., D.J. Lee, K.L. Wang, J.P. Hsu, R.R. Juang and K.J. Hwang (2007) Biofouling in membrane bioreactor, *Separation Science and Technology*, 41(7):1345–1370, https://doi.org/10.1080/01496390600633782.

Randrianarison, G., and M.A. Ashraf (2017) Microalgae: a potential plant for energy production, *Geology, Ecology, and Landscapes*, 1(2):104–120, https://www.tandfonline.com/doi/full/10.1080/24749508.2017.1332853.

Rees, W.E. (1992) Ecological footprints and appropriate carrying capacity: What urban economist leaves out, *Environment and Urbanization* 4(2):121–130.

Rees, W.E. (1997) Urban ecosystems: The human dimension, *Urban Ecosystems* 1(1):63075.

Rees, W.E. (2014) Avoiding Collapse: An agenda for sustainable degrowth and relocalizing the economy, Canadian Centre for Policy Alternatives, Vancouver BC, Canada https://www.policyalternatives.ca/sites/default/files/uploads/publications/BC%20Office/2014/06/ccpa-bc_AvoidingCollapse_Rees.pdf.

Register, R. (1987) *Ecocity Berkeley: Building cities for a healthy future*, North Atlantic Books.

Register, R. (2006) *Ecocities: Rebuilding cities in balance with nature.* Gabriola Island, BC: New Society.

Resilience Alliance (2015) Key concepts, http://www.resalliance.org/key-concepts.

Rich, Nathaniel (2018) Losing Earth: The decade we almost stopped climate change, *New York Times Magazine*, special issue, August 1, https://www.nytimes.com/interactive/2018/08/01/magazine/climate-change-losing-earth.html.

Richardson, Y., M. Drobek, A. Julbe, J. Blin, and F. Pinta (2015) Biomass gasification to produce syngas, Chapter 8 in *Recent Advances in Thermo-chemical Conversion of Biomass*, Elsevier, pp. 209–246, DOI: 10.1016/B978-0-444-63289-0.00008-9.

Rinaldi, G., D. McLarty, J. Brouwer, A. Lanzini, and M. Santarelli (2015) Study of CO_2 recovery in a carbonate fuel cell tri-generation plant, *Journal of Power Sources*, accessed September 2017, https://labs.wsu.edu/cleanenergy/documents/2015/08/powersources2015.pdf/.

Roberts, P.N ., P.W. Newton, and L. Pearson (Eds.) (2014) *Resilient Sustainable Cities: A Future*, Routledge, NY.

Roddy, D.J., and C. Manson-Whitton, (2012) Biomass gasification and pyrolysis, *Comprehensive Renewable Energy*, 5:133–153.

Rogers, P., and A Rosenthal (1988) The imperatives of nonpoint source pollution control, In *Political, Institutional and Fiscal Alternatives for Nonpoint Pollution Control Programs*, (V. Novotny, ed.). Marquette University Press, Milwaukee, WI.

Rubin E, H. Mantripragada, A. Marks, P. Versteeg, J., and Kitchin (2012) The outlook for improved carbon capture technology. *Progress in Energy and Combustion Science* 38, 630–671.

Ryan, J. (2017) China is about to bury Elon Musk in batteries, *Blumberg*, June 2, 2017, accessed 5/2018, https://www.bloomberg.com/news/articles/2017-06-28/china-is-about-to-bury-elon-musk-in-batteries.

Samson, P. (2016) Evolution of the atmosphere, composition, structure and energy, Syllabus, http://www.globalchange.umich.edu/globalchange1/current/lectures/Perry_Samson_lectures/evolution_atm/index.html, University of Michigan.

Sands, K. (2014) Personal communication, Metropolitan Milwaukee, Sewerage District, Milwaukee, WI.

Saufi, S.M., and A.F. Ismail (2004) Fabrication of carbon membranes for gas separation: A review, *Carbon* 42(2):241–259.

Schiemenz, K., and B. Eichler-Löbermann (2010) Biomass ashes and their phosphorus fertilizing effects on different crops, *Nutrient Cycling in Agrosystems* 87(3):471–482, https://link.springer.com/article/10.1007/s10705-010-9353-9.

Schiemenz K., J. Kern, H.M. Paulsen, S. Bachmann, and B. Eichler-Löbermann (2011) Phosphorus Fertilizing Effects of Biomass Ashes. In: *Insam H., Knapp B. (eds) Recycling of Biomass Ashes*. Springer, Berlin, Heidelberg, https://doi.org/10.1007/978-3-642-19354-5_2.

Schoon, N., F. Seath and L. Jackson (2013) *One Planet Living: The Case for Sustainable Consumption and Production in Post-2015 Development Agenda*, Beyond 2015, Bond, and BioRegional, http://sustainabledevelopment.un.org/content/documents/5483bioregional3.pdf.

Sharrer, M.J., K. Rishel, and S/T. Summerfelt (2010) Evaluation of a membrane biological reactor for reclaiming water, alkalinity, salt, phosphorus, and protein contained in a high-strength aquacultural wastewater, *Bioresource Technology* 101:4322-4330.

Shell Global (2018) *Shell Scenarios, Sky: Meeting the Goals of the Paris Agreement*, Shell Oil Co., https://www.shell.com/promos/meeting-the-goals-of-the-paris-agreement/_jcr_content.stream/1521983847468/5f624b9260ef2625f319558cbb652f8b23d331933439435d7a0fc7003f346f94/shell-scenarios-sky.pdf, accessed March 2018.

Shepard, W. (2017) No joke: China is building 285 eco-cities, here is why, *Forbes*, Sep. 1, 2017, accessed February 2018, https://www.forbes.com/sites/wadeshepard/2017/09/01/no-joke-china-is-building-285-eco-cities-heres-why/#1ecbf0a32fe8.

Shin, C., McCarty, P.L., Kim, J., Bae, J. (2014) Pilot-scale temperate-climate treatment of domestic wastewater with a staged anaerobic fluidized membrane bioreactor (SAF-MBR). *Bioresour. Technol.* 159, 95–103.

Shindo, R., and K. Nagai (2014) *Gas Separation Membranes*, DOI 10.1007/978-3-642-36199-9_134-1, Springer, Berlin, Heidelberg.

Siedlecki, M., W. de Jong, and A.H.M. Verkooijen (2011) Fluidized bed gasification as a mature and reliable technology for the production of bio-syngas and applied in the production of liquid transportation fuels: A review, *Energies 2011* 4: 389–434; DOI:10.3390/en4030389, https://www.mdpi.com/1996-1073/4/3/389/htm.

Siegrist, H., Salzgeber, D., Eugster, J., Joss, A., 2008. Anammox brings WWTP closer to energy autarky due to increased biogas production and reduced aeration energy for N-removal, *Water Sci. Technol.* 57: 383–388.

Smil, V. (2000) Phosphorus in the environment: Natural flows and human interferences, *Annual Review of Energy and the Environment* 25: 53–88.

Smith, S.R. (2002) Management of non-hazardous solid wastes, *Encyclopedia of Life Support Systems, UNESCO*, Paris, France, EOLSS Publishers Co. Ltd.

Smith, A., S. Skerlos, and L. Raskin (2013) Psychrophilic anaerobic membrane bioreactor treatment of domestic wastewater, *Wat. Res.* 47: 1655–1665.

Solar Power Authority (2017) Facts about solar power, accessed July 2017 https://www.solarpowerauthority.com/25-facts-about-solar-power/.

Solow, R.M. (1971) The economic approach to pollution and its controls, *Science* 173:497–503.

South Florida Water Management District (2017) *South Florida Environmental Report 2017*, West Palm Beach, Florida, https://www.sfwmd.gov/science-data/sfer.

South Florida Water Management District (2019) Lake Okeechobee, *South Florida Environmental Report* 2019, West Palm Beach, Florida, https://www.sfwmd.gov/science-data/sfer.

Stach, E. et al. (1982) *Coal Petrology*, Gebrűder Bornthaeger, Berlin, Stuttgart.

Stahre P. (2008*) Blue Green Fingerprints in the City of Malmö, Sweden*, VA SYD, Malmö.

Stuckey, D. (2012) Recent developments in anaerobic membrane reactors, *Bioresource Technology* 122:137–48, DOI: 10.1016/j.biortech.2012.05.138.

Steen I. (1998) Phosphorus availability in the 21st century: Management of a non-renewable resource, *Phosphorus and Potassium* 217: 25–31.

Steffen, W., et al. (2018) Trajectories of the Earth System in the Anthropocene. *PNAS*, 2018, DOI: 10.1073/pnas.1810141115, https://www.sciencedaily.com/releases/2018/08/180806152040.htm.

Svoboda, E., E. Mika and S. Berthil (2008) Americas Top 50 Green Cities, *Popular Science*, 115 (33):8252–8259, https://www.popsci.com/environment/article/2008-02/americas-50-greenest-cities.

Sýkorová, E., J. Wanner, and O. Beneš (2014) Analysis of Phosphorus Recovery by Struvite Precipitation from Sludge Water in Selected Wastewater Treatment Plants, *Chemicke Listy* 108(6):610–614, https://www.researchgate.net/publication/288722649_Analysis_of_Phosphorus_Recovery_by_Struvite_Precipitation_from_Sludge_Water_in_Selected_Wastewater_Treatment_Plants/references.

Takaoka, M. et al. (2015) A biomass power generation system combined sewage incineration, *Proc. Water and Energy 2015: Opportunities for Energy and Resource Recovery in the Changing World*, Session 7, WEF, EWA, JSWA, Washington, DC, June 7–10.

Takizawa, K. (2013) Molten Carbonate Fuel Cells, *Energy Carriers and Conversion Systems, Vol. II, Encyclopedia of Life Support Systems*, http://www.eolss.net/sample-chapters/c08/E3-13-10-06.pdf.

Tanner M. (2010) Projecting the scale of the pipeline network for CO2-EOR and its implications for CCS infrastructure development, Office of Petroleum, Gas, & Biofuels Analysis U.S. Energy Information Administration, http://www.eia.gov/workingpapers/co2pipeline.pdf, accessed February 2017.

Tanthapanichakoon, W., and S.W. Jian (2012) Bioethanol Production from Cellulose and Biomass-Derived Syngas, *Engineering Journal* 16(5), http://engj.org/index.php/ej/article/view/356/243.

Tchobanoglous, G., H. Thiesen, and S.A. Vigil (1993) *Integrated Solid Waste Management: Engineering Principles and Management Issues*, McGraw-Hill, New York.

Tchobanoglous, G., and F. Kreith (2003) *Handbook of Solid Waste Management*, McGraw-Hill, New York.

Tchobanoglous, G., F. Burton, and D. Steusel (2003) *Metcalf and Eddy, Wastewater Engineering: Treatment and Reuse*, 4th ed. McGraw-Hill, New York.

Themelis, N.J., Y.H. Kim, M.H. Brady (2002) Energy recovery from New York City solid wastes, *ISWA journal: Waste Management and Research* 20:223–233, http://www.betalabservices.com/PDF/Energy%20recovery%20from%20New%20York%20City%20solid%20wastes.pdf.

Thompson, M., R. Eliis, and A. Vildavsky (1990) *Cultural Theory*, Westview Press, Bolder CO.

Thompson, R.S. (2008) Hydrogen Production by Anaerobic Fermentation Using Agricultural and Food Processing Waste Utilizing a Two-Stage Digestion System, Paper 208, *All Graduate Theses and Dissertations*, Utah State University, Logan.

Tilley, E., L. Ulrich, C. Lüthi, P. Reymond and C. Zurbrügg (2014) *Compendium of Sanitation Systems and Technologies*, 2nd ed. Swiss Agency for Development and Cooperation, Duebendorf, Switzerland: Swiss Federal Institute of Aquatic Science and Technology (Eawag), http://www.eawag.ch/fileadmin/Domain1/Abteilungen/sandec/schwerpunkte/sesp/CLUES/Compendium_2nd_pdfs/Compendium_2nd_Ed_Lowres_1p.pdf, Pdf accessed Dec. 2916.

Torcellini, P., and S. Plees (2012) Defining net zero energy buildings, Chapter 1, Building Design + Construction, March 2012, *Zero and Net Zero Energy Buildings*, accessed January 2015, http://twgi.com/downloads/NetZeroWhitePaper_BDC.pdf.

Tse H.T., S. Luo, J. Li, and Z. He (2016) Coupling microbial fuel cells with a membrane photobioreactor for wastewater treatment and bioenergy production, *Bioprocess Biosyst Eng* 39(11): 1703–1710, https://doi.org/10.1007/s00449-016-1645-2.

Tum, S., C. Kinoshita, Z. Zhang, D. Ishmura, and J. Zhouu (1998) An experimental investigation of hydrogen production from biomass gasification, *International Journal of Hydrogen Energy* 23(8):641–648.

Tyler, T. (2016) It is all about energy – power generation through heat recovery in Hartford, *NEWEA Journal*, Fall 2–16, pp. 23–27.

Ulrich, L. (2015) Hydrogen fuel cell cars return for another look, *New York Times*, April 16, http://www.nytimes.com/2015/04/17/automobiles/hydrogen-fuel-cell-cars-return-for-another-run.html?_r=0.

UN Water (2014) *Water and Energy*, United Nation World Water Development Report 2014, vol. 1, UNESCO, Paris, http://unesdoc.unesco.org/images/0022/002257/225741E.pdf, accessed November 2017.

United Nations (2015a) Water for Life 2005-2015. International Decade for Action, UNDESA, http://www.un.org/waterforlifedecade/scarcity.s\ignorespaceshtml.

United Nations (2015b) Transforming Our World: The 2030 Agenda for Sustainable World, A/Res/70/1, http://www.un.org/ga/search/view_doc.asp?symbol=A/RES/70/1&Lang=E.

United Nations (2017) *World Population Prospects*, DESA Population Division, New York, https://population.un.org/wpp/.

University of Illinois Extension (2018) Composting for the Homeowner, accessed November 2018, https://m.extension.illinois.edu/homecompost/science.cfm.

U.S. Department of Energy (2004) *Fuel Cell Handbook* 7th ed. Office of Fossil Energy National Energy Technology Laboratory, Morgantown, West Virginia, https://www.netl.doe.gov/sites/default/files/netl-file/FCHandbook7.pdf.

U.S. Department of Energy (2008) *Carbon Dioxide Emissions from the Generation of Plants in the United States*, also published by US EPA, Washington, DC.

U.S. Department of Energy (DOE) (2010a) *Report of the Interagency Task Force on Carbon Capture and Storage*, U.S. Department of Energy and U.S. Environmental Protection Agency report August 2010.

U.S. Department of Energy (2010b) *Carbon Dioxide Enhanced Oil Recovery - Untapped Domestic Energy Supply and Long-Term Carbon Storage Solution*, National Energy Technology Laboratory, Washington, DC, https://www.netl.doe.gov/file%20library/research/oil-gas/CO2_EOR_Primer.pdf.

U.S. Department of Energy (2013) *Solid Oxide Fuel Cells: Technology Program Plan*, https://www.netl.doe.gov/File%20Library/Research/Coal/energy%20systems/fuel%20cells/Program-Plan-Solid-Oxide-Fuel-Cells-2013.pdf, Office of Fossil Energy, Washington, DC.

U.S. Department of Energy (2014a) *Innovative Concepts for Beneficial Reuse of Carbon Dioxide*, Office of Fossil Energy, https://energy.gov/fe/innovative-concepts-beneficial-reuse-carbon-dioxide-0.

U.S. DOE (2014b) DOE Hydrogen and Fuel Cells Program Record, US Department of Energy, https://www.hydrogen.energy.gov/pdfs/14005_hydrogen_production_status_2006-2013.pdf.

U.S. Department of Energy (2016a) *Central Versus Distributed Hydrogen Production*, Office of Energy Efficiency & Renewable Energy, US Department of Energy, Washington, DC https://energy.gov/eere/fuelcells/central-versus-distributed-hydrogen-production.

U.S. Department of Energy (2016b) *2015 Wind Technologies Market Report*, Office of Scientific and Technical Information Oak Ridge, https://www.energy.gov/sites/prod/files/2016/08/f33/2015-Wind-Technologies-Market-Report-08162016.pdf, accessed July 2017.

U.S. Department of Energy (2016c) *Environment Baseline, Volume 1: Greenhouse Gas Emissions from the U.S. Power Sector*, Office of Energy Policy and Systems Analysis, https://www.energy.gov/sites/prod/files/2017/01/f34/Environment%20Baseline%20Vol.%201--Greenhouse%20Gas%20Emissions%20from%20the%20U.S.%20Power%20Sector.pdf.

U.S. Department of Energy (2018) *Hydrogen Production: Electrolysis*, Office of Energy Efficiency and Renewable Energy, Washington, DC, accessed October 2018, https://www.energy.gov/eere/fuelcells/hydrogen-production-electrolysis.

UN Environment (2017) *More bang for the buck: Record new renewable power capacity added at lower cost*, United Nations Environmental Program, http://web.unep.org/newscentre/more-bang-buck-record-new-renewable-power-capacity-added-lower-cost.

UNIDO (1996) *Fertilizer Manual*, pp. 207-256, United Nations Industrial Development Organization, Kluver Academic Publishers, Dordrecht, The Netherlands.

US EIA (2011) *Today in Energy, Energy Information Agency*, accessed March 2015, http://www.eia.gov/todayinenergy/detail.cfm?id=3590.

US EIA (2013) A*nnual Energy Review,* Energy Information Administration, http://www.eia.gov/totalenergy/data/annual/?src=Total-f5#summary.

US EIA (2017) *Annual Energy Outlook 2017 with projections to 2050*, Energy Information Administration Washington, DC, https://www.eia.gov/outlooks/aeo/pdf/0383(2017).pdf.

US EIA (2017) How much carbon dioxide is produced per kilowatt hour when generating electricity with fossil fuels? Energy Information Administration, accessed January 2017, https://www.eia.gov/tools/faqs/faq.cfm?id=74&t=11.

U.S. Environmental Protection Agency (1991) *Guidance for Water Quality-Based Decisions: The TMDL Process*. Assessment and Watershed Protection Division, U.S. EPA, Washington, DC.

US EPA (2000) *Constructed Wetlands Treatment of Municipal Wastewater Manual*, EPA A/625/R-99/010, Office of Research and development, Cincinnati, OH https://nepis.epa.gov/Exe/tiff2png.cgi/30004TBD.PNG?-r+75+-g+7+D%3A%5CZYFILES%5CINDEX%20DATA%5C95THRU99%5CTIFF%5C00001401%5C30004TBD.TIF.

US EPA (2003) *Cooling Summertime Temperatures: Strategies to Reduce Urban Heat Islands*, US EPA 430-F-03-014, Environmental Protection Agency, Washington, DC, https://www.epa.gov/sites/production/files/2014-06/documents/hiribrochure.pdf.

US EPA (2007) *Total Maximum Daily Loads with Stormwater Sources: A Summary of 17 TMDLs*, EPA 841-R-07-002, US Environmental Protection Agency, Office of Wetlands, Oceans and Watersheds, Washington, DC, www.epa.gov/owow/tmdl/techsupp.html.

US EPA (2009) *Stormwater Wet Pond and Wetland Management Guidebook*, EPA 833-B-09-001, US Environmental Protection Agency, Washington, DC https://www3.epa.gov/npdes/pubs/pondmgmtguide.pdf.

US EPA (2010) *U.S. Anaerobic Digester: Status Report,* October 2010, accessed March 2014, http://www.epa.gov/agstar/documents/digester_status_report2010.pdf.

US EPA-CHPP (2011) *Opportunities for Combined Heat and Power Wastewater Treatment Facilities: Market Analysis and Lessons from the Field*, Report to US EPA by Eastern Research Group and Resources Dynamics Corporation, https://www.epa.gov/sites/production/files/2015-07/documents/opportunities_for_combined_heat_and_power_at_wastewater_treatment_facilities_market_analysis_and_lessons_from_the_field.pdf, accessed June 2017.

US EPA (2013) *Energy efficiencies for water and wastewater utilities*, US Environmental Protection Agency, Washington, DC, http://water.epa.gov/infrastructure/sustain/energyefficiency.cfm.

US EPA (2014a) *Municipal Solid Waste Generation, Recycling, and Disposal in the United States: Facts and Figures 2012*, US Environmental Protection Agency, Washington DC, accessed June 2015, http://www.epa.gov/osw/nonhaz/municipal/pubs/2012_msw_fs.pdf.

US EPA (2014b) *Municipal Solid Waste Landfills Economic Impact Analysis for the Proposed New Subpart to the New Source Performance Standards*, U.S. Environmental Protection Agency Office of Air and Radiation, Research Triangle Park, NC, accessed July 2017, https://www3.epa.gov/ttnecas1/regdata/EIAs/LandfillsNSPSProposalEIA.pdf.

US EPA (2015) *Fuel and Carbon Dioxide Emissions Savings Calculation Methodology for Combined Heat and Power Systems,* Combined Heat and Power Partnership, accessed June 2017, https://www.epa.gov/sites/production/files/2015-07/documents/fuel_and_carbon_dioxide_emissions_savings_calculation_methodology_for_combined_heat_and_power_systems.pdf.

US EPA (2016) *Energy Recovery from Waste*, accessed March 2017, https://archive.epa.gov/epawaste/nonhaz/municipal/web/html/index-11.html.

U.S. EPA (2017) Inventory of U.S. Greenhouse Gas Emissions and Sinks: 1990–2015. EPA 430-P-17-001, Washington, DC, https://www.epa.gov/ghgemissions/overview-greenhouse-gases.

US EPA (2018) *Advancing Sustainable Materials Management: 2015 Fact Sheet,* accessed October 2018, https://www.epa.gov/sites/production/files/2018-07/documents/2015_smm_msw_factsheet_07242018_fnl_508_002.pdf.

US EPA (2018a) *Facts and Figures about Materials, Waste and Recycling, National Overview: Facts and Figures on Materials, Wastes and Recycling,* https://www.epa.gov/facts-and-figures-about-materials-waste-and-recycling/national-overview-facts-and-figures-materials.

US EPA (2018b) *Greenhouse Gas Reporting Program (GHGRP),GHGRP 2015: Waste* https://www.epa.gov/ghgreporting/ghgrp-2015-waste.

USGCRP (2018a) *Fourth National Climate Assessment, Volume II: Impacts, Risks, and Adaptation in the United States*, U.S. Global Change Research Program, Washington, DC, https://nca2018.globalchange.gov/.

USGCRP (2018b) *Second State of the Carbon Cycle Report (SOCCR2): A Sustained Assessment Report* [Cavallaro, N., G. Shrestha, R. Birdsey, M.A. Mayes, R.G. Najjar, S.C. Reed, P. Romero-Lankao, and Z. Zhu (eds.)]. U.S. Global Change Research Program, Washington, DC, https://doi.org/10.7930/SOCCR2.2018, https://carbon2018.globalchange.gov/.

US Green Buildings Council (2014) *LEED*, accessed November 2014, http://www.usgbc.org/leed.

Vacccari, D.A. (2009) Phosphorus: A looming crisis, *Scientific American* 300:54–59, DOI:10.1038/scientificamerican0609-54.

Valkenburg, C., M.A. Gerber, C.W. Walton, S.B. Jones, B.L. Thompson, and D.J. Stevens (2008) *Municipal Solid Waste (MSW) to Liquid Fuel Synthesis: Volume 1. Availability of Feedstock Technology*. PNNL 18144, Pacific Northwest National Laboratory, Battelle http://www.pnl.gov/main/publications/external/technical_reports/PNNL-18144.pdf.

Van de Graaf, A.A, A. Mulder, P. De Bruijn, M.S.M. Jetten, L.A. Robertson, and J.G. Kuenen (1995), Anaerobic oxidation of ammonium is a biologically mediated process, *Applied and Environmental Microbiology*, 61(4):1246–1251.

Vandenberg, M.P and J.M. Gilligan (2017) *Beyond Politics: The Private Governance Response to Climate Change*, Cambridge University Press, Cambridge, UK, http://www.cambridge.org/fr/academic/subjects/politics-international-relations/political-economy/beyond-politics-private-governance-response-climate-change#kCS3osRFhcZFd5Gy.99.

Van Lier, J.B ., N. Mahmoud, and G. Zeeman (2008) Anaerobic Wastewater Treatment, Chapter 16 in *Biological Wastewater Treatment: Principles, Modeling and Design* (M. Henze, M.C.M. van Loosdrecht, G.A. Ekama, and D. Brdjanovic, eds.), IWA Publishing. London.

van Loosdrecht, M.C.M., and S. Salem (2006) Biological treatment of sludge digester liquids, *Wat. Sci. & Technol.* 53(12):11–20, DOI: 0.2166/wst.2006.401.

Van Niftrik L., and M.S.M Jetten (2012) Anaerobic ammonium oxidizing bacteria: Unique microorganisms with exceptional properties, *Microbiology and Molecular Biology Reviews* 76(3):585–596, https://www.ncbi.nlm.nih.gov/pmc/articles/PMC3429623/, DOI: 10.1128/MMBR.05025-11.

Veneman, P.L., and B. Stewart (2002) *Greywater Characterization and Treatment Efficiency*, Massachusetts Department of Environmental Protection, Bureau, Boston: University of Massachusetts at Amherst, http://archives.lib.state.ma.us/handle/2452/69133.

Veolia (2017) *Billund Biorefinery: Resource Recovery for the Future,* October 2017, accessed October 2018, https://www.veoliawatertechnologies.com/en/media/articles/billund-biorefinery-0.

Vergini, S., A. Aravantiinou, and I. Manariotis (2015) Harvesting of freshwater and marine microalgae by common flocculants and magnetic microparticles, *Journal of Applied Phycology* 28(2), DOI:10.1007/s10811-015-0662-x.

Verstraete, W., P. Van de Caveye, and V. Diamantis (2009) Maximum use of resources present in domestic "used water," *Bioresource Technol.* 100:5537–5545.

Verstraete, W., B.Bundervolt, and B. Eggermont (2010) *Zero Waste Water: Short-cycling of Wastewater Resources for Sustainable Cities of the Future*, 2nd Xiamen International Forum on Urban Environment, Lab. of Microbial Ecology and Technology, (LabMET), University of Ghent, Belgium, www.LabMET.UGent.be.

Viader, P. (2017) Comparison of phosphorus recovery from incineration and gasification sewage sludge ash, *Water Sci. and Technol.* 75(5):1251–1260 DOI: 10.2166/wst.2016.620.

Vijayakumar, J., G., Anderson, S. Gent, and A. Rajendran (2013) Calculation of biomass capacity of algae based on their chemical composition, 2013 ASABE Conference, Kansas City, Missouri, https://www.researchgate.net/publication/271421394_Calculation_of_biomass_capacity_of_Algae_based_on_their_elemental_composition.

Vymazal, J. (2005) Horizontal sub-surface flow and hybrid constructed wetland systems for wastewater treatment, *Ecological Engineering* 25:478–490.

Vymazal, J., and I. Kröpfelová (2008) *Wastewater Treatment in Constructed Wetlands with Horizontal Sub-Surface Flow*, Springer Verlag, Dortrecht, The Netherlands.

Wakerley, D.W., M.F. Kuehnel, K.L. Orchard, K.H. Ly, T.E. Rosser and E. Reisner (2017) Solar-driven reforming of lignocellulose to H_2 with a CdS/CdOx photocatalyst, *Nature Energy*, vol. 2, Article number: 17021 DOI:10.1038/nenergy.2017.21.

Walker, B. (1995) *Conserving Biological Diversity through Ecosystem Resilience*, Wiley Online Library, https://doi.org/10.1046/j.1523-1739.1995.09040747.x.

Walker, B., C.S. Holling, S.R. Carpenter, and A. Kinzig (2004) Resilience, adaptability and transformability in social–ecological systems, *Ecology and Society* 9(2): 5. [online] URL: http://www.ecologyandsociety.org/vol9/iss2/art5/.

Wan, C., F. Yu, Y. Zhang, Q. Li, and J. Wooten (2013) Material balance and energy balance analysis for syngas generation by a pilot-plant downdraft gasifier, *Journal of Biomassed Materials and Bioenergy* 7(6):690–695.

Wang, J., J.G. Burken, X., Zhang (2005) Engineered struvite precipitation: Impacts of component-ion molar ratios aH, *Journal of Environmental Engineering* 131(10):1433–1440, https://doi.org/10.1061/(ASCE)0733-9372(2005)131:10(1433).

Wang, Y., T. Liu., L. Lei and F. Chen (2016) High temperature solid oxide $H2O/CO2$ co-electrolysis for syngas production, *Fuel Processing Technology* 161, https://www.research gate.net/deref/http%3A%2F%2Fdx.doi.org%2F10.1016%2Fj.fuproc.2016.08.009.

Wanner, J. (2014) History of activate sludge: 100 years and counting, IWA Conference Activated Sludge Conference, http://www.iwa100as.org/history.php.

Watts, R,G. ed. (2013) *Engineering Response to Climate Change*, 2nd ed. CER Press, Taylor & Francis, Boca Raton, ISBN 9781138074118 - CAT# K3397, https://www.crcpress.com/Engineering-Response-to-Climate-Change-Second-Edition/Watts/p/book/9781138074118.

WBCSD (2014) *World, Food, and Energy Nexus Challenges*, World Business, World Business Council for Sustainable Development. Geneva, Switzerland, accessed November 2017, http://www.gwp .org/globalassets/global/toolbox/references/water-food-and-energy-nexus-challenges-wbcsd-2014.pdf.

WEF (2007) Biological Nutrient Removal Processes, Chapter 22 in *Operation of Biological Wastewater Treatment Plants*, Water Environment Federation, Alexandria, VA.

Wei, X, et al. (2018) CO_2 Mineral Sequestration in Naturally Porous Basalt, *Environ. Sci. Technol. Lett.*, 5 (3):142–147. DOI: 10.1021/acs.estlett.8b00047.

Weigandt, H., M. Bertau, W. Hübner, F. Bohndick, and A. Bruckert (2013) RecoPhos: Full-scale fertilizer production from sewage sludge ash, *Waste Management* 33(3):540-544 https://www .sciencedirect.com/science/article/pii/S0956053X12003091.

Welch, J., G. Dreckermann, and D. Doerr (2013) Organic Waste to Energy, webinar, accessed January 1–3, 2015, http://www.appliedsolutions.org/Portals/_Appleseed/documents/Webinars/Applied %20Solutions%20biodigester%20webinar%20slides.pdf.

Werle, S., and M. Dudziak (2014) Analysis of organic and inorganic contaminants in dried sewage sludge and by-products of dried sewage sludge gasification, *Energies* 7:462–476, www.mdpi.com/ journal/energies.

Wheeler, S.M., and T. Beatley (2014) *Sustainable Urban Development Reader*, Routledge.

Whitlock, R. (2018) An interview with Tony Leo and Kurt Goddard of FuelCell, *Renewable Energy Magazine*, https://www.renewableenergymagazine.com/interviews/supporting-the-emerging-fcev-market-an-interview-20160212.

Williams, A. (2017) *China's Urban Revolution: Understanding Chinese Eco-cities*, Bloomsbury Academic Publishers, London.

Windley, B.F. (2015) Geologic history of Earth, Revised 2015, *Encyclopedia Britanica*, https://www .britannica.com/science/geologic-history-of-Earth.

Woetzel, J. et al. (2009) *Preparing for China's Urban Billion*, McKinsey Global Institute, San Francisco, http://www.mckinsey.com/global-themes/urbanization/preparing-for-chinas-urban-billion.

Woodard, C. (2009) Iceland strides toward a hydrogen economy, *Christian Science Monitor*, February 9, http://www.csmonitor.com/Technology/Energy/2009/0212/iceland-strides-toward-a-hydrogen-economy.

World Economic Forum (2014) *Towards the Circular Economy: Accelerating the Scale-up Across Global Supply Chains*, Geneva, Switzerland, retrieved July 2017, http://www3.weforum.org/docs/ WEF_ENV_TowardsCircularEconomy_Report_2014.pdf.

World Health Organization (WHO) (2006) Guidelines for the safe use of wastewater, excreta and greywater, https://www.who.int/water_sanitation_health/sanitation-waste/wastewater/ wastewater-guidelines/en/, retrieved September 7, 2010.

Wyman, C.E., S.R. Decker, J.W. Brady, et al. (2005) *Hydrolysis of Cellulose and Hemicellulose*, Public Text, DOI: 10.1201/9781420030822.ch43, https://www.researchgate.net/publication/251337925_43_ Hydrolysis_of_Cellulose_and_Hemicellulose.

WWF (2014) *The Ten Principles of One Planet Living*, World Wildlife Fund, accessed November 2014, http://wwf.panda.org/what_we_do/how_we_work/conservation/one_planet_living/about_opl/principles/.

Yamamoto, S., J.B. Alcauskas, and T.E. Crozier (1976) Solubility of methane in distilled water and seawater, *J. Chem. Eng. Data*, 21(1): 78–80, DOI: 10.1021/je60068a029.

Yang, H., Z. Xu1, M. Fan, R. Gupta, R.B. Slimane, A.E. Bland, and I. Wright (2008) Progress in carbon dioxide separation and capture: A review, *Journal of Environmental Sciences* 20:14–27.

Zarebska, A., K.V. Christensen and B. Norddahl (2012) The application of membrane contactors for ammonia recovery from pig slurry, *Procedia Engineering* 44:1642–1645 https://doi.org/10.1016/j.proeng.2012.08.895.

Zeverenhoven, R., and P. Kilpinen (2002) Control of Pollutants in Fleu Gases and Fuel Gases, Chapter 2, Espo/Turku, Finland, users.abo.fi/rzevenho/gasbook.html, accessed September 2017.

Zhang, Q., L. Dor, L. Zhang, W. Yang, and W. Blasiak (2011) Performance Analysis of Municipal Solid Waste gasification with Steam in a Plasma Gasification Melting Reactor, Royal Institute of Technology, Stockholm, Sweden, accessed February 2012, http://www.eer-pgm.com/_uploads/dbsattachedfiles/july_2011.pdf.

Zhang, Z., J. Zhang, J. Zhao, and S. Xia (2015) Effect of short-time aerobic digestion on bioflocculation of extracellular polymeric substances from waste activated sludge. *Environmental Science and Pollution Research, Environ Sci Pollut Res* 22(3):1812–1818. https://doi.org/10.1007/s11356-013-1887-3.

Zhang, X. (2018) Current status of stationary fuel cells for coal power generation, *Clean Energy* 2(2):126–139, https://doi.org/10.1093/ce/zky012.

Zhu, X, (2009) Micro/nanoporous membrane-based gas–water separation in microchannel, *Microsyst Technol.* 15(9): 1459–1465, https://doi.org/10.1007/s00542-009-0903-5.

Zitomer, D., and P. Adhikari (2005) Extra methane production from municipal anaerobic digesters, *BioCycle: Energy*, pp. 64–66, September 2005.

Zitomer, D.H., P. Adhikari, C. Heisel, and D. Dinen (2008) Municipal anaerobic digesters for codigestion, energy recovery, and greenhouse gas reductions, *Water Environment Research* 80(3):229–237.

Zitomer, D., N. Ferguson, K. McGrady, and J.Schilling (2001) Anaerobic co-digestion of aircraft fluid and municipal wastewater sludge, *Water Environment Research* 73(6):645–654.

Index